HETEROJUNCTION BAND DISCONTINUITIES
Physics and Device Applications

HETEROJUNCTION BAND DISCONTINUITIES
Physics and Device Applications

Edited by

Federico Capasso

AT&T Bell Laboratories
600 Mountain Avenue
Murray Hill, NJ 07974, USA

Giorgio Margaritondo

Department of Physics
and Synchrotron Radiation Center
University of Wisconsin
Madison, WI 53706, USA

1987

NORTH-HOLLAND
AMSTERDAM · OXFORD · NEW YORK · TOKYO

© Elsevier Science Publishers B.V., 1987

All rights reserved. No part of this publication may be reproduced, stored in a retrieval system, or transmitted, in any form or by any means, electronic, mechanical photocopying, recording or otherwise, without the prior permission of the publisher, Elsevier Science Publishers B.V. (North-Holland Physics Publishing Division), P.O. Box 103, 1000 AC Amsterdam, The Netherlands.
Special regulations for readers in the USA: This publication has been registered with the Copyright Clearance Center Inc. (CCC), Salem, Massachusetts. Information can be obtained from the CCC about conditions under which photocopies of parts of this publication may be made in the USA.
All other copyright questions, including photocopying outside of the USA, should be referred to the publisher.

ISBN: 0444 87060 1

Published by:
North-Holland Physics Publishing
a division of
Elsevier Science Publishers B.V.
P.O. Box 103
1000 AC Amsterdam
The Netherlands

Sole distributors for the USA and Canada:
Elsevier Science Publishing Company, Inc.
52 Vanderbilt Avenue
New York, NY 10017
USA

Library of Congress Cataloging-in-Publication Data

Heterojunction band discontinuities

    Includes bibliographies and indexes.
    1. Semiconductors – Junctions. 2. Photoelectronic devices. 3. Energy-band theory of solids. 4. Energy gap (Physics)
I. Capasso, Federico, 1949-   . II. Margaritondo, Giorgio, 1946-
III. Title: Band discontinuities and device applications.
QC611.6.J85H49 1987   530.4'1   87-14123
ISBN 0-444-87060-1

Printed in The Netherlands

*The Editors dedicate their work
to Paola Capasso and to
Giuseppe and Maria Luisa Margaritondo*

Introduction

For many years, heterojunctions have been one of the fundamental research areas of solid-state science. The interest in this topic is stimulated by the wide applications of heterojunctions in microelectronics. Devices such as heterojunction bipolar transistors, quantum well lasers and heterojunction FETs, already have a significant technological impact. This impact is likely to increase in the future, and heterojunction devices promise to play a revolutionary role in microelectronics.

The results produced by heterojunction research have been described by many articles in professional journals. At regular intervals, the most important achievements have been overviewed in monographic presentations. Some of these books are classic in solid-state physics, e.g., *Heterojunctions and Metal–Semiconductor Junctions* by Milnes and Feucht. Until now, however, no comprehensive presentation was available for the results obtained after the 1970s. Our book eliminates this gap.

The book consists of two parts: (I) Band Discontinuities: Measurements and Theory; and (II) Physics of Heterojunction Devices: Band-Gap Engineering. Each chapter is written by one of the top experts in the field. For controversial issues, the different points of view are presented – whenever possible, by one of their main supporters. Thus, rather than offering a unilateral presentation, the book realistically reflects this active research area, including its controversies.

The results obtained in recent years are both abundant and important in heterojunction research. This is the consequence of the exceptional progress in epitaxial growth techniques, such as molecular beam epitaxy and metallo-organic chemical vapor deposition, in the design of novel devices, and in the experimental and theoretical study of microscopic heterojunction interface properties. We hope that the dissemination of the recent results will stimulate further interest in this fascinating field, and further exploitation of the properties of semiconductor heterostructures.

We express our sincere thanks to all the authors who contributed to the book. We would also like to thank AT&T Bell Laboratories, the National Science Foundation, the Office of Naval Research and the Wisconsin Alumni Research Foundation for their support of our personal research efforts in this area.

CONTENTS

Introduction vii

Part I. Band Discontinuities: Measurements and Theory. 1

1. The theory of heterojunction band lineups
   J. Tersoff 3

2. The problem of heterojunction band discontinuities
   G. Margaritondo and P. Perfetti 59

3. Trends in semiconductor heterojunctions
   A.D. Katnani 115

4. Interface contributions to heterojunction band discontinuities: X-ray photoemission spectroscopy investigations
   R.W. Grant, E.A. Kraut, J.R. Waldrop and S.P. Kowalczyk 167

5. Measurements of band discontinuities using optical techniques
   G. Duggan 207

6. The direct optical determination of $GaAs/Al_xGa_{1-x}As$ valence-band offsets
   D.J. Wolford, T.F. Keuch and M. Jaros 263

7. Band discontinuities in HgTe–CdTe superlattices
   J.P. Faurie and Y. Guldner 283

8. Measurement of energy band offsets using capacitance and current measurement techniques
   S.R. Forrest 311

9. Measurement of band offsets by space charge spectroscopy
   D.V. Lang 377

Part II. Physics of Heterojunction Devices: Band-Gap Engineering .. 397

10. Band-gap engineering and interface engineering: from graded-gap structures to tunable band discontinuities
    F. Capasso 399

11. Modern aspects of heterojunction transport theory
    K. Hess and G.J. Iafrate 451

12. Hot-electron injection and resonant-tunneling heterojunction devices
    S. Luryi 489

13. Physics of quantum well lasers
    N.K. Dutta 565

14. Physics and applications of excitons confined in semiconductor quantum wells
    D.S. Chemla and D.A.B. Miller 595

Author index 625

Subject index 647

PART I

BAND DISCONTINUITIES: MEASUREMENTS AND THEORY

CHAPTER 1

THE THEORY OF HETEROJUNCTION BAND LINEUPS

J. TERSOFF

IBM Thomas J. Watson Research Center
Yorktown Heights, NY 10598, USA

Heterojunction Band Discontinuities: Physics and Device Applications
Edited by F. Capasso and G. Margaritondo
© *Elsevier Science Publishers B.V., 1987*

Contents

1. Introduction 5
2. The band lineup problem 7
   2.1. Defining the problem 7
   2.2. Issues in the band lineup problem 9
3. Theories of band lineups: a critical review 11
   3.1. The role of theory 11
   3.2. Numerical interface calculations 12
   3.3. Model theories and reference levels 15
   3.4. The electron affinity rule and related approaches . 21
   3.5. Role of interface dipoles 27
4. A recent approach to the band lineup problem 28
   4.1. Charge transfer and interface dipoles 28
   4.2. Linear response theory: the self-consistent lineup . 29
   4.3. The neutral lineup 36
   4.4. Calculating the neutral lineup 41
   4.5. Comparison with experiment 44
5. Connection between lineups and Schottky barriers 45
   5.1. Relationship to Schottky barriers 45
   5.2. Implications for theory 50
6. Conclusions 52
   6.1. A field in flux 52
   6.2. Ab initio calculations 52
   6.3. Model theories and empirical rules 53
   6.4. Prospects for predictive accuracy 54
References 56

1. Introduction

The last few years have witnessed tremendous growth in the field of semiconductor heterojunctions. It is now possible to routinely make well characterized high-quality interfaces, which are finding applications in novel device structures. The crucial parameters which determine the electronic behavior of the heterojunction interface are the valence and conduction band-edge discontinuities. Since these are not independent, they are referred to collectively as the band lineup. These discontinuities play a role analogous to the p- and n-type Schottky barriers at a metal–semiconductor interface, in that they determine the barrier for hole or electron transport across the interface, and act as a boundary condition in calculations of band bending and interface electrostatics.

Because heterojunctions represent such simple and ideal systems, with bulk-like geometry and bonding, one might expect that accurate theoretical calculations of band lineups would be straighforward. However, the theoretical understanding of heterojunction band lineups remains controversial. Moreover, until recently, reliable band lineup data were scarce, making it particularly difficult to assess available theories. This chapter attempts to give an overview of the field, as well as a detailed discussion of one particular approach which has led to some successful predictions.

The organization of this chapter is as follows:

Section 2.1 attempts to define just what the band lineup problem is, and to draw a clear distinction between the lineup, and related interface properties. Section 2.2 identifies the issues in the band lineup problem, which will provide the major focus for the rest of the chapter.

Theories of band lineups are reviewed from a broad perspective in sect. 3. Section 3.1 categorizes the various types of theories, and the respective contributions which they can make to our understanding of the band lineup problem. Section 3.2 focuses on detailed numerical calculations of the microscopic electronic structure of interfaces, and their use as a tool for predicting band lineups.

Other than such ab initio calculations, all theories of band lineups can be described as "model" theories, in that they make severe approximations in describing the interface, in order to focus more narrowly on the physical mechanisms which are believed to be relevant. Section 3.3 discusses such model theories collectively, along with some strictly empirical rules for band lineups. All of these theories and empirical rules have in common a similar form, based on the idea of a "reference level". Regardless of the physical model, the mere fact that the theory has this form has important implications, which are discussed.

The model theories of interest here are all founded on one of two basic ideas. These two ideas are the subject of sections 3.4 and 3.5. The first of these sections concerns the class of theories which assume that the vacuum level outside the surface of each semiconductor can serve as a reference level for lining up the bands in the two semiconductors. These theories are lumped here under the term "electron affinity rule", although historically this term refers to a particular one of these theories.

The theories of the "electron affinity rule" type represent the oldest class of band lineup theories. The assumptions implicit in this approach are analyzed, and a few specific theories based on this idea are described in more detail. In particular, we consider what conditions must be satisfied for this approach to provide good predictive accuracy. Section 3.5 merely introduces a very different approach, which is the subject of sect. 4.

The approach of using the vacuum level as a reference is clearly appropriate for a highly inert interface, such as that between two rare-gas solids. The opposite limit is a metal–metal interface, where the Fermi level serves as the appropriate reference level, and the vacuum level does not. In section 4.1 these two limits are considered, and the role of interface dipoles for metal–metal interfaces is discussed. Section 4.2 addresses the question of which limit is appropriate for semiconductor heterojunctions, and concludes that the metal–metal case provides the more powerful paradigm.

To obtain predictions of band lineups, it is still necessary to identify some energy which plays a role like that of the Fermi level in the metal–metal interface. Section 4.3 addresses the physical meaning of such a level in the case of a semiconductor, where it is referred to as the "neutrality level". Section 4.4 describes a specific procedure for calculating this energy. The predictions of this approach are compared with experiment in sect. 4.5.

Since the metal–metal interface served as a paradigm for the semiconductor–semiconductor interface, we might expect that this approach could be extended to deal with the metal–semiconductor interface as well. In fact, historically, the metal–semiconductor interface was studied from this viewpoint long before semiconductor heterojunctions were. The relationship between the band lineup problem and the Schottky barrier problem is analyzed in section 5.1. In this section is shown that both Schottky barriers and band lineups can be predicted from the same, calculated "neutrality level". Moreover, band lineups can instead be predicted directly from measured Schottky barrier heights. The success of these predicted relationships has important implications for the assessment of theories of both heterojunction band lineups and Schottky barrier heights, which are briefly considered in section 5.2.

Finally, section 6 summarizes the conclusions of this chapter.

2. The band lineup problem

2.1. Defining the problem

At the level of detail appropriate for device modeling, the electronic structure of a heterojunction interface is very simple, as is shown in fig. 1. If the "band bending" associated with ionized dopant impurities or with

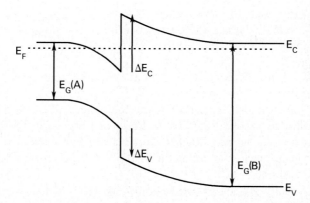

Fig. 1. Schematic band structure of a semiconductor heterojunction, showing position-dependent conduction-band minimum, E_c, and valence-band maximum, E_v, versus position, for an interface between two n-type semiconductors. Note the valence-band discontinuity ΔE_v and the conduction-band discontinuity ΔE_c, with arrows denoting sign convention (upward direction positive). On the right-hand side, band bending results from ionized impurities, and on the left from accumulation of charge in the conduction band (which falls below the Fermi level E_F). E_g is the bulk band gap in the respective semiconductor.

simply the energy at which that band edge would fall, if the local electrostatic potential were extended to infinity to form an ideal bulk semiconductor.

The position-dependent band edges defined in this way vary smoothly in the respective semiconductors. However, because the band gaps of the two semiconductors are in general different, there must be a discontinuity in one or both band edges at the interface. Figure 1 shows the local valence- and conduction-band edges, E_v and E_c, as a function of position in the interface region, along with the Fermi level E_F. As shown in fig. 1, the valence-band discontinuity at the interface is given by

$$\Delta E_v(A, B) \equiv E_v(B) - E_v(A). \tag{2.1}$$

The conduction-band discontinuity ΔE_c is defined similarly. These discontinuities are related by

$$\Delta E_c = \Delta E_g + \Delta E_v, \tag{2.2}$$

where E_g is the band gap, and the sign conventions are illustrated in fig. 1. Either band-edge discontinuity is therefore sufficient to specify the band lineup, which is in general independent of doping.

If the band lineup is known, then for a given doping profile one can calculate the band bending and charge density using standard techniques. Thus knowledge of the band lineup is the central requirement for device design and analysis.

The view of interface electron structure usually employed for device design, as in fig. 1, is, however, simplistic, in that atomic-scale features are neglected. The electronic structure at the interface could, in principle, be quite complicated, with large local fields which are very different from anything in the bulk semiconductors. Fortunately, with respect to transport properties, this neglect of atomic-scale features in the electronic structure is usually well justified.

The band offset gives a barrier to transport which has an exponentially large effect on the current, by limiting thermionic emission etc. On the other hand, the local potential disturbance right at the interface may give a locally higher or lower barrier, but since the width of this extra barrier is only a few Å, the carriers can tunnel through this region, so the current is not greatly affected. The local properties of the interface, while interesting, are therefore usually not very important for device modeling. (An exception arises when there are localized interface states in the forbidden gap.)

Experiments designed to measure the band lineup must therefore look at atoms far enough from the interface to avoid the local (atomic-scale) disturbance, but close enough to be inside the range of appreciable band bending. Since these two length scales could, in principle, overlap, the band lineup cannot be defined with absolute rigor except in the case of perfectly intrinsic semiconductors at absolute zero temperature. However, in practice, this poses no problem. At reasonable doping levels the length scale for band bending is hundreds of Å, or at the least many tens of Å. The localized disturbance at the interface is generally confined to one or two atomic layers in each semiconductor, as will be discussed below. The two length scales are thus, in general, very well separated, and there is, in principle, no problem in defining the band lineup with high accuracy.

In some cases, such as at the left of the junction between n-type semiconductors in fig. 1, there is a region of charge accumulation. In that case there is charge associated with filling of the local conduction bands, rather than with ionized impurities. As long as this accumulation charge density is sufficiently small, it gives band-bending on a relatively long length scale, just like the band-bending from ionized impurities. In principle, it might happen that the charge density was so high that the associated band bending was on a nearly atomic length scale. In that case the band lineup might become ill-defined. However, we do not know any case where this problem occurs.

In addition to its technological importance, the band lineup problem holds a special place in basic interface physics, since it is one of the cleanest, simplest and most well-defined interface problems. Atomic positions differ little from their bulk values, and all atoms are in nearly bulk-like environments. The closely analogous Schottky barrier problem is relatively messy and intractable by comparison, because of uncertainties in the microscopic structure and chemistry of the interface, as well as because of the difficulty of performing reliable spectroscopic measurements of the semiconductor in the presence of a metal.

2.2. Issues in the band lineup problem

In the discussion above, the band lineup problem amounted simply to determining the band-edge discontinuities at a given interface. However, in practice the problem is more complicated, and there are a number of issues which are foci of current research.

One important issue is the identification of factors affecting the band lineup. Beyond merely tabulating band lineups for specific cases, it is also important to have a general understanding of what factors are relevant. In this way one gains a deeper understanding, and at the same time avoids unnecessary work by identifying whole classes of interfaces as being essentially similar. For example, it is important to know whether and how the band lineup depends on factors such as interface orientation, strain, or the ideality of the interface (e.g., the presence of point defects or dislocations). Many experiments have helped to resolve these points, while model theories often avoid the problem by simply assuming an answer, albeit with some plausible justification.

Another issue, although not strictly speaking a theoretical question, has been establishing the accuracy and reliability of experimental determinations of band lineups. Here again, the problem of identifying relevant factors arises. Many experiments are performed with samples which deviate substantially from the ideal of an atomically abrupt epitaxial interface. Some interfaces are graded rather than abrupt; some thin film samples exhibit poor crystallinity. Although there is some evidence that the lineup is relatively insensitive to defects at the interface, device scientists are naturally somewhat suspicious of band lineup determinations based on samples which are not of device quality.

The most fundamental theoretical issue in the band lineup problem, as well as the most controversial, has been the role of interface dipoles. Such dipoles are therefore discussed here with some care. In particular, sect. 3 tries to put the different theories, which make apparently opposite assumptions about interface dipoles, into a unified context, which clarifies the underlying relationships.

Because the interface dipole can only be defined with respect to some reference interface, the problem may be formulated so that the dipole is more or less important, depending on the reference interface. Successful theories of band lineups may therefore be formulated either from the viewpoint that such dipoles are negligible, or that they are the central factor determining band lineups. However, the latter approach has the advantage that it allows metal heterojunctions, semiconductor heterojunctions, and metal–semiconductor junctions to be treated on an equal footing. In contrast, the approach of formulating the problem so that interface dipoles are negligible is limited to interfaces involving semiconductors and insulators.

Aside from understanding the physical mechanisms determining the band lineup, the major theoretical issue here is obtaining quantitative predictions of band lineups. Several theoretical approaches have been reasonably successful in calculating band lineups. However, there are also some empirical approaches which may yield predictions as accurate as any theory, without requiring any calculation or providing any fundamental understanding. Two such empirical approaches are described in sect. 3.3.

3. Theories of band lineups: a critical review

3.1. The role of theory

Theory is not a monolithic enterprise, but rather includes a variety of methods and approaches for attacking various aspects of the problem at hand. Most of the work on heterojunctions can be divided broadly into three categories.

The first category consists of numerical calculations of the complete electronic structure of a specified interface [1–4]. If the Hamiltonian is specified exactly (which among other things requires knowing or calculating the atomic positions), then, in principle, one need only solve the Schrödinger equation to obtain the complete electronic structure of the interface, including the band lineup. In practice, the accuracy of this approach is limited by approximations made to the Hamiltonian, and by numerical precision. However, even if the calculations gave the band lineup exactly, the problem of extracting a qualitative physical understanding would remain. Such numerical calculations of band lineups are discussed below in sect. 3.2.

The second category of theoretical work, which is the main focus of this chapter, involves qualitative analysis of the problem, leading to model theories [5–14] which retain only the factors believed to be essential to the band lineup problem. In such theories one is faced with two distinct problems: first, to correctly identify the essential physics of the lineup problem; and second, to devise a technique, based on this understanding, for actually calculating band lineups for real materials. Because this second step may involve more or less quantitative inaccuracy, it is difficult to assess the correctness of physical arguments based only on the agreement of lineup predictions with experiment. Section 3.3. discusses some general features [15] common to most model theories of band lineups. Sections 3.4 and 3.5 then discuss the two main classes of model theories in greater detail.

Finally, there is a third major category of work which falls outside the scope of this chapter. This consists of analyses of other electronic properties of the interface, *given* the band lineup. Such properties could include, among other things, quantum well exciton energies and luminescence spectra, impurity states near interfaces, transport properties, etc.

3.2. Numerical interface calculations

With the advent of modern high-speed computers, it became possible to attack many interesting problems in electronic structure by directly solving the Schrödinger equation to determine the energies and wavefunctions of all the electrons in the system. This section summarizes the contribution of such microscopic calculations [1–4] to our understanding of band lineups, and attempts to put these calculations in a broader perspective.

Such an approach of calculating the full electronic structure from first principles has the major advantage that, in principle, it makes no assumptions or approximations, but rather approaches the problem without prejudice. On the other hand, the results must then be treated much like experimental results: the physical significance of the results is not obvious, but must be inferred by subsequent analysis.

If such calculations were exact, or nearly so, they would be doubly valuable. First, they would give reliable numerical values for the band lineups at specific interfaces. Second, they would provide "data" (the complete electronic structure of the interface) from which one might extract a simple qualitative picture of the physical mechanism at work, or against which one could at least test simple models which claimed to identify the essential physics of the problem.

In reality, these calculations involve many approximations. While, in principle, the atomic positions can be calculated by minimizing the total energy, heterojunction calculations to date have assumed essentially ideal atomic geometries. Also, these calculations have generally involved approximating the atomic core with a pseudopotential. We discuss separately the earlier calculations [1,2] using empirical forms for the pseudopotential, and the more recent calculations [3,4] using norm-conserving pseudopotentials, which are generally more accurate. Finally, the problems associated with the local-density approximation used for correlation and exchange are discussed below.

Unfortunately, many in this field were disappointed by the imperfect agreement with experiment achieved by the earlier interface calculations [15]. This may be attributable in part to unrealistic hopes for extreme numerical accuracy. Another important reason is no doubt that only a few calculations had been reported, so one could not focus on trends while disregarding modest quantitative discrepancies.

Table 1 summarizes selected [16] experimental results for band lineups, along with results of a number of different theoretical approaches, including the empirical pseudopotential method (column labeled "EP"; other columns in table 1 are discussed below). The empirical pseudopotential calculations differ from experiment by only about 0.2 eV in two cases, but in the other two, which involve ZnSe, the discrepancy is quite large (0.5–1.0 eV).

Whether the discrepancies between calculations and experiments were due to fundamental problems with the approach, or merely to inadequate numerical accuracy, has been unclear. Moreover, no clear qualitative picture of band lineups emerged from such calculations. As a result, microscopic calculations, though widely quoted, had relatively little impact on the field of heterojunction band lineups.

More recently, however, Van de Walle and Martin have reported a series of calculations for a number of heterojunction interfaces [3,4]. These calculations appear to be significantly more accurate than the earlier ones, as can be seen in table 1 (column labeled SCIC). This improvement was seen to stem, at least in the case of AlAs–GaAs, from the use of norm-conserving pseudopotentials [4]. However, the most important gain over previous studies is not the modest improvement in accuracy, but rather the large number of interfaces treated in a uniform way, and the attempt to extract important auxiliary information, such as the effect of strain [3] and orientation [3,4]. These results revive the hope that microscopic calculations will play a central role in elucidating the heterojunction band lineup problem.

Table 1
Theoretical results for valence-band discontinuities ΔE_v in eV, for selected interfaces [16] from various approaches: self-consistent interface calculations (SCIC) [4], atom-superposition reference surface approach (ASRS) [5], empirical pseudopotential method (EP) [2], Harrison tight-binding theory (TB) [9], Tersoff interface-dipole theory (ID) [12], Schottky barrier rule ($\Delta\phi_{bp}$[Au]) [12,14], and experiment [16]. Theoretical results are for a unstrained interface.

Interface	SCIC	ASRS	EP	TB	ID	$\Delta\phi_{bp}$	Exp.
AlAs/GaAs	0.37	0.60	0.25	0.04	0.55	0.44	0.50
InAs/GaSb	0.38	0.58	–	0.52	0.43	0.40	0.51
GaAs/InAs	–	–	–	0.32	0.00	0.05	0.17
Si/Ge	0.53 *	0.60	–	0.38	0.18	0.25	0.20
ZnSe/GaAs	1.59	1.48	2.0	1.05	1.20	0.82	0.96
ZnSe/Ge	2.17	2.07	2.0	1.46	1.52	1.27	1.52
GaAs/Ge	0.63	0.59	0.35	0.41	0.32	0.45	0.53

* Unstrained.

It is not appropriate to go into great detail here in describing such calculations. Instead we mention the basic approximations involved, and then summarize the results obtained.

The earlier calculations [1,2] used empirical ionic pseudopotentials, while the more recent calculations [3,4] used the norm-conserving pseudopotentials developed by Hamann et al. [17], which have given very good agreement with all-electron calculations. Other than this, there are only three crucial approximations which enter: the use of "ideal" atomic positions (except for the effects of strain); the finite basis set used to represent the wavefunctions and, more generally, the finite numerical accuracy; and the local-density approximation (LDA) used for correlation and exchange.

These approximations may have a significant effect on the calculated lineups. There are to our knowledge no estimates of the magnitude of the interface dipole associated with displacements of interface atoms from their "ideal" positions. However, LDA is known to give large errors in the calculated bulk band gaps.

If the band gap values are wrong there must be a corresponding error in the band-edge discontinuities. It is usually assumed that the calculated discontinuity is more reliable for the valence band than for the conduction band, and ΔE_c is inferred from the calculated ΔE_v by using eq. (2.2) with the experimental band gaps. However, Carlsson [18] has argued that the error in the LDA band gaps comes from errors in both valence and conduction bands, with the valence-band error in fact being the larger of the two, being typically 0.5 eV.

If this is correct, then the local-density approximation may be a major source of quantitative inaccuracy in numerical calculations of band lineups. Still, one expects substantial cancellation between the LDA errors in the two semiconductors. The cancellation is probably worst in cases where the semiconductors have rather different dielectric constant ϵ_∞, so those may be the cases where LDA give the largest errors.

With these reservations, the results of calculations of ΔE_v for a number of lattice-matched interfaces are given in the first column of table 1. These interfaces were selected by Kroemer [16] as cases where the band lineups are relatively well-established experimentally. [The calculations are for nonpolar (110) interfaces of zincblende-structure semiconductors.]

Experimental discontinuities are also given in table 1. However, in comparing theory and experiment, it is important to bear in mind the possible large uncertainties in the experimental values.

As can be seen in table 1, the calculated values [3,4] for elemental and III–V semiconductors agree with experiment to within 0.1 or 0.2 eV. For the two interfaces involving a II–VI semiconductor with a less polar semiconductor, the discrepancies are more than twice as large. A similar trend is seen in the earlier empirical pseudopotential calculations [2]. Whether the greater discrepancy in those two mixed cases is due to non-ideal atomic positions in the real interface, inaccuracy in the experiment, limitations of the theoretical approach, or other factors, is not known. It is for example possible that the column-VI elements pose a particularly demanding problem in terms of pseudopotentials and numerical convergence.

Van de Walle and Martin also considered the dependence of ΔE_v on the orientation [3,4] and strain [3] at the interface. They found that, in the cases studied, if one considers the center of gravity of the valence maximum (i.e. the average over the three Γ'_{25} states), then the strain has no significant effect on the lineup between these average valence maxima in the respective semiconductors. However, strain at the interface will in general split the states at the valence maximum, so the highest-lying valence state in each semiconductor is shifted up in energy, and the difference between the shifts in the respective semiconductors alters the discontinuity which would be seen in transport experiments (although presumably not that seen in core-level spectroscopy).

Those authors also found that the AlAs–GaAs lineup did not depend upon interface orientation [4]. In addition, they developed a model approach to the band lineup problem [5], which is discussed in section 3.4 below.

3.3. Model theories and reference levels

While detailed calculations such as those described above are helpful and important, model theories permit one to focus more sharply on the essential physics. Such theories involve a minimum of computation, because they reduce the problem to a few relatively simple factors.

The disadvantage of such an approach is that one has no direct knowledge of whether the approximations made to simplify the problem are qualitatively or quantitatively correct. Instead one relies on physical arguments or numerical estimates. Moreover, the quantitative accuracy or inaccuracy of a model theory is not a guarantee of the correctness or incorrectness of the basic physical model. There are always auxiliary assumptions made in extracting numerical results, which may lead to inaccuracy in implementing a fundamentally sound model. Conversely, since many physically distinct quantities are nonetheless strongly correlated, a theory which focuses on the wrong physical quantity may still give a surprisingly good description of experimental results.

There are certain elements which most model theories have in common. It is therefore appropriate to discuss some of these general aspects, before concentrating on specific theories in the sections below.

All popular model theories and empirical rules for band lineups are based on "reference levels". The role of a reference level is to put all semiconductors on a common absolute energy scale [19]. One merely lines up the reference levels in the respective semiconductors. Then the relative energies of the valence maxima in the respective semiconductors can be determined trivially. Of course, there is no guarantee that there even exists a set of reference levels which, when used in this way, will give the true band lineups. However, if such a set of levels exists, then such an approach makes determining band lineups trival, once the reference levels are determined.

The simplest example of a reference level is the Fermi level in a metal. At a metal–metal heterojunction, the true "lineup" can be determined simply by aligning E_F in the respective metals. If one wants to know, for example, the discontinuity in the bottom of the valence band at the interface, it is only necessary to know the position of the valence minimum relative to the Fermi level for the respective bulk metals. No interface calculation is required, and yet the result is exact. In fact, it would be exceedingly difficult to perform a numerical calculation for the interface which would give this "lineup" as accurately as the reference-level scheme, in conjunction with bulk calculations to determine the valence minimum with respect to the Fermi level.

Of course, the same is, strictly speaking, true of a heterojunction. That is, the lineup of band edges *deep in the bulk* of the two semiconductors is found simply by aligning the two Fermi levels. The difference here is that, because the dipole associated with free carriers is so spatially diffuse at reasonable doping levels, for the heterojunction we are also interested in the lineup of features at distances much smaller than the depletion width. Unlike the lineup of features deep in the bulk, the band lineup which we are interested in cannot be shown rigorously to depend only on a well-defined bulk thermodynamic quantity.

More generally, consider any reference level E_r which is to be aligned in the respective materials. We make the crucial assumption that E_r is a well-defined property of the bulk semiconductor. The implications of this assumption are explored below.

For specificity, we consider the problem of determining the valence-band discontinuity between two semiconductors, although the valence-band maximum could be replaced with any feature whose lineup across the interface is desired. Then we need to know only the valence-band maximum relative to E_r in the bulk, which we denote as

$$E_v^r \equiv E_v - E_r. \tag{3.1}$$

At a heterojunction between materials A and B, the reference-level approach consists of assuming that E_r is the same on both sides of the interface, i.e.

$$\Delta E_r(A, B) \equiv E_r(B) - E_r(A) = 0. \tag{3.2}$$

In this case, by combining eqs. (3.1) and (3.2), one trivially obtains

$$\Delta E_v(A, B) = \Delta E_v^r(A, B). \tag{3.3}$$

Thus the interface band lineup is obtained strictly in terms of *bulk* properties of the semiconductors, specifically E_v^r, which are presumably much easier to calculate than interface properties. Of course, we have not yet shown that a reference level with these properties can or does exists. Various proposals for such reference levels are discussed in this section, and two particularly important examples are analyzed in more detail in subsequent sections.

The most familiar example of a reference level used in the context of semiconductor heterojunctions is the "electron affinity rule" [6], which may be rephrased as an ionization potential rule. This is simply the case where the reference level E_r is taken as the vacuum level E_0 outside the surface. The situation is shown in fig. 2. If the ionization potential of semiconductor A is $\phi(A)$, then the valence maximum (relative to the vacuum level) is

$$E_v^r(A) \equiv E_v^0(A) = -\phi(A). \tag{3.4}$$

Here the superscript zero simply denotes the particular choice of reference level, i.e. E_r taken as the vacuum level. Then from eq. (3.3) one obtains

$$\Delta E_v(A, B) = -\Delta \phi(A, B), \tag{3.5}$$

as shown in fig. 2. [Hereafter we will frequently abbreviate differences such as $\Delta E_v(A, B)$ by ΔE_v, with the suffix (A, B) being understood.]

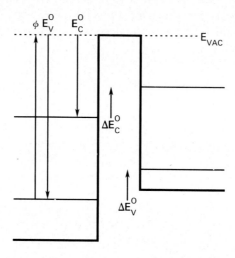

Fig. 2. Schematic picture of electronic structure of two semiconductor surfaces, separated by a small vacuum region, showing quantities relevant to the electron affinity rule. The bold line is the potential associated with the two semiconductors, the lighter horizontal lines show the valence-band maximum and the conduction-band minimum versus position. The ionization potential is ϕ, and E_{VAC} is the vacuum level (dashed line). The valence maximum and conduction minimum, relative to the vacuum level, are E_v^0 and E_c^0. Valence- and conduction-band discontinuities in the absence of any interaction between the surfaces are ΔE_v^0 and ΔE_c^0. Arrows indicate sign convention, with up positive.

For historical reasons, this approach was first formulated using the electron affinity, which specifies the position of the conduction minimum relative to the vacuum level. The same analysis then yields the conduction band difference as the difference in the two affinities. Obviously, if the band gap is known, the two formulations are equivalent using eq. (2.2), and we refer to the "electron affinity rule" in deference to tradition, while actually focusing on the ionization potential for purposes of discussion or calculation.

So far this is rigorously correct, because as long as the semiconductors remain out of contact and charge neutral, the vacuum level remains as a well-defined external reference. However, this is not yet a theory of band lineups, since we have assumed that the semiconductors are not in contact. Changes might take place upon bringing the two semiconductors together, which would render this analysis invalid.

The affinity rule is discussed in detail in sect. 3.4; there we consider the conditions necessary eq. (3.5) to remain approximately valid when the two semiconductors are brought into contact. Another recent reference-level approach which has been particularly successful in giving accurate band lineups is discussed in sect. 4. The affinity rule is used here only to illustrate how simply a theory works when based on reference levels.

The mere existence of a reference level for lining up the bands in semiconductors has important implications. Since the band edges of all semiconductors can then be put on a common absolute energy scale, the band lineups of different pairs of semiconductors are not independent. In particular, from eq. (3.3) above one immediately obtains the important rule of *transitivity*, as discussed by Kroemer [15]:

$$\Delta E_v(A, B) + \Delta E_v(B, C) = \Delta E_v(A, C). \tag{3.6}$$

The existence of a *bulk* reference level also implies that the band lineup is independent of interface orientation.

While it is not possible to prove a general rule by a few specific examples, the present evidence tends to support the validity of the transitivity rule, and the lack of dependence of lineup on interface orientation. (We exclude from the discussion cases where one orientation or growth order leads to poor sample quality, as for heteropolar semiconductors grown on homopolar substrates [15], or where there is a "built-in" dipole at the interface, as at an ideal Ge–GaAs(100) interface [20].) Both experiments [21] and calculations [3,4] indicate that the lineup is independent of orientation in those cases studied, and the transitivity rule is also well satisfied in experimental [15,22,23] and theoretical [4] tests.

If the transitivity rule holds, and the lineup is orientation-independent, then one reference level for each semiconductor must suffice for predicting band lineups. (If the lineup depends on orientation, one might still get by with one reference level for each surface of each semiconductor.)

One expects in general that any reference-level picture may be applied to strained interfaces, but that one must then use the position of the valence maximum relative to the reference level in the strained bulk. Van de Walle and Martin found [3] in numerical calculations of Si–Ge interfaces that a reference level sufficed to describe the alignment of the respective (three-fold degenerate) valence maxima, *except* that one must then add the effect of strain in lifting the degeneracy, and raising the highest of the strain-split valence states. In other words, strain did not shift the valence states significantly relative to the reference level *on average*, but strain-induced splitting represented a significant correction.

If one bulk reference level for each semiconductor suffices for predicting band lineups, as experiments and calculations seem to suggest, then a table of such levels would be of great value even in the absence of any theoretical understanding. Such a table was begun by Katnani and Margaritondo [23], and further extended by Margaritondo [24]. Since the reference levels are defined only up to an arbitrary constant, Katnani and Margaritondo chose the reference level at the valence maximum in germanium, so $E_v^r(\text{Ge}) = 0$. Each time the band lineup for an interface between germanium and another semiconductor is measured, that semiconductor can be placed in the table by giving its valence maximum relative to that of germanium.

As lineup measurements are never perfectly accurate, it is important to have a test of internal consistency in such a table. For example, Katnani and Margaritondo also measured interfaces with silicon. The difference between the valence-band discontinuity for silicon versus germanium on a given semiconductor should then equal the discontinuity at the silicon–germanium interface. Their data obeys this rule to within about 0.15 eV, consistent with the expected experimental accuracy.

Such a table can be further improved by including all available data for band lineups, and determining the optimum choice of reference levels by statistical methods. If enough data are available, then the problem is highly overdetermined. In that case, the ability to fit the entire data set with only one level per semiconductor provides a strong test of the principle of using such levels. Moreover, the statistical accuracy with which the levels are determined could exceed the accuracy of the original lineup measurements.

Finally, we mention two particularly interesting recent reference-level schemes. It was recently suggested that there should be a correlation between Schottky barrier heights and band lineups, whereby the Fermi-level position at an interface with a metal serves as the reference level for a given semiconductor [12,14]. Then the p-type Schottky barrier $\phi_{bp} = E_F - E_v$ acts as $-E_v^r$ in eq. (3.3) above, and the lineup is given by

$$\Delta E_v = -\Delta \phi_{bp}. \tag{3.7}$$

This relationship gives very good predictions of band lineups [14,24]. The accuracy and theoretical basis of eq. (3.7) is discussed in section 5.2.

The other recent scheme referred to above was introduced by Zunger [25], and was also discussed by Langer and Heinrich [26]. There, the set of deep levels associated with transition-metal impurities serves as the reference level. In the spirit of the electron affinity rule, it has been argued that these deep levels provide a measure of some "internal vacuum level".

However, such an interpretation is by no means necessary for the success of this scheme.

In fact, Tersoff and Harrison [27] have pointed out that the impurity levels should be "pinned" relative to the dangling-bond level or the "neutrality level", rather than free-floating, as the "internal vacuum level" interpretation assumes. This gives a clear physical explanation of the success of the scheme in predicting band lineups, since these lineups are also determined (according to this approach) by aligning the neutrality levels, as discussed in sect. 4. More important, such an explanation accounts for the fact the these impurity levels also correlate very closely with Schottky barrier heights where an affinity-rule picture is almost universally agreed to be inappropriate.

Of course, the whole attraction of an empirical reference level scheme is that its utility is not dependent on a complete physical understanding.

3.4. The electron affinity rule and related approaches

The electron affinity rule is the prototypical reference-level theory, and as such was discussed briefly in sect. 3.3 above. Here the term "affinity rule" is sometimes used to refer collectively to the entire class of theories based on the same physical principle, rather than just the theory based on the experimentally measured electron affinity. It is hoped that this usage will not lead to confusion.

In its oldest form [6], the electron affinity rule simply states that the conduction-band discontinuity at a heterojunction is given by the difference in electron affinity of the respective semiconductors. This rule can be equivalently phrased in terms of the ionization potential (the energy to remove an electron from the valence maximum) and the valence-band discontinuity as in sect. 3.3 above. Since the electron affinity and ionization potential differ simply by the band gap, the choice of formulation is a matter of convenience.

The physical picture is simple. If we bring two neutral semiconductors together from infinity, it is natural to take the vacuum level as the zero of energy. As long as the semiconductors are not allowed to interact, the valence maximum of semiconductor A falls at an energy $-\phi(A)$, where ϕ is the ionization potential. The difference in the positions in energy of the two valence maxima is simply $\phi(A) - \phi(B)$, as shown in fig. 2. *If* this remains true when the heterojunction is formed, then we can write

$$\Delta E_v(A, B) = -\Delta\phi(A, B),$$

which was eq. (3.5) above. In that case the electron affinity rule or the ionization potential rule, eq. (3.5), will correctly predict the band lineup, given an accurate knowledge of the electron affinity or the ionization potential.

Unfortunately this does not give us a bulk reference level. The ionization potential is not a bulk property, but depends on the surface. Different crystal faces have different ionization potentials. The ionization potential also depends in general on the structure of the surface, and is affected by surface relaxations and reconstructions which may be entirely irrelevant to the structure of the heterojunction interface.

Another problem with the affinity rule is that there may be charge transfer at the interface, which would alter the lineup. This effect is discussed in detail later.

Putting aside these objections for a moment, the most fruitful and illuminating approach is to ask what are the necessary and sufficient conditions for eq. (3.5) to remain true when the interface is formed. The energy of the local valence maximum depends only on local properties, in particular the charge density and the local electrostatic potential. In the bulk of either semiconductor, the valence maximum etc. will therefore not be affected by any change far away, except via the long-ranged electrostatic potential.

A redistribution of charge at the interface can set up a dipole, which uniformly shifts the electrostatic potential on one side of the interface, relative to the other side. Since the electrostatic potential is a linear function of the charge density, the electron affinity rule will hold *exactly*, if the interface charge density is exactly equal to a linear superposition of the respective surface charge densities.

Let us make this a little more formal, in order to get a more precise idea of when and how well the electron affinity rule should work. First we let A and B denote, not merely two semiconductors, but two semiconductor surfaces of some specified orientation and structure. If we use the experimental electron affinity, then A and B should refer to the actual structure of the surface on which the measurement was performed. We denote the charge density of the surface A by $\rho(A)$, and similarly for B. The true interface charge density is denoted $\rho(I)$. Then, as discussed above, the affinity rule is rigorously correct and exact if $\rho(I) = \rho(A) + \rho(B)$. It is then convenient to define the difference, i.e. the amount by which this relationship is violated, by

$$\delta\rho = \rho(A) + \rho(B) - \rho(I). \tag{3.8}$$

This represents a thin sheet of charge, with zero monopole moment. In general, $\delta\rho$ may be expected to have a finite dipole moment. In that case, the affinity rule will be violated by precisely the potential drop associated with this dipole sheet $\delta\rho$.

It is sometimes useful to speak of an "interface dipole". This dipole is just $\delta\rho$, or the potential drop associated with $\delta\rho$, depending upon the context. It is important to stress from the outset that this dipole is not uniquely defined. Rather, it is only defined relative to the reference interface whose charge density is $\rho(A) + \rho(B)$. Since $\rho(A)$ and $\rho(B)$ refer to reference surfaces which may in principle be chosen arbitrarily, the dipole may be arbitrarily small or large in any given case depending on the choice of reference surfaces.

As long as we use experimental values for the electron affinity, it is very difficult to assess the validity of the affinity rule. This is because the actual surfaces are in general reconstructed in a complicated way, whereas the interface is not, so that the difference $\delta\rho$ has a very complicated structure, and it is difficult to draw useful generalizations.

This problem can be avoided by using a model surface instead of the real surface. Until recently, such improvements on the electron affinity rule constituted the major avenue of theoretical work in this field [8,9]. In particular, for the affinity rule to be applicable we want $\delta\rho$ to have a small dipole moment. The only simple way to assure this is to have $\delta\rho$ itself be small, which requires that the surfaces be chosen so that $\rho(A)$ and $\rho(B)$, when combined, give a good approximation to $\rho(I)$. Since the structure of the interface is nearly ideal, it is natural then to choose $\rho(A)$ and $\rho(B)$ to correspond to the respective ideal surfaces, or some approximation thereto.

It is worth stressing that there is no guarantee a priori that there exists any choice of reference surfaces which will make $\delta\rho$ have zero dipole moment for all possible pairs of semiconductors. However, if the lineups obey the transitivity rule given by eq. (3.6), then we know that one reference level per semiconductor suffices to give all lineups correctly, and so there necessarily exists some reference surface for which the affinity will work. The question is then whether there is a simple prescription for generating this reference surface.

There remain at least three potential problems. First, there is more than one ideal surface. Any low-Miller-index surface might seem a natural choice, and yet these may have different affinities. However, for homopolar materials such as diamond-structure semiconductors, the affinity may be relatively insensitive to orientation. In particular, if the surface charge density can be well approximated by a superposition of spherical charge densities, the affinity is guaranteed to be independent of orientation [8].

For polar materials, the situation is more complicated. It seems most attractive to concentrate on non-polar faces, e.g. (110) for zincblende structures. However, if the real heterojunction in question has a polar interface with a built-in dipole, such as ideal Ge–GaAs(100), then one should in principle use the affinities for the corresponding ideal polar surfaces. On the other hand, the electrostatic energy may drive intermixing, which would reduce or eliminate the built-in dipole [20]. Such polar interfaces are discussed elsewhere [20], and are not considered further here.

The second problem is that the interface atoms may be displaced somewhat from their ideal positions. For example, at a Ge–ZnSe(110) interface, since the Se atom has a smaller covalent radius that the Zn, one might expect small displacements which would place the Ge closer to the Se, and further from the Zn, than for an ideal interface. Unfortunately, no estimates of the size of such effects are available.

The third problem concerns electronic rearrangements at the interface. Even if the atomic positions remain unchanged when the two surfaces are brought together to form an interface, there may be rearrangement of the electronic charge, in particular charge transfer, which will contribute a net dipole, and hence shift the band lineup. Traditionally, it has been assumed that such shifts are small, and early estimates [9] tended to confirm this assumption, although Frensley and Kroemer [8] suggested that the interface dipole might in fact be important.

More recently, Tejedor and Flores [11] and Tersoff [12] argued that such dipoles were critically important. These approaches based on the central role of the interface dipole are discussed in the following section.

If we assume that interface dipoles can in fact be neglected, we are still left with the highly nontrivial problem of calculating the ionization potential of the appropriate ideal semiconductor surface. As noted above, the best choice is not, in general, the real surface, nor even an exact calculation for the electronic structure of a surface with ideal nuclear positions. Rather, the best reference surface is one chosen so that the linear superposition of two surfaces will be as near as possible to the true charge density of the interface. In fact, if a "reference surface" satisfying this condition can be found, then by construction there will be no effect from interface dipoles, and an electron affinity rule *based on this hypothetical surface* will give accurate band lineups.

This fact highlights the difficulty in establishing whether interface dipoles are important. There is no way to define the dipole except as the deviation from some reference system, such as the reference interface obtained by linear superposition of the two reference surfaces. If the transitivity condition is satisfied, then one can in principle always find a reference system such that the dipole is zero. Thus a theory emphasizing dipoles may be consistent with a theory which neglects dipoles but chooses the reference surface in a clever way.

There is in fact a natural choice for the reference surface. It has long been noted that a superposition of free-atom charge densities provides a rather good starting approximation to the true charge density of most solid-state systems. Thus, if we construct the reference surface by superposition of atom charge densities, it may turn out that the dipole is small, perhaps even negligible. There is still a minor problem, that the atom-superposition charge density is not an exact description of the bulk, but this can be dealt with in various ways.

Two approaches based on such reference surfaces are worth discussing in further detail. These are due to Harrison [9] and to Van de Walle and Martin [5]. Harrison used a tight-binding approach which takes as its starting point the free-atom term values plus interatomic matrix elements. The valence maximum is then calculated on the absolute energy scale determined by the free-atom term values, i.e. relative to the vacuum level. The use of free-atom term values suggests that the reference surface is a nonpolar surface generated by superposition of atom charge densities, although within the tight-binding approach the reference surface is never specified explicitly.

The Harrison approach [9] is reasonably successful in predicting band lineups. Results of this tight-binding method are given in the column labeled TB in table 1. The typical errors are seen to be perhaps 0.2–0.4 eV. Since the tight-binding approximation, and neglect of relativistic effects, are already enough to give errors of this magnitude, it is impossible to say whether the accuracy limitations are attributable to the neglect of dipoles for these reference interfaces, or merely to limitations in the tight-binding approach.

[In fact, Harrison and Tersoff [28] have shown that the effect of interface dipoles can be easily included in the tight-binding approach, in a manner analogous to that described in sect. 4. This appears to give at most a slight improvement in accuracy. An analysis of the magnitude of relativistic corrections suggests that these may be the dominant source of error in the tight-binding approach as implemented [28], masking the smaller changes associated with inclusion of interface dipoles. While the relativistic effects could be included in the tight-binding scheme in a straightforward manner, this has not yet been done.]

Van de Walle and Martin [5] took an ab initio approach to the problem of choosing a model surface. They constructed the atom-superposition charge density for a nonpolar (110) zincblende surface, and used it to determine the average electrostatic potential in the crystal bulk. The bulk bandstructure calculations for the respective semiconductors could then be placed on a common absolute scale (i.e. relative to the vacuum level) by setting the average electrostatic potential in the bulk, which would otherwise be arbitrary, so as to be equal to that obtained with the reference surface.

The resulting lineups were, overall, about equal in accuracy to the full self-consistent-field interface calculations, and rarely differed from the full calculation by more than 0.2 eV. These results are given in the column marked ASRS in table 1. This suggests that the atom-superposition reference surface is indeed a choice which successfully minimizes the need to include dipole corrections.

Finally, it is worthwhile to mention two more approaches within the context of the electron affinity rule, those of Frensley and Kroemer [8] and of Freeouf and Woodall [10]. Frensley and Kroemer pointed out that, if the total charge density could be represented as a superposition of spherical charges, and if these were sufficiently localized that the charge density in the interstitial positions were negligible, then the average electrostatic potential in the interstitial positions would be equal to the electrostatic potential at infinity (for the reference surface obtained by superposition of the spherical charge densities). They concluded that this average interstitial potential could be taken as a reference level. This might remain a good approximation, even if the conditions for it to hold rigorously are not satisfied. Moreover, it has the clear advantage that only a single calculation for the actual bulk potential is required. The reference surface is implicit, but is never used.

The predictions made using the Frensley–Kroemer approach were somewhat disappointing [15]. However, in view of the severe approximations made in performing the calculations, it is impossible to determine from those calculations how well the scheme would work in principle. It would be very interesting to examine the lineups which would be obtained by using as a reference level the interstitial potential determined by a state-of-the-art calculation. In fact, in the few cases where this has been done, the results are rather encouraging [25].

While we concluded above that using experimental electron affinities made it impossible to assess the validity of the affinity rule, we never addressed the question of whether such a procedure gave good lineup predictions in practice. While Kroemer has criticized such use of the electron affinity rule [15], Freeouf and Woodall [10] have argued that experimental affinities in fact give rather good results for band lineups. However, in view of the scatter among published affinity data, it is difficult to make any definitive assessment of this approach.

Finally, it is worth mentioning the "common anion rule", which states simply that the valence-band discontinuity between two semiconductors with the same anion will be "very small" [8]. The physical picture behind this rule is simply that the states at the valence maximum are associated primarily with the anion. Within the tight-binding theory of Harrison [9], it is easily seen that this rule should be accurate *if* interface dipoles can be neglected. Thus, the rule may be regarded a special case of the electron affinity rule. Nevertheless, it is far less valuable, because it applies to only a small subset of possible interfaces. More important, this rule is known to fail for AlAs–GaAs [29], and it has recently been suggested [30] that the rule is least accurate for lattice-matched heterojunctions, which are the heterojunctions of greatest practical interest.

3.5. *Role of interface dipoles*

As discussed above, it seems that one can calculate lineups with some success by choosing a reference surface for each semiconductor so as to minimize the need for including additional dipoles when the interface is formed. However, such an approach has its limitations. It requires a clever guess; and quantifying how good that guess is, is in principle no easier than calculating the dipole for an arbitrary reference interface.

A completely different approach is to focus on the dipole, and how it depends on the electron affinity difference (or, equivalently, on the choice of reference interface). If this dependence is sufficiently simple, one may be able to solve the problem without appealing to a particular choice of reference surface.

The best example of this is the metal–metal interface. At a metal–metal junction in equilibrium, the Fermi levels in the bulk of the two metals must be at the same energy level. If one imagines starting with a discontinuity in the local Fermi level, charge will then flow so as to form a dipole. This flow will continue until the dipole is just large enough that the two Fermi levels be identical.

Choosing either the atom-superposition charge density or the actual surface as the reference surfaces gives a "lineup" which leaves a substantial misalignment of the Fermi levels in the respective metals. In fact, using the original electron affinity rule based on experimental affinities gives an error in the lineup which, in this case, is precisely the contact potential difference. That approach therefore does not provide a good approximation for the metal–metal heterojunction.

On the other hand, by focusing on the dipole the problem can be solved simply and accurately. Any mismatch in the Fermi levels leads to an unbounded flow of charge from one metal to the other, which in turn gives an infinite dipole whose sign is such as to raise the Fermi level in the metal whose Fermi level had been lower. Thus, the only stable lineup is where the Fermi levels coincide, regardless of the starting point or the reference surface.

For semiconductors, the situation is not so trivial. Section 4 below is in fact devoted primarily to generalizing this rather trivial treatment of metals to include semiconductors as well. The net result is a theory which does not depend sensitively on the choice of the starting point or the reference surface, and which yields rather accurate predictions of band lineups. Moreover, the approach also gives Schottky barrier heights at the same time, whereas there has been no success in finding reference surfaces which give quantitative accuracy for Schottky barriers, any more than for metal–metal junctions.

Finally, it is worth mentioning that charged defects or impurities can also contribute a dipole at the interface, which would alter the apparent band discontinuity. While reasonable numbers of defects (well under a monolayer) have been shown experimentally [31] and theoretically [32] to have little effect on the lineup, dopant impurities can be concentrated at the interface in sufficient numbers to give an appreciable local dipole, altering the effective lineup [33]. Such extrinsic interface dipoles will not be considered in this chapter.

4. A recent approach to the band lineup problem

4.1. Charge transfer and interface dipoles

Whenever two dissimilar materials are brought together, there will in general be some charge transfer, which will lead to an interface dipole. If one material is much more electronegative (attractive to electrons) than the other, then electrons will flow from the less electronegative material to the more electronegative material. Here we are interested in determining how such charge transfer dipoles affect heterojunction band lineups. However, it is helpful to consider first the extreme limits of metallic and insulating behavior.

For metals, the simplest measure of electronegativity is the workfunction. A metal with a large workfunction has a Fermi level which is very low (very attractive to electrons) relative to the vacuum level. When two metals are brought together, charge flows from the less electronegative metal to the more electronegative one. Since metals have free carriers, there is nothing to impede this flow except the electrostatic interaction among the electrons.

As long as the bulk Fermi level is lower in one metal than in the other, charge will flow. However, this charge transfer results in a dipole which raises the Fermi level in the more electronegative metal, relative to the other metal. When this dipole exactly cancels the difference in workfunction (i.e. the difference in electronegativity), the interface will be in equilibrium and no further charge will flow.

Thus the dipole, for the metal–metal junction, is exactly equal to the mismatch in electronegativity (i.e. workfunction) in the respective metals. This mismatch, and thus the dipole, still of course depends on the reference surface used, which affects the workfunction and hence the position of the Fermi level relative to the vacuum level. However, the final self-consistent lineup is completely independent of the reference surface, and is easily obtained simply by aligning the respective bulk Fermi levels.

Note that, just as for the affinity rule, the lineup may be described by a reference level. The critical difference is that now the reference level is truly a bulk quantity (the neutrality level), whereas for the affinity rule, the reference level was in principle surface-dependent (although in practice this could be circumvented within some level of approximation).

The opposite extreme from a metal–metal interface is an interface between two highly insulating, nonpolarizable solids, e.g. two rare-gas solids. In that case, either the ideal surface or the superposition of free-atom charge densities forms a very good reference surface. We expect little charge rearrangement, and hence an extremely small interface dipole relative to the free surfaces, because the electrons are so tightly bound.

A semiconductor should lie somewhere in between these two extremes. In order to quantify where on the scale between metals and insulators the semiconductors of interest lie, we require a more quantitative analysis of the interface dipole. Such a description is the subject of the next two sections.

4.2. Linear response theory: the self-consistent lineup

The most natural and powerful way to include the effect of interface dipoles at the heterojunction interface is through linear response theory.

This was first done by Tejedor and Flores [11], and we follow their analysis, with some modifications, in this section.

We begin by introducing an assumption, which will be justified in the next section. That assumption is that there exists a "neutrality level" for each semiconductor, which is characteristic of the bulk semiconductor, and which plays a role analogous to the Fermi level in a metal. Specifically, we assume that if, before the semiconductors are brought together, the two neutrality levels coincide, then there will be no charge transfer when the interface is formed. If the neutrality levels do not line up, then the semiconductor with the lower neutrality level is in effect more electronegative than the other, and some charge transfer will occur. For the time being, this is all we need to know about the neutrality level. The physical meaning of such a neutrality level for a semiconductor is the subject of the next section.

Let us denote the neutrality level in a semiconductor E_n, and its position relative to the vacuum level E_0, by E_n^0. As before the valence maximum relative to the vacuum level is called E_v^0, with the superscript 0 indicating that the quantity is given relative to the vacuum level. Similarly, E_v^n is the valence maximum relative to the neutrality level. Then for semiconductors A and B, in the absence of any interface dipole, the valence band discontinuity

$$\Delta E_v \equiv E_v(B) - E_v(A) \tag{4.1}$$

becomes simply

$$\Delta E_v \to \Delta E_v^0, \tag{4.2}$$

where $\Delta E_v^0 \equiv E_v^0(B) - E_v^0(A)$. These relationships are illustrated in fig. 3a.

We assume that the reference surfaces A and B have been chosen so that, *except* for the problem of charge transfer, the superposition of the reference surfaces would give a good description of the interface, and $\delta\rho$ would be negligible. However, if there is a small mismatch in the two neutrality levels, then there will be a small additional dipole δ linear in this mismatch,

$$\delta = -\alpha[E_n(B) - E_n(A)]. \tag{4.3}$$

Figure 3b shows schematically the lineup at the interface when this dipole in included. Note that for convenience we use δ to refer to the potential drop associated with the interface dipole. Thus the susceptibility α is

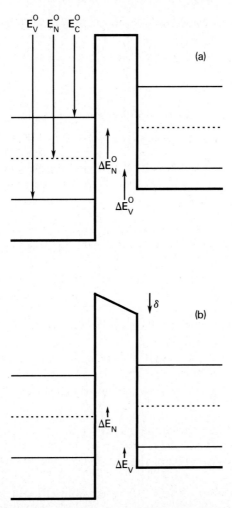

Fig. 3. Qualitative relationship between interface dipole and band lineup. (a) Schematic picture of electronic structure of two semiconductor surfaces, separated by a small vacuum region, as in fig. 2. Here the "neutrality level" relative to the vacuum level, E_n^0, is also shown, along with the associated discontinuity ΔE_n^0 for the case of noninteracting surfaces. (b) Same as above after surfaces interact to form true heterojunction, including interface dipole δ. Picture is still drawn with a gap between the surfaces for visual clarity. Actual discontinuities in E_n and E_v are denoted by ΔE_n and ΔE_v.

dimensionless, and is proportional to the usual suceptibility (which refers to the induced charge density).

The mismatch in the neutrality levels at the interface,

$$\Delta E_n \equiv E_n(B) - E_n(A), \qquad (4.4)$$

depends on the dipole, as shown in fig. 3b:

$$\Delta E_n = \Delta E_n^0 + \delta = \Delta E_n^0 - \alpha \Delta E_n. \qquad (4.5)$$

The problem must then be solved self-consistently, and the solution is

$$\Delta E_n = (1 + \alpha)^{-1} \Delta E_n^0, \qquad (4.6)$$

i.e. any mismatch ΔE_n^0 in the starting neutrality levels is reduced by a factor of $1 + \alpha$ due to the charge-transfer dipole δ. Similarly, we can solve for the dipole δ,

$$\delta = -\frac{\alpha}{1 - \alpha} \Delta E_n^0. \qquad (4.7)$$

Remember, however, that E_n^0 depends on the choice of reference surface, and hence so does the dipole δ.

What we are really interested in here is the valence-band discontinuity ΔE_v, which can be written

$$\Delta E_v = \Delta E_v^0 + \delta. \qquad (4.8)$$

Instead of giving the valence maximum relative to the vacuum level, it is sometimes convenient to give it relative to E_n. That way, even though E_v^0 and E_n^0 are surface-dependent, the difference $E_v^n \equiv E_v^0 - E_n^0$ is a true bulk quantity.

Then a little algebra suffices to give the desired result,

$$\Delta E_v = \frac{\alpha}{1 + \alpha} \Delta E_v^n + \frac{1}{1 + \alpha} \Delta E_v^0. \qquad (4.9)$$

This equation has two important limits, to which we have already referred qualitatively. For a metal, any mismatch ΔE_n in the neutrality level leads to unbounded charge transfer and hence an infinite dipole, so $\alpha \to \infty$, and the self-consistent lineup can be obtained from eq. (4.9) as

$$\lim_{\alpha \to \infty} \Delta E_v = \Delta E_v^n. \qquad (4.10)$$

This is equivalent to $\Delta E_n = 0$, i.e. to lining up the neutrality levels. On the other hand, for a rare-gas solid we expect a negligibly small susceptibility α, so the appropriate limit is

$$\lim_{\alpha \to 0} \Delta E_v = \Delta E_v^0. \tag{4.11}$$

This corresponds to the electron affinity rule.

It is worth noting that, regardless of the value of α, we can define a reference level such that aligning the reference levels of the respective semiconductors will give the self-consistent lineup. Specifically, if we defined the level

$$E_r = \frac{\alpha}{1+\alpha} E_n + \frac{1}{1+\alpha} E_0, \tag{4.12}$$

then the self-consistent lineup, eq. (4.9), corresponds to $\Delta E_r = 0$, i.e. one simply lines up this reference level in the respective semiconductors. Thus the success of empirical reference levels gives us no hint of which physical limit is appropriate.

In the limit $\alpha \to 0$ it is necessary that the reference surface be carefully chosen, as discussed for the affinity rule in sect. 3.4. For $\alpha \to \infty$ the neutrality level must be known accurately. To deal with the intermediate case, both these conditions must be satisfied. As we show below, the cases of most interest lie near the $\alpha \to \infty$ limit, so while an accurate neutrality level is required, the results are not very sensitive to the choice of reference surface.

We have assumed that the dipole and hence the susceptibility could be treated as scalar quantities, whereas in fact the dipole may have a complicated spatial distribution, and the susceptibility is a nonlocal operator which is not even diagonal in reciprocal space. Fortunately, this more complicated behavior does not appear to affect any of the conclusions here [34].

There remains one crucial question, which is the magnitude of the susceptibility α. This can in principle be calculated directly; Tejedor and Flores, for example, estimated that $\alpha \simeq 2.5$ for typical III–V semiconductors. In that case a situation intermediate between the affinity rule and the neutrality rule is obtained, with deviations (they estimated) of typically about 0.2 eV from the strong neutrality rule, eq. (4.10).

However, we believe that it is both more intuitive and more accurate to use a simple thought experiment [12,35] to estimate the susceptibility. We begin by considering a homojunction, in which two ideal surfaces of a given semiconductor are brought together to form an interface, which is simply the perfect bulk. We then make the semiconductor on one side more electronegative, i.e. we decrease E_v^0 and E_n^0 on that side. We can imagine doing this by adding a constant attractive background potential in that semiconductor, i.e. by changing the zero Fourier component of the pseudopotential. This is, however, entirely equivalent to adding a dipole sheet at the surface. The spatial arrangement of the dipole depends on how we let the additional background potential go from attractive in the bulk to zero in the vacuum.

If we shifted the potential, and hence E_n^0 and E_v^0, by an amount V_x, this corresponds to adding a surface dipole $\delta_x = V_x$. When the interface is formed, the resulting valence-band discontinuity is, from eq. (4.6),

$$\Delta E_v = \frac{1}{1+\alpha} V_x, \qquad (4.13)$$

using the facts that $\Delta E_v^n = 0$ and $\Delta E_v^0 = V_x$.

However, this situation also corresponds exactly to embedding a sheet dipole in the ideal bulk semiconductor. From dielectric response theory, we know that embedding a sheet dipole δ_x leads to a net screened dipole of precisely

$$\delta = \delta_x/\epsilon, \qquad (4.14)$$

where ϵ is the static long-wavelength dielectric constant. Since the screening charge in a semiconductor is almost completely confined to a region of a few Å around the charge being screened, the observable valence-band discontinuity in this case corresponds simply to the net screened dipole, and it is not possible in practice to separate out the contributions from the initial dipole δ_x and the screening dipole $\delta = -[\alpha/(1-\alpha)]\delta$.

By equating δ and δ_x in eq. (4.14) with ΔE_v and V_x in eq. (4.13), the screened and unscreened valence-band discontinuity in this thought experiment, we obtain the simple result

$$\alpha + 1 = \epsilon, \qquad (4.15)$$

i.e. α is just a dimensionless version of the usual bulk dielectric susceptibility. For typical elemental and III–V semiconductors, $\epsilon \sim 10$. If we consider the general lineup result, eq. (4.9), as a weighted average of the two limits, eqs. (4.10) and (4.11), the actual lineup corresponds to 10% of the affinity-rule lineup plus 90% of the neutrality-rule lineup. If this reasoning is correct, then clearly the neutrality rule represents the best zeroth-order approximation.

In fact, the correction to the neutrality-rule lineup is just the difference between eq. (4.10) and eq. (4.11),

$$\Delta E_v - \Delta E_v^n = \frac{1}{\alpha}\left(\Delta E_v^0 - \Delta E_v\right). \tag{4.16}$$

Since the affinity rule is typically accurate to about 0.5 eV, we can conclude that $\Delta E_v^0 - \Delta E_v$ is no more than 0.5 eV in magnitude, in which case the correction to the neutrality lineup is of order 0.05 eV or less. Thus it appears that the neutrality rule alone is quite adequate for quantitative calculation of lineups:

$$\Delta E_v \simeq \Delta E_v^n. \tag{4.17}$$

Moreover, any change in the reference surfaces will have only a weak effect on the calculated lineup, since this change in ΔE_v^0 is to be divided by the dielectric constant in eq. (4.9). In fact, given the uncertainties involved in determining E_v^0, it is not clear that there is anything to be gained by including the correction given by eq. (4.16), i.e. by using the full result, eq. (4.9), instead of eq. (4.17).

One should bear in mind that, with a good choice of reference surface, the affinity rule can also provide reasonably accurate predictions for heterojunctions. The greatest advantage of the neutrality rule is that it can handle semiconductor–semiconductor, metal–semiconductor, and metal–metal interfaces in a simple, accurate and unified way. This point is discussed further in sect. 5.

In conclusion, it should be quite adequate for the present purpose to use the straight neutrality rule, eq. (4.17), to calculate band lineups. The errors due to neglecting the correction given by eq. (4.16) are small compared to any reasonable estimate of the accuracy with which the neutrality level can be calculated, or of the accuracy of the rule itself (which, at the least, is limited by the accuracy with which the transitivity rule is obeyed). These points are discussed further in the next section, which addresses the physical interpretation of the neutrality level, which was simply assumed to exist in the discussion above.

4.3. The neutral lineup

We have not yet addressed the question of how to determine the neutral lineup, or whether there really exists a set of reference levels (the "neutrality levels" E_n), with the property that aligning the neutrality level in the two semiconductors in fact gives the neutral lineup. That question is the subject of this section.

The question of the neutrality level can be answered rather simply within the context of tight-binding theory. Harrison and Tersoff [28] have shown that, in the bond-orbital approximation, the semiconductor's sp^3 hybrid energy (the dangling-bond energy) plays the role of the neutrality level.

This result is intuitively satisfying. The hybrid energy gives a measure of the electronegativity which is specifically appropriate for the solid. As discussed above, charge transfer will lead to an interface dipole which nearly cancels the electronegativity difference, i.e. which equalizes the hybrid energies on the two sides of the interface.

While the tight-binding approach gives perhaps a clearer physical picture than any other model, in applying it to real semiconductors there are several sources of quantitative inaccuracy [28]. We therefore emphasize a rather different approach in this chapter.

Tejedor and Flores [11] showed that, in a one-dimensional model appropriate in the limit of small band gap, the neutral lineup was obtained by aligning the centers of the band gaps of the two semiconductors. Thus, at least in this special case, the center of the band gap could serve as the neutrality level in sect. 4.2.

These ideas were applied [11] to a few real semiconductors with reasonable success. However, those authors did not calculate enough lineups for a convincing test of the theory.

Unfortunately, that ingenious and promising theory went largely unnoticed in the semiconductor community. More recently, these principles were re-discovered independently by the present author [12], who suggested a convenient ansatz for calculating the neutrality level for real semiconductors [36].

Although the result of Tejedor and Flores regarding the neutral lineup in their one-dimensional model was obtained analytically, it can actually be obtained by inspection without any calculation, once the underlying physical picture is understood. We therefore emphasize here the qualitative origin of interface dipoles. For a more formal approach, the reader is referred to ref. [11].

First, let us return to the thought experiment of the homojunction mentioned above. In that case the neutral lineup was trivially the bulk lineup, with no band discontinuity. That is, if a (nonpolar) plane in the bulk is arbitrarily designated as the interface, then there is no dipole across that "interface", nor is there net charge induced by the interface in any unit cell.

On the other hand, when the semiconductor on one side was shifted in energy to give a band discontinuity, this induced a dipole. We estimated that dipole using a result of dielectric theory, without actually looking at the microscopic origin of the dipole. Considering the origin of that dipole now will help us generalize to the case of a true heterojunction.

Figure 4a shows schematically the electronic structure of our hypothetical interface. The bands on the left are shifted down in energy by an

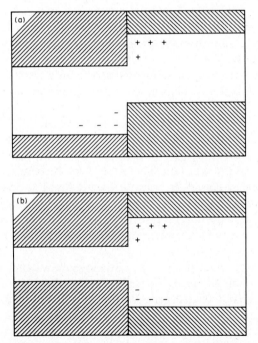

Fig. 4. Two simple examples of relationship between band lineup and interface dipole. Figures may be viewed as describing a one-dimensional model, or as referring to a single interface wavevector $k_{\|}$. Crosshatching shows projected bulk bands versus position. Symbols + and − are intended to suggest the net local charge which is due to the presence of gap states. (a) A single semiconductor in which a band discontinuity is artificially induced, e.g., by imposing an external step potential or embedding a dipole sheet. (b) An interface between two semiconductors, both with "symmetric" valence and conduction bands, as described in text, but with unequal band gaps.

amount V_x, but otherwise the two semiconductors are assumed to be identical. Consider an electron at the top of the valence band of the semiconductor on the right. That electron feels a potential barrier of height V_x preventing it from moving into the left semiconductor. Nevertheless, the electron is not strictly confined to the right side, but rather decays exponentially into the semiconductor on the left. This "tunneling" results in a finite charge density on the left due to states near the upper valence maximum, even though on the left these are "forbidden" states in the (local) band gap.

It is not hard to imagine that this tunneling leads to an excess electron density on the left, and a deficit on the right, due to the extra electrons tunneling over. However, the origin of this charge is actually somewhat subtle. In particular, there is also a net charge associated with tunneling of empty conduction states, as suggested pictorially in fig. 4a.

Of course, there is really no charge coming directly from empty states. But because of a sum rule, the distribution of the empty states affects the distribution of the filled states, and hence the charge density [37,38]. In order to understand how these factors lead to an interface dipole, we must first discuss the sum rule, and how it comes into play when gap states are present.

There is no need for a formal development here. Instead we confine ourselves to describing a few simple results. Put simply, the sum rule states that at any site, the integrated local density of states (including both filled and empty states) is a constant. Therefore, if we introduce a state in the gap by changing the potential or boundary condition, that state must take its spectral weight from the local valence and conduction bands. Qualitatively speaking, states at the top of the gap take their spectral weight primarily from the conduction band, while states near the bottom of the gap come primarily from the valence band.

Now consider a state in the gap, just above the valence maximum. That state takes most of its weight from the valence band, but also a little from the conduction band. If that state is filled, this amounts to filling, not only the local valence band, but also a small fraction of a state in the local conduction band. Thus a filled gap state leads to a net negative charge.

If the filled state lies higher in the gap, it has more conduction character, and so leads to a greater net negative charge. Conversely, an empty state high in the gap leads to a small net positive charge locally, whereas an empty state low in the gap leads to a larger net positive charge. This effect has been analyzed in detail by Appelbaum and Hamann [37] and by Kallin and Halperin [38].

Now we return to the example of the homojunction which was discussed above. In fig. 4a there will be states near the top of the valence band on the right, which tunnel into the band gap on the left. In general, the Fermi level must lie above (or very near) the valence maximum on the right, so these states will be occupied, leading to a net charge transfer to the left as discussed above. This charge is suggested visually by the (-) symbols in fig. 4a. Similarly, there are states near the conduction minimum on the left tunneling into the gap on the right. The Fermi level must lie below (or very near) the conduction minimum on the left, so these states are empty, leading to a charge deficit on the right, as suggested by the (+) symbols. [While the positive charge density arises because of a reduced spectral weight in the valence band, it seems most graphic to associate this charge with the empty gap states, which are fractionally valence-like. This is, however, just a book-keeping convention.]

In principle, the Fermi level might fall well above the lower conduction minimum in fig. 4a, as in the accumulation layer pictured in fig. 1. In reality, though, this only occurs because of the exceedingly small density of states associated with the conduction minimum in some heteropolar semiconductors. This effect does not, in general, occur in one dimension. In three dimensions, there is a picture like fig. 4a associated with every wavevector $k_\parallel$ parallel to the interface, and only for an almost negligible subset (associated with $k_\parallel \simeq 0$) does the Fermi level ever fall above the projected band edge. Thus, fig. 4a still applies in the case of an accumulation layer.

The net dipole in fig. 4a is clearly of the same sign as inferred earlier from dielectric screening, and in fact the discussion above is, in that simple case, just another way of looking at the dielectric response. Now we consider a slightly less trivial example, as the first step towards treating a general heterojunction.

Consider the somewhat artificial case of a semiconductor where the conduction band has exactly the opposite dispersion from the valence band, i.e. in the bandstructure $E(k)$ the conduction band looks just like the valence band, only "upside down" in energy. In that case, if we join two such semiconductors with different band gaps as shown in fig. 4b, there will be no net charge transfer or dipole if the centers of the two gaps fall at the same energy. This is guaranteed by symmetry for this artificial bandstructure, assuming that the atomic structure of the interface does not break the left–right reflection symmetry. As suggested in fig. 4b, the valence-band discontinuity tends to lead to charge transfer from left to right, as might be expected. What is less intuitive, is that the conduction-band discontinuity tends to give right-to-left charge transfer. (This occurs indirectly, via the influence of these gap states on the occupied states, as discussed above.) It may be helpful to think of this as transfer of holes from left to right.

Figure 4b thus represents the first nontrivial case of the "neutral lineup" which was discussed in section 4.2. An obvious choice for the reference level E_n in this case would be the center of the band gap. It is natural then to ask to what extent this can be generalized to two arbitrary semiconductors.

In one dimension, there is only one energy in the band gap which is "special". This is the branch point E_B in the complex bandstructure, which separates the more valence-like gap states from the more conduction-like ones. Therefore, a natural choice for the effective midgap energy is this branch point, which is illustrated in fig. 5. In fact, this energy is the natural division between the valence and conduction bands, as discussed by Kohn and Rehr [39].

Because E_B represents the energy at which the states cross over from being more valence-like to being more conduction-like, it seems reasonable

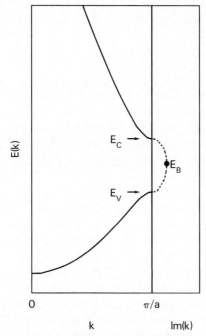

Fig. 5. Complex band structure in one dimension. Left panel shows usual free-electron-like energy bands $E(k)$, with band gap at the edge of the Brillouin zone. Right panel shows energy versus $\text{Im}(k)$ for $\text{Re}(k) = \pi/a$. The energy where $\text{Im}(k)$ is maximal is the branch point E_B.

that when the branch points are aligned on the two sides, the valence- and conduction-band tunneling will approximately cancel. This is exact in the small-gap limit, where the branch point becomes arbitrarily close to the center of the gap. Unfortunately, an exact analysis of the general case would probably be no simpler than doing a full microscopic interface calculation.

In conclusion, then, we take as an ansatz that the branch point E_B plays precisely the role required for the neutrality level discussed in sect. 4.2. Then if by convention we measure E_B relative to the valence maximum, ie. $E_B = E_n - E_v$, we can write

$$E_v^n = -E_B. \tag{4.18}$$

This becomes rigorously correct in certain limits, specifically a small gap or a band structure which is symmetric with respect to energy about the center of the gap.

Since, in applying this to real semiconductors, we will have to make further approximations, it is difficult to be certain how accurate this assumption is. Real semiconductors are not so far from the small-gap limit, so this approach certainly should not go too far wrong.

4.4. Calculating the neutral lineup

For one-dimensional semiconductors, we now have a working procedure for calculating band lineups. In sect. 4.2 we concluded that it sufficed to align the "neutrality levels" in the two semiconductors, and in sect. 4.3 we suggested that the branch point E_B provided a good approximation for the neutrality level.

It is not obvious, however, how we should apply this in three dimensions. One approach, taken by Tejedor and Flores [11], is to break down the three-dimensional problem into an infinite set of one-dimensional problems, one for each value of $k_\parallel$ of the two-dimensional interface wavevector. Then one may choose some finite sampling of wavevectors, and sum the dipoles associated with each sample wavevector to obtain the total interface dipole.

Unfortunately, this does not immediately give us a reference level. In fact, if taken to its logical conclusion, this reduces to a microscopic interface calculation. However, by making some approximations regarding the dipole associated with each interface wavevector, a reference level may be extracted. Also, since the orientation appears explicitly in this approach, one does not obtain a bulk reference level, unless one shows, either analytically or by examples, that the net result is orientation-independent.

In the interests of simplicity, we may choose to *assume* that there exists a bulk reference, and then develop a procedure for calculating it by analogy with the one-dimensional case [13,36]. Such an approach is reasonable, but can never be rigorous; its ultimate justification lies in its success in calculating band lineups for real materials. However, the underlying ideas regarding the role and physical origin of the interface dipole are quite firmly founded.

In sect. 4.3 and elsewhere [13,36], the property which suggested using the branch point as the neutrality level, was the fact that it represented the energy at which the gap states crossed over from being more valence-like to being more conduction-like. In order to generalize this to three dimensions, we need some way to characterize an "average" gap state at a given energy.

One simple way to do this is through a real-space Green function. We begin with the Green function $g(r, r', E)$ characterizing propagation with an energy E from r to r',

$$g(r, r', E) = \sum_{nk} \frac{\psi^*_{nk}(r)\, \psi_{nk}(r')}{E - E_{nk}}. \tag{4.19}$$

Here ψ_{nk} is the usual Bloch state of the real wavevector k and the band n, and E_{nk} is its energy. Details of the wavefunction within a unit cell are not of interest here. It is therefore convenient to consider propagation by a lattice vector R, and average over the unit cell. Then using Bloch's theorem, eq. (4.19) simplifies to

$$G(R, E) \equiv \int dr\, g(r, r+R, E) = \sum_{nk} \frac{\exp(ik \cdot R)}{E - E_{nk}}. \tag{4.20}$$

G can be decomposed into valence- and conduction-band contributions, G_v and G_c, by performing the sums over these sets of bands separately. For a given value of R, at energies high (low) in the gap, the conduction (valence) band contribution to G dominates. One can then locate the energy at which these two contributions are of equal magnitude. In analogy with the branch point we call this energy E_B, i.e.

$$|G_v(R, E_B)| = |G_c(R, E_B)|. \tag{4.21}$$

If the result is relatively independent of R over the physically relevant range, one may speak of an effective midgap energy E_B without specifying R. The interested reader is referred to ref. [13] for a detailed discussion of the R-dependence and its implications.

It is tempting to associate $G(R, E)$ with lineups at interfaces oriented perpendicular to R. However, this is probably not justified. The spirit of the present work is to *assume* that there is a bulk neutrality level, which may be defined without reference to the details of the interface. Calculating G along the $\langle 110 \rangle$ direction, with R a few lattice constants in magnitude, has proven to be an effective way of locating this energy [13]. If the pinning depends sensitively on the interface, either through its orientation or its structure, then further work would be required to incorporate these effects within the present framework.

Equation (4.20) requires as input only the semiconductor band structure. It is therefore worth considering in some detail how the band structures were calculated in applying this theory. Perhaps the best approach would be to use the empirical pseudopotential method, where the band structure is constrained to fit experimental data such as optical photoemission spectra. However, this method entails problems of nonuniqueness in the fitting procedure, and results are not available for all semiconductors. Therefore for convenience and consistency the band structures used in calculating E_B were calculated from first principles, using a linearized augmented plane-wave method [40]. These all-electron calculations used the frozen-core approximation, and made no shape approximations in the charge or potential.

Correlation and exchange were included using the Wigner local density functional. As always with this approach, the calculated band gaps were too small. To correct this problem, the conduction bands were rigidly shifted in energy so as to give the correct gap, following Baraff and Schluter [41]. This approach to the band-gap problem is rather successful in reproducing the experimental bands for indirect-gap semiconductors, because the error from the local density approximation is relatively constant across the conduction bands (Technical problems associated with direct-gap semiconductors are discussed in ref. [13].)

The band structure calculations were scalar-relativistic, i.e. they did not include spin–orbit splitting. This correction can be included perturbatively, as was done for the antimonides and II–VI's [13,30] Including this effect would also give slight corrections to E_B ($\sim -\frac{1}{6}\Delta$, where Δ is the spin–orbit splitting of the valence maximum) for other semiconductors as well, if it were included.

4.5. Comparison with experiment

The central conclusion of sect. 4.2 was conclusion (4.17), which states that, for the semiconductors of interest, an excellent estimate of the true band lineup can be obtained simply by aligning the neutrality levels in the respective semiconductors. In sect. 4.3 this neutrality level was related via eq. (4.18) to the branch point E_B in the complex bandstructure in one dimension, and sect. 4.4 described an ansatz, eq. (4.21), for calculating the neutrality level for real semiconductors.

The calculated neutrality levels are given, relative to the valence maximum, in table 2. Using the notation of sect. 4.2, this would be called $-E_v^n$. However, to emphasize that this represents a specific approximation for the neutrality level, which was obtained by analogy to the branch point in the complex bandstructure, we called this effective midgap energy E_B.

Results for several theoretical approaches to band lineups were given in table 1, and are illustrated in figure 6. In particular, the column labeled ID gives the result obtained in the present approach of aligning the neutrality levels according to eqs. (4.17) and (4.18) i.e.

$$\Delta E_v \simeq -\Delta E_B. \qquad (4.22)$$

Table 2
Semiconductor "midgap" energy E_B, and measured Fermi-level positions at metal–semiconductor interfaces, relative to valence-band maxima (eV), from refs. [12,30].

	E_B	E_F(Au) (Ref. [42])	E_F(Al) (Ref. [42])
Si	0.36	0.32	0.40
Ge	0.18	0.07	0.18
AlP	1.27		
GaP	0.81	0.94	1.17
InP	0.76	0.77	
AlAs	1.05	0.96	
GaAs	0.50	0.52	0.62
InAs	0.50	0.47	
AlSb	0.45	0.55	
GaSb	0.07	0.07	
InSb	0.01	0.00	
ZnSe	1.70	1.34	1.94
MnTe	1.6 *		
ZnTe	0.84		
CdTe	0.85	0.73	0.68
HgTe	0.34		

* See ref. [30] for approximations.

This is also shown in fig. 6a. Predictions for other interfaces can be obtained from table 2 with eq. (4.22).

The present approach is seen to be accurate to about 0.1 eV, although it would be difficult to argue that this level of accuracy should be expected a priori. Several other theoretical approaches are seen to work also reasonably well. A comparison to a much larger base of experimental data has been given by Margaritondo [43], who concluded that the most accurate theoretical approach is the one described here, eq. (4.22).

Nevertheless, because of the great difficulty in measuring band lineups precisely, the experimental values are subject to uncertainties which may be just as large as those in the theoretical predictions. It is therefore risky to assess the merits of the respective theories based on the precise level of agreement with experiment. It is even less justified to use such comparisons to decide the validity of the physical picture underlying a given theory, since there are sources of quantitative error in any theory which are unrelated to the essential physical mechanism envisioned.

We have already alluded to a connection between Schottky barriers and semiconductor heterojunctions. If we equate the neutrality level E_B in the semiconductor with the neutrality level (the Fermi level) in a metal, we can obtain a rule relating Schottky barriers to band lineups,

$$\Delta E_v = -\Delta \phi_{bp}, \tag{4.23}$$

where ϕ_{bp} is the p-type Schottky barrier height (for an Au contact) on a given semiconductor. In table 1 and fig. 7 it is seen that this rule predicts band lineups more accurately than any theory to date, except perhaps that described in sect. 4. This is confirmed by Margaritondo's analysis [24] with a larger data base. The physical picture behind this rule, and its important implications for the theoretical understanding of both band lineups and Schottky barriers, is the subject of sect. 5.

5. Connection between lineups and Schottky barriers

5.1. Relationship to Schottky barriers

Consider the heterojunction interface shown schematically in fig. 4b. If we let the band gap of the semiconductor on the left shrink to zero, we have a metal–semiconductor interface, i.e. a Schottky barrier. The p-type barrier height is by definition just the energy difference between the metal Fermi level E_F and the semiconductor valence-band maximum at the interface,

$$\phi_{bp} \equiv E_F - E_v. \tag{5.1}$$

It is important to remember here that fig. 4 really refers to a one-dimensional model. In three dimensions, a zero-gap semiconductor is qualitatively different from a metal. However, in one dimension there is no such distinction. We first discuss the problem in one dimension for simplicity, and then draw the correspondence to three dimensions.

If we let the band gap in a one-dimensional model go to zero, the valence and conduction edges and the branch point or neutrality level all collapse to the Fermi level. Then the valence-band discontinuity ΔE_v becomes exactly the same as the p-type Schottky barrier height ϕ_{bp}.

More formally, denoting the metal and semiconductor by M and S, we can write

$$E_n(M) = E_v(M) = E_F. \tag{5.2}$$

Thus $E_v^n(M) = 0$, and also $\Delta E_v(S, M) = E_F(M) - E_v(S) \equiv \phi_{bp}$. Combining these with eq. (4.17) gives the result

$$\phi_{bp} = -E_v^n(S). \tag{5.3}$$

In other words, the Schottky barrier height gives us the position of the neutrality level relative to the valence-band maximum. (Using the n-type barrier gives the level relative to the conduction-band minimum.)

From eq. (4.17), the neutrality level is all we need to calculate band lineups. Substituting the neutrality level, eq. (5.3), derived from the Schottky height into band lineup formula (4.17) gives the simple relationship

$$\Delta E_v = -\Delta \phi_{bp}, \tag{5.4}$$

between band lineups and Schottky barriers.

In going from one dimension to three dimensions we simply adopt eq. (5.4) without worrying about the fact that the Fermi level no longer corresponds also to the valence maximum and conduction minimum. In other words, we simply line up the "neutrality level" in the two materials, where for metals the neutrality level is E_F, and for semiconductors we use E_B.

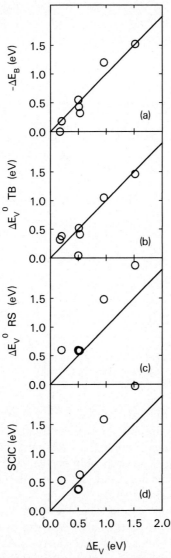

Fig. 6. Results of various theories for the valence-band discontinuity ΔE_v, plotted versus experimental results, for the set of interfaces selected by Kroemer [16]. The diagonal line represents perfect agreement. Numerical values are given in table 1. (a) Neutrality rule $\Delta E_v = -\Delta E_B$. (Tersoff [12].) (b) Tight-binding version of electron affinity rule. (Harrison [9].) (c) Atom-superposition reference surface approach to electron affinity rule. (Van de Walle and Martin [5].) (d) Self-consistent interface calculation. (Van de Walle and Martin [4].)

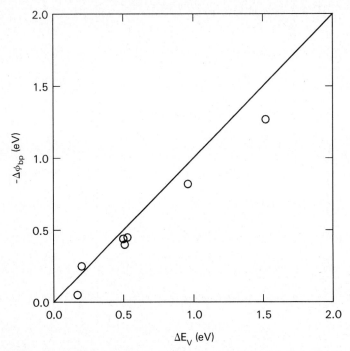

Fig. 7. Prediction of the Schottky-barrier rule, eq. (5.4), for band discontinuities, versus experiment, for selected [16] interfaces as in fig. 6.

Since Schottky barriers are relatively easy to measure, and have been tabulated for a large number of semiconductors, eq. (5.4) allows us to trivially obtain band lineup predictions for most interfaces. This approach has the advantage that it depends only on the validity of eq. (4.17) and its metal–semiconductor analog, and not upon the specific approximation used to calculate the neutral level.

The results, as summarized in table 1 and fig. 7, are at least as accurate as any theory to date. In particular, it is worth noting that this approach gives a very accurate lineup for the GaAs–AlAs junction, where until recently ΔE_v was erroneously believed to be very small [29].

Using eq. (5.3), in conjunction with formula (4.18) for the semiconductor neutrality level, gives the result

$$\phi_{bp} = E_B, \tag{5.5}$$

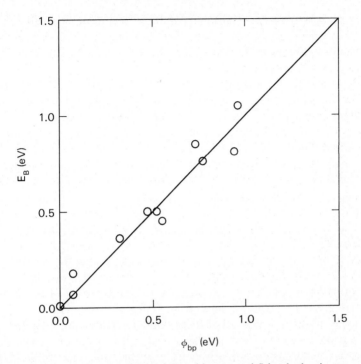

Fig. 8. Comparison of bulk neutrality level E_B with measured Schottky barriers ϕ_{bp} for Au contacts, from table 2. The diagonal line represents perfect agreement.

which should also permit us to calculate Schottky barrier heights. In fact, historically, this approach to Schottky barrier heights [36,44] preceded the corresponding analysis of heterojunctions. Moreover, ansätze (4.18) and (4.21) for the neutrality level have a particularly clear motivation in the case of Schottky barriers [13,36]. Unlike the semiconductor case, the importance of interface dipoles at metal–semiconductor interfaces is almost universally acknowledged, although the microscopic origin of those dipoles remains somewhat controversial [14].

Results of eq. (5.5) can be compared with experiment in table 2 and fig. 8. For Au contacts, the theory and experiment are seen to agree to within about 0.1 eV. This lends considerable credibility to the approach of sect. 4.4 for calculating the neutrality level.

If we are interested in relatively high accuracy, there is a further point we must worry about. In the case of semiconductor heterojunctions, we found that the neutrality rule, eq. (4.17), was not exact, and in principle a correction, given by eq. (4.16), should be added. However, since affinity rule (4.11) works reasonably well for heterojunctions, it was concluded that correction (4.16) was entirely negligible on the scale of accuracy which we can reasonably hope for here.

For metals, the affinity rule fails badly. Therefore, we cannot assume that, for the metal–semiconductor interface, the corrections to the neutrality rule are negligible. In fact, if they were, then from eq. (5.5) the barrier height would be strictly independent of the metal. However, experimentally it is observed that, while the barrier height is much less sensitive to the metal than one would expect from affinity rule (4.11) (or more precisely, from the analogous model for Schottky barriers) the absolute magnitude of the change in barrier height can be a substantial fraction of an eV.

Table 2 gives experimental values of ϕ_{bp} for contacts of both Au and Al. The difference is typically about 0.1 eV, which is not negligible on the scale of interest here. The dependence of barrier height on metal can in fact be quantified using linear response theory, just as was done for the heterojunction in sect. 4.2. This dependence, which determines the so-called pinning-strength parameter S, is given by an equation similar to eq. (4.16). For more details interested readers are referred to Ref. [35].

For the present purpose, it suffices that the experimental barrier height gives a good measure of the neutrality level, up to an additive term which, for a given metal, depends relatively little on the semiconductor (at least for the more covalent semiconductors). In fact, if we pick Au contacts (for which the data are most plentiful), the correspondence between E_B and ϕ_{bp} is seen in table 2 and fig. 8 to be close indeed.

Thus we see that it requires only a trivial extension of the heterojunction theory to deal with Schottky barriers. Then the calculation of a single number, E_B, for each semiconductor, gives predictions of both band lineups and Schottky barriers.

5.2. Implications for theory

Result (5.4) provides a uniquely convenient and accurate way of predicting band lineups. However, this relationship has important implications quite aside from its predictive value. The theories of both band lineups and Schottky barrier heights have been controversial in recent years. The fact that band lineups and barrier heights are so closely connected experimentally argues strongly for a theoretical approach which deals with both of these on a unified footing, and gives the relationship between Schottky barriers and band lineups as a natural result.

As discussed in sect. 3.4, the class of theories based on the electron affinity rule may provide a very reasonable description of band lineups, although the accuracy of such approaches remains somewhat limited to date. Nevertheless, such a perspective is clearly inadequate for dealing with Schottky barriers, and so such approaches miss a very interesting and important part of the problem.

Freeouf and Woodall [10] have argued that Schottky barriers can in fact be understood in terms of the electron affinity rule, which in this context is the original Schottky model of Schottky barriers. To obtain reasonable results without invoking interface dipoles, however, they had to assume that the relevant metal at the interface is a particular reaction product, e.g. anion clusters or a silicon-rich silicide. Because it requires a microscopic knowledge of interface structure and chemistry, it is difficult to assess the merits of such a model. However, virtually all other workers in the Schottky barrier field concur that Fermi-level "pinning" is caused by electronic states in the band gap, and the only controversial issue is the nature and origin of these states.

So far we have suggested that the correlation between barriers and lineups favors a neutrality-level theory over an affinity-rule theory for understanding heterojunctions. In addition, though, this correlation is important as a test of theories of Schottky barriers. While the theoretical understanding of Schottky barriers has long been focused on an intrinsic neutrality level [44], more recently there has been much interest in a possible extrinsic mechanism for Schottky barrier formation [45].

In particular, it has been suggested that at a metal–semiconductor interface, Fermi-level "pinning" may be due to gap states associated with point defects [45]. Such defects might occur as a result of energy released during interface formation. While defects might certainly play some role at a metal–semiconductor interface, the hypothesis that they actually determine the barrier height raises several problems [14]. The point of interest here is that it cannot readily explain the correlation between barrier heights and band lineups.

If Schottky barrier heights are determined by Fermi-level pinning associated with defect levels, then they should be unrelated to heterojunction band lineups. This is true because, at device-quality heterojunctions, there are very few defects, and the Fermi level is not even pinned. Moreover, it has been shown in at least one case that Fermi level pinning does not affect the band lineup at a heterojunction [31], and simulations suggest that this will be true in general for realistic densities of defects [32].

6. Conclusions

6.1. A field in flux

The study of semiconductor heterojunctions and their band lineups is a relatively new field, which is developing rapidly at this time. New experimental results, and new theoretical models and calculations, continue to appear regularly. As a result, any attempt to assess the status of the field must be regarded as provisional.

While the subject of this chapter is the theory of band lineups, one must rely to some extent on experimental results in attempting to judge the accuracy of the various theories. In the past, the reliability of experimental data has itself been something of a problem. If the experimental data are in doubt, any assessment of the theories becomes particularly risky.

A sobering example is the case of AlAs–GaAs, which was perhaps the most thoroughly studied interface, and the one where the lineup was believed to be best understood both experimentally and theoretically [15]. Yet recently it was convincingly demonstrated that the accepted lineup was completely incorrect [29]. Since AlAs–GaAs is one case where various theories give substantially different results, this experimental revision has had a significant (and perhaps even exaggerated) impact on the perceived reliability of the respective theories.

Even with the wider availability of reasonably accurate lineup data, the theory of band lineups remains an open and controversial field. No single model is universally accepted. Basic issues, such as the role of charge-transfer dipoles at the interface, are still debated.

Yet despite the variety of theoretical approaches to the band lineup problem, and the lack of concensus among theorists as to which approach is best, there is good reason for optimism. As was illustrated in fig. 6, several different approaches are now capable of predicting band lineups with reasonable accuracy, and at least one approach gives accuracy of typically around 0.1 eV, comparable to the accuracy of the experimental data. Moreover, there are several promising empirical schemes for predicting band lineups, as well as a predicted correlation with Schottky barrier heights which is remarkably well obeyed experimentally.

6.2. Ab initio calculations

In principle, it should be possible to calculate the full electronic structure of a heterojunction from first principles. Such a calculation would yield the lineup, along with much more information. While such calculations [2–4] have been reasonably successful, especially recently, they have not yet achieved the accuracy of the best model theories.

This is a bit disturbing, since the full calculations ought to work *better* than model theories. Whether the limited accuracy of the full microscopic calculations is due to numerical problems, or to inherent limitations in the local density functional approach, or to a completely different problem, is not clear.

If the primary problem in ab initio calculations is simply one of numerical precision, then it is not so difficult to see how model theories might be more accurate. For example, as was mentioned in sect. 3.3, it would be relatively difficult to calculate the "lineup" at a metal–metal interface with anything close to the accuracy which can be obtained by performing two bulk calculations, and aligning the resulting Fermi levels. On the other hand, model theories make certain uncontrolled approximations, so their accuracy cannot be systematically improved.

Quite aside from predicting band lineups, ab initio calculations have a major role to play in providing a microscopic picture of the interface, including local charge rearrangements. While such calculations have not yet had much impact on our qualitative understanding of the mechanisms determining the band lineup, the approach remains promising in this respect. In particular, it has already helped clarify the manner in which strain affects the band lineup [3], as well as providing further evidence [3,4] for the orientation-independence of the band lineup.

6.3. Model theories and empirical rules

Several model theories of band lineups were discussed in sections 3 and 4. While none of these describe the data perfectly, perhaps the most noteworthy feature is how well some of them work.

As discussed in sect. 3, some theories emphasize the role of charge-transfer dipoles at the interface, while others appear to neglect this effect completely. This seeming contradiction was seen to be largely a matter of how the dipole is handled. Even when dipoles are apparently neglected, one must choose the "reference surface" so as to insure that the dipoles will in fact be small.

While both methods give reasonable results, the approach of emphasizing dipoles appears to be ultimately more powerful. That approach has yielded the most accurate lineup predictions to date [12,43]. More important, it can quantitatively describe metal–metal, metal–semiconductor, and semiconductor–semiconductor junctions, all within a single unified approach, using only one reference level for each material. The only other model with this scope is that of Freeouf and Woodall [10], which requires severe and highly controversial assumptions about the structure and chemistry at the interface.

Among theories which neglect interface dipoles, that of Harrison [9] has proven notably accurate, especially in view of its simplicity. Nevertheless, the tight-binding approach can never be expected to give overall accuracy better than a few tenths of an eV. Also, Harrison and Tersoff [28] pointed out that, even within the tight-binding context, it is better justified (and just as simple) to include the effect of interface dipoles.

Another approach in which the interface dipole is not included explicitly, but rather is minimized by an appropriate choice of reference surface, is that of Van de Walle and Martin [5]. Besides being reasonably accurate, this approach is particularly notable in that the reference surface discussed in sect. 3.4 is constructed explicitly.

The effect of interface dipoles has been included in a simple way by Tejedor and Flores [11], Tersoff [12], and Harrison and Tersoff [28]. This approach appears to offer a significant improvement in accuracy over the affinity-rule type of approach. However, as we stressed above, the most important advantage of this approach is that is can also handle metal–semiconductor and metal–metal interfaces within the same simple approach.

The interface-dipole approach predicted a simple connection between band lineups and Schottky barrier heights, which may provide the most accurate and convenient means to date for predicting Schottky barrier heights. Another promising scheme, proposed by Zunger [25] and subsequently by Langer and Heinrich [26], involved using transition-metal impurity levels as reference levels to predict band lineups. Finally, if many band lineups are know, others can be predicted using the assumption of transitivity. The most powerful and convenient way of doing this is by compiling a set of reference levels which give the best fit to the available data base, following Katnani and Margaritondo [23].

6.4. Prospects for predictive accuracy

The goal of understanding heterojunction band lineups, and the goal of predicting them with high accuracy, are to a large extent separate issues. The theory of band lineups is often viewed as merely a tool for predicting lineups. However, lineups can always be measured, presumably with more accuracy than they can be calculated. The most valuable contribution theory can make, therefore, is a qualitative understanding of the mechanisms which determine the band lineup, and an identification of the relevant factors determining the interface electronic properties.

Even so, in the end one generally resorts to comparison with experiment to determine how believable a given theory may be. It is therefore desirable to estimate what the fundamental accuracy limits of the respective theories are, so that we can tell whether discrepancies between theory and experiment indicate a shortcoming in the theory, or merely a lack of precision due to approximations which are in principle avoidable.

Such an analysis for each theoretical approach would be beyond the scope of this chapter. Nevertheless, we can identify one problem which plagues many very different approaches. That is the problem of treating exchange and correlation among valence electrons. In ab initio calculations, one invariably uses the local density approximation for correlation and exchange. This approximation gives large errors, of the order of 1 eV, in the band gaps, so we must expect some error also in the band-edge discontinuities. It is usual to assume that the valence-band discontinuity ΔE_v will be rather accurate, while the conduction-band discontinuity will be unreliable. However, the errors in ΔE_v due to this problem are actually entirely unknown at present. Even in a tight-binding approach, the choice of approximation (e.g. local density or Hartree–Fock) for exchange and correlation in the atomic calculation has a significant effect on the predicted lineups, perhaps as much as a few tenths of an eV in some cases.

From this perspective, it is perhaps surprising that so many theories work so well. Certainly there is no reason at present to expect that theoretical predictions of band linueups can be systematically improved well beyond the present level of almost 0.1 eV accuracy. Moreover, this is already close enough to the experimental accuracy, that while individual cases are clearly out of line with experiment, any improvement in the overall acuracy of the best available theory would not be terribly meaningful until more precise data becomes available.

For sheer predictive accuracy, the best schemes would seem to be those based on experimental inputs. A particularly appealing scheme is that of compiling a table of reference levels which provide the best overall fit to the available data. Since this approach is statistical, it is to some extent insensitive to errors in individual measurements. Ultimately, however, the best way to determine a band lineup is to measure it.

Acknowledgments

It is a pleasure to thank the many people who have helped teach me about heterojunctions, or who have contributed to this chapter by their comments and encouragement. I am particularly grateful to F. Capasso, D.R. Hamann, W.A. Harrison, G. Margaritondo, R.M. Martin, S.T. Pantelides, C.G. Van de Walle, and W.I. Wang.

References

[1] G.A. Baraff, J.A. Appelbaum and D.R. Hamann, Phys. Rev. Lett. 38 (1977) 237.
[2] W.E. Pickett, S.G. Louie and M.L. Cohen, Phys. Rev. B 17 (1978) 815;
J. Ihm and M.L. Cohen, Phys. Rev. B 20 (1979) 729;
W.E. Pickett and M.L. Cohen, Phys. Rev. B 18 (1978) 939.
[3] C.G. Van de Walle and R.M. Martin, J. Vac. Sci. & Technol. B 3 (1985) 1256
[4] C.G. Van de Walle and R.M. Martin, Mater. Res. Soc. Symp. Proc. 63 (1986) 21. These results are also summarized in ref. [5].
[5] C.G. Van de Walle and R.M. Martin, J. Vac. Sci. & Technol. B 4 (1986) 1055.
[6] R.L. Anderson, Solid-State Electron. 5 (1962) 341.
[7] J.A. Van Vechten, J. Vac. Sci. & Technol. B 3 (1985) 1240.
[8] W.R. Frensley and H. Kroemer, Phys. Rev. B 16 (1977) 2642.
[9] W.A. Harrison, J. Vac. Sci. & Technol. 14 (1977) 1016.
[10] J.L. Freeouf and J.M. Woodall, Surf. Sci. 168 (1986) 518; Appl. Phys. Lett. 39 (1981) 727.
[11] C. Tejedor and F. Flores, J. Phys. C 11 (1978) L19;
F. Flores and C. Tejedor, J. Phys. C 12 (1979) 731.
[12] J. Tersoff, Phys. Rev. B 30 (1984) 4874. For details of the calculation, and a revised value for GaAs, see ref. [13].
[13] J. Tersoff, Surf. Sci. 168 (1986) 275.
[14] J. Tersoff, J. Vac. Sci. & Technol. B3 (1985) 1157.
[15] For a review see H. Kroemer, in: Proc. NATO Advanced Study Institute on Molecular Beam Epitaxy and Heterostructures, Erice, Sicily, 1983, eds by L.L. Chang and K. Ploog (Martinus Nijhoff, The Hague, 1984).
[16] Reference [15] selects several heterojunctions as being particularly appropriate for comparing critically theories, because their lineups are accurately known. However, the experimental value quoted for AlAs/GaAs is obsolete. For a recent re-evaluation of AlAs/GaAs see ref. [29].
[17] D.R. Hamann, M. Schluter and C. Chiang, Phys. Rev. Lett. 43 (1979) 1494.
[18] A.E. Carlsson, Phys. Rev. B 31 (1985) 5178.
[19] J. Tersoff, Phys. Rev. Lett. 56 (1986) 675.
[20] W.A. Harrison, R. Grant, E. Kraut, and D.J. Waldrup, Phys. Rev. B 18 (1978) 4402;
K. Kunc and R.M. Martin, Phys. Rev. B 24 (1981) 3445.
[21] W.I. Wang, T.S. Kuan, E.E. Mendez and L. Esaki, Phys. Rev. B 31 (1985) 6890.
[22] A.D. Katnani and R.S. Bauer, Phys. Rev. B 33 (1986) 1106.
[23] A.D. Katnani and G. Margaritondo, Phys. Rev. B 28 (1983) 1944.
[24] G. Margaritondo, Surf. Sci. 168 (1986) 439.

[25] A. Zunger, Ann. Rev. Mater. Sci. 15 (1985) 411; see also
M.J. Caldas, A. Fazzio and A. Zunger, Appl. Phys. Lett. 45 (1984) 671; and
A. Zunger, Phys. Rev. Lett. 54 (1985) 849.
[26] J.M. Langer and H. Heinrich, Phys. Rev. Lett. 55 (1985) 1414; see also
J.M. Langer and H. Heinrich, Physica B+C 134 (1985) 444, and references therein.
[27] J. Tersoff and W. Harrison, Phys. Rev. Lett. (25 May 1987).
[28] W.A. Harrison and J. Tersoff, J. Vac. Sci. & Technol. B 4 (1986) 1068.
[29] A review of recent work on AlAs–GaAs is given by G. Duggan, J. Vac. Sci. & Technol. B 3 (1985) 1224.
[30] J. Tersoff, Phys. Rev. Lett. 56 (1986) 2755.
[31] P. Chiaradia, A.D. Katnani, H.W. Sang Jr and R.S. Bauer, Phys. Rev. Lett. 52 (1984) 1246.
[32] A. Zur and T.C. McGill, J. Vac. Sci. & Technol. B 2 (1984) 440; see also
A. Zur, T.C. McGill and D.L. Smith, Phys. Rev. B 28 (1983) 2060.
[33] F. Capasso, A.Y. Cho, K. Mohammed and P.W. Foy, Appl. Phys. Lett. 46 (1985) 664.
[34] R.M. Martin has emphasized (private communication) that an arbitrary dipole or other charge disturbance will be exactly screened by ϵ *only* if the components of the dielectric matrix which are off-diagonal in reciprocal space may be neglected. This condition is always satisfied if the disturbance is sufficiently smooth. In the present thought experiment, the fictitious dipole δ_x referred to in eq. (4.14) must be confined to a length scale of the same order as the bondlength, in order to reasonably describe a heterojunction. In this case the off-diagonal dielectric response might introduce some quantitative modification. However, the off-diagonal matrix elements appear in the few examples studied to give screening stronger than ϵ rather than weaker, which would tend to strengthen the point made here.
[35] J. Tersoff, Phys. Rev. B 32 (1985) 6968.
[36] J. Tersoff, Phys. Rev. Lett. 52 (1984) 465.
[37] J.A. Appelbaum and D.R. Hamann, Phys. Rev. B 10 (1974) 4973.
[38] C. Kallin and B.I. Halperin, Phys. Rev. B 29 (1984) 2175; and references therein.
[39] W. Kohn, Phys. Rev. 115 (1959) 809;
J.J. Rehr and W. Kohn, Phys. Rev. B 9 (1974) 1981; and Phys. Rev. B 10 (1974) 448.
[40] D.R. Hamann, Phys. Rev. Lett. 42 (1979) 662.
[41] G.A. Baraff and M. Schluter, Phys. Rev. B 30 (1984) 3460.
[42] S.M. Sze, Physics of Semiconductor Devices (John Wiley & Sons, New York, NY, 1969). The relationship $\phi_{bp} = E_g - \phi_{bn}$ is used where appropriate in extracting ϕ_{bp} from this tabulation.
[43] G. Margaritondo, Phys. Rev. B 31 (1985) 2526.
[44] F. Yndurain, J. Phys. C 4 (1971) 2849;
C. Tejedor, F. Flores and E. Louis, J. Phys. C 10 (1977) 2163.
[45] W.E. Spicer, I. Lindau, P.R. Skeath, C.Y. Su and P.W. Chye, Phys. Rev. Lett. 44 (1980) 420;
W.E. Spicer, P.W. Chye, P.R. Skeath, C.Y. Su and I. Lindau, J. Vac. Sci. & Technol. 16 (1979) 1422.

CHAPTER 2

THE PROBLEM OF HETEROJUNCTION BAND DISCONTINUITIES

Giorgio MARGARITONDO

University of Wisconsin-Madison, WI, USA

and

Paolo PERFETTI

Istituto di Struttura della Materia del CNR, Frascati, Italy

Heterojunction Band Discontinuities: Physics and Device Applications
Edited by F. Capasso and G. Margaritondo
© *Elsevier Science Publishers B.V., 1987*

Contents

1. Introduction to the problem 61
2. The data base 64
   2.1. Transport measurements 65
   2.2. Photoemission measurements 66
   2.3. Optical measurements 68
   2.4. Accuracy and reliability 69
   2.5. Presentation of the data base 71
3. A first level of approximation 76
   3.1. Underlying accuracy limits of the general-purpose theories 77
   3.2. Overview of general-purpose discontinuity theories 80
      3.2.1. Model derived from the electron affinity rule 80
      3.2.2. The empirical deep-level model 85
      3.2.3. The Harrison tight-binding model 87
      3.2.4. The Frensley–Kroemer model 90
      3.2.5. The dielectric electronegativity approach 91
      3.2.6. The midgap-energy approach 92
      3.2.7. Other models 94
      3.2.8. Is there a "best" linear model? 95
   3.3. Correlations between Schottky barriers and heterojunctions 99
   3.4. Empirical solutions 103
4. Beyond the first level of approximation 104
   4.1. Controlling the band discontinuities 107
5. Final remarks 109
Note added in proof 110
References 111

1. Introduction to the problem

Heterojunctions devices are attracting increasing attention due to their superior flexibility. This flexibility originates from the use of two different semiconductors with two different sets of parameters. Thus, a heterojunction interface offers many more controllable parameters than, for example, p–n junctions or Schottky barriers. In principle, the control of these parameters enables the designers to tailor the heterojunction device characteristics and fit the desired functions. In practice, however, controlling the parameters and their influence on the device characteristics is a difficult task. Achieving such control is the goal of most research programs on heterojunctions.

The most important parameters which can be controlled in a heterojunction interface are the minimum forbidden gaps of the two semiconductors, E_g^A and E_g^B. The difference $\Delta E_g = E_g^A - E_g^B$ gives rise to the valence band discontinuity ΔE_v and to the conduction band discontinuity ΔE_c, which play a leading role in determining the transport and optical properties of the interface and, therefore, the device performance. How is ΔE_g shared between ΔE_c and ΔE_v? This is perhaps the most important problem in heterojunction research [1].

The importance of the band discontinuities became apparent in the very early days of heterojunctions research. The first fundamental step was the formulation of Anderson's *Electron affinity rule* in 1962 [2]. Since then, a huge research effort has been made to clarify the nature of ΔE_v and ΔE_c. A fundamental component of this effort is the use of surface-sensitive experimental techniques [3], e.g., photoemission spectroscopy with or without synchrotron radiation. In recent years, the advent of these techniques in heterojunctions research stimulated a renewed theoretical interest in this problem, and the formulation of several new discontinuity models [4–23].

Recent efforts have produced substantial progress towards the final understanding of the nature of the discontinuities. It should be emphasized, however, that the final solution to this problem is still elusive. In fact, the recent notorious setback concerning $Al_xGa_{1-x}As$–GaAs heterojunctions is a clear warning against excessive optimism. One important result of the recent research is the clarification of the complexity of the problem, and of the need for more advanced theoretical tools to solve it.

The definition of the problem is provided by practical aspects of device applications. For these applications, the control of ΔE_v and ΔE_c must achieve an accuracy better than kT at room temperature, i.e., and accuracy of ± 0.25 meV. Therefore, the ultimate goal of the research in this field is the formulation of a theoretical approach capable of predicting ΔE_v and ΔE_c with the above-mentioned accuracy.

At present, theory is very far from that accuracy. Therefore, it is necessary to define intermediate goals in this research program, and use them to discuss the current theories. From a quantitative point of view, two other factors besides kT are important in this field. The most obvious is the magnitude of the two gaps. This ranges from a few tenths of an eV to 2–3 eV. The second factor is the magnitude of the contributions to ΔE_v and ΔE_c arising from the non-ideal character of the interface. This influences the peculiar microscopic charge distribution at the interface. "Realistic" factors influencing the charge distribution are, for example, chemisorption-bond dipoles, the microdiffusion of atoms across the interface, the presence of defects and/or contaminants and the interface reconstruction. All these factors can, in principle, contribute to the microscopic electrostatic dipole and, therefore, influence the discontinuities.

A full theoretical description of the "realistic" microscopic charge distribution requires extremely sophisticated approaches – which are not yet available. Even if such approaches were available, they would require a complete characterization of the atomic and electronic structure of the interface. Such characterization, necessary to provide the input data for theory, would be somewhat beyond our present capabilities. Therefore, current heterojunction theories are intrinsically unable to describe in detail the microscopic interface charge distribution. This limits their accuracy in predicting the band discontinuities. The magnitude of this underlying accuracy limit corresponds to the magnitude of the dipole contributions to ΔE_v and ΔE_c which are related to the microscopic interface charge distribution.

We shall see that the *average* accuracy limit of discontinuity models which neglect the microscopic charge distribution is of the order of 0.15 eV. Specific models can overcome this average accuracy limit in the case of specific interfaces. Their average accuracy for all interfaces, however, does not. We emphasize that 0.15 eV is still, in general, smaller than ΔE_g. Therefore, models which neglect the microscopic charge distribution are not necessarily worthless, and could provide a first-order understanding of the

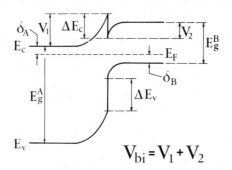

Fig. 1. Schematic diagram of a p–n heterojunction interface. E_v, E_c and E_F are, respectively, the valence- and conduction-band edges and the Fermi level. The other parameters are defined in the text.

band lineup. On the other hand, 0.15 eV is much larger than kT at room temperature. This means that the final solution to this problem will require significant theoretical advances in providing a realistic microscopic description of the interface.

The above elementary considerations provide the background for our review of the problem of band discontinuities. This chapter will consider the problem primarily from the point of view of the intermediate goal, i.e., understanding and describing ΔE_v and ΔE_c with an accuracy of the order of 0.15 eV. We will introduce the data base which has been accumulated in recent years and discuss its use for the achievement of the above intermediate goal. Then we will consider the current theoretical models in the light of the same goal. Finally, we will discuss the realistic prospectives of moving beyond the intermediate goal and towards the final solution of the problem, considering both theoretical and experimental aspects.

Treatment of the discontinuity problem requires a clear definition of the relevant physical quantities. The definition is based on fig. 1, which shows a simplified picture of the interface of a p–n heterojunction. Relevant physical quantities, besides E_g^A, E_g^B, ΔE_v and ΔE_c, are:
(1) The barriers, V_1 and V_2, due to the band bending at the two sides of the interfaces.
(2) The corresponding *built-in potential*, V_{bi}.
(3) The distances between the Fermi level and the band edges, δ_A and δ_B.

It is important to unambiguously define the *sign* of ΔE_v and ΔE_c. Throughout this chapter we shall use the following convention: for the interface A–B, a *positive* ΔE_v means that the valence-band edge of semiconductor A is at a lower energy than that of semiconductor B, and vice versa if ΔE_v is negative. For the same interface, a positive ΔE_c means that the conduction-band edge of semiconductor A is at a higher energy than that of semiconductor B, and vice versa if ΔE_c is negative. Therefore, for the specific case shown in fig. 1, the interface A–B would have positive ΔE_v and ΔE_c, while the interface B–A would have negative ΔE_v and ΔE_c. To avoid ambiguities, in the remainder of this chapter we shall list the two components of each heterojunction starting from the semiconductor with the larger gap, i.e., for the A–B interface, $E_g^A > E_g^B$. We shall also define ΔE_g as the difference between the gap of the first semiconductor and that of the second semiconductor, so that $\Delta E_g > 0$. Finally, we shall consider δ_A and δ_B to be positive quantities. The definitions based on the p–n heterojunction case shown in fig. 1 can be easily generalized to p–n, n–n and p–p interfaces.

Several simple equations link the above relevant quantities. Here is a list of the most frequently used:

$$\Delta E_g = E_g^A - E_g^B, \tag{1}$$

$$\Delta E_v + \Delta E_c = \Delta E_g, \tag{2}$$

$$\Delta E_v + V_{bi} + \delta_A + \delta_B = E_g^A, \tag{3}$$

$$\Delta E_c - V_{bi} - \delta_A - \delta_B = -E_g^B. \tag{4}$$

Equation (2) shows that, once ΔE_v is determined from theory or experiment, ΔE_c is determined too. Equations (3) and (4) emphasize the direct link between discontinuities and junction barriers. Both for heterojunctions and metal–semiconductor junctions, the junction barriers are determined by the mechanism which establishes the interface position of the Fermi level. Thus, there is a potential link between the heterojunctions discontinuities and Schottky barriers.

2. The data base

The chances of solving the problem of band discontinuities – or any other problem in science – are only as good as the corresponding data base. The progress recently made in developing a large and reliable data base on heterojunction band discontinuities is reason for optimism. This progress was made possible by the refinement of existing experimental techniques, by the improved control of the interface preparation processes, by the elimination of some critical systematic errors, and by the introduction of new experimental techniques in this field, e.g., photoemission spectroscopy.

Other chapters of this book analyze in detail the different experimental methods currently used to measure heterojunction band discontinuities. In this chapter we will give only a short description of some of the most widely used approaches.

2.1. Transport measurements

The original background of this class of techniques was provided by eqs. (3) and (4). These equations reduce the problem of measuring the band discontinuities to that of estimating the built-in potential and combining it with the bulk Fermi level positions determined by the doping of the two semiconductors. Therefore, any technique capable of measuring the built-in potential V_{bi} can, at least in principle, be used to measure ΔE_v and ΔE.

The most widely used method for measuring V_{bi} is the study of capacitance–voltage characteristics. This approach is based on the assumption that the $C(V)$ function has the form:

$$C \propto (V_{bi} - V)^{-1/2}, \tag{5}$$

and, therefore, that a plot of C^{-2} versus V gives a straight line, intercepting the V-axis exactly at $V = V_{bi}$. Other methods for measuring V_{bi} are based on the study of current–voltage characteristics [1]. Apparently, both methods are straightforward and of simple implementation. The reality, however, is much more complex.

An excellent analysis of the limits of these techniques was recently presented by Kroemer [24]. In particular, he emphasized the critical and commonly neglected effects of impurity gradients and interface charges on the C^{-2}–V plots. He also emphasized the severe limitation of the I–V methods, except for heterojunctions with unusual band lineups, e.g., those for which the conduction-band edge of one semiconductor is at a lower energy than the valence-band edge of the other. We do agree with Kroemer's reservation about the C–V and I–V techniques, and with his emphasis on their limitations.

Not all the measurements of V_{bi} based on the C–V characteristics are affected by the above problems. Kroemer et al. [25] developed a powerful

C–V profiling method which yields the interface charge and ΔE_c for a n–n heterojunction with known doping profile, coupled to a Schottky barrier. At present, this is probably the most reliable transport method for determining band discontinuities.

Another advanced electrical approach is the "charge transfer method" recently developed by Wang and co-workers [26,27]. This method is based on the study of the dependence of ΔE_c of the charge trapped at an interface triangular potential well.

2.2. Photoemission measurements

Photoemission probes the local density of electronic states [3]. Therefore, it is a direct method for measuring the band edges which are features of the local density of states. The principles of this method are illustrated in fig. 2.

This approach requires the in situ growth of a thin overlayer of semiconductor B on top of semiconductor A. The extreme surface sensitivity of the

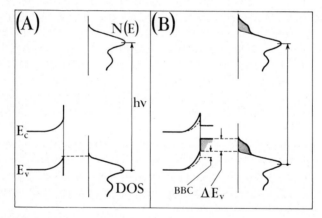

Fig. 2. Photoemission measurement of the valence band discontinuity ΔE_v. (A) The first part of the experiments is performed on the clean surface of the substrate. Upon absorption of photons of energy $h\nu$, photoelectrons are created. Their distribution in energy, $N(E)$, reflects in first approximation the density of electronic states (DOS) of the substrate, shifted in energy because of the band bending and also shifted by $h\nu$. Notice that the position in energy of E_v *on the surface* can be measured by subtracting $h\nu$ from the position of the leading edge of $N(E)$. (B) A thin overlayer of the second semiconductor is deposited. The two valence-band edges of the two semiconductors give rise to a double-edge structure in $N(E)$. From the double edge, the discontinuity ΔE_v can be directly deduced. If the double edge is not visible because ΔE_v is too small, ΔE_v can be deduced indirectly. This requires measuring the position of the leading edge of $N(E)$ before and after deposition, and correcting the substrate result for the overlayer-induced band-bending change (BBC). In turn, the BBC can be deduced from the overlayer-induced shift in energy of spectral features in $N(E)$.

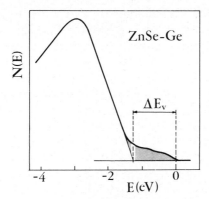

Fig. 3. An example of a double edge in the photoemission spectrum of a heterojunction. The interface consisted of a thin overlayer of amorphous Ge deposited on cleaved ZnSe (see ref. [28]).

photoemission probes requires the contamination level of substrate and overlayer to be kept very low – typically, less than $\frac{1}{100}$ equivalent monolayer at the interface. This, however, is a desirable features for *any* interface measurement, including discontinuity measurements with other methods.

When bombarded with ultraviolet or soft X-ray photons, the system emits photoelectrons. Due to the short mean free-path of electrons inside the two semiconductors, the photoelectrons originate from the overlayer and from the substrate region immediately close to the interface. Their measured distribution in energy is, at least in first approximation an image of the local density of states at the interface.

The presence of the valence-band discontinuity corresponds to a double edge in the distribution of the photoelectrons. This double edge can be used to study and measure ΔE_v. Figure 3 shows the double-edge structure in the case of the interface between amorphous Ge and a cleaved ZnSe substrate [28].

The above approach is affected by several problems. The relation between photoelectron distribution and local density of states is not direct, and in fact is quite complex [3,29]. The short mean free path of the electrons requires the overlayer to be thin, typically limited to a few monolayers. However, both theory and experiment indicate that even a few monolayers are sufficient to establish the basic valence-band features of the overlayer, including its leading edge. A more serious problem is that many interfaces have a small ΔE_v, for which the double-edge structure cannot be resolved.

Even without a resolved double-edge structure, the photoemission spectra contain enough information to derive ΔE_v [29]. The leading spectral edge of the clean substrate gives its valence-band edge. The leading spectral edge of the overlayer-covered substrate gives the valence-band edges of the overlayer. The distance in energy between the two edges corresponds to ΔE_v, except for the overlayer-induced change in the position of the substrate valence-band edge. This change corresponds to the change in band bending visible in fig. 2. Since the band-bending changes cause a rigid shift of all substrate valence-band features, its magnitude can be estimated from the analysis of the corresponding photoemission spectral features.

The above mentioned is often implemented by using the shift of substrate core-level peaks to estimate the changes in band bending. Of course, this approach is not as direct as that based on the double-edge structure – and this can affect its reliability. The accuracy of the photoemission measurements of ΔE_v is affected by several factors which change, to some extent, from experiment to experiment. These include, for example, the accuracy in determining the valence-band edges from the spectral edges, and overlayer-induced changes in the core-level peaks other than the rigid shifts due to changes in band bending. In the majority of the experiments the accuracy does not exceed 0.1 eV. The contribution of Grant et al. (ch. 7) to this book discusses methods to achieve significantly better accuracies, which require a full theoretical study of the density of states near the valence-band edges.

The internal photoemission technique recently proposed by Heiblum et al. [30] to measure ΔE_c is also based on the photoelectric effect. This technique is an elegant extension of the internal photoemission technique widely used to measure Schottky barriers at metal–semiconductor interfaces.

2.3. Optical measurements

These techniques are based on the study of the optical properties of alternating layers of two semiconductors [31]. They have been widely applied to interfaces between GaAs and $Al_xGa_{1-x}As$. Molecular beam epitaxy (MBE) enables the experimentalists to fabricate multi-layered structures of these two materials. As shown in fig. 4, the presence of ΔE_c and ΔE_v implies the creation of quantum wells for electrons and holes. The quantized energy levels associated with each well depend on the corresponding discontinuity, on the width of the well and on the effective mass.

Processes involving the localized quantum-well states influence the opti-

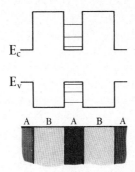

Fig. 4. Periodic structure consisting of alternating thin layers of the two semiconductors A and B. This structure has quantum wells for electrons and holes. The quantum levels of each well depend on its depth, i.e., on the valence- or conduction-band discontinuity. By analyzing the optical processes related to these levels it is possible to deduce ΔE_v and ΔE_c.

cal properties of the system. They introduce series of peaks both in the infrared absorption spectra and in the photoluminescence spectra [31–34]. From the position in energy of the peaks in each series it is possible to retrieve the parameters of the well and in particular the value of ΔE_c or ΔE_v.

In principle, the accuracy of these estimates of ΔE_v and ΔE_c is higher than that of other techniques and in particular that of photoemission techniques. The past experience, however, suggests some prudence. For example, in the early 1970s, the results of infrared measurements [31] suggested the "15/85%" rule for the sharing of the ΔE_g value of $Al_xGa_{1-x}As$–$GaAs$ between valence- and conduction-band discontinuity – which most authors now believe to be wrong. Another obvious limitation of these approaches is that they require the fabrication of high-quality multi-layer structures with MBE, and can only be applied to nearly ideal interfaces with excellent crystal quality.

2.4. Accuracy and reliability

It is not an easy task to make realistic estimates of the accuracy of the discontinuity measurements. The quoted accuracies by the different authors are often in sharp contrast to reality. For example, the discontinuities measured with C–V and I–V methods are often given with meV accuracy. Yet, according to some of the best experts in this field, most of the measurements with those methods are totally inaccurate. Likewise, optical-method measurements are often given with meV accuracy. The notorious case of $Al_xGa_{1-x}As$–GaAs, however, shows that for several years this method was in error by tenths of an eV! The quoted accuracies of photoemission experiments are, in some cases, as good as 30–40 meV. Yet, the photoemission measurements of ΔE_v for GaAs–Ge range between 0.24 and 0.7 eV! These sobering facts suggest prudence in accepting the authors' claims about accuracy.

One important point is the distinction between the accuracy achieved in measuring the quantities necessary to estimate ΔE_v and ΔE_c, and the final accuracy in estimating the discontinuities. For example, the optical methods require measuring the position of spectral lines, and this can be done with excellent accuracy. This, however, may not be the major accuracy limit for the corresponding estimates of ΔE_c and ΔE_v – it is necessary to estimate carefully the uncertainty introduced by the specific model adopted to link the experimental spectra and the discontinuities, and by the choice of the corresponding parameters. Similar problems often exist for other measurement methods.

On the borderline between accuracy problems and reliability is the crucial question of the control of the heterojunction preparation procedure. Great progress has been made with the advent of MBE and with the general improvement of the vacuum conditions and of the interface cleanliness. Still, the growth conditions change substantially from experiment to experiment, and many parameters are not communicated and sometimes not even controlled by the authors.

How much can the interface growth conditions influence ΔE_v and ΔE_c? In essence, answering this question is equivalent to estimating the effects of the growth conditions on the microscopic charge distribution at the interface. The growth conditions are one of the potential factors not accounted for by theories which neglect the microscopic interface charge distribution. We have mentioned that the average accuracy limit is of the order of 0.15 eV in predicting ΔE_v. This also gives the maximum magnitude of the contribution of the growth conditions, except for pathologic cases, e.g., badly contaminated interfaces.

Of course, 0.15 eV is a huge quantity in device applications. On the one hand, this magnitude emphasizes the necessity for a strict control of the interface growth conditions in applied research. On the other hand, it suggests that a manipulation of those conditions can provide some flexibility in controlling the discontinuities – as was demonstrated in recent experiments (see sect. 4.1).

The general improvement in the control of the growth conditions makes the discontinuities measured today more accurate and reliable than those measured ten or even five years ago. On the other hand, the accuracy must still be estimated with prudence and not necessarily by relying on the authors' claims. In our opinion, a prudent estimate of the typical accuracy of each measurement of ΔE_v or ΔE_c is ± 0.1 eV.

This level of accuracy may be disappointing for scientists who need better accuracies for their research – and were probably under the impression that such accuracies could be reached with existing techniques. Indeed they can, if the experimentalists are extremely careful with their interface preparation procedure and with the data analysis. The typical experiment, however, is not likely to exceed 0.1 eV in accuracy.

One final but important remark is that certain interfaces have been studied very extensively by several groups. While the typical accuracy of each individual measurement of ΔE_v and ΔE_c may not be better than the above limits, the average of the different results may give an estimate with better accuracy.

2.5. Presentation of the data base

Our discussion of the problem of band discontinuity will be based, hereafter, on experimental data. Specifically, we collected measurements of ΔE_v and ΔE_c obtained from different experimental methods on a variety of interfaces. The results of these measurements are listed in table 1.

We collected the data without eliminating a priori anyone of the experimental results known to us, except those which were obtained more than 4 to 5 year ago. The reason for this time limit is that, in our opinion, the accuracy in measuring ΔE_v and ΔE_c has dramatically improved in recent years. In particular, this is a result of better control of the interface preparation process. Therefore, we eliminated old results, since many of them were obtained on heavily contaminated interfaces. The time limit of 4 to 5 years is undoubtly arbitrary – it removes some perfectly good results and retains some questionable ones. On the whole, however, it does provide better accuracy than an indiscriminate use of all the results ever obtained.

Except for the above time limit, we did not eliminate any other data known to us. We feel that the selection of subsets of "good" data, based on criteria which are often different for different authors, risks producing a biased data base and a biased assessment of the theoretical progress. Of course, the data base of table 1 can be used by the readers to extract subsets of data according to their own rule to define "good" data.

Table 1
Experimental band discontinuities *.

Interface	ΔE_v	ΔE_c	$\Delta E_c/\Delta E_g$ (%)	Exp. method	Ref.	Comments
Si–Ge	0.17	(0.27)		PH	[29]	a-Ge or c-Ge deposited on Si(111)
	0.17	(0.27)		PH	[29]	a-Si deposited on Ge(111)
	0.4	(0.04)		PH	[39]	a-Si or c-Si deposited on Ge(111)
	0.39	(0.05)		CV	[83]	Ge on Si (also IV)
	0.28	*0.16*				*Average values*
AlAs–Ge	0.95	(0.58)		PH	[35]	a-Ge deposited on AlAs(110)
	0.9	(0.63)		PH	[86]	c-Ge on AlAs
	0.78	(0.75)		PH	[101]	AlAs on Ge(100)
	0.86	*0.67*				*Average values*
AlAs–GaAs	0.45	(0.4)	(47)	CT	[27]	MBE, (100) orientation
	0.4	(0.45)	(53)	PH	[69]	GaAs on AlAs(110)
	0.15	(0.7)	(82)	PH	[69]	AlAs on GaAs(110)
	0.38	(0.47)	(55)	PH	[101]	AlAs on GaAs(100), GaAs on AlAs(100)
	0.34	*0.51*	*60*			*Average values*
$Al_xGa_{1-x}As$–GaAs	0.113	0.160	60	IP	[30]	$x = 0.22$
	0.122	0.211	63	IP	[30]	$x = 0.27$
	0.175	0.310	64	IP	[30]	$x = 0.39$
	0.370	0.160	30	IP	[30]	$x = 0.45$
	0.376	0.164	30	IP	[30]	$x = 0.49$
	(0.19)	(0.32)	63	CV	[42]	$x = 0.4$
			62	CV	[42]	$x = 0.37$
	(0.16)	(0.21)	57	IR	[44]	$x = 0.3$
	(0.18)	(0.19)	51	IR	[45]	$x = 0.3$
	0.210	(0.33)	(61)	OT	[26]	$x = 0.5$
			65	IV	[40]	$x = 0.54 – 1$
			67	CV	[52]	$x = 0.09 – 0.41$
	(0.08)	(0.13)	60	OT	[53]	$x = 0.17$
	(0.17)	0.30	63	IV	[54]	$x = 0.38$
	(0.13)	0.17	57	IV	[54]	$x = 0.24$
			60	IV	[54]	$x = 0.38$
	0.19	(0.28)	62	CV	[55]	$x = 0.15 – 0.3$

System	ΔE_v	ΔE_c	%	Method	Ref	Notes
Al$_x$In$_{1-x}$As–InP	0.49	(0.11)	70	IV	[56]	$x = 0.3$
	0.4	(0.26)	65	IV	[56]	$x = 0.57$, 0.7 and 1
AlSb–GaSb	0.05	(0.10)	59	PL	[32]	$x = 0.13$
GaAs–Si	0.35		61	IV	[75]	$x = 0.26$, (100) and (311)
GaAs–Ge	0.47	0.25 (0.13)	66	CV	[25]	$x = 0.3$
	0.53	0.28 (0.19)	(64)	IV	[76]	$x = 0.35$
	0.42	0.28	60	IV	[79]	$x = 0.24$
	0.65		75	PL	[33]	$x = 0.26 - 0.36$
	0.24	(0.20)	(48)	OT	[80]	$x = 0.3$
	0.46	0.18	(62)	PL	[34]	$x = 0.2 - 0.3$
			59			*Average value*
	0.23	−0.52		PL	[62]	$x = 0.48$, staggered gaps
	0.59	(0.5)		OT	[74]	Superlattice
	0.54	(0.19)		PH	[29]	a-Si on GaAs(110)
	0.68	(0.33)		PH	[29]	a-Ge on GaAs(110)
	0.34	(0.21)		PH	[46]	c-Ge on GaAs(100)
	0.53	(0.15)		PH	[41]	c-Ge on GaAs(110)
	0.7	(0.26)		PH	[47]	c-Ge on GaAs(110)
	0.25	(0.03)		PH	[47]	a-Ge on GaAs(110)
	0.48	(0.34)		PH	[48]	c-GaAs on Ge(110)
	0.55	(0.22)		PH	[64]	c-Ge on GaAs(110)
	0.56	(0.45)		PH	[65]	GaAs on Ge(110)
	0.60	(0.09)		PH	[66]	p-Ge on n-GaAs(110)
	0.60	(0.14)		PH	[66]	p-Ge on p-GaAs(110)
	0.44	0.050		IV	[67]	c-Ge on GaAs(100)
		0.39		OT	[70]	c-Ge on GaAs(110)
		(0.15)		PH	[71]	c-Ge on GaAs(110)
		(−0.02)		PH	[72]	c-Ge on GaAs(110)
		(0.43)		PH	[73]	a-Ge on GaAs(110)
		(0.20)		PH	[81]	c-Ge on GaAs(111)Ga
		(0.13)		PH	[81]	c-Ge on GaAs(100)Ga
		(0.12)		PH	[81]	c-Ge on GaAs(110)
		(0.08)		PH	[81]	c-Ge on GaAs(100)As
		(0.08)		PH	[81]	c-Ge on GaAs($\bar{1}\bar{1}\bar{1}$)As
	0.49	(0.24)		PH	[101]	GaAs on Ge(100)
		0.19				*Average values*

Table 1 (continued)

Interface	ΔE_v	ΔE_c	$\Delta E_c/\Delta E_g$ (%)	Exp. method	Ref.	Comments
GaAs–InAs	0.17	(−0.09)		PH	[58]	GaAs on InAs(100), staggered gaps
GaP–Ge	0.80	(0.77)		PH	[29]	a-Ge deposited on GaP(110)
GaP–Si	0.80	(0.33)		PH	[36]	a-Si or c-Si deposited on GaP(110)
GaSb–Ge	0.20	(−0.20)		PH	[29]	a-Ge on GaSb(110), staggered gaps
GaSb–Si	0.05	(−0.49)		PH	[29]	a-Si on GaSb(110), staggered gaps
InAs–Ge	0.33	(0.27)		PH	[29]	a-Ge on InAs(110)
InAs–Si	0.15	(0.01)		PH	[29]	a-Si on InAs(110)
InP–In$_x$Ga$_{1-x}$As	0.37	0.23	38	CV	[57]	$x = 0.53$
In$_x$Al$_{1-x}$As–In$_y$Ga$_{1-y}$As	(0.20)	0.50	71	IP	[51]	$x = 0.52, y = 0.53$
	(0.29)	(0.43)	60	PL	[60]	$x = 0.52, y = 0.53$
	(0.22)	(0.50)	70	PL	[63]	$x = 0.52, y = 0.53$
	(0.40)	0.33	45	CV	[68]	$x = 0.52, y = 0.53$
	(0.53)	0.19	(26)	CV	[78]	$x = 0.52, y = 0.53$ (Also IV, OT)
	(0.50)	0.22	(31)	CV	[77]	$x = 0.52, y = 0.53$ ($T > 170$ K)
	0.36	0.36	51			Average values
InP–(In,Ga), (As,P)			39	CV	[57]	Different quaternary compositions, lattice-matched with InP, with $E_g = 0.80$–1.20 eV
			65	PL	[59]	Different quaternary compositions, lattice-matched with InP, with $E_g = 0.70$–1.12 eV
	0.08	0.16	(67)	PL	[85]	
			57			Average value
InP–Ge	0.64	(−0.04)		PH	[29]	a-Ge on InP(110) (staggered gaps?)
InP–Si	0.57	(−0.41)		PH	[29]	a-Si on InP(110), staggered gaps
In$_x$Ga$_{1-x}$P–GaAs	0.08	0.59	88	CV	[88]	$x = 0.48$
InSb–Ge	0.0	(−0.50)		PH	[29]	a-Ge on InSb(110)
InSb–Si	0.0	(−0.94)		PH	[29]	a-Si on InSb(110)
CdS–Ge	1.75	(0.00)		PH	[29]	a-Ge on cleaved CdS
CdS–Si	1.55	(−0.24)		PH	[29]	a-Si on cleaved CdS, staggered gaps
CdSe–Ge	1.30	(−0.23)		PH	[29]	a-Ge on cleaved CdSe, staggered gaps

The problem of heterojunction band discontinuities

Heterojunction	ΔE_v	(ΔE_c)	Method	Ref.	Notes
CdSe–Si	1.20	(−0.57)	PH	[29]	a-Si on cleaved CdSe, staggered gaps
CdTe–Ge	0.85	(−0.08)	PH	[29]	a-Ge on cleaved CdTe (staggered gaps?)
CdTe–Si	0.75	(−0.42)	PH	[29]	a-Si on cleaved CdTe, staggered gaps
CdTe–α-Sn	1.1	(0.26)	PH	[61]	α-Sn on CdTe(111)
CdTe–HgTe	0.040	(1.25)	IR	[43]	Superlattice
	~0	(−1.29)	IV	[82]	
	0.35	(0.94)	PH	[102]	
	0.13	*1.16*			*Average values*
ZnS–Cu$_2$S	1.4	1.35	IV	[84]	Also OT
ZnSe–Ge	1.40	(0.51)	PH	[28,29]	a-Ge and c-Ge on cleaved ZnSe
	1.52	(0.39)	PH	[49]	c-Ge on ZnSe(110)
	1.29	(0.62)	PH	[49]	c-ZnSe on Ge(110)
	1.40	*0.51*			*Average values*
ZnSe–Si	1.25	(0.22)	PH	[29]	a-Si on cleaved ZnSe
ZnSe–GaAs	1.10	(0.13)	PH	[49]	Annealed c-ZnSe on GaAs(110)
	0.96	(0.27)	PH	[49]	c-ZnSe deposited on hot GaAs(110)
	1.03	*0.20*			*Average values*
ZnTe–Ge	0.95	(0.64)	PH	[29]	a-Ge on cleaved ZnTe
ZnTe–Si	0.85	(0.30)	PH	[29]	a-Si on cleaved ZnTe
PbTe–Ge	0.35	(−0.42)	PH	[50]	a-Ge on PbTe(100), staggered gaps
GaSe–Ge	0.83	(0.55)	PH	[38]	a-Ge on cleaved GaS
GaSe–Si	0.74	(0.20)	PH	[38]	a-Si on cleaved GaSe
CuBr–GaAs	0.85	(0.74)	PH	[86]	
CuBr–Ge	0.7	(1.6)	PH	[86]	
CuInSe$_2$–Ge	0.48	(−0.25)	PH	[37]	a-Ge on fractured CuInSe$_2$, staggered gaps
CuInSe$_2$–Si	0.00	(−0.21)	PH	[37]	a-Si on fractured CuInSe$_2$
CuGaSe$_2$–Ge	0.62	(−0.33)	PH	[37]	a-Ge on fractured CuGaSe$_2$, staggered gaps
ZnSnP$_2$–GaAs	0.13	(−0.03)	OT	[87]	Staggered gaps

* A positive sign for the valence-band discontinuity of the A–B heterojunction means that the valence-band edge of B is above that of A; a positive sign for the conduction-band discontinuity of the same heterojunction means that the conduction-band edge of B is below that of A. As a general rule, data not in parentheses were directly quoted by the corresponding publications while those in parentheses were estimated, e.g., from eq. (2). Experimental methods used: photoemission (PH); internal photoemission (IP); IR absorption (IR); luminescence (LU); C–V characteristics (CV); I–V characteristics (IV); charge transfer (CT); other methods (OT). The prefixes a and c refer to amorphous and crystalline materials, respectively.

3. A first level of approximation

The first theory of the heterojunction band discontinuity was formulated more than twenty years ago by Anderson [2], and it is an extension of the Schottky model for metal–semiconductor interfaces [1]. New theories have been formulated with increasing frequency in recent years [4–23]. Most of these theories have a very ambitious goal: given any two semiconductors A and B, predict the discontinuities for the corresponding heterojunction A–B. Specifically, they try to identify some general rule linking the properties of the two semiconductors with the properties of the corresponding discontinuities.

In Anderson's approach [2], the relevant property of each semiconductor is its electron affinity. The philosophy of this model is schematically explained by fig. 5. The basic prediction of the model is that the valence-band discontinuity is given by the difference between the electron affinities of the two semiconductors,

$$\Delta E_c = \chi_B - \chi_A. \tag{6}$$

Equation (6) shows that the Anderson model is a general-purpose theory in the sense discussed above. Given any two semiconductors, one substitutes the corresponding electron affinities in eq. (6) to find ΔE_c.

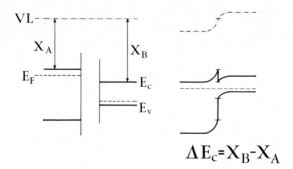

Fig. 5. Schematic explanation of the electron affinity model [2]. The model ignores the specific microscopic charge distribution created by the formation of the interface. Thus, when the two semiconductors are joined together and the Fermi level becomes the same throughout the system, the conduction-band discontinuity is simply given by the difference of the electron affinities.

For a generic general-purpose model, eq. (6) must be replaced by

$$\Delta E_v = f(\xi_B, \xi_A), \tag{7}$$

which states that the discontinuity depends on the values of a given physical quantity ξ for the two semiconductors. All current general-purpose models [2,5-19] use a specific, *linear* form for eq. (7),

$$\Delta E_v = \xi_B - \xi_A. \tag{8}$$

For example, eq. (8) gives Anderson's model when each ξ is the forbidden gap width minus the electron affinity. Below, we shall discuss other theories which use the general form of eq. (8) to express ΔE_v.

The linear character of all current general-purpose theories is an extremely important point. It gives us the opportunity to estimate the limits of this entire class of theories a priori, without analyzing each single theory. This approach was originally inspired by the experimental test of the transitivity of ΔE_v [89], and fully developed by Katnani and Margaritondo [29]. An up-to-date version of this analysis is presented here, based on the extended data base of table 1.

The analysis of the limits of the "linear" theories is more important than the mere assessment of the value of these theories. The reduction of ΔE_v to a form like eq. (8) prevents the theory from estimating in detail the effects of the "realistic" microscopic charge distribution at the interface. This is, therefore, an approximation common to all current general-purpose discontinuity theories. An assessment of their underlying accuracy limit also provides a good idea of the effect of the peculiar charge distribution at each interface. This enables us to divide the problem into a "macroscopic" part and a "microscopic" part, along the lines explained in the introduction. The "macroscopic" part can be solved by a linear model, while the "microscopic" cannot. The first step, therefore, must be a quantitative assessment of the accuracy level achievable by the "linear" models, i.e., of the accuracy limits implicit in considering only the "macroscopic" part of the problem.

3.1. Underlying accuracy limits of the general-purpose theories

The linear character expressed by eq. (8) has several elementary but important consequences. The first consequence of the linear character is *commutativity*, i.e.,

$$\Delta E_v^{A-B} + \Delta E_v^{B-A} = 0. \tag{9}$$

In practical terms, this rule can be referred to two different interface preparation processes: deposition of semiconductor B on substrate A or vice versa.

An extension of commutativity is the rule of transitivity, which applies to the interfaces corresponding to all possible pair combinations of a given group of semiconductors. Calling A_i the ith semiconductor of a group of n materials

$$\sum_{i,j=1}^{n} \Delta E_v^{A_i - A_j} = 0. \tag{10}$$

For example, given the three semiconductors A, B and C, eq. (10) requires that

$$\Delta E_v^{A-B} + \Delta E_v^{B-C} + \Delta E_v^{C-A} = 0. \tag{11}$$

The predictions of eqs. (9) and (11) can be tested using the data base of table 1. These are the results:

(1) Equation (9) was tested for the case of AlAs–GaAs and ZnSe–Ge. Scientists at Rockwell International grew these interfaces [69] by using both possible sequences for each of them, minimizing the changes in the other growth conditions. The deviations from zero, i.e., from the predictions of eq. (9), where 0.27 eV and 0.23 eV. The corresponding average magnitude of the discrepancy is, therefore, 0.14 and 0.12 eV per interface. A similar test for Ge–Si would nominally give a smaller average magnitude, but the corresponding data were taken in different laboratories and on interface grown under widely varying conditions [29,39,83]. The test for AlAs–GaAs was repeated by Katnani and Bauer for the (100) orientation, finding a negligible deviation [101].

(2) Equation (11) was tested for thirteen groups of three semiconductors, each containing Ge, Si and one of the following semiconductors: GaAs, GaP, GaSb, InAs, InP, InSb, CdS, CdSe, CdTe, ZnSe, ZnTe, GaSe and $CuInSe_2$. The average magnitude of the deviation from zero, i.e., from the prediction of eq. (11), was 0.09 eV per group, corresponding to 0.03 eV per interface.

(3) Equation (11) was also tested for the two groups (GaAs, Ge, InAs) and (GaAs, Si, InAs). The average magnitude of the deviation was 0.05 eV per interface.

(4) A similar test for the two groups (GaAs, Ge, ZnSe) and (GaAs, Si,

ZnSe) also gave an average magnitude of the deviation of 0.05 eV per interface.

(5) A test of eq. (11) for the group (GaAs, Ge, AlAs) gave an average magnitude of the deviation of 0.03 eV per interface. The test was also performed by Katnani and Bauer [101], obtaining a value of 0.01–0.02 eV per interface.

(6) A test of eq. (11) for the group (GaAs, Ge, CuBr) gave an average magnitude of the deviation of 0.21 eV per interface.

In summary, the tests derived from the assumption of linearity suggest accuracy limits ranging from values below the experimental accuracy to values of the order of 0.2 eV.

A second input for the estimate of the accuracy limits of all linear theories is the scattering of the data produced by different experiments on the same interface. Some of this scattering is the product of the experimental uncertainty. However, scatterings larger than, conservatively speaking, 0.1 eV can be assumed to be real. The most likely cause for them are the differences in the interface preparation process, which result in different "realistic" microscopic charge distributions – not accounted for by any linear model.

Three sets of data from table 1 are particularly relevant for this estimate of the accuracy limit: Si–Ge, GaAs–Ge and $In_{0.52}Al_{0.48}As$–$In_{0.53}Ga_{0.47}As$. The large set of data on $Al_xGa_{1-x}As$–GaAs is not equally valuable due to the scattering in composition. The standard deviations of the above three sets of data are 0.11, 0.14, and 0.14 eV, respectively. As 0.1 eV is a conservative estimate of the experimental accuracy, these values are beyond it.

A third estimate of the accuracy limits of linear models is provided by the results of experiments which tried to modify the band discontinuity by changing the microscopic structure of the interface. This was done by a controlled contamination of the interface, with contamination levels of the order of a monolayer. These experiments will be discussed later in this chapter (see sect. 4.1). We anticipate here that the intralayers modify the discontinuity by an amount that ranges from zero, in the case of GaAs–Ge, to 0.5 eV for hydrogen at the SiO_2–Si interface. These intralayer effects belong to the "microscopic" part of the discontinuity problem, which cannot be effectively described by linear models.

On the average, the estimated accuracy limits are below 0.1 eV from the first kind of tests, 0.1–0.15 eV from the second kind of tests and 0.1–0.25 eV from the third kind of tests. The latter kind of tests were primarily performed with ionic semiconductors. Therefore, we can set the accuracy limit estimated from all tests at 0.1–0.15 eV. In the following discussions, we shall use conservatively the upper limit, 0.15 eV. Notice that this is larger than the typical experimental accuracy. Particularly important in that regard are the experiments on the intralayer effects, which measured *changes* in ΔE_v rather than absolute values. The corresponding experimental accuracy is much better than the typical 0.1 eV value for absolute discontinuity measurements.

3.2. Overview of general-purpose discontinuity theories

The above-mentioned underlying accuracy limit common to all current general-purpose discontinuity models has two important implications. First, this limit is much too high to be acceptable for applied research and technology. Even after allowing for excessive pessimism, the limit is much higher than the ~ 0.025 eV level necessary for technology. Thus, no one of the general-purpose discontinuity theories currently available is likely to be satisfactory for applied research on discontinuities.

The second important implication is that the limit is still much smaller than the typical difference between the two forbidden gaps. This means that the general-purpose theories are not necessarily worthless from a fundamental point of view. As we already argued, they can provide a first-approximation understanding of the nature of the discontinuities. This step is extremely important – without such an understanding, it is impossible to move into high-accuracy theories as required for applications.

Therefore, each one of the current general-purpose theories could, in principle, have fundamental validity in the solution to this problem. It is important, therefore, to assess the specific accuracy of each theory. This will be done for some of the most important general-purpose models after discussing their general features.

3.2.1. Models derived from the electron affinity rule

The electron affinity rule [2], schematically explained in fig. 5, has inspired a number of recent discontinuity models. In essence, the fundamental limitations of the rule are related to its inability to describe the microscopic dipole present at the interface, which contributes to the discontinuities.

From an empirical point of view, the electron affinity is measured by extracting electrons from the semiconductors, and, therefore, it is influenced by the interfaces between the surface and vacuum. The electron affinity rule uses a linear combination of the results of such measurements to describe all the factors contributing to the discontinuities. It is certainly not clear a priori that this strategy can produce results of reasonable accuracy.

A practical problem affecting the electron affinity rule is the difficulty in measuring the electron affinities, and the consequent lack of a reliable data

Table 2
Semiconductor parameters.

Material	E_g, minimum forbidden gap * (eV, room temperature)	Electron affinity ** (eV)
Si	1.107	4.01 (Ref. [1])
Ge	0.67	4.13
α-Sn	0.08	
AlAs	2.2	
$Al_xGa_{1-x}As$	for $x < 0.37$: $1.35 + 1.255x$; for $x > 0.37$: $1.73 + 0.312x + 0.033x^2$ (Refs. [30,94])	
AlSb	1.6	3.6
GaAs	1.35	4.07
GaP	2.24	3.61
GaSb	0.67	4.08
InAs	1.27	4.9
$In_xGa_{1-x}As$	0.75, $x = 0.53$ (Ref. [57])	
$In_xAl_{1-x}As$	1.47, $x = 0.52$ (Ref. [51])	
InP	1.27	4.48
$In_xGa_{1-x}P$	1.90, $x = 0.48$ (Ref. [88])	
InSb	0.165	4.60
CdS	2.42	4.79
CdSe	1.74	4.95
CdTe	1.44	4.28
ZnS	3.54	3.9
ZnSe	2.58	4.09
ZnTe	2.26	3.53
PbTe	0.25	
HgTe	0.15 (inverted gap)	
CuBr	2.94	
Cu_2S	0.8	
GaSe	2.05	3.35
$CuInSe_2$	0.9	
$CuGaSe_2$	0.96	
$ZnSnP_2$	1.45 (ref. [86])	

* From: CRC Handbook of Chemistry and Physics, except where indicated otherwise.
** From ref. [6], except where indicated otherwise.

base. This point must be kept in mind when considering the commonly accepted electron affinity values listed in table 2.

The problem of the microscopic dipole is also present for the Schottky theory of metal–semiconductor interfaces. Mailhiot and Duke recently treated this problem with a local-density jellium model [4]. The Schottky barrier for an n-type semiconductor can be written:

$$V_{bn} = (\phi_m - \chi_s) + V_{dipole}, \tag{12}$$

where ϕ_m is the metal work function and χ_s is the electron affinity of the semiconductor. Equation (12) differs from the Schottky model because of dipole term:

$$V_{dipole} = \Delta\phi_{ms} - (\Delta\phi_m^s - \Delta\phi_s^s), \tag{13}$$

where $\Delta\phi_{ms}$ is the contribution due to the self-consistent valence electron charge rearrangement at the contact, and $\Delta\phi_m^s$, $\Delta\phi_s^s$ are the dipole contributions to the work functions. The central result of Mailhiot and Duke's model is that the different terms in V_{dipole} tend to cancel each other. As a result, the magnitude of V_{dipole} does not exceed, typically, 50 meV.

The above result implies that the Schottky model is valid *in first approximation*, at least with a jellium approach. The failures of the Schottky model, therefore, should be attributed to changes in the microscopic structure of the interface which cannot be described with the above model – such as interface relaxation of the atomic positions or microdiffusion processes.

The extension of the above results to the case of heterojunction band discontinuities is quite straightforward. The electron affinity rule is established in this way as a first-approximation result, due to the relatively small magnitude of the dipole correction.

The electron affinity rule is also the starting point for the discontinuity model developed by Tejedor and Flores [5]. This approach, however, should not be considered as a model derived from the rule. Its background is closely related to the model proposed by Tersoff [16], discussed below. Tejedor and Flores propose a modified version of eq. (6), which includes the interface dipole factors neglected by the electron affinity rule,

$$\Delta E_v = \Delta E_g - (\chi_B - \chi_A) - \Delta_J - \alpha\Delta. \tag{14}$$

The two correcting terms are Δ_J and $\alpha\Delta$. The first term incorporates all dipole contributions which can be estimated by treating the interface as a metal–metal junction. The second term describes the effects of the flow of charge upon creation of an intimate contact between the two semiconductors.

To estimate the latter terms, Tejedor and Flores use the concept of "charge neutrality level", which they had previously introduced to treat metal–semiconductor interfaces [103]. The definition of this level is based on the fact that when the metal–semiconductor interface is formed, interface states are created in the forbidden gap and there is a lack of states in the semiconductor valence band. The charge neutrality level is the energy below which these two factors cancel each other, giving a local charge neutrality. The charge neutrality level is an intrinsic property of each semiconductor, corresponding to Tersoff's "midgap energy" [16].

In the case of a heterojunction interface, the "charge-flow" dipole term $\alpha \Delta$ depends linearly on the misalignment of the two charge neutrality levels at the interface, Δ. Calling ϕ_0^A and ϕ_0^B the distance between each charge neutrality level and the corresponding valence band edge, eq. (14) can be easily remanipulated obtaining:

$$\Delta E_v = (1 + \alpha)^{-1} \left[\Delta E_g - (\chi_B - \chi_A) - \Delta_J + \alpha \left(\phi_0^B - \phi_0^A \right) \right]. \tag{15}$$

Table 3 shows the results of Tejedor and Flores' calculations for four heterojunctions, together with experimental results extracted from table 1. While the correlation between theory and experiment appears to be good, the cases for which a comparison can be made are too few to assess the merit of this approach with respect to other theories. We already mentioned, however, the strong interest of the theoretical background of this model, and its relation to the highly successful approach developed by Tersoff [16]. The condition of local charge neutrality was recently used by

Table 3
Results of the model by Tejedor and Flores [5] for (110) interfaces.

Interface	ϕ_0's (eV)	Δ (eV)	ΔE_v (eV)	ΔE_v (Exp.) * (eV)
GaAs–Ge	0.55, 0.17	−0.03	0.35	0.49
GaP–Si	0.77, 0.00	0.23–0.60	1.00–1.37	0.80
AlSb–GaSb	0.74, 0.61	0.16	0.29	0.4
InAs–GaSb	0.45, 0.61	0.20	0.04	

* Experimental results from table 1.

Priester et al. [104] within a tight-binding framework to calculate the band offset of the GaAs(110)–Ge interface, obtaining $\Delta E_v = 0.62$ and 0.68 eV with two different parametrization schemes.

Again within the general framework of the electron affinity rule is an approach based on the extension of the Freeouf–Woodall effective work function model for Schottky barriers [6]. The model treats metal–semiconductor interfaces with the standard Schottky picture, but it replaces the pure metal phase of the interface with an anion-rich interface phase. The Schottky-model equations, e.g., $V_{bn} = (\phi_m - \chi_s)$ for n-type semiconductors, is valid as long as ϕ_m is interpreted as the work function of this anion-rich phase. Experimental support for this model is provided by the correlation between the measured Au Schottky barriers for III–V's and II–VI's and the anion work function.

The effective work function model can be extended to the case of heterojunction interfaces. The interface formation is assumed to produce local disruption and the creation of an anion-rich interface phase. If the anion element is different for the two semiconductors, the interface phase is likely to contain both of them. The position of the Fermi level with respect to the band edges is determined by the work function of the interface phase. This parameter cancels out when one evaluates the position of the valence-band edges relative to each other, i.e., ΔE_v. All that is left is the electron affinity of each semiconductor, and ΔE_v is given by eq. (6).

We already mentioned the basic difficulty in testing the electron affinity rule and the models which are related to it. The electron affinity measurements are strongly influenced by the characteristics of the interface between the material and vacuum – and most of the available data were taken in times when the advanced surface characterization and the use of ultrahigh vacuum was not a routine practice. A reliable test requires measurements of electron affinities and band discontinuities in the same experiment, under well-characterized conditions of surface and interface cleanliness.

Such tests were performed by Zurcher and Bauer [48] and by Niles and Margaritondo [105]. In both cases, the results were negative, i.e., the electron affinity rule failed. The discrepancy between theory and experiment reported by Niles and Margaritondo [105] for ZnSe–Ge is 0.77 eV. They also argued that the result is also negative for the effective work function model and in general for other approaches which are directly linked to the electron affinity rule. These experiments indicate that the electron affinity rule does not provide a reasonable theoretical background for explaining band lineups.

The failure of the electron affinity rule increases the emphasis on other linear models. The rule, as all linear models, expresses the discontinuities in terms of the absolute position in energy of the valence-band edges of the two semiconductors. Other models follow the same route, trying to identify a reference energy level (the vacuum level for the electron affinity rule), then to estimate the absolute position in energy of the valence band edges, E_v^A and E_v^B, and finally to predict ΔE_v from the absolute positions,

$$\Delta E_v = E_v^B - E_v^A. \tag{16}$$

The most notable examples are the empirical deep-core-level approach proposed independently by Zunger and co-workers [8] and by Langer and Heinrich [7], the empirical approach by Katnani and Margaritondo [29] which will be discussed in sect. 3.4, Harrison's LCAO model [9,10], the Frensley–Kroemer model [14], Van Vechten's ionization potential model [15] and Tersoff's midgap-energy model [16].

3.2.2. The empirical deep-level model

The empirical deep-level model was independently formulated by Langer and Heinrich [7] and by Alex Zunger and his co-workers [8]. The approach is based on the empirical observation that a given deep level of a given transition-metal impurity, *referred to the vacuum level*, occurs at similar energies in different semiconductors in the same family. This is emphasized by fig. 6, which shows the binding energy with respect to the vacuum level of several different impurity levels. Shown in the figure are acceptor levels in several III–V compounds and donor level in several II–VI compounds.

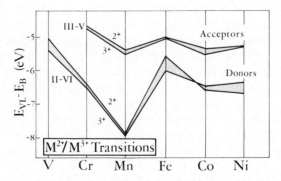

Fig. 6. Binding energy, E_B, of different transition-metal deep impurities in III–V and II–VI semiconductors, referred to the position of the vacuum level, E_{VL} (ref. [8]). Notice that the binding energy is very similar for the same impurity in different materials of the same family.

Table 4
Valence-band edge position deduced by Langer and Heinrich [7] with the deep-impurity model. All values in eV.

III–V		II–VI	
GaP	0 (reference)	ZnSe	0 (reference)
GaAs	0.33	CdTe	0.80
InP	0.17	CdSe	0.09
AlAs	−0.12	CdS	−0.37
		ZnS	−0.56

In essence, the above observations show that the deep impurity levels are not pinned to either band edge, but to the vacuum level, when the host material is replaced by another one in the same family. The immediate consequence is that the impurity levels can be used as substitute reference energies instead of the vacuum level. Thus, this approach can be used to derive empirical absolute energy positions of the valence-band edge for semiconductors in the same family. Table 4 shows the E_v positions so derived by Langer and Heinrich for several III–V compounds (referred to GaP) and for several II–VI compounds (referred to ZnSe).

The most relevant success of this approach is the correct prediction of the partition of ΔE_g between valence-band and conduction-band discontinuity for the $Al_xGa_{1-x}As$–GaAs interfaces. Table 1 showed that, in contrast to earlier estimates, the average of many recent experiments gives $\Delta E_c/\Delta E_g = 0.59$. This implies that ΔE_v is *not* small for these systems. Such a result is not easy to understand if one uses the simplistic assumption that two materials from the same family, with the same structure and with the same anion component, should give a heterojunction with small ΔE_v.

The above result is clearly predicted by the deep-impurity approach, if one assumes for the Fe^{2+} acceptor level in $Al_xGa_{1-x}As$ the same properties discussed above for transition-metal impurity levels in binary III–V's. the experimental base is provided by photocapacitance data [7]. The results are shown in fig. 7, as position of the band edges versus x, referred to the Fe^{2+} level. From these data, one predicts $\Delta E_c/\Delta E_g = 0.64$ for the AlAs–GaAs heterojunction, close to the average experimental result (0.60) from table 1, and also close to the average experimental result for $Al_xGa_{1-x}As$–GaAs (0.59).

The above results on AlAs–GaAs are one of the many examples of failure of the "common anion rule", which was commonly used a few years ago to estimate band discontinuities. The experimental evidence indicates

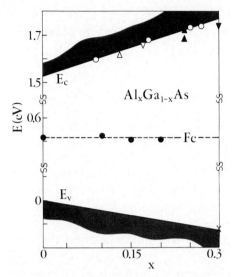

Fig. 7. Position in energy of the band edges of $Al_xGa_{1-x}As$ with respect to the Fe^{2+} deep impurity level, as a function of the composition parameter x (see ref. [7], and references therein). From these results, $\Delta E_c/\Delta E_g = 0.64$ is predicted for $Ga_xAl_{1-x}As$–GaAs interfaces.

that this rule, like the electron affinity rule, should no longer be considered a reasonable theoretical framework to treat band lineups.

3.2.3. The Harrison tight-binding model

The empirical approach discussed in the previous section does provide the necessary input for eq. (16), but only a limited understanding of the physical background leading to the band lineup. The problem remains of finding a theory capable of calculating the valence-band edge positions with respect to some absolute reference level.

Harrison emphasized [9,10] that the simplest meaningful band calculations which place all systems on the same scale are tight-binding calculations based upon universal parameters. Thus, this approach is best suited to find E_v terms for different semiconductors, whose difference define the "natural" band lineups and, in first approximation [eq. (16)], the ΔE_v's. The tight-binding result is immediate and quite simple. Given a binary semiconducting compound, the valence-band maximum is

$$E_v = \tfrac{1}{2}(\epsilon_a + \epsilon_c) + \left[\tfrac{1}{4}(\epsilon_a - \epsilon_c)^2 + (4E_{xx})^2\right]^{1/2}, \qquad (17)$$

where ϵ_a and ϵ_c are the cation and anion atomic energies, and the matrix element E_{xx} depends on the nearest cation–anion distance, d:

$$4E_{xx} \approx -1.28\frac{\hbar^2}{md^2}, \tag{18}$$

where m is the free electron mass. In the earliest version of this approach [9], Harrison used Herman–Skillman term values for ϵ_c and ϵ_a, while in the most recent version [10] he used Hartree–Fock term values. Table 5 shows the calculated E_v terms with this approach. Also shown in table 5 are the E_v's calculated by Vogl et al. [11] with different values for the matrix element E_{xx}.

Corrections to the above tight-binding approach were proposed by Chen et al. [12] and by Bechstedt et al. [13]. In the first case, the tight-binding formalism was modified to introduce the spin–orbit interaction, extending the work originally developed by Chadi [90]. Chen et al. [12] argued that the spin–orbit correction is not negligible when the atomic number of the constituent atoms is larger than ~ 40. Their calculated E_v's are also listed in table 5. Bechstedt et al. [13] argued that the expression for ΔE_v should not only include the tight-binding E_v terms [see eq. (16)], but also the relaxation self-energies of holes at the valence-band maxima. In fact, ΔE_v is the difference between the initial and final state energies of the two-step process, excitation of an electron from E_v^A to the vacuum level and then decay of the electron from the vacuum level to E_v^B. Including the hole relaxation self-energies, eq. (16) becomes

$$\Delta E_v = E_v^B - E_v^A + \Sigma_v^B - \Sigma_v^A. \tag{19}$$

The calculated values of the relaxation self energies are included in table 5.

Harrison recently discussed the possible limits of the tight-binding approach [10]. That such limits must exist is implicit in the fact that this is a linear model, and, therefore, subject to the underlying accuracy limit of all linear models. The primary cause of this limit is, again, the specific charge distribution of the "non-ideal" interface. There is, however, a dipole term which is not accounted for by the above-mentioned tight-binding model even for the most "ideal" interfaces [10].

For example, if one takes a (111) interface between Si and Ge and neglects possible effets of lattice mismatch, the difference in electronegativity between the two elements produces a polarization of the Si–Ge bonds and a dipole correction to the "natural" value of ΔE_v. The calculated value

Table 5
Calculated tight-binding valence-band edge positions. *

Material	Tight-binding E_v			E_v, including spin–orbit (Ref. [12])	Σ_v, relaxation hole self-energy (Ref. [13])
	HF (Refs. [9,10])	HS Refs. [9,10])	(Refs. [11,13])		
C	−15.18	−15.91			4.32
Si	−9.35	−9.50	−8.24		3.15
Ge	−8.97	−9.12	−7.97		3.08
Sn	−8.00	−8.04	−7.27		2.75
SiC	−12.59	−12.56			
BN	−15.93	−16.19			
BP	−11.56	−11.81			
BAs	−11.00	−11.17			
AlN	−14.67	−13.84			
AlP	−10.22	−10.03	−9.40		3.09
AlAs	−9.67	−9.57	−8.67	−9.54	3.06
AlSb	−8.76	−8.67	−8.05	−8.80	2.81
GaN	−14.59	−13.66			
GaP	−10.21	−10.00	−9.24		3.09
GaAs	−9.64	−9.53	−8.76	−9.49	3.06
GaSb	−8.67	−8.69	−7.96	−8.85	2.89
InN	−14.34	−13.00			
InP	−10.03	−9.64	−8.97	−9.55	2.93
InAs	−9.46	−9.21	−8.62	−9.15	2.91
InSb	−8.61	−8.41	−7.76	−8.61	2.75
BeO	−18.05	−16.27			
BeS	−12.74	−12.05			
BeSe	−11.86	−11.19			
BeTe	−10.77	−10.00			
MgTe	−9.81	−9.33			
ZnO	−17.19	−15.58			
ZnS	−12.00	−11.40	−11.23		2.98
ZnSe	−11.06	−10.58	−10.21		2.93
ZnTe	−9.87	−9.50	−9.36		2.82
CdS	−11.89	−11.12	−10.85		2.80
CdSe	−10.97	−10.35	−9.98	−10.42	2.76
CdTe	−9.80	−9.32	−9.04		2.66
CuF	−20.32	−18.41			
CuCl	−14.05	−13.11			
CuBr	−12.67	−11.90			
CuI	−11.20	−10.62			
AgI	−11.15	−10.49			

* HF = using Hartree–Fock term values; HS = using Herman–Skillman term values.

of this correction for the Si–Ge interface is 0.034 eV – small, but not necessarily negligible. Polar interfaces are likely to have bigger corrections, and the dipole contributions increase further for interfaces with charged extended defects, e.g., due to microdiffusion. Thus, our previous estimates of the magnitude of these effects appears reasonable.

3.2.4. The Frensley–Kroemer model

This model is a pseudopotential approach to the estimate of the valence-band edge positions [14]. The reference energy in each semiconductor is the interstitial potential $\overline{V}_i$, which is the average of the electrostatic potentials at the midpoints between adjacent atoms. The calculated valence-band edge positions with respect to $\overline{V}_i$ are shown in table 6. These results were obtained from pseudopotential band structure calculations including the exchange interaction via a Slater approximation.

If one assumes that the interstitial potentials are equal in the two semiconductors, then, in this model, ΔE_v is simply given by eq. (16), with the E_v's derived from table 6. In general, however, they will not. The difference between the two interstitial potentials gives the dipole potential, V_{dipole}, which contributed to ΔE_v,

$$\Delta E_v = E_v^B - E_v^A + V_{\text{dipole}}. \tag{20}$$

Frensley and Kroemer estimated the dipole potential by taking the difference of the "electronegativity potentials" of the two semiconductors. The "electronegativity potential" of a material is, by definition, proportional to average Phillips electronegativity of its component elements.

Table 6
E_v terms calculated by Frensley and Kroemer [14].

Material	E_v * (eV)	Material	E_v * (eV)
Si	−3.16	InP	−4.58
Ge	−3.25	InAs	−4.38
AlAs	−3.96	ZnS	−5.34
AlSb	−3.94	ZnSe	−5.07
GaP	−4.12	ZnTe	−4.74
GaAs	−3.96	CdS	−5.42
GaSb	−3.89	CdTe	−4.90

* Relative to the "interstitial potential".

A remarkable result of this approach is its linear character, even including the dipole corrections. In fact, the dipole correction itself is the difference between the two terms, each determined by one of the two semiconductors. Thus, eq. (20) has the same structure as eq. (8). However, a comparison with experimental ΔE_v's shows that this is not an effective way to treat the dipole correction – since it decreases, instead of increasing, the average accuracy of Frensley–Kroemer's model.

3.2.5. The dielectric electronegativity approach

A third, interesting approach within the framework established by eq. (16) was proposed by Van Vechten [15], and was based on the semi-empirical dielectric approach he originally developed in 1969. The presentation of this approach includes a detailed and clear discussion of the meaning of the quantities we call conduction and valence band discontinuities. The ΔE_v and ΔE_c we are interested in are those corresponding to potential steps capable of producing rectification at a heterojunction diode. This eliminates steps which are not abrupt on the scale of the carrier diffusion length, and the corresponding components of the band discontinuities. An example of these "non-abrupt" steps are those due to surface depolarization fields.

To eliminate the non-abrupt components from the picture, Van Vechten proposes to define the "effective" valence-band discontinuity as

$$\Delta E_v = \int_{-\delta}^{+\delta} \frac{\partial E_v}{\partial z} dz, \tag{21}$$

where the integrand is the gradient of the valence-band edge position along the direction perpendicular to the interface, and the integral is extended to a region of width 2δ on both sides of the interface ($z = 0$), larger than the metallurgical junction but small enough that the effects of electrostatic gradients due to surfaces or dopants is negligible. This effective ΔE_v should be rigorously given by eq. (16), if one interprets the ΔE_v's terms as ionization potentials.

To estimate the ionization potentials, Van Vechten proposed a reoptimization of his 1969 semi-empirical formula:

$$E_v = E_v^h \left[1 + \left(C/E_v^h\right)^2\right]^{1/2}, \tag{22}$$

where C is the electronegativity difference of the material and E_v^h, the homopolar equivalent of E_v, is the value of the ionization potential given by a material with the same density but with all atoms equal, so that $C = 0$. E_v^h is assumed to follow a simple exponential variation with the density, i.e., with the lattice constant, d. A fit with the experimental values for Si and Ge originally gave

$$E_v^h = 5.17(d/d_{Si})^{-1.3077} \text{ eV}. \tag{23}$$

The experimental evidence, however, suggest that the above values of the exponent overestimates the dependence of E_v^h on d. Van Vechten suggested to extract better values of the empirical parameters in eq. (23) by fitting four "reliable" ΔE_v's. A practical test of this approach [91] suggests that accuracy problems may arise – the formula tries to estimate relatively large quantities (the E_v's), but the input for the fit are small differences between those quantities. Without a good method to estimate the empirical parameter, it is impossible to assess the accuracy of this particular approach.

3.2.6. The midgap-energy approach

This approach belongs to a fundamental class of semiconductor interface models, which can be traced back to Heine's theory of metal-induced gap states for metal–semiconductor interfaces [92]. We already discussed an important model in this class, the Tejedor–Flores model. The major contribution to the current interest in these theories has been given by the recent work of Tersoff, both for Schottky barriers and for semiconductor–semiconductor interfaces [16,93].

For metal–semiconductor interfaces, the basic idea of Tersoff's approach is that the metal work functions have tails well inside the semiconductor, corresponding to states inside the gap. Metallic screening by these states tends to pin the Fermi level so as to maintain local charge neutrality. Tersoff argued that the charge neutrality conditions requires occupancy of the metal-induced gap states with mainly valence-like character, and non-occupancy of those with prevailing conduction-like character. Thus, the interface E_F must be close to the "midgap energy", E_B, for which the character of the gap states changes from valence-like to conduction-like. From these arguments the direct relation between this approach and the Tejedor–Flores–Louis model [103] should be clear, in particular as far as the "charge neutrality level" and the midgap energy are concerned.

The Schottky barrier produced by Tersoff's mechanism, e.g., for n-type semiconductors is approximately equal to the distance between the midgap energy and the bottom of the conduction band. Tersoff calculated the

Table 7
Midgap energies calculated by Tersoff [16,107].

Material	E_B * (eV)	Material	E_B * (eV)
Si	0.36	AlP	1.27
Ge	0.18	GaP	0.81
AlAs	1.05	InP	0.76
GaAs	0.50	ZnSe	1.70
InAs	0.50	MnTe	1.6
AlSb	0.45	ZnTe	0.84
GaSb	0.07	CdTe	0.85
InSb	0.01	HgTe	0.34

* Measured from E_v.

position of E_B for several semiconductors using energy bands obtained with the linearized augmented-plane-wave method. The results are shown in table 7, referred to the top of the valence band.

Tersoff also extended this theoretical line to heterojunction interfaces. His main conclusion is that the midgap energies of the two semiconductors must be close in energy to each other. In fact, a misalignment of the two E_B's corresponds to a "restoring" dipole, caused by the occupancy or non-occupancy of the gap states induced by the band discontinuity, i.e., by the tailing of the conduction and (or) valence-band wave functions of one side of the junction into the gap of the other side, or (and) vice-versa. This dipole tends to restore the "canonical" lineup, with the two E_B's aligned. Therefore, the midgap energy acts as a natural reference point in the band lineup. The data of table 7 can be interpreted, after changing sign, as the position of E_v with respect to this natural reference point. The value of ΔE_v is then given by eq. (16). As we shall discuss later, this simple approach achieves remarkable accuracy in predicting the valence-band discontinuity.

It is obvious that Tersoff's approach predicts a link between Schottky barriers and heterojunction band discontinuities – although this result is not unique to the midgap-energy approach. This link can be used to eliminate the need to estimate the midgap-energy positions. These positions can be empirically extracted, rather than calculated, from experimental Schottky barrier heights.

We will argue below that the theories in this class are the only current general-purpose discontinuity models which are not in disagreement with existing experimental data. One should emphasize, however, that this approach has been criticized by several authors. Very recently, Kane analyzed the possible charge transfers underlying the concept of charge neutrality level, arguing that one should include a dominant two-level contribution arising from junction bond breaking [106,107].

3.2.7. Other models

Models in the general category defined by eq. (8) have been proposed by several other authors [17–19]. Among those authors was Nussbaum who presented arguments in favor of the continuity across the junction of the intrinsic position of the Fermi level [18] this would imply a near-equipartition of ΔE_g between ΔE_v and ΔE_c. In a recent evolution [17], the same author developed a discontinuity model based on the concept of position-dependent band edges and on its consequences on the chemical potential. The main conclusion of this approach is that

$$\Delta E_v = \tfrac{1}{2}\Delta E_g - \tfrac{1}{2}kT \log\left(\frac{N_{cA}N_{vA}}{N_{cB}N_{vB}}\right), \tag{24}$$

where N_{cA}, N_{cB}, N_{vA} and N_{vB} are the bulk densities of states of the conduction and valence bands of the two semiconductors near the band edges.

This approach is of course a linear theory since eq. (24) has a form similar to eq. (16), if one takes

$$E_v = \tfrac{1}{2}E_g - \tfrac{1}{2}kT \log(N_c N_v), \tag{25}$$

with obvious meaning of the symbols (and redefining the densities of states so that they are dimensionless). This term, however, should not be immediately interpreted as the position of the valence-band edge with respect to a standard reference level. Table 8 shows the values of N_c and N_v for different semiconductors.

A model calculation of the interface dipoles was recently proposed by Ruan and Ching [108] as a method to estimate band discontinuities. The conceptual framework of this theory is somewhat similar to that of the midgap-energy or charge-neutrality-point approach [5,16]. However, these authors directly calculate the interface dipoles due to tunneling states using an iterative process and a square-well model. Interestingly enough, this approach cannot be classified as a linear model, since it does not have the

Table 8
Bulk densities of states used in the Unlu–Nussbaum model [17].

Material	N_c (10^{19} cm^{-3})	N_v (10^{19} cm^{-3})	Material	N_c (10^{19} cm^{-3})	N_v (10^{19} cm^{-3})
Si	2.8	1.04	InP	0.052	1.26
Ge	1.04	0.61	InSb	0.0043	0.62
GaAs	0.047	0.70	CdSe	0.11	0.74
GaP	1.83	1.14	CdTe	0.13	0.55
GaSb	0.021	0.62	ZnSe	0.31	0.87
InAs	0.0056	0.62	ZnTe	0.22	0.078

general form of eq. (8). The accuracy achieved by this model is remarkable, considering the simplicity of the calculation scheme.

3.2.8. Is there a "best" linear model?

This question can be addressed from two different points of view. First, one can try to identify which one of the "linear" models is conceptually more correct than the others. Second, one can empirically find which model achieves the best accuracy in predicting the discontinuities.

The first approach must necessarily be pragmatic. Since all models are based on approximations, the problem is to identify those approximations which have the best conceptual and practical background. The debate in this area is very much alive. A strongly controversial question is the role of the interface dipoles. This role is central in certain theories, notably Tersoff's approach [16] and also Tejedor–Flores' [5]. The role is neglected or minimized in other approaches, e.g., the tight-binding models in general [9–11]. Arguments have been presented by theorists on both sides, but the issue is still very much open.

The empirical approach based on the accuracy test is somewhat more conclusive – but it still is affected by several problems. The most fundamental limitation is that this test cannot provide final evidence in favor of a given model. First, there is no guarantee that a conceptually wrong model will not reach excellent accuracy for a certain set of interfaces by luck or coincidence. Even more important, the accuracy of a model can be limited by the practical approximations used to calculate its predictions rather than by its conceptual weaknesses.

The second problem is the data base to use in the experiments. Should this base be limited to "good" interfaces, e.g., those with good lattice

Table 9a
Comparison between experimental valence-band discontinuities and the theoretical predictions of different linear models *.

Interface	Exp.	EA	FK	TBHF	TBHS	TBV	TER	UN
Si–Ge	0.28	0.32	0.09	0.38(0.31)	0.38(0.31)	0.27(0.20)	0.18	0.21
AlAs–Ge	0.86	–	0.71	0.70(0.72)	0.45(0.47)	0.70(0.72)	0.87	–
AlAs–GaAs	0.34	–	0.00	0.03(0.03)	0.04(0.04)	−0.10(−0.10)	0.35	–
AlSb–GaSb	0.4	0.45	0.05	0.09(0.17)	−0.02(0.06)	0.09(0.17)	0.38	–
GaAs–Ge	0.49	0.62	0.71	0.67(0.69)	0.41(0.43)	0.79(0.81)	0.32	0.41
GaAs–Si	0.05	0.30	0.62	0.29(0.38)	0.03(0.12)	0.52(0.61)	0.14	0.21
GaAs–InAs	0.17	0.75	−0.42	0.18(0.03)	0.32(0.17)	0.14(−0.01)	0.00	0.51
GaP–Ge	0.80	1.05	0.87	1.24(1.23)	0.88(0.87)	1.27(1.26)	0.63	0.78
GaP–Si	0.80	0.73	0.78	0.86(0.92)	0.50(0.56)	1.00(1.06)	0.45	0.57
GaSb–Ge	0.20	−0.05	0.64	−0.30(−0.11)	−0.43(−0.24)	−0.01(0.18)	−0.11	0.07
GaSb–Si	0.05	−0.51	0.55	−0.68(−0.42)	−0.81(−0.55)	−0.28(−0.02)	−0.29	−0.13
InAs–Ge	0.33	1.37	1.13	0.49(0.66)	0.09(0.26)	0.65(0.82)	0.32	−0.09
InAs–Si	0.15	1.05	1.04	0.11(0.35)	−0.27(−0.05)	0.38(0.62)	0.14	−0.30
InP–Ge	0.64	0.95	1.33	1.06(1.21)	0.52(0.67)	1.00(1.15)	0.58	0.37
InP–Si	0.57	0.63	1.24	0.68(0.90)	0.14(0.36)	0.73(0.85)	0.40	0.16
InSb–Ge	0.0	−0.04	–	−0.36(−0.03)	−0.71(−0.38)	−0.21(0.12)	−0.17	−0.18
InSb–Si	0.0	−0.35	–	−0.74(−0.28)	−1.09(−0.69)	−0.48(−0.08)	−0.35	−0.39
CdS–Ge	1.75	2.41	2.17	2.92(3.10)	2.00(2.28)	2.88(3.16)	–	0.88
CdS–Si	1.55	2.09	2.08	2.54(2.79)	1.62(1.97)	2.61(2.96)	–	0.67
CdSe–Ge	1.30	1.89	2.04	2.00(2.32)	1.23(1.55)	2.01(2.33)	–	0.54
CdSe–Si	1.20	1.57	1.95	1.62(2.01)	0.85(1.24)	1.74(2.13)	–	0.33
CdTe–Ge	0.85	0.92	1.65	0.83(1.25)	0.20(0.62)	1.07(1.49)	–	0.41
CdTe–Si	0.75	0.60	1.56	0.45(0.94)	−0.18(0.31)	0.80(1.29)	–	0.21
CdTe–α-Sn	1.1	–	–	1.80(1.86)	1.28(1.34)	–	–	–
ZnSe–Ge	1.40	1.87	1.82	2.09(2.24)	1.46(1.61)	2.24(2.39)	1.52	1.47
ZnSe–Si	1.25	1.55	1.73	1.71(1.93)	1.08(1.30)	1.97(1.19)	1.34	1.26
ZnSe–GaAs	1.03	1.25	1.11	1.42(1.55)	1.05(1.18)	1.45(1.58)	1.20	1.05

Interface								
ZnTe–Ge	0.95	0.99	1.49	0.90(1.16)	0.38(0.64)	1.39(1.65)	0.66	0.84
ZnTe–Si	0.85	0.67	1.40	0.52(0.85)	0.00(0.33)	1.12(1.45)	0.48	0.64
GaSe–Ge	0.83	0.60	–	–	–	–	–	–
GaSe–Si	0.74	0.28	–	–	–	–	–	–
CuBr–GaAs	0.85	–	–	3.03	2.37	–	–	–
CuBr–Ge	0.7	–	–	3.70	2.78	–	–	–
Average accuracy								
IV's and III–V's		0.28	0.41	0.28(0.34)	0.37(0.25)	0.27(0.28)	0.14	0.23
All data		0.30	0.49	0.37(0.48)	0.45(0.28)	0.39(0.52)	0.16	0.32

* EXP = experimental values; EA = electron affinity rule [2]; FK = Frensley–Kroemer values [14]; TBHF = Harrison's tight-binding values with Hartree–Fock terms [9] [relaxation-corrected values [13] in parentheses, as for all other tight-binding calculations]; TBHS = Harrison's tight-binding values with Herman–Skillman terms [10]; TBV = tight-binding values from Vogl et al. [11]; TER = Tersoff's values [16]; UN = Unlu–Nussbaum values [17].

Table 9b
Comparison between measured and calculated ΔE_v values by Ruan and Ching [108].

Interface	Exp.	Theory	Interface	Exp.	Theory
Si–Ge	0.28	0.22	CdS–Ge	1.75	1.87
AlAs–Ge	0.86	0.92	CdS–Si	1.55	1.44
AlAs–GaAs	0.34	0.36	CdSe–Ge	1.30	1.35
GaAs–Ge	0.49	0.51	CdSe–Si	1.20	1.11
GaAs–InAs	0.17	0.18	CdTe–Ge	0.85	0.92
GaP–Ge	0.80	0.95	ZnSe–Ge	1.40	1.56
GaP–Si	0.80	0.61	ZnSe–Si	1.25	1.18
InAs–Si	0.15	0.10	ZnSe–GaAs	1.03	0.96
InP–Ge	0.64	0.67			

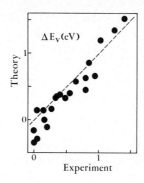

Fig. 8. Comparison between measured ΔE_v's (from table 1) and the predictions of Tersoff's model (refs. [16,106], and tables 7 and 9). The dashed line corresponds to perfect agreement between theory and experiment.

matching, no microdiffusion, etc.? We already debated this point and the severe risks in limiting the data base, due to the arbitrary character of the selection criteria. Other authors, however, disagree with our conclusions [24].

All of the above caveats must be kept in mind while considering the results of the accuracy tests of different linear models. These results are summarized in table 9. We limited table 9a to models which numerically predict a number of interfaces large enough to make the accuracy test meaningful.

From the averaged accuracies of table 9a it is possible to draw the following conclusions. From an empirical point of view, Tersoff's model [16] is the most accurate. One should notice, however, that this model extends to a more limited number of interfaces than most other models. In particular, Tersoff's model is much more accurate than the electron affinity rule [8], whose validity had already been severely challenged by direct experimental tests [48,105]. The accuracy of Tersoff's approach is emphasized by fig. 8.

The spin–orbit and relaxation corrections to the tight-binding models do not significantly modify their accuracy. The magnitude of these corrections, however, could be taken as an indication that these effects are not negligible and should be included in a realistic theoretical treatment of the discontinuities.

An important general conclusion from table 9a is that the accuracy decreases *for all models* when the data base is expanded from group IV and III–V compounds to all interfaces, involving those made with strongly ionic materials. This suggests that the microscopic interface dipoles are strong contributors to the discontinuities, at least for interfaces involving ionic compounds.

Table 9b shows that the simple approach proposed by Ruan and Ching achieves better accuracy than the linear models. We consider this as additional evidence for the validity of the conceptual background of the class of the theories including Tersoff's [16] and Tejedor–Flores models [5]. In fact, it indicates that this conceptual background can yield better accuracy when the hypothesis of linearity and its intrinsic accuracy limit are removed. We shall present in the next section additional evidence in favor of this class of theories.

3.3. Correlations between Schottky barriers and heterojunctions

The question of a possible correlation between Schottky barriers and discontinuities has direct relevance to the conceptual soundness of some of the linear discontinuity models. An intuitive but naive way to predict such correlation is illustrated by fig. 9. The assumption here is that the position

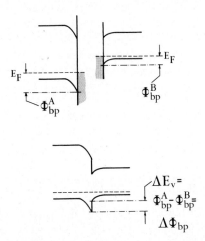

Fig. 9. An intuitive although naive way to explain the possible relation between heterojunction band discontinuities and Schottky barriers. The top part of the figure shows the Schottky barrier obtained by interfacing two different p-type semiconductors with the same metal. The ϕ_{bp}'s are the corresponding Schottky barrier heights, and correspond to the interface distances in energy between the top of the valence band of each semiconductor and the Fermi level. Equation (26) is obtained assuming that these distances do not change if one considers the heterojunction interface involving the two semiconductors instead of the two semiconductor–metal interfaces. This means that the Fermi level pinning position is the same for a given semiconductor in both kinds of interfaces.

of E_F relative to the band edges does not change when the metal side of each Schottky barrier is eliminated and the two semiconductors are brought together. One has, then:

$$\Delta E_v = \Delta \phi_{bp}, \tag{26}$$

where $\Delta \phi_{bp}$ is the difference between the two Schottky barrier heights (assuming that the semiconductors are both p-type). The corresponding equation for the Schottky barriers on n-type semiconductors is:

$$\Delta E_v = \Delta E_g - \Delta \phi_{bn}. \tag{27}$$

More rigorously, the above correlation is predicted by Tersoff's general midgap-energy theory of Schottky barriers and heterojunctions [16,93]. It would also be predicted by a combination of the electron affinity rule for heterojunctions (or a theory related to it) with the Schottky model for Schottky barriers. The Schottky model, however, does not work for covalent semiconductors. Furthermore, direct experimental evidence exists against the electron affinity rule and related models [48,105].

A positive test of eqs. (26) and (27), therefore, is also a positive test of the midgap-energy approach and, in general, of the theories in its class. Furthermore, eqs. (26) and (27), after a positive test, could be used to

Table 10
Measured Au Schottky barried heights for n-type and p-type materials [1,6]. All values in eV.

Material	ϕ_{bn}	ϕ_{bp}	$\langle \phi_{bp} \rangle$, average interface distance between E_F and E_v
Si	0.81		0.30
Ge	0.45		0.22
AlAs		0.9	0.9
GaAs	0.90	0.5	0.47
GaP	1.30	0.96	0.95
GaSb	0.60	0.1	0.08
InSb		0.1	0.1
InP	0.49		0.78
CdS		1.63	1.63
CdSe	0.49	1.21	1.23
CdTe	0.60	0.78	0.81
ZnSe	1.36	1.31	1.26
ZnTe		0.65	0.65

predict heterojunctions band discontinuities. The input would be measured Schottky barrier heights rather than calculated terms. This is important, since it eliminates the additional uncertainty caused by the practical approximation in calculating theoretical terms, e.g., the midgap-energy positions in Tersoff's approach [16].

A correlation between Schottky barrier heights and heterojunction band discontinuities, as predicted by eq. (28), was found by Eizenberg et al. [11]. The correlation was found for Mo–Al$_{1-x}$Ga$_x$As and GaAs–Al$_{1-x}$Ga$_x$As interfaces. We will describe the results of an extensive test of eq. (28), based on Schottky barriers between different semiconductors and Au [1,6]. This metal was selected because it corresponds to the most extensive set of Schottky barrier data. Another advantage is that the Au interfaces are less likely to be affected by contamination than other interfaces. Table 10 shows experimentally measured Schottky barriers for Au on n-type and p-type semiconductors, ϕ_{bn} and ϕ_{bp}. These parameters correspond to the distance in energy at the interface between the Fermi level and E_c or E_v. Therefore, it is straightforward to extract from the p-type and n-type data an average interface position of the Fermi level with respect to E_v, $\langle \phi_{bp} \rangle$. Equation (26) can be rewritten in terms of this parameter,

$$\Delta E_v = \Delta \langle \phi_{bp} \rangle. \tag{28}$$

Table 11
Correlation between heterojunction valence-band discontinuities and Au Schottky barrier heights.

Interface	$\Delta \langle \phi_{bp} \rangle$ (eV)	Experimental ΔE_v (eV)	Interface	$\Delta \langle \phi_{bp} \rangle$ (eV)	Experimental ΔE_v (eV)
Si–Ge	0.08	0.28	CdS–Ge	1.41	1.75
AlAs–Ge	0.68	0.86	CdS–Si	1.33	1.55
AlAs–GaAs	0.44	0.33	CdSe–Ge	1.01	1.30
GaAs–Ge	0.25	0.49	CdSe–Si	0.93	1.20
GaAs–Si	0.17	0.05	CdTe–Ge	0.59	0.85
GaP–Ge	0.73	0.80	CdTe–Si	0.51	0.75
GaP–Si	0.65	0.80	ZnSe–Ge	1.04	1.40
GaSb–Ge	−0.14	0.20	ZnSe–Si	0.95	1.25
GaSb–Si	−0.22	0.05	ZnSe–GaAs	0.79	1.03
InP–Ge	0.56	0.64	ZnTe–Ge	0.57	0.95
InP–Si	0.48	0.57	ZnTe–Si	0.49	0.85
InSb–Ge	−0.12	0.00			
InSb–Si	−0.20	0.00			

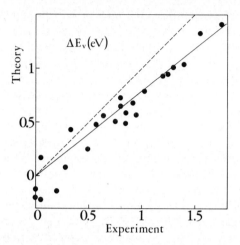

Fig. 10. Experimental test of eq. (28). The correlation between heterojunction valence-band discontinuities and Schottky barrier heights is evident. The dashed line corresponds to perfect agreement between theory [eq. (28)] and experiment. The solid line emphasizes the better fit given by eq. (31).

The right-hand and left-hand terms of eq. (28) are listed in table 11 for 24 different interfaces. The correspondence between the two columns is quite clear, and it is further emphasized by fig. 10. The average magnitude of the discrepancy between ΔE_v and $\Delta \langle \phi_{bp} \rangle$ is 0.23 eV, i.e., better than the accuracy of most linear models and not much above the underlying accuracy limit arising from linearity. These facts indicate that there is, indeed, a correlation between ΔE_v and the Schottky barrier heights, as predicted by the midgap-energy approach.

One interesting element in fig. 10 is the systematic character of the deviations. Most data points are on the same side of the "perfect agreement" line. The simplest modifications of eq. (28) which could account for the systematic discrepancy are an additive constant or a slope different from 1. While no theoretical justification has been provided for an additive constant, a modification of the slope can be explained within the framework of the midgap-energy approach [95,109].

Equation (28) is valid only in the strong-Fermi-level-pinning limit. Deviations from such limit, while essentially negligible for heterojunctions, may be significant for Schottky barriers. In general, the Schottky barrier can be written as

$$\phi_{bp} = E_B + S\Delta\chi, \tag{29}$$

where E_B is the midgap energy, $\Delta\chi$ is the difference between the metal and semiconductor electronegativities, and S is the pinning strength parameter. Only for $S = 0$, i.e., for the strong pinning limit, $\phi_{bp} = E_B$, and Tersoff's model [16] gives eq. (28).

In general, however, $S \neq 0$. In fact [110], S is linear in the indirect band gap, and, therefore, rather linear in $(\epsilon_\infty)^{-1}$, the reciprocal of the optical dielectric constant. Thus, it should be roughly proportional to the "average" band gap, and hence linear with E_B.

If we, therefore, substitute $(a + bE_B)$ for S, and take $\Delta\chi = -c$, we find

$$\phi_{bp} = (1 - A)E_B - B, \tag{30}$$

where $A = bc$ and $B = ac$. Since, according to the midgap-energy approach [5], ΔE_v is given by the difference between the E_B's of the two semiconductors, we obtain

$$\Delta\phi_{bp} = (1 - A)\Delta E_v, \tag{31}$$

which replaces eq. (28).

The constants a and b are both positive. In the case of Au c is also positive, and so is A. Thus, the slope, $1 - A$, is smaller than one, in agreement with the experimental findings.

A least-square fit of eq. (31) to the data of table 11 gives $A = 0.21$. The results were shown in fig. 10 (solid line). The fit is clearly better than the predictions of eq. (28) (dashed line). The average magnitude of the discrepancy between theory and experiment decreases from 0.23 eV in the case of eq. (28) to 0.10 eV in the case of eq. (31).

3.4. Empirical solutions

The fundamental understanding of the nature of the band lineup requires the development of complete theoretical models. The practical task of *predicting* band discontinuities, however, can be accomplished by using semi-empirical approaches. Equation (31) is a good example of one of these approaches, since it can be used to estimate band discontinuities from measured Schottky barrier values.

The entire class of linear models, i.e., models based on eq. (8), can of course be optimized by using the experimental data. This approach was first proposed by Katnani and Margaritondo [29,96]. The expanded data base of table 1 makes possible an optimization of the original terms. The philoso-

Table 12

Material	Empirical position in energy of the valence-band edge (in eV) for eq. (16).	Material	Empirical position in energy of the valence-band edge (in eV) for eq. (16).
Ge	(Reference)	CdS	−1.74
Si	−0.16	CdSe	−1.33
α-Sn	+0.22	CdTe	−0.88
AlAs	−0.78	ZnSe	−1.40
AlSb	−0.61	ZnTe	−1.00
GaAs	−0.35	PbTe	−0.35
GaP	−0.89	HgTe	−0.75
GaSb	−0.21	CuBr	−0.87
InAs	−0.28	GaSe	−0.95
InP	−0.69	$CuInSe_2$	−0.33
InSb	−0.09	$CuGaSe_2$	−0.62
		$ZnSnP_2$	−0.48

phy of this approach is simply to estimate from the experimental ΔE_v's the best values for the E_v terms to be used in eq. (16). The reference level for these empirical valence-band edge positions is irrelevant, and it was arbitrarily set to coincide with the top of the valence band in Ge.

The optimization of the terms was achieved with a least-squares fitting procedure. The results are shown in table 12. By inserting the above E_v terms in eq. (16) one can predict ΔE_v for any pair of semiconductors. This procedure appears capable of reaching the underlying accuracy limits of all linear models. The terms in table 12 are an optimization of *all* the linear discontinuity models. We emphasize, however, that this approach is *not* a theoretical model, except for the use of the hypothesis of linearity, eq. (8).

4. Beyond the first level of approximation

The main conclusion of the previous discussion is that all current general-purpose theories are inadequate in predicting the band discontinuities with the accuracy required for applications. These theories can have fundamental validity, and are a necessary first step in the solution of the problem but they cannot overcome the underlying accuracy limit. Although there is some uncertainty about the magnitude of this limit, all reasonable values are much larger than kT at room temperature.

In the future, therefore, the theoretical research in this area must find new avenues to predict ΔE_v and ΔE_c with better accuracy. There are essentially two major approaches which can be used. First, one can try to develop realistic theoretical calculations of the microscopic electronic structure of the interface. Second, one can empirically identify the factors which contribute to the discontinuities and are not accounted for by the current linear models – and estimate the magnitude of their contribution. This second approach has an additional advantage. It identifies practical ways of changing the value of ΔE_v and ΔE_c, which can be used to control them. Recent breakthroughs in this area may open up a tremendous range of opportunities in heterojunction technology.

Detailed calculations of the electronic structure of an interface can be performed with a variety of theoretical schemes, e.g., tight-binding, cluster models, etc. The most effective and advanced method is the self-consistent pseudopotential, implemented with iterative computer calculations. Self-consistent pseudopotential calculations of the electronic structure of the semiconductor–semiconductor interfaces have been published in recent years [20–23].

The above calculations, however, are very scarce. Furthermore, there is no evidence that they reach a better accuracy than linear models in predicting the band discontinuities. The problem, in our opinion, is once again describing a "realistic" interface. An idealized interface, i.e., atomically sharp and, therefore, free of microdiffusion, alloying, formation of interface compounds, etc., would still have a complex microscopic structure which is not accounted for by linear models. The self-consistent pseudopotential approach can, within reasonable limits, describe this kind of structure. However, it cannot – or at least not yet – describe all of the factors contributing to ΔE_v and ΔE_c at a realistic interface.

Efforts are underway to extend the self-consistent pseudopotential approach to localized defects. This is a good step towards the description of the more realistic semiconductor–semiconductor interfaces. However, the complete description of all the possible factors contributing to the discontinuities is beyond the reasonable limits attainable even with self-consistent pseudopotential calculations. For example, the exact atomic positions at the interface could be one such factor. The total-energy minimization is now capable of predicting those positions for some "good" interfaces – but it would be totally unrealistic to expect it to predict all the different atomic positions for an interface full of different kinds of defects.

Even within the above-mentioned reasonable limits, self-consistent pseudopotential calculations and other kinds of theoretical approaches are very

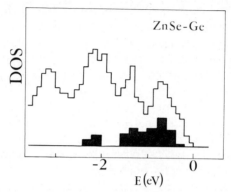

Fig. 11. The density of electronic states of the interface layer of the ZnSe–Ge (110) heterojunction, predicted by Pickett and Cohen [21]. The shaded area shows the contribution from interface states associated with Ge–Zn interface bonds.

valuable when they describe the local electronic structure. This structure is, potentially, an important element both in determining the band lineup and because it influences the physical properties of the interface. Figure 11 provides a classic example. The local ZnSe–Ge electronic structure calculated with the pseudopotential approach [21] predicts the existence of a strong density of local states exactly in the same energy region where the band discontinuity occurs. Therefore, what we call "band discontinuity" is locally "made of" interface states. This information cannot be ignored in estimating the effects of band discontinuities, e.g., on the transport properties of the interface. Experimental evidence for the existence of the ZnSe–Ge interface states was recently provided by photoemission experiments [97].

The limitations of the theoretical tools emphasize the importance of an integrated approach, coupling those tools to empirical observations. The experiments on realistic interfaces can provide two inputs of invaluable importance for the theory. First, they can identify the presence or absence of a particular kind of interface effect which can contribute to the discontinuities, e.g., microdiffusion of a particular species. The new kinds of microscopy, such as atomic-resolution TEM and tunneling microscopy are particularly promising from this point of view. Second, they can empirically test the relevance of these effects to the band lineup. Many effects can affect a priori the band lineup, but some, or most, of them could be unimportant in practice. The magnitude of the different contributions to ΔE_v and ΔE_c are virtually impossible to estimate with a purely theoretical approach, and an empirical solution is both necessary and desirable.

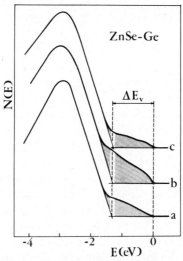

Fig. 12. ZnSe–Ge photoemission spectra showing the valence-band discontinuity [28]. These spectra were taken on unannealed, amorphous Ge overlayers (curve a) and on two different annealed Ge overlayers, exhibiting low-energy electron diffraction spots (curves b and c). These spectra suggest that the order or disorder of the overlayer is not an important factor in ΔE_v for this particular system.

A typical example of this last approach is provided by the interface order or disorder. Suppose we measure ΔE_v for an interface between a given substrate and a given overlayer. Does the result change if the overlayer is changed from amorphous to ordered? The empirical results in this area suggest a different magnitude of this effect for different overlayers, and the necessity of a test on a case-by-case basis. For example, careful analysis of the ZnSe–Ge band discontinuity by photoemission spectroscopy ruled out [28] strong disorder-related contributions to ΔE_v, as shown in fig. 12.

4.1. Controlling the band discontinuities

Recently, the empirical study of factors influencing the band lineup has produced some exciting developments, which may lead to the control of the discontinuities. The quantitative limit of this control is essentially set by the magnitude of the underlying accuracy limit of the general-purpose linear models. The same effects which limit the accuracy of those models can be used to modify the value of ΔE_v and ΔE_c. Thus, on the average one should not expect a flexibility of more than 0.1–0.15 eV in controlling the discontinuities. This is an average limit, however, which does not preclude larger controllability for specific interfaces.

The control of the band lineup at the $Al_{1-x}Ga_xAs$–GaAs interface was achieved by Capasso and co-workers using controlled local doping [98]. This approach is described in Capasso's chapter (ch. 10, this volume). Changes in ΔE_v up to several tenths of an eV were achieved at Wisconsin, Frascati and Caltech by introducing ultra-thin intralayers between the two components of the interface [99,112,113].

This method was inspired by the extensive work done by Brillson et al. [100] on the effects of reactive metal intralayers on the Schottky barrier parameters. The intralayers are known to strongly influence the microdiffusion processes, and lead to changes in the Schottky barrier heights. Likewise, a ultra-thin intralayer can influence the interface dipoles and, therefore, modify the band lineup at a heterojunction interface. In fact, such modifications can be caused by a variety of intralayer-induced effects: formation of interface bonds with their own dipole moments, saturation of dangling bonds, activation or inhibition of microdiffusion processes, etc.

Strong intralayer-induced modifications in the heterojunction have been observed in several systems. Figure 13 shows, for example, the effects of a ultra-thin aluminum intralayer between ZnSe and Ge on the band lineup. The intralayer causes ΔE_v to increase by 0.2–0.3 eV. Such a modification is large enough to be directly visible in the photoemission spectra. Similar effects were observed for a variety of systems. Al intralayers increase the valence-band discontinuity by 0.1–0.2 eV for CdS–Ge and by 0.1–0.3 eV for CdS–Si. Likewise, Cs intralayers increase ΔE_v by 0.2–0.3 eV for SiO_2–Si, by 0.2 eV for CdS–Si and by 0.3–0.4 eV for GaP–Ge. Hydrogen intralayers decrease ΔE_v for CdS–Si while they do not produce detectable effects for GaP–Ge. In general, intralayer effects are negligible for GaAs–Ge.

Some of the most striking intralayer results were observed in the case of controlled hydrogen or deuterium contamination of SiO_2–Si [112,113]. The modifications of ΔE_v have magnitudes up to 0.5 eV. Furthermore, depending on the interface preparation technique, the hydrogen intralayer can cause either an increase or a decrease of ΔE_v. These results suggest that the intralayer effects are an interplay of several different mechanisms. This may enhance the flexibility in controlling the band lineup, but it could also make it difficult to provide a simple explanation for the phenomena.

Quaresima and co-workers [112] adopted a simple model to calculate the dipole term due to the formation of chemical bonds involving the in-

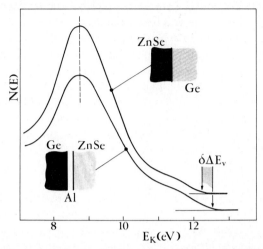

Fig. 13. Photoemission spectra showing the change in the ZnSe–Ge band discontinuity, $\delta\Delta E_v$, caused by a 2 Å thick Al intralayer. The main peak is primarily due to the ZnSe substrate. Its alignment in energy for the two curves guarantees that the ZnSe valence band edges are also aligned. The leading spectral edges of the two curves are due to Ge. Their position in energy is different for the interfaces with and without Al intralayer. This corresponds to a ΔE_v increase of 0.2–0.3 (see ref. [99]).

tralayer. The model used Sanderson's approach to calculate charge transfers upon formation of the bonds [114]. In spite of its simplicity and neglect of other effects, this model is remarkably successful in quantitatively explaining a number of intralayer-induced effects. This, at least, indicates that the phenomena cannot be entirely explained by microdiffusion processes.

The experimental observation of intralayer-induced effects is, of course, only a first step towards their practical exploitation. The feasibility of exploiting them has not yet been demonstrated. Extensive R&D must test the stability of the intralayer under realistic operating conditions as well as their direct and indirect influence on the device operation. In parallel, new kinds of intralayers should be explored. The possibility of controlling heterojunction band lineups in practical devices is so appealing that it entirely justifies and stimulates these extensive research programs.

5. Final remarks

Better experimental and theoretical methods have greatly improved our understanding of the problem of band discontinuities. While this brings us quite close to a good fundamental knowledge of their nature, we are still far from results suitable for technological applications.

The large and reasonably reliable data base produced by photoemission spectroscopy and by other methods enabled us to clarify the limits of all general-purpose linear theoretical approaches to band discontinuity estimates. We also extracted an approximate value for the corresponding underlying accuracy limit. These conclusions clearly put technological applications beyond the reach of any current general-purpose discontinuity model. From a fundamental point of view, some significant step was made towards the identification of the linear model with the soundest theoretical foundations. Tersoff's approach [16] is the front runner, since it does produce the best accuracy, and since its prediction of a correlation between ΔE_v and the Schottky barrier heights has been experimentally verified.

A theoretical model capable of reaching the necessary accuracy for applied research must still be developed. The most promising avenue appears an integrated approach, in which self-consistent pseudopotential calculations are coupled with empirical estimates of the factors contributing to the band lineup, and to a detailed picture of the atomic and electronic structure of the "realistic" interface. One important by-product of this approach is the identification of methods to control the band lineup. Significant results have already been obtained in this area, and they show promise of the development of an entire new branch of device engineering.

Note added in proof

In recent months, there have been several important developments in this rapidly expanding field. Some of the most interesting are:
- A study of the pressure dependence of the band offset in an InAs–GaSb superlattice [115].
- A theory of band offsets based on acoustic deformation potentials, which is directly related to the Tersoff–Heine–Flores models [116].
- The explanation of the failure of the common-anion rule in terms of the cation d contributions to the valence band [117].
- A local-density calculation of the GaAs–AlAs offset by Massida and co-workers [118].
- A test of the linearity of the band offset for II–VI heterojunction interfaces [119].
- Measurements of the valence-band discontinuities for ZnSe–MnSe [120] and GaAs–Si [121]. ΔE_v was 0.17 eV in the first case, and 0.23–0.55 eV for GaAs–Si, depending on the substrate temperature.
- An explanation of the intralayer-induced band lineup changes by Durán and co-workers [122].

Other interesting recent articles are refs. [123–125].

References

[1] A.G. Milnes and D.L. Feucht, Heterojunctions and Metal–Semiconductor Junctions (Academic Press, New York, 1972).
[2] R.L. Anderson, Solid-State Electron. 5 (1962) 341.
[3] For a recent review on surface sensitive techniques see:
G. Margaritondo and J.E. Rowe, Electron spectroscopy, in: Treatise of Analytical Chemistry, Vol. 8, eds. I.M. Kolthoff and P.J. Elving (Wiley, New York, 1986) Part I, ch. 17; and
G. Margaritondo and J.H. Weaver, Photoemission spectroscopy of valence state, in: Methods of Experimental Physics, Vol. 22, eds. R.L. Park and M.J. Lagally (Academic Press, New York, 1985) ch. 4.
Some of the first experiments in this area are described in ref. [73].
[4] C. Mailhiot and C.B. Duke, Phys. Rev. B 33 (1986) 1118.
[5] C. Tejedor and F. Flores, J. Phys. C 11 (1978) L19.
[6] J.L. Freeouf and J.M. Woodall, Appl. Phys. Lett. 39 (1981) 727; Surf. Sci. 168 (1986) 518.
[7] J.M. Langer and H. Heinrich, Phys. Rev. Lett. 55 (1985) 1414; and Physica B+C 134 (1985) 444.
[8] M.J. Caldas, A. Fazzio and Alex Zunger, Appl. Phys. Lett. 45 (1984) 671;
Alex Zunger, Ann. Rev. Mater. Sci. 15 (1985) 411; Solid State Physics, Vol. 39, eds. H. Ehrenreich and D. Turnbull (Academic Press, New York, 1986) sect. VI.29, p. 275; Phys. Rev. Lett. 54 (1985) 849.
[9] W.A. Harrison, J. Vac. Sci. & Technol. 14 (1977) 1016.
[10] W.A. Harrison, J. Vac. Sci. & Technol. B 3 (1985) 1231.
[11] P. Vogl, H.P. Hjalmarson and J.D. Dow, J. Phys. & Chem. Solids 44 (1983) 365.
[12] Z.H. Chen, S. Margalit and A. Yariv, J. Appl. Phys. 57 (1985) 2970.
[13] F. Bechstedt, R. Enderlein and O. Heinrich, Phys. Status Solidi b 126 (1984) 575.
[14] W.R. Frensley and H. Kroemer, Phys. Rev. B 16 (1977) 2642.
[15] J.A. Van Vechten, J. Vac. Sci. & Technol. B 3 (1985) 1240; Phys. Rev. 182 (1969) 891.
[16] J. Tersoff, Phys. Rev. B 30 (1984) 4875.
[17] H. Unlu and A. Nussbaum, private communication and to be published.
[18] A. Nussbaum, The theory of semiconducting junctions, in: Semiconductors and Semimetals, Vol. 15, eds. R.K. Willardson and A.C. Beer (Academic Press, New York, 1981) ch. 2, and references therein.
[19] O. Von Ross, Solid-State Electron. 23 (1980) 1069.
[20] W.E. Pickett, S.G. Louie and M.L. Cohen, Phys. Rev. Lett. 39 (1977) 109.
[21] W.E. Pickett and M.L. Cohen, Phys. Rev. B 18 (1978) 939.
[22] M.L. Cohen, Adv. Electron. & Electron Phys. 51 (1980) 1, and references therein.
[23] J. Pollman, Festkörperprobleme 20 (1980) 117.
[24] H. Kroemer, Surf. Sci. 132 (1983) 543.
[25] H. Kroemer, W.-Y. Chien, J.S. Harris and D.D. Edwall, Appl. Phys. Lett. 36 (1980) 295.
[26] W.I. Wang, E.E. Mendez and F. Stern, Appl. Phys. Lett. 45 (1985) 639.
[27] W.I. Wang and F. Stern, J. Vac. Sci. & Technol. B 3 (1985) 1280.
[28] G. Margaritondo, C. Quaresima, F. Patella, F. Sette, C. Capasso, A. Savoia and P. Perfetti, J. Vac. Sci. & Technol. A 2 (1984) 508.
[29] A.D. Katnani and G. Margaritondo, Phys. Rev. B 28 (1983) 1944.
[30] M. Heiblum, M.I. Nathan and M. Eizenberg, Appl. Phys. Lett. 47 (1985) 503.

[31] R. Dingle, W. Weigmann and C.H. Henry, Phys. Rev. Lett. 33 (1974) 827;
R. People, K.W. Wecht, K. Alavi and A.Y. Cho, Appl. Phys. Lett. 43 (1983) 118;
A.C. Gossard, W. Brown, C.L. Allyn and W. Wiegmann, J. Vac. Sci. & Technol. 20 (1982) 694.
[32] N.H. Meynadier, C. Delalande, G. Bastard, M. Voos, F. Alexandre and J.L. Lievin, Phys. Rev. B 31 (1985) 5539.
[33] P. Dawson, G. Duggan, H.I. Ralph, K. Woodbridge and G.W. 't Hooft, Superlattices and Microstruct. 1 (1985) 231.
[34] R.C. Miller, A.C. Gossard and D.A. Kleinman, Phys. Rev. B 32 (1985) 5443.
[35] M.K. Kelly, D.W. Niles, E. Colavita, G. Margaritondo and M. Henzler, Appl. Phys. Lett. 46 (1985) 768.
[36] P. Perfetti, F. Patella, F. Sette, C. Quaresima, C. Capasso, A. Savoia and G. Margaritondo, Phys. Rev. B 30 (1984) 4533.
[37] M. Turowski, G. Margaritondo, M.K. Kelly and R.D. Tomlinson, Phys. Rev. B 31 (1985) 1022.
[38] R.R. Daniels, G. Margaritondo, C. Quaresima, P. Perfetti and F. Levy, J. Vac. Sci. & Technol. A 3 (1985) 479.
[39] P.H. Mahowald, R.S. List, W.E. Spicer, J. Woicik and P. Pianetta, J. Vac. Sci. & Technol. B 3 (1985) 1252.
[40] D. Arnold, A. Ketterson, T. Henderson, J. Klein and H. Morkoc, Appl. Phys. Lett. 45 (1984) 1237.
[41] E.A. Kraut, R.W. Grant, J.R. Waldrop and S.P. Kowalczyk, Phys. Rev. Lett. 44 (1980) 1620.
[42] T.W. Hickmott, P.M. Solomon, R. Fischer and H. Morkoc, J. Appl. Phys. 57 (1985) 2844.
[43] Y. Guldner, G. Bastard, J.P. Vieren, M. Voos, J.P. Faurie and A. Million, Phys. Rev. Lett. 51 (1983) 907.
[44] R.C. Miller, D.A. Kleinman and A.C. Gossard, Phys. Rev. B 29 (1984) 7085.
[45] R.C. Miller, A.C. Gossard, D.A. Kleinman and O. Munteanu, Phys. Rev. B 29 (1984) 3470.
[46] A.D. Katnani, P. Chiaradia, H.W. Sang Jr, P. Zurcher and R.S. Bauer, Phys. Rev. B 31 (1985) 2146.
[47] W. Mönch, R.S. Bauer, H. Gant and Murschall, J. Vac. Sci. & Technol. 21 (1982) 498.
[48] P. Zurcher and R.S. Bauer, J. Vac. Sci. & Technol. A 1 (1983) 695.
[49] S.P. Kowalczyk, E.A. Kraut, J.R. Waldrop and R.W. Grant, J. Vac. Sci. & Technol. 21 (1982) 482.
[50] F. Cerrina, R.R. Daniels and V. Fano, Appl. Phys. Lett. 43 (1983) 182;
F. Cerrina, R.R. Daniels, Te-Xiu Zhao and V. Fano, J. Vac. Sci. & Technol. B 1 (1983) 570.
[51] R. People, K.W. Wecht, K. Alavi and A.Y. Cho, Appl. Phys. Lett. 43 (1983) 118.
[52] H. Okumura, S. Misawa, S. Yoshida and S. Gonda, Appl. Phys. Lett. 46 (1985) 377.
[53] O.J. Glembocki, B.V. Shanabrook, N. Bottka, W.T. Beard and J. Comas, Appl. Phys. Lett. 46 (1985) 970.
[54] J. Batey, S.L. Wright and D.J. DiMaria, J. Appl. Phys. 57 (1985) 484.
[55] M.O. Watanabe, J. Yoshida, M. Mashita, T. Nakanishi and A. Hojo, J. Appl. Phys. 57 (1985) 5340.
[56] D. Arnold, A. Ketterson, T. Henderson, J. Klem and H. Morkoc, J. Appl. Phys. 57 (1985) 2880.

[57] S.R. Forrest, P.H. Schmidt, R.B. Wilson and M.L. Kaplan, Appl. Phys. Lett. 45 (1984) 1199.
[58] S.P. Kowalczyk, W.J. Schaffer, E.A. Kraut and R.W. Grant, J. Vac. Sci. & Technol. 20 (1982) 705.
[59] P.E. Brunemeier, D.G. Deppe and N. Holonyak Jr, Appl. Phys. Lett. 46 (1985) 755.
[60] J.S. Weiner, D.S. Chemla, D.A.B. Miller, T.H. Wood, D. Sivco and A.Y. Cho, Appl. Phys. Lett. 46 (1985) 619.
[61] S. Takatani and Y.W. Chung, Phys. Rev. B 31 (1985) 2290.
[62] E.J. Caine, S. Subbanna, H. Kroemer, J.L. Merz and A.Y. Cho, Appl. Phys. Lett. 45 (1984) 1123.
[63] D.F. Welch, G.W. Wicks and L.F. Eastman, J. Appl. Phys. 55 (1984) 3176.
[64] P. Chiaradia, A.D. Katnani, H.W. Sang Jr and R.S. Bauer, Phys. Rev. Lett. 52 (1984) 1246.
[65] R.S. Bauer, P. Zurcher and H.W. Sang, Appl. Phys. Lett. 43 (1983) 663.
[66] S.P. Kowalczyk, R.W. Grant, J.W. Waldrop and E.A. Kraut, J. Vac. Sci. & Technol. B 1 (1983) 684.
[67] J.M. Ballingall, C.E.C. Wood and L.F. Eastman, J. Vac. Sci. & Technol. B 1 (1983) 675.
[68] R. People, K.W. Wecht, K. Alavi and A.Y. Cho, Appl. Phys. Lett. 43 (1983) 118.
[69] J.R. Waldrop, S.P. Kowalczyk, R.W. Grant, E.A. Kraut and D.L. Miller, J. Vac. Sci. & Technol. 19 (1981) 573.
[70] W. Mönch and H. Gant, J. Vac. Sci. & Technol. 17 (1980) 1094.
[71] E.A. Kraut, R.W. Grant, J.R. Waldrop and S.P. Kowalczyk, Phys. Rev. Lett. 44 (1980) 1620.
[72] R.S. Bauer and J.C. McMenamin, J. Vac. Sci. & Technol. 15 (1978) 1444.
[73] P. Perfetti, D. Denley, K.A. Mills and D.A. Shirley, Appl. Phys. Lett. 33 (1978) 66.
[74] C. Tejedor, J.M. Calleja, F. Meseguer, E.E. Mendez, C.A. Chang and L. Esaki, in: Proc. Int. Conf. on the Physics of Semiconductors, San Francisco, CA, 1984, eds. J.A. Chadi and W.A. Harrison (Springer, Berlin, 1985) p. 559.
[75] W.I. Wang, T.S. Kuan, E.E. Mendez and L. Esaki, Phys. Rev. B 31 (1985) 6890.
[76] C.M. Wu and E.S. Yang, J. Appl. Phys. 51 (1980) 2261.
[77] S.R. Forrest and O.K. Kim, J. Appl. Phys. 52 (1981) 5838.
[78] S.R. Forrest and O.K. Kim, J. Appl. Phys. 53 (1982) 5738.
[79] J. Batey, S.L. Wright, D.J. Dimaria and T.N. Theis, J. Vac. Sci. & Technol. B 3 (1984) 653.
[80] Y.L. Liu, R.J. Anderson, R.A. Milano and H.J. Cohen, Appl. Phys. Lett. 40 (1982) 967.
[81] J.R. Waldrop, E.A. Kraut, S.P. Kowalczyk and R.W. Grant, Surf. Sci. 132 (1982) 513.
[82] T.F. Kuech and J.O. McCaldin, J. Appl. Phys. 53 (1982) 3121.
[83] M. Maenpaa, T.F. Kuech, M.-A. Nicolet, S.S. Lau and D.K. Sadana, J. Appl. Phys. 53 (1982) 1076.
[84] P.P. Gorbik, V.N. Komashchenko and G.A. Fedorus, Fiz. & Tekh. Poluprovodn. 14 (1980) 1276 [Sov. Phys.-Semicond. 14 (1980) 753].
[85] R. Chin, N. Holonyak Jr, S.W. Kirchoefer, R.M. Kolbas and E.A. Rezek, Appl. Phys. Lett. 34 (1979) 862.
[86] J.R. Waldrop, R.W. Grant, S.P. Kowalczyk and E.A. Kraut, J. Vac. Sci. & Technol. A 3 (1985) 835.
[87] E.A. Patten, G.A. Davis, S.J. Hsieh and C.M. Wolfe, IEEE Electron Device Lett. EDL-6 (1985) 60.
[88] S.J. Hsieh, E.A. Patten and C.M. Wolfe, Appl. Phys. Lett. 45 (1984) 1125.

[89] J.R. Waldrop and R.W. Grant, Phys. Rev. Lett. 26 (1978) 1686.
[90] D.J. Chadi, Phys. Rev. B 16 (1977) 790.
[91] G. Margaritondo, Superlattices and Microstruct. 2 (1986) 173.
[92] V. Heine, Phys. Rev. A 138 (1965) 1689.
[93] J. Tersoff, Phys. Rev. Lett. 52 (1984) 465.
[94] H. Temkin and U.G. Keramidas, J. Appl. Phys. 51 (1980) 3269.
[95] J. Tersoff, unpublished results.
[96] A.D. Katnani and G. Margaritondo, J. Appl. Phys. 54 (1983) 2522.
[97] G. Margaritondo, F. Cerrina, C. Capasso, F. Patella, P. Perfetti, C. Quaresima and F.J. Grunthaner, Solid State Commun. 52 (1984) 495.
[98] F. Capasso, A.Y. Cho, K. Mohammed and P.W. Foy, Appl. Phys. Lett. 46 (1985) 664; F. Capasso, K. Mohammed and A.Y. Cho, J. Vac. Sci. & Technol. B 3 (1985) 1245.
[99] D.W. Niles, G. Margaritondo, P. Perfetti, C. Quaresima and M. Capozi, Appl. Phys. Lett. 47 (1985) 1092;
D.W. Niles, G. Margaritondo, E. Colavita, P. Perfetti, C. Quaresima and M. Capozi, J. Vac. Sci. & Technol. A 4 (1986) 962.
[100] L.J. Brillson, Surf. Sci. Rep. 2 (1982) 123.
[101] A.D. Katani and R.S. Bauer, Phys. Rev. B 33 (1986) 1106.
[102] S.P. Kowalczyk, J.T. Cheung, E.A. Kraut and R.W. Grant, Phys. Rev. Lett. 56 (1986) 1605.
[103] C. Tejedor, F. Flores and E. Louis, J. Phys. C 10 (1977) 2163.
[104] C. Priester, G. Allan and M. Lannoo, Phys. Rev. B 33 (1986) 7386.
[105] D.W. Niles and G. Margaritondo, Phys. Rev. B 34 (1986) 2923.
[106] E.O. Kane, Phys. Rev. B 33 (1986) 4428; J. Vac. Sci. & Technol. B 4 (1986) 1051.
[107] E.O. Kane, J. Vac. Sci. & Technol. (1987) in press.
[108] Y.-C. Ruan and W.Y. Ching, J. Appl. Phys. 60 (1986) 4035.
[109] J. Tersoff and G. Margaritondo, unpublished.
[110] J. Tersoff, Phys. Rev. B 32 (1985) 6968.
[111] M. Eizenberg, M. Heiblum, M.I. Nathan and N. Braslau, Appl. Phys. Lett. 49 (1986) 422.
[112] P. Perfetti, C. Quaresima, C. Coluzza, C. Fortunato and G. Margaritondo, Phys. Rev. Lett. 57 (1986) 2065.
[113] P.J. Grunthaner, F.J. Grunthaner, M.H. Hecht and N.M. Johnson, unpublished.
[114] R.T. Sanderson, Inorganic Chemistry (Reinhold, New York, 1967); Chemical Bond and Bond Energy (Academic Press, New York, 1971).
[115] L.M. Claessen, J.C. Maan, M. Altarelli, P. Wyder, L.L. Chang and L. Esaki, Phys. Rev. Lett. 57 (1986) 2556.
[116] M. Cardona and N.E. Christensen, Phys. Rev. B 35 (1987) 6182.
[117] A. Zunger, private communication.
[118] A.J. Freeman, Phys. Rev. B (1987) in press.
[119] T.M. Duc, C. Hsu and J.P. Faurie, Phys. Rev. Lett. 58 (1987) 1127.
[120] H. Asonen, J. Lilja, A. Vuoristo, M. Ishiko and M. Pessa, private communication.
[121] R.S. List, J. Woicik, P.H. Mahowald, I. Lindau and W.E. Spicer, private communication.
[122] J.C. Durán, A. Muñoz and F. Flores, Phys. Rev. B35 (1987) 7721.
[123] D.D. Coon and H.C. Liu, J. Appl. Phys. 60 (1986) 2893.
[124] H. Unlu and A. Nussbaum, IEEE Trans. Electr. Devices ED-33 (1986) 616.
[125] D.V. Lang, M.B. Panish, F. Capasso, J. Allam, R.A. Hamm, A.M. Sergent and W.T. Tsang, Appl. Phys. Lett. 50 (1987) 736.

CHAPTER 3

TRENDS IN SEMICONDUCTOR HETEROJUNCTIONS

A.D. KATNANI

IBM Corporation
System Technology Division
Endicott, NY 13760, USA

Heterojunction Band Discontinuities: Physics and Device Applications
Edited by F. Capasso and G. Margaritondo
© Elsevier Science Publishers B.V., 1987

Contents

Introduction	117
1. Heterojunction interface characteristics	117
1.1. Definition	117
1.2. History	118
2. Experimental procedure	122
2.1. Preparation techniques	122
2.2. Experimental approach	124
2.3. Photoemission measurement of the valence-band discontinuity	126
2.3.1. Synchrotron radiation measurements	126
2.3.2. XPS measurements	128
2.3.3. Accuracy of measurements	129
3. Nature of the valence-band edge discontinuity	132
3.1. Theoretical discontinuity models	132
3.1.1. Anderson's model	132
3.1.2. Frensley and Kroemer's model	133
3.1.3. Harrison's model	133
3.1.4. Tersoff's model	134
3.2. Experimental work	135
3.2.1. Consequences of linearity	135
3.2.1.1. Overlayer ordering	136
3.2.1.2. Dependence on surface orientation	136
3.2.1.3. Commutativity rule	137
3.2.1.4. Transitivity rule	138
3.2.2. Theoretical predictions versus experimental data	139
4. Interfacial effects and the band edge discontinuity: the GaAs/Ge interface	144
5. Potential barrier height	153
6. Effects of an interlayer on the band edge discontinuities	157
7. Summary and conclusions	160
References	163

*Scientific knowledge is a body of
statements of varying degrees of certainty –
some most uncertain, some nearly sure,
none absolutely certain.*

R.P. Feynman

Introduction

Over the past two decades semiconductor heterojunction has evolved into one of the most active areas of research and development. One pressing question in this field is concerned with the origin of the band lineup at semiconductor heterojunctions and the interfacial Fermi level position. The search for an answer to this question has motivated a large number of research groups. The ability to predict and measure band discontinuities to within an accuracy of kT, and to control their value over the difference between the band gaps of the semiconductors forming the junction (ΔE_g) is of great technological importance. This prompted the need to understand the nature of band lineup and its relationship to interfacial dipoles and/or intrinsic properties of the semiconductors forming the junction.

A large amount of experimental and theoretical work has been accumulated to address the above questions. A review of the available data seems appropriate at this time. This chapter examines the experimental techniques used to measure the valence-band discontinuity, particularly, photoemission spectroscopy. The predictions of the available theoretical models will be examined in the light of experimental band discontinuity measurements. Furthermore, the relationship between the band discontinuity and the Fermi-level position will be examined. Finally, the effect of interfacial microscopic properties and foreign adatoms on band discontinuities will be discussed.

1. Heterojunction interface characteristics

1.1. Definition

Figure 1 shows a schematic energy band diagram of two semiconductors forming a heterojunction. The transport properties of all heterojunction devices strongly depend on three interface characteristics [1–5]: band

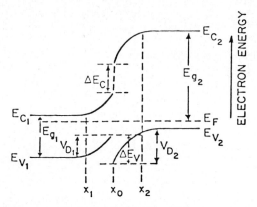

Fig. 1. Schematic of the energy bands for a hypothetical heterojunction. The two semiconductors have a band gap E_{g1} and E_{g2}, respectively.

discontinuities, interface states and potential barrier height. The change in the forbidden gap across the interface is distributed between a valence-band discontinuity (ΔE_v), and a conduction-band discontinuity (ΔE_c). These discontinuities may form barriers for the charge carriers crossing the interface and dramatically influence the operation of heterojunction devices [1–3]. Thus, the possibility of separately biasing electrons versus holes or of confining one or both classes of carriers without applied voltage could be achieved. Also, devices with greater mobility and ballistic transport could be designed. Interface states, including defect states, also influence the heterojunction device behavior by acting as charge traps or recombination centers [4,5]. Finally, the position of the Fermi level (E_F) at the interface determines the barrier heights on the two sides of the interface (V_{D1} and V_{D2}). Actually, it is possible to view the heterojunction as a double Schottky barrier; thus, as discussed below, leading experimentalists and theorists to search for a correlation between ΔE_v and E_F.

1.2. History

Scientists in the late 1950s entertained the possibility of using two different semiconductors to form a heterojunction as opposed to a homojunction. Gubanov [6] formulated a theoretical description of the capabilities of such junctions. Later, Shockley [7] and Kroemer [8] pointed out the potential applications of such a junction in device technology. They showed that for some classes of devices these "heterojunctions" have several advantages over homojunctions. For example, Shockley [7] suggested that a wide-gap emitter would increase the injection efficiency in a transistor. Also, Kroemer [8] showed that, by using heterojunctions, the threshold for an injection laser can be lowered to room temperature. Later Anderson [9] and Oldham and Milnes [4] took a pioneering step by actually growing such junctions and studying their electrical and transport properties. At the same time, major experimental efforts were directed toward developing a room-temperature injection laser using heterojunctions [10].

Knowledge of the band schemes was necessary to interpret and understand the experimental measurements, and especially the distribution of the band-gap difference between the valence- and conduction-band discontinuities. This led Anderson to formulate a rule to calculate band discontinuities. Anderson's model [9] estimated the conduction-band discontinuity as the difference between the electron affinities of the semiconductors forming the junction. The rule is, in essence, based on the continuity of the vacuum level across the junction, and it is an extension of the Schottky barrier model. Since the early 1960s and until the 1970s the electron affinity rule was the only widely accepted method to estimate the band edge discontinuity and, therefore, analysis of experimental data was largely based on its predictions. Attempts to test the validity of the electron affinity rule [11] using transport techniques were inconclusive because these techniques provided only indirect "macroscopic" estimates of the band edge discontinuity, as will be discussed below. Furthermore, these measurements relied heavily on specific assumptions about the distribution of dopants at the interface and about the spatial distribution of the interface states.

The difficulty in measuring the electron affinity with reasonable accuracy combined with the inadequacy of using the electron affinity, which is a surface property, to describe the interfacial band edge discontinuity necessitated the search for an alternative model. The capability of growing heterojunctions with good crystal quality enabled Dingle in 1973 to deduce an empirical rule to estimate the band edge discontinuity [12]. Dingle's rule was based on photoluminescence measurements performed on heterojunction systems involving III–V semiconductors. The rule gained wide acceptance for heterojunctions involving semiconductors whose electron affinities were either unknown or inaccurate.

The early 1970s brought about increased interest in heterojunctions for various areas of device applications. In fact, the realization of a room-temperature heterostructure laser demonstrated the feasibility of a new generation of devices whose performance would have been difficult with the conventional homojunction devices. Such growth in interest created a pressing need for development of experimental and theoretical tools to describe and understand the band edge discontinuity. Furthermore, neither the electron affinity rule nor the Dingle rule could provide the accuracy or the wide applicability needed. In addition, the electron affinity rule was not based on fundamentally sound theoretical ground, and Dingle's rule was simply an empirical rule deduced from indirect measurements of the band edge discontinuity. In fact, the electron affinity rule has been widely criticized by a number of authors, among them Kroemer who refuted the rule in details [13].

Up until the late 1970s, no serious theoretical attempts to describe the band edge discontinuity had been made. Besides the electron affinity rule, the early theoretical work on heterojunctions primarily involved solving the electrostatic equations for the charge distribution on the two sides of the interface. These calculations tried to explain the transport properties of the heterojunctions [1] and the effect of the interface states [5]. This delay in theoretical involvement is probably due to the level of complexity of heterojunctions. The theoretical description of the Schottky barrier problem was already a concern with the involvement of one semiconductor. However, the technological and fundamental importance of heterojunction systems stimulated a number of theoreticians to attempt to tackle this complexity. In 1977, a number of theoretical calculations of the interface electronic properties were reported. These calculations used two different approaches. In the first approach, the band energy positions for the free semiconductors were calculated with respect to a common energy reference. To reduce the problem, the complexity of the interface between the two semiconductors forming the junction is ignored. Then, the band edge discontinuity is simply calculated by lining up the common energy reference. The second approach calculated the local electronic structure of the interface in detail, giving the band discontinuities as a by-product. Examples of the first approach are the potential-matching model of Frensley and Kroemer [14], the tight-binding approach of Harrison [15], the continuous intrinsic Fermi-level model by Adams and Nussbaum [16], the continuous conduction-band model by Von Ross [17] and the induced gap state model by Tersoff [18]. A short description of some of these theories will be given later. Examples of the second approach are the self-consistent calculations of Baraff et al. [19] and Pickett et al. [20]; the cluster approach of Swarts et al. [21] and the tight-binding approach of Pollman and Pantelides [22]. In principle, some of the second type of models can be developed to any degree of accuracy; therefore, they are ideal methods to estimate the discontinuities. However, the practical current accuracy is still unsatisfactory, and these models involve complicated calculations and hours of expensive computer time.

Due to the nature of the preparation techniques and the experimental probes, early experimental work on heterojunction transport properties did not provide detailed information about heterojunction interface characteristics. The development of liquid-phase epitaxy to yield crystals with reasonable quality enabled a number of researchers to use photoluminescence to estimate the band edge discontinuity. Later, the development of molecular-beam epitaxy, an ultrahigh vacuum compatible technique, enabled scientists to use surface sensitive techniques like photoemission [23–28], Auger electron spectroscopy and electron energy loss spectroscopy [29–31] to gain insight into the microscopic electronic structure of the interface and to obtain a local measurement of ΔE_v and E_F. The most extensive results were obtained using photoemission spectroscopy with synchrotron radiation. For example, studies of the interface states and, in general, of the evolution of the local electronic structure during the interface formation were made possible by the use of angle-resolved synchrotron radiation photoemission [32,33].

The first results on measurements of the valence-band discontinuity using photoemission spectroscopy were reported in 1978 [23–25]. Subsequent experimental investigations have primarily addressed the problem of the distribution of the gap difference between valence-band and conduction-band discontinuities and the interfacial Fermi level position [25–28,34–36]. The results of these experiments were used to test the validity of the predictions of the different discontinuity models. The "microscopic" experimental studies also investigated other heterojunction interface characteristics, such as the evolution of the electronic structure during the interface formation, and the resulting interface states [32,33,37–39]. For example, the interface electronic states were detected and investigated for Si/Ge [37,38] and ZnSe/Ge [39] interfaces.

The above historical introduction described the progress in the experimental and theoretical tools used to study the band edge discontinuity. Another fundamental parameter of heterojunction devices is the interfacial Fermi-level position. The theoretical problems concerning Fermi-level pinning and potential-barrier formation are similar to those found for metal–semiconductor interfaces. The Schottky barrier for metal–semiconductor interfaces, which is the equivalent of the heterojunction potential barriers, has been widely studied [40–43]. Experimental results demonstrated that for many metal–semiconductor interfaces, the Fermi-level pinning position is obtained at small metal coverages and is independent of the metal overlayer. This observation led Spicer and co-workers [40] to propose in 1979 the "defect model" for Fermi-level pinning at III–V/metal interfaces, which relates this effect to native defects at the semiconductor surface created during the interface formation. Several theoretical studies recently tried to understand the nature of the Fermi-level-pinning defects. Daw and Smith [44] identified the defects as vacancies created at the surface during the metal deposition. However, Dow and Allen [45] argued that anti-site defects are energetically more favorable. In general, the nature of local defects at metal–semiconductor interfaces and their role in pinning E_F remains a rather controversial issue.

A number of experiments [27,46] suggested that the defect model could be extended to certain types of semiconductor–semiconductor interfaces. The extension of the defect model to heterojunction interfaces would have implications on the band edge discontinuity and the density of interfacial defect states. In fact, in a recent study, Katnani and Margaritondo suggested a correlation between the band edge discontinuity and the Fermi-level pinning position [27]. However, the nature of defects and their role in determining band discontinuity is controversial. Chiaradia et al. [47] showed that defects do not occur in sufficient density at the GaAs/Ge(100) interface to influence either the Fermi-level position or the band discontinuity. Furthermore, they did not find a correlation between ΔE_v and E_F. Also, Brugger et al. [48] argued against the presence of such defects at the GaAs/Ge(110) interface. In addition, Zur et al. found that a large density of states is necessary to pin the Fermi level [49]. All current experimental and theoretical work addressing the question of the Fermi-level pinning position, if anything, leads to more controversy. Nonetheless, this question is vital to heterojunction research.

2. Experimental procedure

2.1. Preparation techniques

Much of the progress in the development of many heterojunction devices and in the fundamental understanding of heterointerfaces is attributed to the emergence of a number of high crystal quality growth techniques. One of the first and most widely used techniques is liquid-phase epitaxy (LPE), which was invented by Nelson in 1963 [50]. Later, Panish introduced the sliding boat LPE to grow double heterostructures [51]. Two other growth techniques were introduced in the early 1970s, metal-organic chemical vapor deposition (MOCVD) and molecular beam epitaxy (MBE) [52,53]. The quality, background doping or impurities and control over the thickness of the grown crystal are among the parameters that make one technique advantageous over the others. A comparison of these techniques is beyond the scope of this chapter.

For any crystal growth technique, the quality of the starting substrate surface influences the quality of the interface and the grown crystal. There are several preparation methods for the starting substrate surface. Among them are chemical polishing, vacuum annealing, sputtering followed by annealing and in situ cleaving. There are numerous reports discussing the effect of surface contamination on the quality of the grown crystal and the interface [54,55].

Another parameter that affects the quality of the grown crystal and the interface is the substrate temperature during growth. For example, an amorphous Ge overlayer and a negligible amount of interdiffusion is obtained when Ge is deposited on a room-temperature GaAs substrate. When the substrate temperature is raised to 320–360°C, an epitaxial Ge overlayer and an abrupt interface are obtained [56]. Further increases in the substrate temperature result in a diffused interface. Therefore, for two semiconductors to be compatible, their crystallinity temperatures have to be within 50°C of each other, and both must be lower than the temperature at which interdiffusion occurs. The effect of growth conditions on the heterojunction characteristics, particularly the band edge discontinuity, will be discussed later in the text.

The majority of basic studies used two preparation methods. The first one involves interfaces between in situ cleaved crystals and a semiconductor overlayer deposited by simple evaporation methods. The studies in this case are limited to cleavage surface orientations only and the results depended on the quality of the cleaved surface. The other method involves interfaces grown with MBE. In this case both semiconductors are grown in situ. A thick buffer layer is usually grown to eliminate any problems due to the original substrate or those that may arise due to prior treatment. Furthermore, the number of surface orientations available for study is limited by the quality of crystal growth on that surface orientation. In fact, MBE combined with surface techniques offers a golden opportunity to combine interfacial information gained from surface techniques with device characteristics gained from electrical and optical techniques. This opportunity has not been fully exploited yet because of the costs involved.

2.2. Experimental approach

The two major experimental approaches used to investigate semiconductor–semiconductor interfaces fall into two categories. The first category is transport (optical and electrical) techniques, such as thermionic emission above barriers [57], capacitance–voltage [58,59], internal photoemission [60], and photoluminescence [12,61,62], and the second category is surface-sensitive techniques, such as Auger electron spectroscopy [29,31], electron energy loss spectroscopy [30] and photoemission spectroscopy [X-ray photoelectron spectroscopy (XPS) and soft XPS (SXPS) or synchrotron radiation] [25–28,34–39]. Transport techniques provide indirect information about the interface characteristics, while surface-sensitive techniques provide direct information about the microscopic properties of the interface. The former techniques are performed on thick overlayer or device-like structures while the latter require thin overlayers. There is a debate concerning the stability of the band edge discontinuity during the transition from the thin layer regime to the thick layer one. Although it is a legitimate concern, there is growing evidence in the literature of the insensitivity of the band edge discontinuity to the overlayer thickness above 5–10 Å [48,63].

The accuracy of each technique is also of great importance. Any technique must be able to measure the band edge discontinuity to better than kT. Unfortunately, there is no current technique that can achieve such accuracy. Electrical and optical measurements could be claimed to achieve accuracies within 10 meV. However, this is by no means the accuracy of the measured band edge discontinuity because the band edge discontinuity is not a direct outcome of the measurement but an outcome of a model calculation of the data. Therefore, the accuracy of these measurements are not limited by the accuracy of the actual measurement but by the accuracy of the model and the parameters involved in it. For example, although photoluminescence measurements achieve an accuracy within 10 meV, there are a number of issues raised about the way the band edge discontinuity is extracted from the data. The type of well and the boundary conditions used in addition to the large inaccuracy of electron and hole masses are among the issues that deteriorate the accuracy of the actual measurement. However, it is beyond the scope of this chapter to discuss the limitations of the various techniques in the first category [64].

For the second category, SXPS and XPS measure the density of occupied states and, therefore, one could measure the valence-band edge of the semiconductor. Furthermore, the surface sensitivity of this technique makes it possible to track the valence-band edges of both semiconductors. This technique could provide a direct measurement of the band edge discontinuity. In which case, the accuracy of the measurements is model-independent and is limited by the measurement itself. Unfortunately, the accuracy of the techniques in this category is much worse than that of photoluminescence measurements. Accuracies of the order of 100–250 meV are reported in the literature. The origin of these accuracies will be discussed later in the text.

The information gained from SXPS is not limited to providing a direct measurement of the band discontinuity, but also provides information about the initial stages of interface formation and about the chemistry of the interface. In fact, the capabilities of the MBE synchrotron radiation combination have not been fully realized. This combination provides a link between measurements performed on device-like structures and those performed on thin overlayers. Furthermore, the information gained from such techniques is insightful for thin superlattices and for two-dimensional electron gases.

Photoemission involves the photo-excitation of an electron from an occupied state, valence-band or core level, to the continuum [65],

$$E_{kin} = h\nu - E_b - \Phi, \tag{1}$$

where E_{kin} and E_b are the kinetic and binding energy of the excited photoelectron, $h\nu$ is the energy of the incident photon and Φ is the work

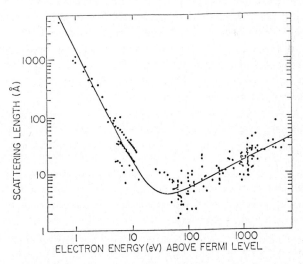

Fig. 2. The inelastic mean free path of excited electrons plotted as a function of their kinetic energy.

function. The photons are produced by either X-ray excitation from Al or Mg targets or synchrotron radiation. Synchrotron radiation has several advantages over conventional photon sources; among them the wide energy range, high luminosity and high resolution [66]. The wide energy range proves advantageous since the inelastic mean free path (IMFP) of the emitted photoelectron depends on its kinetic energy [67]. Figure 2 shows a plot of IMFP as a function of photoelectron kinetic energy. Therefore, electron mean free paths from 5 to 60 Å are possible with synchrotron radiation. This, in essence, provides a nondestructive depth profiling tool within that region. The IMFP for photoelectrons excited with Al Kα and Mg Kα radiations is 10–40 Å. It is apparent that synchrotron radiation is a more suitable technique to monitor the evolution of interface formation.

2.3. Photoemission measurement of the valence-band discontinuity

Two equivalent approaches are used to measure the valence-band discontinuity depending on the source of radiation [27,68]. For synchrotron radiation, the photon energy can be chosen to maximize the cross section for valence-band excitation. While for X-ray radiation the cross section for valence-band excitation is very small. Therefore, to obtain valence-band spectra with reasonable S/N ratio is time consuming. For synchrotron radiation measurements, the top of the valence band is readily defined by extrapolating the valence-band edge and ΔE_v can be measured. For XPS measurements the top of the valence band is indirectly measured by measuring the position of a core level with known binding energy with respect to the top of the valence band. This measurement requires that at least the position of one core level has to track the top of the valence band on each side of the junction upon interface formation.

2.3.1. Synchrotron radiation measurements

In synchrotron radiation measurements the valence-band spectra from the clean substrate surface and from the covered-surface at consecutive overlayer thickness are recorded. The surface sensitivity of the technique allows one to track the evolution of the substrate valence-band electronic features into the overlayer ones. Consequently, this enables one to observe at what coverage the overlayer semiconductor valence-band electronic features are fully developed. For example, fig. 3 shows a typical evolution of the GaAs valence-band spectra taken at $h\nu = 130$ eV as a function of Ge coverage for c(8 × 2). The arrows point out those features of the clean spectrum under-

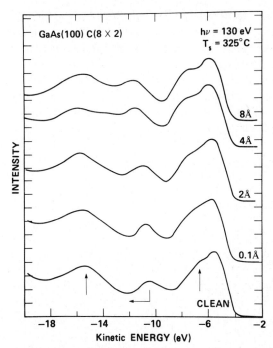

Fig. 3. The evolution of the valence-band spectra of GaAs with increasing Ge coverage. Notice that at coverages above 5 Å the spectral features of the Ge valence band are dominant.

going the most prominent changes. It is clear that the Ge valence-band spectral features (e.g., disappearance of heteropolar gap) are fully developed at Ge coverages between 4 Å and 8 Å, in agreement with theoretical predictions [63] and similar to those observed for GaAs/Ge(110) interfaces [27].

Unfortunately, the energy distance between the top of the valence band of the clean substrate and that of the fully developed overlayer does not correspond to the valence-band discontinuity because the top of the valence band of the clean substrate changes upon the interface formation. This change in position is attributed to the change in band bending, which is due to charge rearrangement and results from the arrival of the overlayer atoms. Experimental measurements show that the change in band bending saturate at coverages ranging from less than a monolayer to two monolayers depending on the substrate and overlayer materials [40,43]. The overlap between the valence-band spectra of both semiconductors in some cases makes it difficult to determine the top of the valence band of the substrate at intermediate overlayer coverages. A method to estimate the change in band bending is necessary to locate the interfacial position of the top of the substrate valence band. Fortunately, most experimental observations suggest that the cation core level tracks the top of the valence band for compound semiconductors [27,40]. Therefore, the change in band bending can be estimated from the observed shift of the cation core level position upon the interface formation. Hence ΔE_v is given by

$$\Delta E_v = E_v^1 - \left(E_v^2 - \beta\right), \tag{2}$$

where the $E_v^{1,2}$ are the tops of the valence band of the fully developed overlayer and clean substrate, respectively, and β is the change in band bending.

Figure 4 shows a plot of the difference between the shift in the valence-band edge and the Ga(3d) as a function of Ge coverage. Below a certain coverage, the valence-band structure of the overlayer is not fully developed. Clearly, at small coverages this difference should be zero because the shift in both the valence-band edge and Ga(3d) is due mainly to the change in band bending. The dashed line indicates the coverage region where the shift in the valence-band edge is composed of two components. One component is due to the change in band bending, and the other is due to the development of the Ge overlayer electronic structure. Notice that the difference is practically constant in the range 6–8 Å (4 to 6 monolayers). At this stage, the difference is fully indicative of ΔE_v alone.

A direct measurement of ΔE_v can be readily obtained from the valence-band spectra for certain classes of heterojunctions which exhibit a large ΔE_v. Figure 5 shows the valence-band spectra obtained from 3 monolayers of Ge on ZnSe. The spectra clearly indicate the valence-band edges of both semiconductors. The ability to resolve both edges facilitates direct measurement of ΔE_v and eliminates the need to rely on core level tracking the top of the valence band to estimate the change in band bending.

2.3.2. XPS measurements

In XPS measurements the energy distance between a substrate and an overlayer core level is measured. The energy position of these core levels with respect to their respective valence-band maxima is determined a priori. As shown in fig. 6, ΔE_v is then given by [68]

$$\Delta E_v = E_b^{c1} - E_b^{c2} + \Delta E_b^{c1-c2}, \tag{3}$$

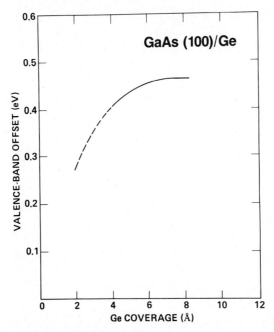

Fig. 4. The difference between the shift of the top of the valence band and the shift of the Ga(3d) is plotted as a function of Ge coverage. Notice that this difference is constant once the electronic spectral features of the Ge valence band are fully developed.

where E_b^{c1} and E_b^{c2} are the binding energies of the substrate and overlayer core levels with respect to the top of the valence band, and ΔE_b^{c1-c2} is the energy distance between these two core levels. The assumption here is that the chosen core level has to track its respective valence band during the interface formation. Figure 7 shows a typical XPS spectra obtained from AlAs/GaAs(100) and AlAs/Ge(100) interfaces. In this case, one needs the binding energy of Al(2p), Ge(3d) and Ga(3d) relative to the top of the valence band of AlAs, Ge and GaAs, respectively, and assumes that those core levels track their respective valence bands during the interface formation. Excitation from the valence band of both AlAs and GaAs has very low cross section in addition to the low resolution attainable at the energies obtained with the X-ray sources.

2.3.3. Accuracy of measurements

As shown above, ΔE_v is the sum of experimentally measured numbers. The accuracy is, therefore, the square root of the sum of the squares of the

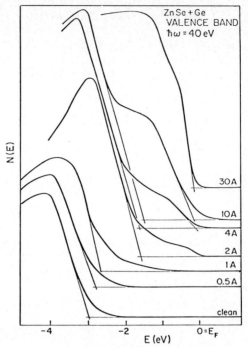

Fig. 5. The evolution of the valence-band spectra of ZnSe with increasing Ge coverage. A direct measurement of the valence-band discontinuity is possible from this figure.

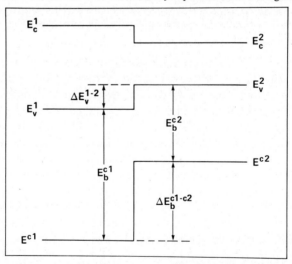

Fig. 6. Schematic of the energy band diagram near the interface. The band discontinuity in this case is deduced from the energy position of the cation core levels relative to their respective valence-band maxima and relative to each other.

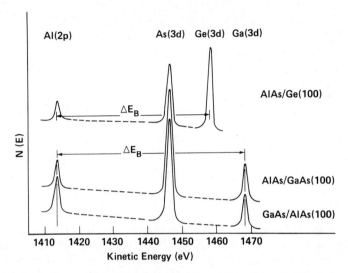

Fig. 7. Core-level spectra obtained from three MBE-grown interfaces. As discussed in the text the band discontinuity is related to the measured distance in energy between the cation core levels.

accuracy of each number. Both methods would have nearly the same accuracy limits. However, synchrotron radiation measurements have the advantage by being more surface sensitive and having higher cross section for the valence band. Therefore, a priori knowledge of the binding energy of the semiconductor cation core levels relative to their respective valence-band maxima is not required. Also, the higher surface sensitivity of synchrotron radiation makes it easier to separate surface from bulk contributions to the core level spectra. In turn, this provides more precise position measurement of the core level. By measuring both the top of the valence band and the change in band bending, synchrotron radiation measurement minimizes errors that might arise from assuming that both cation core levels follow their respective valence bands when XPS is used.

For an ideal case, the accuracy in determining the top of the valence band is the main contributor to the overall accuracy for both methods. The accuracy of the linear extrapolation method of determining the top of the valence band is 100 meV. However, a better accuracy can be achieved by fitting the experimental valence-band spectra to the theoretically calculated density of states [68]. Besides the major effort involved in doing so, theoretical calculations are not available for most semiconductors. Nevertheless, if a precise measurement of the top of the valence band is available, one would expect 40 meV and 50 meV accuracy limits for synchrotron radiation and XPS measurements, respectively. Note that synchrotron radiation measurement, eq. (2), involves two energy separations while XPS measurement, eq. (3), involves three.

3. Nature of the valence-band edge discontinuity

3.1. Theoretical discontinuity models

To obtain insight into the nature of the band edge discontinuity one relies on the combination of experiment and theory. Theory provides predictions based on certain geometrical and structural assumptions. Then a comparison between experimental results and theoretical predictions provides a verification of those assumptions and can help identify the origin of the physical property under consideration. Unfortunately, in the case of heterojunction interfaces, detailed theoretical calculations are complex, costly and have not yet been developed to the accuracy needed. Therefore, in this case one relies on semiquantitative theories. Such theories use first principles analysis which can also shed some light onto the origin of band edge discontinuity.

As discussed earlier, a number of semiquantitative approaches during the last two decades were proposed to describe the band edge discontinuity. Such approaches include: Anderson's electron affinity rule [9], Harrison's LCAO model [15], Frensley–Kroemer pseudopotential model [14], Adam–Nussbaum continuous-intrinsic-E_F rule [16], the continuous conduction-band edge rule by Von Ross [17] and the induced gap state model by Tersoff [18]. The models of Adam and Nussbaum and Von Ross suffer fundamental and practical shortcomings which have been discussed by a number of authors [69–71]. Below a short description of the essential features of the other models.

3.1.1. Anderson's model

This model was developed by Anderson in 1962 [9]. In a heterojunction, the vacuum level is parallel to the band edges and continuous throughout the junction in the absence of dipole layers. Under these assumptions, the conduction-band discontinuity can be expressed as the difference between the electron affinities of the two semiconductors forming the heterojunction, i.e.,

$$\Delta E_c = \chi_1 - \chi_2 \tag{4}$$

where χ_1 and χ_2 are the electron affinities of the two semiconductors 1 and 2. This model is, in essence, an extension of the Schottky barrier model used for metal–semiconductor interfaces. There are fundamental questions about the validity of the electron affinity rule, which also suffers from some practical drawbacks. One example is the large experimental uncertainty in measuring the electron affinity, which affects the accuracy in predicting ΔE_c.

3.1.2. Frensley and Kroemer's model

This model was developed by Frensley and Kroemer in 1977 [14]. Frensley and Kroemer argued that the average interstitial potential, derived from the electrostatic part of the total periodic potential for the bulk semiconductor, is continuous throughout the interface in the absence of dipole layers. On this basis, they calculated the valence-band maximum with respect to the average interstitial potential using the self-consistent pseudopotential approach for the bulk semiconductors. The valence-band discontinuity is given by

$$\Delta E_v = E_v^1 - E_v^2, \tag{5}$$

where E_v^1 and E_v^2 are the valence-band positions of the two semiconductors as given by the calculations. The model was later corrected to include interface dipoles by considering the charge transfer between the two semiconductors. The correction was determined from the relative electronegativities of the constituents of the semiconductors. Frensley and Kroemer showed that the dipole correction is small and does not depend on the semiconductor surface orientation.

3.1.3. Harrison's model

This model was developed by Harrison in 1977 [15]. In this model, the valence-band maxima are calculated by the linear combination of atomic orbital (LCAO) approach, for the bulk semiconductors. The valence-band maximum for the semiconductor is given by

$$E_v = \tfrac{1}{2}\left(e_p^c + e_p^a\right) - \left[\tfrac{1}{4}\left(e_p^c - e_p^a\right)^2 + V_{xx}^2\right]^{1/2}, \tag{6}$$

where e_p^a and e_p^c are the p-atomic energy states of the anion and cation of the semiconductor. In the case of an elemental semiconductor, $e_p^a = e_p^c$. V_{xx} is the interatomic matrix element given by

$$V_{xx} = b/d^2, \qquad (7)$$

where b is a constant determined by fitting the values to the theoretical bands of Si and Ge, and d is the interatomic distance. The valence-band discontinuity is again given by

$$\Delta E_v = E_v^1 - E_v^2, \qquad (8)$$

where E_v^1 and E_v^2 are the calculated valence-band maxima of the two semiconductors.

The dependence of V_{xx} on the interatomic distance allows for empirical corrections to account for lattice relaxation at lattice-mismatched interfaces. For example, an adjustment of the valence-band maximum of one semiconductor relative to the other is possible by assuming that both semiconductors have the same interatomic distance at the interface. Such a correction, in most cases, is found to improve the agreement for lattice-mismatched interfaces [27,34]. Specific examples will be discussed later.

3.1.4. Tersoff's model

Tersoff proposed this model in 1984. He suggested that the mid-gap energy position for a semiconductor should be the reference point to calculate the band discontinuities. The mid-gap energy position is calculated by minimizing the interface dipole that results from the band discontinuity states. In simple terms, the mid-gap energy position is obtained by assuring charge neutrality at the interface and, consequently, zero interfacial dipole. The calculation requires the knowledge of the band structure of the semiconductor under consideration. The fact that the mid-gap energy position, to first-order approximation, is an intrinsic property of the semiconductor makes Tersoff's model applicable to both metal–semiconductor and semiconductor–semiconductor interfaces. In this model the band discontinuity is given by

$$\Delta E_v = E_B^1 - E_B^2, \qquad (9)$$

where the $E_B^{1,2}$ are the mid-gap energy positions for the semiconductor forming the heterojunction.

A common feature of the above models is that they express the band discontinuities as the difference between two terms characteristic of the two semiconductors. This "linearity" is a powerful simplification and at the same time a limiting factor. For example, all linear models ignore the peculiar microscopic properties of each interface. In fact, most of them give a band discontinuity which is independent of the crystallographic faces involved in the interface and of the general interface morphology. This common feature implies, for example, that the predicted band discontinuities must be the same for different surface orientations of a given substrate combined with a given overlayer. It also implies that the discontinuities are not different for ordered and disordered overlayers. Two other general consequences of the above models' linearity are commutativity and transitivity of the predicted discontinuities. The commutativity rule implies that the valence- (or conduction-) band discontinuity for the interface between a substrate of material A and an overlayer of material B (A/B interface) is equal in magnitude and opposite in sign with respect to that for the B/A interface. The transitivity rule implies, for example, that the sum of the valence- (or conduction-) band discontinuities for the three interfaces formed by different combinations of three given semiconductors is zero, i.e. the valence-band discontinuities for the A/B, B/C, and C/A interfaces add up to zero.

3.2. Experimental work

Extensive experimental work has been performed over the past two decades to verify the consequences of the linearity implied by most of the above semiquantative theories and to assess their accuracy limits. Also, the experiments addressed the question of the interface and its contribution to the band edge discontinuity. In fact, most experimentalists sought a parameter by which they would have control over the band edge discontinuity. The following sections discuss the predictions of the current theories and the consequences of linearity in light of the available experimental work. The effect of the interface and its contribution to the band edge discontinuity will be discussed in the light of experimental data obtained from the GaAs/Ge heterojunction system.

3.2.1. Consequences of linearity

The consequences of linearity include the independence of band edge discontinuity on substrate orientation and overlayer ordering, and the commutativity and transitivity rules. Below follows a discussion on these general consequences in the light of experimental data which illustrates the validity of the general concept of the various models regardless of the actual theoretical method used. In a way, this study is an attempt to provide an answer to the question: Can the band edge discontinuity be expressed as the difference between two terms? At the same time, this study will provide the general accuracy limits of the linear models.

3.2.1.1. Overlayer ordering
The literature is full of examples in favor and against the independence of the band edge discontinuity on overlayer ordering. Margaritondo et al. [72] and Perfetti et al. [37] did not detect significant changes in ΔE_v for ordered and disordered Ge overlayers on Si substrates. A difference in ΔE_v of the order of 0.2–0.3 eV was reported for ordered and disordered Ge overlayers on ZnSe [73] and GaAs [46], while no difference was observed for ordered and disordered ZnSe overlayers on Ge or GaAs [73]. Later, Margaritondo et al. observed no difference in ΔE_v for ordered and disordered Ge overlayers on ZnSe [74].

Systematic data on the effects of overlayer ordering are not yet available. The available data as shown above is contradictory and, if anything, suggests that overlayer ordering does not affect ΔE_v by more than a few tenths of an eV (0.1–0.15 eV in the average) per interface.

This point is relevant because the bulk of the data used for testing theoretical models is obtained from interfaces involving disordered overlayers (overlayer semiconductor deposited on a room-temperature substrate). Therefore, the limited overlayer ordering effects mentioned above do not jeopardize the overall comparison between such data and those models, and do not affect significantly the tests of the models and the corresponding conclusions.

3.2.1.2. Dependence on surface orientation
Fang and Howard [75] and Grant et al. [26] revealed non-negligible substrate surface orientation effects for the Ge/GaAs heterojunction system. For example, Grant et al. [26] measured discrepancies of the order of 0.2 eV between the ΔE_v's of Ge-covered GaAs substrates with different orientations. However, careful examination of the data found by Grant et al. shows that the average discrepancy is within 0.1 eV. Nevertheless, later, Katnani et al. measured the same ΔE_v within ± 0.05 eV for the heterojunction between Ge and different surface orientations of GaAs [76]. Katnani and Margaritondo measured the same ΔE_v's for $CdS(10\bar{1}0)/Ge$ and $CdS(11\bar{2}0)/Ge$ interfaces [27].

Again the limited amount of data to date supports either a negligible or nonexistent dependence of ΔE_v on the starting substrate surface orientation, suggesting that the contribution of interfacial dipoles to the band discontinuities is within 0.15 eV.

3.2.1.3. Commutativity rule

Commutativity is an assumption basic to all measurements performed on quantum well and superlattices. Non-commutativity for such structures would imply an asymmetry in the potential barrier on both sides of the interface. Such asymmetry would yield a lowering of the position of the conduction band of the semiconductor as the superlattice size increases. The absence of such asymmetry in electrical and optical measurements is, in a way, an indirect confirmation of the commutativity of the band discontinuity.

Table 1 summarizes the results of photoemission measurements which address the commutativity of the band discontinuities. As seen from the table, the experimental data are controversial. Most of the data is obtained from experiments performed on III–V/IV heterojunction systems [73,77,78]. A number of problems arise when a III–V semiconductor is grown on a group-IV semiconductor [79,80]; among them are site selectivity, growth temperature compatibility and interdiffusion. For GaAs/Ge heterojunction systems all three problems exist, while for ZnSe/Ge, interdiffusion is a severe problem in addition to site selectivity. The effects caused by these problems might, in some cases, influence the band discontinuity measurements; thus, making it difficult to separate measurements artifacts from physical reality. For example, the broadening of the Ga(3d) and/or the Ge(3d) lineshape caused by interdiffusion at the GaAs/Ge interface will influence the measurement of ΔE_v. Therefore, it is important to understand and to assess the effects of such problems on interface quality, overlayer

Table 1
Test of the commutativity rule.

Heterojunction system	Discrepancy (eV)	Ref.
GaAs/Ge	0.20	[77]
GaAs/Ge	0.15	[78]
ZnSe/Ge	0.23	[73]
Si/Ge	0.00	[27]
AlAs/GaAs(110)	0.25	[84]
AlAs/GaAs(100)	0.05	[81]

quality and measurement technique before contesting band edge discontinuity commutativity on the basis of this data.

Commutativity was tested for heterojunctions formed between elemental semiconductors involving Si and Ge, and compound semiconductors involving AlAs/GaAs. Katnani and Margaritondo found that the band discontinuity for an Si/Ge heterojunction is independent of the growth sequence [27]. Katnani and Bauer demonstrated that the band edge discontinuity for the GaAs/AlAs(100) heterojunction is commutative [81]. This independence of growth sequence is consistent with experimental observations from quantum-well structures involving III–V/III–V semiconductors, contrary to previous results obtained from AlAs/GaAs(110) interfaces which showed non-commutativity [82]. Watanabe et al., using C–V profiling measurements, showed that ΔE_v for the AlGaAs/GaAs heterojunction is independent of the growth sequence [83].

In conclusion, the band discontinuity was found to be commutative for lattice-matched III–V/III–V heterojunctions. In spite of the problems involved with III–V/IV heterojunctions, their band discontinuities are independent of growth sequence within 0.15 eV. The observed commutativity of the band discontinuity for the severely lattice-mismatched heterojunction Si/Ge is interesting and suggests that interfacial dipoles, if present, are independent of the growth sequence.

3.2.1.4. Transitivity rule

Earlier XPS measurements by Waldrop and Grant have shown nontransitivity of the band discontinuity of 0.64 eV for the CuBr–Ge–GaAs heterojunction systems [25]. Katnani and Margaritondo have analyzed eleven different groups of three semiconductors each to test the transitivity rule [27]. Also, Katnani and Bauer showed that the band edge discontinuity for heterojunctions involving AlAs–Ge–GaAs is transitive [81].

The results are summarized in table 2 which lists the sum of the experimental ΔE_v's for the various heterojunction systems used to test the transitivity of the band discontinuity. This sum must equal zero if the transitivity rule holds. In fact, the sum is equal to zero within the experimental uncertainty of each result – except for GaP–Si–Ge and CuBr–Ge–GaAs heterojunction systems where a discrepancy of 0.32 eV and 0.64 eV occurs, respectively. The majority of the listed interfaces in table 2 support the transitivity of the band discontinuity at the most within 0.15 eV. These results again support the conclusions drawn in the previous sections.

Both observations, the commutativity and transitivity of the band dis-

Table 2
Test of the transitivity rule.

Heterojunction system	Discrepancy (eV)	Ref.
GaAs–Ge–Si	0.13	[27]
GaP–Ge–Si	0.32	[27]
GaSb–Ge–Si	0.02	[27]
InAs–Ge–Si	0.03	[27]
InP–Ge–Si	0.10	[27]
InSb–Ge–Si	0.17	[27]
CdS–Ge–Si	0.03	[27]
CdSe–Ge–Si	0.07	[27]
CdTe–Ge–Si	0.07	[27]
ZnSe–Ge–Si	0.02	[27]
ZnTe–Ge–Si	0.07	[27]
AlAs–GaAs–Ge(100)	0.04	[81]
GaAS–ZnSe–Ge or Si	0.03	[27,73]
AlAs–CuBr–GaAs	0.64	[25]
GaAs–InAs–Ge or Si	0.17	[27,87]
InAs–GaSb–Ge or Si	0.03	[27,86]

continuity, demonstrate the insensitivity of the band structure lineup to interfacial dipoles within 0.15 eV. This is consistent with recent theoretical studies by Zur et al. [49] and experimental measurements of Katnani et al. [28].

In conclusion, the above discussion demonstrates that the linearity assumption imposes a general accuracy limit on all theoretical models based simply on intrinsic semiconductor properties. This general accuracy limit is 0.15 eV as deduced from the discrepancy between experimental results and the consequences of linearity. Consequently, all simple models are not expected to achieve an accuracy better than 0.15 eV.

3.2.2. Theoretical predictions versus experimental data

In this section a comparison between the predictions of each theory with the available experimental data is made. This comparison will assess the accuracy limit of each model.

The most widely used band discontinuity model is Anderson's electron affinity rule. Photoemission experiments in which the two electron affinities and ΔE_v were measured in the same system demonstrated the failure of

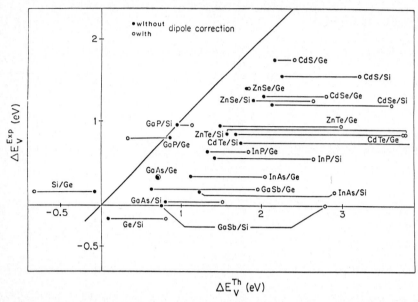

Fig. 8. Comparison between the Katnani–Margaritondo results and the predictions of the Frensley–Kroemer model. ●, without dipole correction; ○ with dipole correction taken into account.

this model in predicting the band discontinuity [84]. One routine difficulty in using Anderson's model is selecting the appropriate electron affinities from the wide range of values found in the literature for each semiconductor.

Katnani and Margaritondo performed a systematic study of several heterojunction interfaces to assess the accuracy of the different semiquantitative models [27]. Figures 8 and 9 show plots of the predictions of the Frensley–Kroemer model and the Harrison models versus the Katnani–Margaritondo experimental results. The solid line corresponds to perfect agreement. The correlation between experimental results and the Frensley–Kroemer model is not excellent, although the model does give reasonable predictions for some interfaces, e.g., GaP/Si. Corrections for the interface dipoles, as suggested by Frensley and Kroemer [14], do not improve the agreement with experimental results. The discrepancy between this theory and experimental data obtained from interfaces involving Si or Ge could possibly arise from errors in the predicted E_v's for Ge and Si. In fact, Katnani and Margaritondo adjusted the predicted valence-band edge

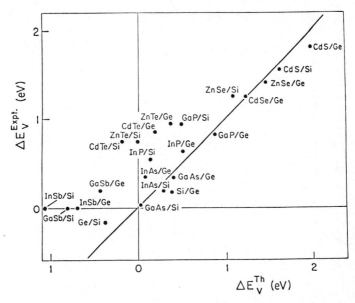

Fig. 9. The predictions of the Harrison model are plotted against the Katnani–Margaritondo experimental data.

energy position by 0.65 eV for Si and by 0.50 eV for Ge to improve the overall accuracy of the model. This empirical correction was deduced from their results. Besides the above comparison, the model predicts ΔE_v's for GaAs/InAs [85], ZnSe/GaAs [73], InAs/GaSb [86] and AlAs/Ge [81] (without the Katnani–Margaritondo correction for the E_v of Ge). The model fails to predict ΔE_v's for AlAs/GaAs [81] and InP/CdS [11]. The overall accuracy of the model as estimated from the above comparison is within 0.30 eV.

The correlation between experimental results and the predictions of Harrison's model is reasonable, as shown in fig. 9. Notice that the model successfully predicts the band discontinuities for lattice-matched interfaces, while it becomes much less accurate for lattice-mismatched interfaces. For example, the predicted ΔE_v's for GaAs/Ge and ZnSe/Ge, which exhibit good lattice matching, give an excellent correlation with the experimental findings. On the contrary, the predictions for InSb/Ge, GaSb/Ge, and CdTe/Ge, all having severe lattice mismatch, are very far from the experimental results. This observation led Katnani and Margaritondo to introduce a simple correction for interface relaxation to compensate for the lattice mismatch. They assumed that the overlayer interatomic distance

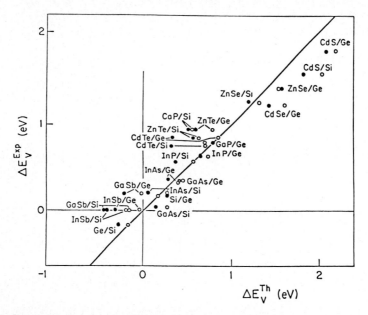

Fig. 10. Comparison between the Katnani–Margaritondo results and the predictions of the Harrison model corrected for the lattice mismatch.

approaches the substrate interatomic distance, d, near the interface. As a result the calculated E_v for the overlayer changes at the interface because of the dependence of the interatomic matrix elements on the interatomic distances. This empirical correction substantially improved the accuracy of Harrison's model. Similar improvements were obtained by replacing the overlayer interatomic distance with the average of substrate and overlayer interatomic distances. Figure 10 shows a comparison between their experimental findings and the predictions of the model after substituting the overlayer interatomic distance with the substrate interatomic distance (open circles), or with the average of the overlayer and substrate interatomic distances (solid circles). The improvement with respect to fig. 9 is evident. Notice in particular that the correction is successful in improving the model for lattice-mismatched interfaces, e.g., for InSb/Ge, GaSb/Ge, CdTe/Ge, CdTe/Si, and ZnTe/Si. In addition, the model gives reasonable predictions for the InAs/GaSb [86], ZnSe/GaAs [73], CdS/InP [11] and GaAs/InAs [85] heterojunctions, while it fails to predict ΔE_v for AlAs/GaAs [83] and AlAs/Ge [81] heterojunction systems. This comparison shows that Harrison's model provides an estimate of ΔE_v within 0.20 eV.

Table 3
Comparison between experimental ΔE_v's and predictions of Tersoff's model.

Heterojunction	ΔE_v^{exp} (eV)	ΔE_v^{th} (eV)	Discrepancy ($\Delta E_v^{\text{exp}} - \Delta E_v^{\text{th}}$)
Si/Ge	0.17	0.18	−0.01
AlAs/Ge	0.78	0.87	−0.09
GaAs/Ge	0.46	0.52	−0.06
InAs/Ge	0.33	0.32	0.01
GaSb/Ge	0.20	−0.11	+0.31
GaP/Ge	0.80	0.63	0.17
InP/Ge	0.64	0.58	0.06
GaAs/Si	0.05	0.34	−0.29
InAs/Si	0.15	0.14	0.01
GaSb/Si	0.05	−0.29	0.34
GaP/Si	0.80	0.45	0.35
InP/Si	0.57	0.40	0.17
AlAs/GaAs	0.38	0.35	0.03
InAs/GaSb	0.51	0.43	0.08
GaAs/InAs	0.17	0.20	−0.03

Margaritondo discussed the correlation between the predictions of Tersoff's model and the Katnani–Margaritondo experimental results [87]. He argued that Tersoff's model approaches the general accuracy limits of all linear models. An overall correlation between available experimental data and predictions of Tersoff's theory is shown in table 3. The model successfully predicts ΔE_v for AlAs/GaAs. However, the data base for comparison in this case is limited because Tersoff's calculation was performed for a limited number of semiconductors. Nevertheless, the above comparison yields an overall accuracy limit of 0.15 eV in agreement with Margaritondo's conclusions [87].

It is clear that the three models which are based on first-principles calculations give reasonable predictions of ΔE_v within 0.15–0.30 eV. However, these models assume different approaches to obtain the semiconductor parameter which determines the band discontinuities. From the early discussion, one finds that Harrison's and Frensley and Kroemer's models are similar in nature. Both models predict the same ΔE_v for lattice-matched III–V/III–V interfaces. Both assume that interfacial dipoles are very small and use the free semiconductor's bulk parameters. On the other hand, Tersoff considered the interfacial dipoles originating from band gap states to play a major role in determining the band discontinuities. It is important to note that the band gap states' dipoles are derived from the free semiconductor's bulk band structure and do not include dipoles resulting from interfacial microscopic effects. In this respect, Tersoff's model is linear and similar to the other two models. In addition to the fundamental soundness of each model (which is controversial), simplicity and wide applicability are factors to be considered when evaluating these models. In this sense, Harrison's model is the simplest and most straightforward (does not require complicated calculations) of the three. This simplicity makes it easy to use and more general.

4. Interfacial effects and the band edge discontinuity: the GaAs/Ge interface

The above discussion revealed deviations from the general predictions of all linear models beyond the combined experimental uncertainty. From the magnitude of these deviations, one concludes that the effects ignored by the linear models are not negligible, but they do not affect each band discontinuity by more than 0.15 eV. If interfacial effects were truly this large and controllable, they could provide a tuning capability of the band edge discontinuities of up to $8kT$ at room temperature. It is, therefore, essential to examine in detail the magnitude of the band discontinuity variations and to understand the origin of possible interfacial contributions.

The effect of the interfacial properties on the band edge discontinuities have been investigated for prototypical lattice-matched, simple to prepare and theoretically well-understood heterojunctions such as GaAs/Ge, ZnSe/Ge, GaAs/ZnSe and GaP/Si [46,73,75,88]. Available experimental results on ΔE_v for the GaAs/Ge interface are consistent within ~ 0.3 eV. Reported experimental values referring to the GaAs/Ge(100) interface ranged from 0.55 eV for Ga-rich and 0.66 eV for As-rich initial surfaces cleaned by sputter-anneal [26], to 0.40 ± 0.1 eV, for an MBE grown c(2 × 8) surface [89]. It is important to understand the various factors that cause such spread in the measured ΔE_v's. Is this spread due to variation in the local microscopic properties of the interface (morphology of the overlayer, variation in the atomic arrangement at the starting surface, surface orientation, etc.) or to different methodologies (preparation procedure, analysis, etc.) used by different authors? The questions reduce to whether the scattering of ΔE_v can be correlated with experimentally controllable chemical or structural parameters.

Katnani and co-workers have performed a systematic study of the interfaces between several GaAs surface orientations and reconstructions and Ge to address the above issues [28,76]. They also studied different surface reconstructions of GaAs(100). For some of the GaAs/Ge(100) interfaces, As_4 has been introduced during the deposition of the overlayer. In fact, the study of the different GaAs(100) surface reconstructions focused on the variations in the interface properties and their effects on ΔE_v.

The GaAs(100) surface exhibits a number of reconstructions, among them 4×6, $c(8 \times 2)$, $c(2 \times 8)$ and $c(4 \times 4)$. The surface stoichiometry as derived from core-level photoemission measurements correlates with the surface ordering. The measured As/Ga ratio increases as the surface ordering changes from 4×6, $c(8 \times 2)$, $c(2 \times 8)$ to $c(4 \times 4)$ [89–94]. The differences among these surface structures and anion to cation ratios serve as a typical example of the variations one would expect to influence the band edge discontinuities. The differences also test the effect of local microscopic properties of the interface on band edge discontinuities.

The upper half of fig. 11 shows the As(3d) lineshape for the $c(2 \times 8)$, $c(4 \times 4)$ and 4×6 surfaces at $h\nu = 106$ eV. Surface and bulk components have been deconvoluted using a Gaussian fit assuming that spin–orbit splitting and branching ratio are the same for all components [92]. The shaded doublet corresponds to the contribution of the As atoms in bulk GaAs for each surface. The shifted peaks are attributed to the rearranged surface As atoms. The $c(4 \times 4)$ exhibits an extra doublet which is attributed to the formation of As–As bonds [93]. Notice that the relative position and intensity of the deconvoluted peaks are different for the different surfaces. The lower half of fig. 11 shows the Ga(3d) lineshape for the three surface reconstructions. Both Ga(3d) and As(3d) lineshapes demonstrate the distinct bonding and stoichiometry variations among the GaAs(100) surface reconstructions. The figure emphasizes the differences among the three surface reconstructions.

The evolution of the core-level spectra during the interface formation emphasizes the differences among the three surface reconstructions. The As(3d) linewidth decreased with increasing Ge coverage. Eventually, it stabilizes and does not change even with very thick Ge overlayers. A signal from As(3d) was still observed in the absence of a significant Ga(3d) signal at very high coverages of Ge (160 Å) for all GaAs(100) starting surfaces. The reduced As(3d) linewidth suggests a more localized anion bonding at the Ge:As surface than in GaAs. This result is consistent with the formation of As lone pairs as seen in GeAs [95,96]. The evolution of the different As(3d) lineshapes to a unique one at thick Ge coverage suggests that the chemistry of the interface should either be different for the distinct starting surfaces or distributed over different effective interface widths.

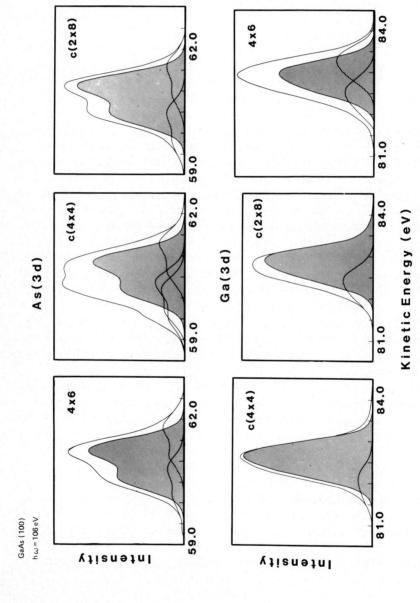

Fig. 11. The upper [lower] curves show the As(3d) [Ga(3d)] lineshape for three different GaAs(100) surface reconstructions.

The evolution of the Ga(3d) lineshape does not indicate a strong chemical activity of the Ga atoms up to 7 Å of Ge coverage. The slight changes in lineshape observed above 7 Å can be related solely to the outdiffusion of small amounts (< 5% of a monolayer) of Ga. The peak position shift indicates a smooth decrease in binding energy that saturates at 4 Å of Ge coverage. This shift is attributed to changes in band bending rather than any metallic clustering.

To gain insight into the evolution of the As and Ga core levels with coverage, Katnani et al. [27] have devised a new method of analysis where the normalized intensity of one component (As or Ge) is plotted versus another (Ga). This scheme is based on the fact that Ge, As and Ga atoms have approximately the same cross-sections and that IMFP for Ge photoelectrons across a Ge layer is the same as that of Ga and As photoelectrons across GaAs and Ge. Therefore, the total intensity (which is proportional to the number of atoms within the attenuation length of photoelectrons) is the same, independent of the Ge coverage. Hence, such a methodology allows one to observe directly the evolution of the core-level signals without calibrating the overlayer thickness. Errors due to coverage uncertainties are avoided. The curves obtained from such an analysis are usable only when one substrate constituent is recognized to be stationary (i.e., not outdiffusing). Metal [40,41] and oxide [40] overlayers have previously been shown to leave, in most cases, the semiconductor cation unmoved (i.e., attenuates exponentially with increasing overlayer coverage). In the epitaxial-abrupt growth regime for Ge on GaAs, this is usually the case for the Ga atoms [28].

Examples of this new methodology are given in fig. 12, where such type of plot is shown for 4×6, $c(8 \times 2)$ and $c(4 \times 4)$ respectively. The "ideal" case (dashed line) occurs if no exchange reaction or interdiffusion takes place. Both As and Ge curves differ considerably from such an abrupt, unmixed ideal interface situation. In fig. 12, the Ge coverage increase corresponds to changes along the curves, as indicated by the arrows. The As curve is always above the ideal line, indicating an As outdiffusion through the Ge overlayer independent of the starting surface stoichiometry. Moreover, all As curves extrapolate to the same amount of As $(27 \pm 2\%)$ independent of the starting surface As concentration.

Distinct differences occur at the onset of the As curves in the three cases shown in fig. 12. For the 4×6 surface, the As curve immediately starts moving upwards, indicating that As is driven to the surface, most probably from the underlying GaAs layers. A simple picture in which Ge atoms first bond to surface Ga, leaving As atoms unburied, would lead to a horizontal

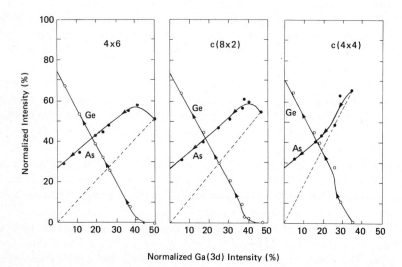

Fig. 12. The normalized intensity of the As(3d) and Ge(3d) core levels is plotted as function of the normalized intensity of the Ga(3d) for the GaAs/Ge interface. This method eliminates errors that might arise due to the uncertainty in measuring the overlayer thickness. The overlayer thickness increases along the arrow.

onset of the As curve, contrary to the experimental finding. On the other hand, the c(4 × 4) surface has a completely different behavior at the onset of the As curve. The As curve initially drops and runs close to the "ideal" case independent of the starting point being different (i.e., higher As signal for this surface compared to the 4 × 6). This trend points out that Ge simply covers the As-saturated c(4 × 4) surface without any major rearrangement at the initial coverage stage. At higher Ge coverage (above ~ 1 ML), an exchange reaction between Ge and As takes place and As floats at the surface [28,95,96]. The case of the c(8 × 2) surface is an intermediate one, as one would expect if surface chemical processes scale with the initial As/Ga ratio. Indeed, the case of the c(8 × 2) bears much analogy with that of 4 × 6, suggesting that differences in surface atomic structure (primitive versus centered) exert less influence on interface formation than differences in surface chemistry (stoichiometry).

All three cases extrapolate to the same end-point, 27 ± 2%, regardless of the starting point. This result demonstrates that the formation of an

As-stabilized Ge surface phase influences the interface growth. The LEED pattern of this Ge:As surface reconstruction on GaAs/Ge(100) is a reproducible two domain 2×1 reconstruction [96,97]. The pattern is the same for all starting GaAs(100) surfaces and it does not change even upon deposition of a thick (~ 2000 Å) layer of Ge. Furthermore, experimental studies associates this LEED pattern with the formation of a $GeAs_x$ compound [95,96]. This commonality of primitive and centered starting structures again suggests that surface stoichiometry is a greater determinant of interface formation behavior than the initial surface bonding structure.

The observation of the As capping layer of the Ge surface for all the interfaces, raised a question about the origin of this layer. Recently, Neave et al. suggested contamination from the environment as the origin of this layer [91]. However, Monch and Gant observed at least one monolayer of As at the free Ge surface of a GaAs/Ge(110) interface where the GaAs surface was prepared by cleavage with no As source in the experimental chamber [29]. The ambient probably does not contribute significant amounts of As because As contamination was not detected on a sputter cleaned Ge surface even when it was left in the experimental chamber for many hours after cleaning. Furthermore, no As is observed on the surface of a Ge buffer layer when grown on a pure Ge substrate [78]. These observations suggest that As keeps floating at the free surface during the Ge deposition on GaAs and that indeed surface As is segregated rather than being an environmental contamination. This attribution is consistent with total energy calculations, which show that As segregates to the Ge free surface to minimize the total energy of the system [98].

The measured ΔE_v is the same (0.47 ± 0.05 eV) regardless of all the variations induced at the interface as presented above. These results suggest that no correlation exists between ΔE_v and the initial As/Ga ratio or between ΔE_v and whether the substrate has a primitive [i.e., 4×6] or centered [i.e., $c(4 \times 4)$, $c(2 \times 8)$, $c(8 \times 2)$] surface reconstruction. It is of interest to consider how this relates to chemical composition and ordering at the final heterojunction interface. In this context, by interface we mean the transition region between the first complete Ge plane and the first complete plane of As or Ga. For an epitaxial heterojunction, this region extends over a few atomic planes [80,99]. In principle, two hypotheses can be formulated: (1) either the ΔE_v is the same because the microscopic properties of the heterojunction are identical, or (2) it is the same in spite of the fact that the microscopic properties are different.

At the present time this question cannot be answered experimentally in a convincing way because extensive modelling of core-level lineshapes, even when possible, does not yield a unique description of the atomic arrangement. However, the evolution of the interface growth suggests that the heterojunction interface might actually be different on an atomic scale among the number of samples covering the different surface reconstructions of GaAs(100) where the same ΔE_v is measured. In fact, just accounting for the As monolayer at the free Ge surface suggests that the evolution goes in the direction of enriching the interface with As when the starting GaAs(100) surface is As-depleted. The amount, ordering and chemical bonding of the segregated As, as well as the evolution of the As(3d) lineshape, tend to emphasize the irrelevance of the chemistry and stoichiometry of the starting surface. In turn, we propose that the driving force for MBE interface formation always yields a unique, low-energy equilibrium structure at the material's transition region, though its atomic extent may vary by as much as a couple of atomic layers. Therefore, if there is any "nanoscopic" dipole contribution to the valence-band discontinuity, it appears to be either very small or to be the same (within ± 0.05 eV).

The segregation of As on the free Ge(100) surface raises a question about the role of As in interface formation. The fact that As is not supplied by the environment (as discussed before) and that the same amount of As is observed regardless of the starting As coverage requires that either As outdiffuses from the bulk or is supplied by anion clusters in the near surface region. One observation consistent with this attribution is Joyce and Foxon's [100] report of GaAs surface enrichment upon annealing MBE films grown with As_4. The depth from which such As is supplied is not clear; nevertheless, it probably will increase the width of the interface if it is supplied from the first few monolayers of GaAs. Therefore, the interface width should decrease if As is intentionally introduced in the ambient during the initial stages of the Ge epitaxial growth. The question then arises as to whether ΔE_v and/or E_F would be affected.

To address the above question, Katnani et al. performed a series of experiments where they introduced an overpressure of As_4 during the MBE deposition of Ge [28]. The results of these experiments point out that the presence of As in the environment during Ge evaporation modifies several parameters in the initial stages of interface formation. The core-level intensities clearly show an As uptake at submonolayer coverages. In fig. 13, the As(3d) core-level signal is shown as a function of the Ga(3d) intensity for GaAs(100) c(4 × 4) substrate with As_4 introduced during the MBE growth of Ge. This type of stationary cation analysis provides the clearest picture of the interface growth process. Comparing fig. 13 with fig. 12, it is obvious that the presence of As_4 during Ge overlayer growth drastically

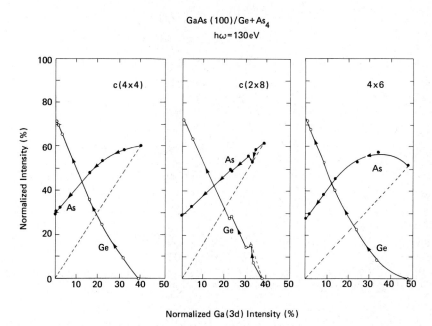

Fig. 13. The effect of introducing As_4 during the interface formation between GaAs and Ge is shown in this plot. Notice the difference between this figure and fig. 12.

alters the onset of the As curve. However, while the initial monolayer formation is modified, in both cases the same amount of As segregates at the final free Ge surface. A similar behavior has been observed for the GaAs(100) 4×6 and $c(2 \times 8)$ surface reconstructions, as shown in the figure.

Another parameter sensitive to the presence of As during the interface formation is the electron affinity χ of the free surface. Katnani et al. [28] measured changes in χ with respect to the initial value by subtracting the change in band bending from the change in work function with increasing coverage of Ge. These quantities are determined experimentally by measuring the Ga(3d) energy position and the secondary electron distribution cut-off [78]. Figure 14 shows the change in the electron affinity, $\Delta \chi$, as a function of Ge coverage for three initial surface reconstructions, namely, 4×6 $c(2 \times 8)$ and $c(4 \times 4)$. The electron affinity referenced to the substrate tends to increase upon heterojunction growth. This behavior can be explained by the dipole associated with the As–Ge bond at the interface.

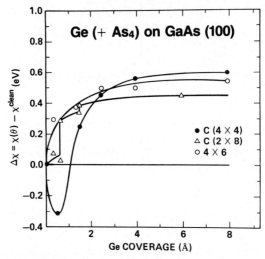

Fig. 14. The changes in the electron affinity with increasing Ge coverage are shown for three surface reconstructions of GaAs(100).

The decrease in χ shown in fig. 14 for the c(4 × 4) surface is attributed to the dipole created by the first monolayer of Ge on top of the As-terminated surface. At a larger coverage, an exchange reaction between Ge and As takes place and the microscopic dipole reverses its orientation.

It is remarkable that in spite of all these modifications induced by the presence of As_4 during interface growth, the band edge discontinuities are

Table 4
ΔE_v dependence for GaAs/Ge on surface orientation and As_4 background.

Surface orientation	Surface structure	ΔE_v (eV)	Surface orientation	Surface structure	ΔE_v (eV)
100	4×6	0.46	211	1×1	0.53
100	c(8×2)	0.46	111	2×2	0.50
100	c(4×4)	0.48	110	1×1	0.43
100	4×6+As_4	0.44	$\overline{1}\overline{1}\overline{1}$	2×2	0.48
100	c(2×8)+As_4	0.47			
100	c(4×4)+As_4	0.48			

by no means affected by such changes in environmental conditions. Table 4 summarizes the experimental values of ΔE_v for the different GaAs/Ge(100) interfaces with and without As_4 during Ge film growth. Also listed in this table are the ΔE_v's for the different GaAs/Ge surface orientations. No differences or trends are detectable within an experimental uncertainty of ± 0.05 eV. This result is interpreted as being caused by strong local electrostatic forces during the initial stages of interface formation. Such forces prevent modification of the built-in potential barrier by introducing As "foreign" species. These measurements provide further support for theories that account for intrinsic properties of semiconductors being the most suitable for describing heterojunction band edge discontinuities.

5. Potential barrier height

Most of the atomic scale theoretical and experimental studies have been concerned with the valence-band discontinuity with little attention given to the Fermi level at the interface. While little is known about the Fermi-level position at heterojunction interfaces, the Fermi level has been extensively studied for semiconductor/metal and oxide interfaces [40,43]. There is a general agreement that for Si/metal interfaces, the Fermi level is pinned by localized interface states. For compound semiconductor/metal interfaces, the defect model proposed by Spicer et al. [38] attributes the final position of E_F to native defect states created during interface formation. The type and origin of these defects is unknown and theoretically controversial [44,45]. The backbone of this model is formed by two experimental observations. First, the final E_F position at the interface on the substrate side is the same regardless of the deposited atom. Second, the final E_F position is achieved at very low coverages of the foreign atom.

The relevant point here is that some of the above results find their counterparts for heterojunction interfaces. Similar to silicon–simple-metal interfaces, localized electronic states have been detected at the ZnSe/Ge [39] and Si/Ge [37,38] interfaces, and have been explained theoretically in terms of chemisorption bonds. Katnani and Margaritondo observed that the Fermi level position at the interface between cleaved III–V semiconductors and Si or Ge coincides with that observed at the interface between those compound semiconductors and metals [27]. Also Monch and Gant observed the same final Fermi-level position for ordered and disordered Ge overlayers on GaAs(110) [101]. In addition, Katnani et al. found that the distance between E_v and E_F changes on going from GaAs to GaP as

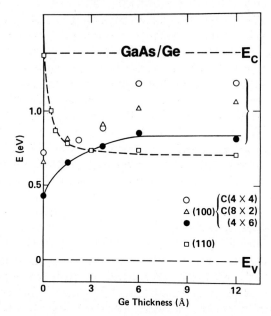

Fig. 15. The position of the Fermi level in the GaAs(100) and (110) band gap is plotted as a function of Ge coverage. Notice the difference in behavior between the GaAs(100) and (110) surfaces.

qualitatively predicted by theoretical calculations based on surface anti-site defects [35]. These observations support the extension of the defect model to semiconductor–semiconductor heterojunctions. In turn, this extension raises interesting questions about the correlation between Fermi-level pinning and the establishment of band discontinuities because the pinning positions of E_F in the two gaps are trivially related to ΔE_v and ΔE_c. In fact, Katnani and Margaritondo found that the difference between the pinning positions for cleaved GaAs and ZnSe is ~ 1 eV, which coincides with the measured ΔE_v for the ZnSe/GaAs interface.

The above discussion raises a serious question about the role of defect states at heterojunction interfaces. Katnani et al. performed systematic, correlated measurements of E_F and ΔE_v at the GaAs/Ge(100) heterojunction interface [102]. Figure 15 shows the evolution of E_F during interface formation between GaAs(100) or (110) surfaces and epitaxial, lattice-matched Ge. The very important observation from this figure is the different initial and final positions of E_F for the different starting surface reconstructions of GaAs(100). This change in E_F position correlates with excess of As at the starting GaAs surface. Notice that the initial position of E_F varies by over 0.3 eV, going from at least a monolayer of surface As for the c(4 × 4), to a half monolayer on the c(8 × 2), to at least a quarter monolayer of Ga on the 4 × 6 starting GaAs(100) surface. This trend is maintained as the interface of GaAs/Ge(100) forms. More important is the fact that Katnani et al. measured the same ΔE_v, within ±0.05 eV, independent of the final E_F positions in fig. 15. The constancy of ΔE_v suggests that the rearrangement of the charge distribution on both sides of the interface does not create a measurable dipole layer at the interface. Furthermore, doping does not influence the potential step at the abrupt, epitaxial interface (i.e., ΔE_v). Also, the origin of the valence-band discontinuity and the Fermi-level final position are different for semiconductor/semiconductor interfaces.

The study by Katnani et al. raised an interesting question as to why the E_F position for cleaved GaAs(110) appears to evolve so differently from GaAs(100) and apparently "pin" independent of overlayer material. As seen in fig. 15, the starting E_F position for the cleaved GaAs(110) is indicative of a flat band condition, i.e., no cleavage steps. Notice the difference in the evolution and in the final position of E_F for the two surfaces' crystallographic orientations. This behavior difference between cleaved (110) heterojunctions and MBE(100) interface formation may just reflect the same differences seen in (110) heterojunctions between the cleaved GaAs substrates [27,29] and the MBE prepared GaAs(110) starting surfaces [103]. This difference might indicate that there are two different mechanisms responsible for the final E_F positions.

The observed correlation between E_F and the surface As concentration of the GaAs(100) raised another question about the role of mobile As in determining the valence-band discontinuity [28,96] and the Fermi-level position [104]. To study the extent of the role of As during interface formation, Katnani et al. intentionally introduced As_4 during expitaxial growth of the Ge overlayer. Figure 16 shows the evolution of E_F during interface formation between GaAs(100) subsurfaces and co-MBE deposition of Ge and As_4. Notice that the initial E_F positions for the clean surfaces reproduce those presented in fig. 15. The final E_F position at the interface between GaAs(100) (4 × 6) surface and Ge moved toward the conduction-band edge by over 0.1 eV when As_4 is introduced. The As-induced change for the c(4 × 4) surface is smaller because that starting surface is already saturated with As and the Fermi level is already nearly degenerate with the Ge conduction band without additional As in the ambient. The final E_F position for the c(2 × 8) surface is consistent with

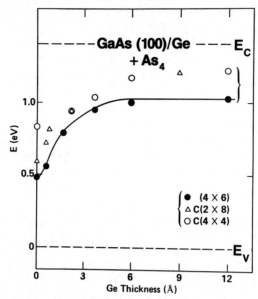

Fig. 16. The effect of introducing As_4 during the GaAs/Ge interface formation is that the Fermi level moves towards the conduction band.

what we expect from the observed trend discussed above. Again, interestingly enough, Katnani et al. measured the same valence-band discontinuity for the GaAs/Ge interfaces prepared in this way and those grown with Ge alone. The role of As as a dopant to change E_F without affecting ΔE_v is supported by these experiments.

The above discussion demonstrates the irrelevance of defect states in determining E_F at the GaAs/Ge(100) interface. In fact, an explanation of the above results in terms of the defect model invokes defect levels at different positions in the gap. Also, it would require an ad hoc redistribution of the charge density on both sides of the interface to maintain the same ΔE_v at the interface. The results do not totally exclude the formation of defects at the MBE-grown GaAs/Ge(100) interface. However, they do suggest that the density of such defects is not sufficient to pin E_F.

Besides the above systematic study, Brugger et al. performed Raman spectroscopic measurements of both ΔE_v and E_F for the GaAs/Ge(110) interface [48]. They concluded that there are no defect states in the gap of GaAs. This result, in fact, contradicts the results of Monch et al. [46]. Also, theoretical calculations by Zur et al. suggested that a large density of defect states is required to pin the Fermi level at heterointerfaces [49].

The interfacial Fermi-level position remains a question of great scientific and technological interest. Clearly, the available data in this area is controversial as discussed. On the one hand, some data shows that the Fermi level is pinned and its position in the gap determines, although to a first order, ΔE_v. On the other hand, there is other experimental evidence which negates such relationship between E_F and ΔE_v. Also the data suggest that the density of interfacial defects is not sufficient to pin the Fermi level. It is important to notice that all data supporting defect or E_F pinning are obtained from cleaved samples. This difference in behavior between cleaved and MBE samples is worth investigating and might explain the observed differences among the data.

6. Effects of an interlayer on the band edge discontinuities

The tunability of the band edge discontinuities remains a question of great fundamental and technological interest. The various studies presented above demonstrate that tunability can not be achieved by varying the intrinsic constituents of the semiconductors forming the junction. This has led a number of authors to study the effect of an interlayer on some prototypical heterojunction interfaces like GaAs/Ge and ZnSe/Ge.

Katnani et al. introduced an Al interlayer at the GaAs/Ge interface [105]. The strong reactivity of Al with As should result in a Ga-rich and perhaps AlAs stabilized interface. If Al atoms outdiffuse, they should act as acceptors in Ge. In addition, AlAs is lattice-matched to $\pm 0.1\%$ with both Ge and GaAs and thus could form a chemically foreign but structurally continuous interface. Therefore, whether Al outdiffuses or stabilizes the As at the interface, it should change the chemistry and the charge redistribution on both sides of the junction compared to GaAs/Ge alone. This study demonstrated the sensitivity of the band edge discontinuities to unintentional local contamination present in the monolayer range at the interface and to the interfacial defect states created during the GaAs/Al interface formation.

Figure 17 shows valence-band spectra for clean MBE-grown GaAs(100) c(4 × 4) and for the Al-covered surfaces at different substrate temperatures. The effect of substrate temperature during deposition is noticeable from the upper two curves. For room-temperature deposition, the valence band shows a contribution from Al metallic states which fill the heteropolar gap at 4.3 eV below the valence-band edge. For deposition at 340°C, the first peak below the valence-band edge narrows, indicating loss of As, and the

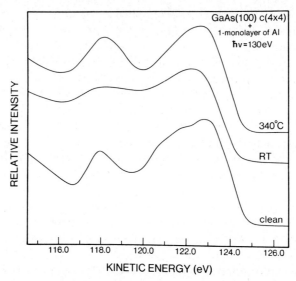

Fig. 17. The effect of the Al interlayer and its deposition temperature on the GaAs surface just before growing the Ge overlayer is illustrated by the valence-band spectra.

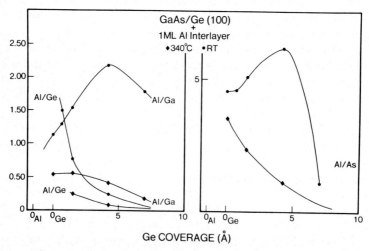

Fig. 18. The changes in the relative intensities of the Al, As and Ge are plotted with increasing Ge coverage during the GaAs:Al/Ge interface formation.

second peak broadens, indicating formation of Al–As bonds. While room temperature deposition results in the formation of small Al clusters, deposition at 340°C results in the formation of AlGaAs alloys.

The above conclusions can also be drawn from core-level spectra. The increase in the Al/Ga and Al/As ratios with increasing coverage of Ge, shown in fig. 18 for room-temperature deposition, indicates the presence of Al clusters from the initial monolayer. The Al is reacting with the topmost GaAs layer, in this case, and does not diffuse into the GaAs. In contrast, for the deposition at 340°C, the Al/Ga and Al/As ratios decrease with increasing Ge coverage, as shown in fig. 18. The Al in this case is completely reacted with and interdiffused with the GaAs surface. For both Al-deposition temperatures, the interface exhibits a small percent of Ga in some bonding configuration different than that of GaAs, as indicated by the broadening of the Ga(3d) lineshape at Ge coverages above 6 Å. Both the broadening of the Ga(3d) lineshape at thick Ge coverages and the presence of a monolayer of As at the free Ge surface, regardless of the Al-deposition temperature, suggest that even if the Al monolayer completely reacts, it does not form a spatially uniform AlGaAs layer. These observations indicate that the Ga–As bond is easier to break than the Al–As bond, as one would expect.

The Ga(3d)/As(3d) ratio is independent of the Al interlayer and its deposition temperature contrary to previous observation from the GaAs(110):Al/Ge system [106]. For both deposition temperatures, As segregates to the free Ge surface in spite of the strong bonding of AlAs. Based on As:Ge (3d)-core intensity ratios, the amount of segregated As is the same, independent of the Al interlayer and the deposition temperature. As for the intrinsic GaAs/Ge heterojunction, the interface formation process is dominated by surface $GeAs_x$ growth, which for the (100) orientation always exhibits the same surface reconstruction, two domains 2×1.

Apparently, the Al interlayer has changed the interfacial chemistry and probably the interfacial structure. However, it did not affect the band edge discontinuities. The constancy of the band edge discontinuity for the GaAs/Ge interface, in spite of the Al interlayer, is surprising. It suggests that the interfacial band lineup is a property of the semiconductors forming the heterojunction. Furthermore, it suggests that the interfacial dipoles are offset by some mechanism to achieve that equilibrium band lineup. This insensitivity in ΔE_v is measured for Al interlayer thickness up to two monolayers (the maximum Al coverage studied).

More important, the effect of the Al interlayer on E_F is of relevance to

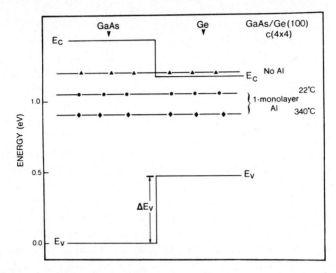

Fig. 19. Schematic of the energy band diagram of the GaAs:Al/Ge interface. The Fermi-level position changes depending on the presence of the Al interlayer and its deposition temperature, while the band discontinuity stayed unaffected.

our previous discussion on the potential barrier height. Figure 19 shows a schematic of the energy band diagram for the GaAs:Al/Ge interface and the final E_F position. The Fermi level moves by 0.15 eV towards the valence band for the room temperature deposition, while it moves by 0.3 eV for the deposition at 340°C relative to its position at the GaAs/Ge interface without the Al interlayer. The presence of the Al interlayer does not pin the Fermi level as it was observed for the GaAs(110)/Al interface [40]. Clearly, the deposition-induced defects created during the GaAs(100)/Al interface formation do not occur in sufficient density to cause a complete pinning of the Fermi level. This observation emphasizes the differences between the GaAs/Ge(100) and (110) interfaces. More important, the constancy of ΔE_v regardless of the interfacial E_F position emphasizes the absence of a simple correlation between ΔE_v and E_F, and supports the conclusions of the previous section.

7. Summary and conclusions

The research discussed here addressed a number of critical issues. First, it showed that the complex problem of determining the band lineup at semiconductor–semiconductor interfaces could be simply reduced to the difference between two numbers obtained either from the intrinsic bulk properties of each semiconductor or from the charge arrangement on each side of the interface. In any case, this approach ignores the interfacial microscopic effects. Although one would expect the interfacial microscopic contributions to be different for different heterojunction systems, the existensive data obtained from GaAs/Ge and the limited data obtained from ZnSe/Ge and GaP/Si heterojunctions support the independence of the band edge discontinuity on the interfacial effects within 0.15 eV.

Different approaches are used to calculate the semiconductor parameter which determines ΔE_v. The overall accuracy of each model is comparable. Harrison's and Tersoff's models approach the accuracy limit imposed on such linear models. However, Tersoff's model was successful in identifying a single mechanism, namely the induced gap states to explain the band lineup for heterojunction and the Schottky barrier for metal–semiconductor interfaces. The model is theoretically controversial, however, experimental results favor it on both accounts. The advantage in favor of Harrison's model is its simplicity and wide applicability.

The second question that this research has raised and attempted to answer is the relationship between the band discontinuity and the Fermi-level position in the gap. Experimental results on this issue are controversial. Some of the data support the existence of such a relationship. In fact, this motivated theoreticians to examine the relationship between the Schottky barrier for metal–semiconductor systems and the band discontinuity for semiconductor–semiconductor interfaces. This research sought a unified theory which could explain both types of junctions. Other data excluded the presence of any relationship between ΔE_v and E_F. In fact, the data here suggests that an internal chemical potential is not a useful concept to derive the band discontinuity. This result is also supported by theoretical calculations which show that a large density of defects is necessary to pin the Fermi level at heterointerfaces. The driving mechanism which determines the Fermi level position and its relation to the interfacial band lineup still remains an interesting and a challenging problem. One of the challenges concerns providing an explanation for the observed differences between MBE and cleaved surfaces.

It is interesting to note that Tersoff's model might provide an explanation for the behavior of E_F observed at the GaAs/Ge(100), if the barrier height on each side of the heterointerface is not affected by the presence of the induced band gap states. Such explanation is possible if the reference energy position and E_F need not be aligned at heterointerfaces. In this way, the band discontinuity could stay constant while E_F could assume any value in the band gap. The case is different for Schottky barriers.

The third question concerns the tunability of the band discontinuity. If the band lineup mechanism is driven by intrinsic semiconductor properties, then tunability could only be achieved by gradually changing the semiconductor intrinsic properties. This change can be achieved by using alloy semiconductors which span a wide energy gap across the variation in composition. On the other hand, if the band lineup is driven by interfacial dipoles and charge rearrangements, then one could control the band discontinuity by using foreign atoms at the interface to rearrange the dipoles and charge. However, data obtained from the Al interlayer at the GaAs/Ge interface excluded the possibility of using this approach to modify the band discontinuity. It is possible that the insensitivity of the band discontinuity to the presence of the foreign atoms at the interface is intrinsic to the GaAs/Ge heterojunction system. It remains to be seen whether such results can be extended to other heterojunction systems.

These questions pose challenging tasks for future research. Improvements in crystal growth techniques and in the spatial and energy resolution of photoemission spectroscopy in the near future will increase the capabilities of the present techniques. The ability to probe small areas (200 μm small spot electrostatic analyzers) will provide researchers with the opportunity to probe the microscopic chemistry and structure of the interface. Also, the high energy resolution achieved by combining multichannel detectors with hemispherical electrostatic analyzers will result in an improvement of the measurement accuracy of photoemission, both XPS and SXPS.

Also, in recent years a number of techniques have been developed or employed to study the structure and the geometry of the interface. Such techniques like transmission electron microscopy, Rutherford backscattering, high-energy electron diffraction, ion channeling and ellipsometry will provide valuable information about the interfacial atomic arrangements. The combined efforts of these techniques and XPS/SXPS may lead to answers for many of the open questions.

From a theoretical point of view, all linear models are limited. The controversy surrounding the origin of the semiconductor parameter which describes the band discontinuity, will increase. The future will see more approaches. The question here is not their accuracy but their fundamental soundness because their accuracy is predetermined by the linearity accuracy limit. The feasible approach to obtain a complete understanding of certain heterojunction systems of interest is to calculate, in detail, the band structure at the interface using the increasing power of high-speed computers and available structural and geometrical information.

*We have found it of paramount importance
that in order to progress we must recognize the
ignorance and leave room for doubt.*

R.P. Feynman

References

[1] A.G. Milnes and D.L. Feucht, Heterojunctions and Metal–Semiconductor Junctions (Academic Press, New York, 1972); and
S.M. Sze, Physics of Semiconductor Devices (Wiley Interscience, New York, 1969).
[2] H. Kroemer, Jpn. J. Appl. Phys. 20 (1981) 9, and Proc. IEEE 70 (1982) 13.
[3] Mamoru Kurado and Jiro Yoshida, IEEE Trans. Electron Devices ED-31 (1984) 467.
[4] W.G. Oldham and A.G. Milnes, Solid-State Electron. 6 (1963) 121; and Solid-State Electron. 7 (1964) 153.
[5] H.C. Card, J. Appl. Phys. 50 (1979) 2822;
J. Jerhot and V. Snejdar, Phys. Status Solidi a 34 (1976) 505;
L.L. Chang, Solid-State Electron. 8 (1965) 721; and
R.S. Meuller and R. Zuleeg, J. Appl. Phys. 35 (1964) 1550.
[6] A.I. Gubanov, Zh. Tekh. Fiz. 20 (1950) 1287; and Zh. Tekh. Fiz. 22 (1952) 729.
[7] W. Shockley, US Patent 2 569 347 (1951).
[8] H. Kroemer, Proc. IRE 45 (1957) 1535; and Proc. IEEE 51 (1963) 1782.
[9] R.L. Anderson, Solid-State Electron. 5 (1962) 341; and
G. Zeiderbergs and R.L. Anderson, Solid-State Electron. 10 (1967) 113.
[10] H. Kressel and H. Nelson, RCA Rev. (March 1969) p. 106.
[11] J.L. Shay, Singurd Wagner and J.C. Phillips, Appl. Phys. Lett. 28 (1976) 31.
[12] R. Dingle, W. Wiegmann and C.H. Henry, Phys. Rev. Lett. 33 (1974) 827.
[13] H. Kroemer, CRC Crit. Rev. Solid State Sci. 5 (1975) 555.
[14] W.R. Frensley and H. Kroemer, Phys. Rev. B 15 (1977) 2642.
[15] W. Harrison, J. Vac. Sci & Technol. 14 (1977) 1016.
[16] M.J. Adams and A. Nussbaum, Solid-State Electron. 22 (1979) 783.
[17] O. Von Ross, Solid-State Electron. 23 (1980) 1069.
[18] J. Tersoff, Phys. Rev. B 30 (1984) 4874.
[19] G.A. Baraff, J.A. Appelbaum and D.R. Hamann, Phys. Rev. Lett. 38 (1977) 237.
[20] W. Pickett, S.G. Louis and M. Cohen, Phys. Rev. Lett. 39 (1977) 109; and Phys. Rev. B 17 (1978) 815.
[21] C.A. Swarts, W.A. Goddard and T.G. McGill, J. Vac. Sci. & Technol. 19 (1981) 551.
[22] J. Pollman and S. Pantelides, J. Vac. Sci. & Technol. 16 (1979) 1498.
[23] P. Perfetti, D. Denley, K.A. Mills and D.A. Shirley, Appl. Phys. Lett. 33 (1978) 66.
[24] R.S. Bauer and J.C. McMenamin, J. Vac. Sci. & Technol. 15 (1978) 1444.
[25] J.R. Waldrop and R.W. Grant, Phys. Rev. Lett. 26 (1978) 1686.
[26] R.W. Grant, J.R. Waldrop and E.A. Kraut, Phys. Rev. Lett. 40 (1978) 656.

[27] A.D. Katnani and G. Margaritondo, Phys. Rev. B 28 (1983) 1944.
[28] A.D. Katnani, P. Chiaradia, H.W. Sang Jr, P. Zurcher and R. S. Bauer, Phys. Rev. B 31 (1985) 2146.
[29] W. Monch and H. Gant, J. Vac. Sci. & Technol. 17 (1980) 1094.
[30] S. Nannarone, F. Patella, P. Perfetti, C. Quaresima, A. Savoia, C.M. Bertoni, C. Calandra and F. Manghi, Solid State Commun. 34 (1980) 409.
[31] Ping Chen, D.J. Bolmont and C.A. Sebenne, Surf. Sci. 132 (1983) 505.
[32] P. Zurcher, G.J. Lapeyre, J. Anderson and D. Frankel, J. Vac. Sci. & Technol. 21 (1982) 476.
[33] D. Denley, K.A. Mills, P. Perfetti and D.A. Shirley, J. Vac. Sci. & Technol. 16 (1979) 1501.
[34] G. Margaritondo, A.D. Katnani, N.G. Stoffel, R.R. Daniels and Te-Xiu Zhao, Solid State Commun. 43 (1982) 163.
[35] A.D. Katnani, G. Margaritondo, R.E. Allen and J.D. Dow, Solid State Commun. 44 (1982) 1231.
[36] A.D. Katnani, R.R. Daniels, Te-Xui Zhao and G. Margaritondo, J. Vac. Sci. & Technol. 20 (1982) 662;
A.D. Katnani, N.G. Stoffel, R.R. Daniels, Te-Xui Zhao and G. Margaritondo, J. Vac. Sci. & Technol. A 1 (1983) 692.
[37] P. Perfetti, N.G. Stoffel, A.D. Katnani, G. Margaritondo, C. Quaresima, F. Patella, A. Savoia, C.M. Bertoni, C. Calandra and F. Manghi, Phys. Rev. B 24 (1981) 6174.
[38] P. Chen, D. Bolmont and C. Sebenne, Solid State Commun. 44 (1982) 1191.
[39] G. Margaritondo, F. Cerrina, F.J. Grunthaner, Solid State Comm. 52 (1984) 495.
[40] W.E. Spicer, P.W. Chye, P. Skeath, C.Y. Su and I. Lindau, J. Vac. Sci. & Technol. 16 (1979) 1422;
W.E. Spicer, I. Lindau, P. Skeath and C.Y. Su, J. Vac. Sci. & Technol. 17 (1980) 1019.
[41] R.H. Williams, R.R. Varma and V. Montegomery, J. Vac. Sci. & Technol. 16 (1979) 1143.
[42] L.J. Brillson, C.F. Brucker, N.G. Stoffel, A.D. Katnani and G. Margaritondo, Phys. Rev. Lett. 46 (1981) 838.
[43] L.J. Brillson, Surf. Sci. Rep. 2 (1982) 123.
[44] M.S. Daw and D.L. Smith, Phys. Rev. B 20 (1979) 5150; and Appl. Phys. Lett. 36 (1980) 690.
[45] J.D. Dow, R.E. Allen, J. Vac. Sci. & Technol. 20 (1982) 659.
[46] W. Monch, R.S. Bauer, H. Gant and R. Murschall, J. Vac. Sci. & Technol. 21 (1982) 498.
[47] P. Chiaradia, A.D. Katnani, H.W. Sang Jr and R.S. Bauer, Phys. Rev. Lett. 52 (1984) 2.
[48] H. Brugger, F. Schaffler and G. Abstreiter, Phys. Rev. Lett. 52 (1984) 141.
[49] A. Zur, T. McGill and D.L. Smith, Surf. Sci. 132 (1983) 456.
[50] H. Nelson, RCA Rev. 1963, p. 603.
[51] M.B. Panish, S. Summski and I. Hayashi, Metall. Trans. 2 (1971) 795.
[52] A.D. Dupuis and P.D. Dapkus, Appl. Phys. Lett. 31 (1977) 466.
[53] A.Y. Cho, R.W. Dixon, H.C. Casey Jr and R.L. Hartman, Appl. Phys. Lett. 28 (1976) 50.
[54] M. Bafleur and A. Munoz Yague, Thin Solid Films 101 (1983) 299;
A. Munoz Yague, J. Piqueras and N. Fabre, J. Electrochem. Soc. 128 (1981) 149.
[55] J. Szuber, Thin Solid Films 105 (1983) 33;
O. Tejayadi, Y.L. Sun, J. Klem, R. Fischer, M.V. Klein and H. Morkoc, Solid State Commun. 46 (1983) 251.
[56] R.S. Bauer and H.W. Sang Jr, Surf. Sci. 132 (1983) 479.

[57] J. Batey, S.L. Wright and D.J. DiMaria, J. Appl. Phys. 57 (1985) 484.
[58] H. Kroemer, W.Y. Chien, J.S. Harris and D.D. Edwall, Appl. Phys. Lett. 36 (1980) 295.
[59] G.B. Norris, D.C. Look, W. Kopp, J. Klem and H. Morkoc, Appl. Phys. Lett. 47 (1985) 423.
[60] M. Heiblum, M.I. Nathan and M. Eizenberg, Appl. Phys. Lett. 47 (1985) 503.
[61] R.C. Miller, D.A. Kleinman and A.C. Gossard, Phys. Rev. B 29 (1984) 7085.
[62] H. Okumura, S. Misawa, S. Yoshda and S. Gonda, Appl. Phys. Lett. 46 (1985) 477.
[63] M. Cohen, Adv. Electron. Electron Phys. 51 (1980) 1.
[64] Geoffrey Duggan, J. Vac. Sci. & Technol. B 3 (1985) 1224.
[65] Photoemission in Solids I and II, eds M. Cardona and L. Ley (Springer, Berlin, 1978); Photoemission and the Electronic Properties of Surfaces, eds B. Feuerbacher, B. Fitton and R.F. Willis (Wiley Interscience, New York, 1978).
[66] C. Kunz, Synchrotron Radiation Techniques and Applications, ed. C. Kunz (Springer, Berlin, 1979).
[67] I. Lindau and W.E. Spicer, J. Electron Spectros. & Relat. Phenom. 3 (1974) 409; and C.R. Brundle, Surf. Sci. 48 (1975) 99.
[68] E.A. Kraut, R.W. Grant, J.R. Waldrop and S.P. Kowalczyk, Phys. Rev. B 28 (1984) 1965.
[69] R.J. Lee, IEEE Trans. Device Lett. EDL-6 (1985) 130.
[70] A.M. Marshak, IEEE Trans. Device Lett. EDL-6 (1985) 128.
[71] O. Von Ross, IEEE Trans. Device Lett. EDL-6 (1985) 126.
[72] G. Margaritondo, N.G. Stoffel, A.D. Katnani, H.S. Edelman and C.M. Bertoni, J. Vac. Sci. & Technol. 18 (1981) 290; and
G. Margaritondo, N.G. Stoffel, A.D. Katnani and F. Patella, Solid State Commun. 36 (1980) 215.
[73] S.P. Kowalczyk, W.J. Schaffer, E.A. Kraut and R.W. Grant, J. Vac. Sci. & Technol. 20 (1981) 705.
[74] G. Margaritondo, C. Capasso, F. Patella, P. Perfetti, C. Quaresima, A. Savioa and F. Sette, J. Vac. Sci. & Technol. A 2 (1984) 508.
[75] F.F. Fang and W.H. Howard, J. Appl. Phys. 35 (1964) 3.
[76] A.D. Katnani, P. Chiaradia, H.W. Sang Jr and R.S. Bauer, J. Vac. Sci. & Technol. B 2 (1984) 471.
[77] R.W. Grant, J.R. Waldrop, S.P. Kowalczyk and E.A. Kraut, J. Vac. Sci. & Technol. B 3 (1985) 1295.
[78] P. Zurcher and R.S. Bauer, J. Vac. Sci. & Technol. A 1 (1983) 695.
[79] W.A. Harrison, J. Vac. Sci. & Technol. 16 (1972) 1492;
W.A. Harrison, E.A. Kraut, J.R. Waldrop and R.W. Grant, Phys. Rev. B 18 (1978) 4402.
[80] H. Kroemer, Surf. Sci. 132 (1983) 543;
S.L. Wright, M. Inada and H. Kroemer, J. Vac. Sci. & Technol. 21 (1982) 534.
[81] A.D. Katnani and R.S. Bauer, Phys. Rev. B 33 (1986) 1106.
[82] E.A. Kraut, R.W. Grant, J.R. Waldrop and S.P. Kowalczyk, Phys. Rev. Lett. 44 (1980) 1620.
[83] M.O. Watanabe, J. Yoshida, M. Mashita, T. Nakanisi and A. Hojo, Ext. Abstr. 16th Int. Conf. on Solid State Devices and Materials, Kobe, 1984 (1984) p. 181.
[84] R.S. Bauer, P. Zurcher and H.W. Sang Jr, Appl. Phys. Lett. 43 (1983) 663.
[85] S.P. Kowalczyk, E.A. Kraut, J.R. Waldrop and R.W. Grant, J. Vac. Sci. & Technol. 21 (1982) 482.

[86] J. Sakaki, L.L. Chang, R. Ludeke, C.A. Chang, G.A. SaiHalasz and L. Esaki, Appl. Phys. Lett. 31 (1977) 211;
L.L. Chang and L. Esaki, Surf. Sci. 98 (1980) 70.
[87] G. Margaritondo, Phys. Rev. B 31 (1985) 2526.
[88] P. Perfetti, F. Sette and C. Quaresima, P. Esso, E. Sevoia F. Patella, and G. Margaritondo, Proc. 17th Int. Conf. on the Physics of Semiconductors, San Francisco, CA, August 1984, eds. J.A. Chadi and W.A. Anderson (Springer, Berlin, 1985) p. 233.
[89] R.Z. Bachrach, R.S. Bauer, P. Chiaradia and G.V. Hansson, J. Vac. Sci. & Technol. 18 (1981) 797; and 19 (1981) 335.
[90] P.K. Larsen and J.F. Van der Veen, Surf. Sci. 126 (1983) 1.
[91] J.H. Neave, P.K. Larson, B.A. Joyce, J.P. Gowers and J.F. Van der Veen, J. Vac. Sci. & Technol. B 1 (1983) 668.
[92] A.D. Katnani, H.W. Sang Jr, P. Chiaradia and R.S. Bauer, J. Vac. Sci. & Technol. B 2 (1984) 471.
[93] P.K. Larson, J.H. Neave, P.J. Dobson and B.A. Joyce, Phys. Rev. B 27 (1983) 1966;
J.F. Van der Veen, P.K. Larson, J.H. Neave and B.A. Joyce, Solid State Commun. 49 (1984) 659.
[94] J. Massies, P. Devoldere and N.T. Linb, J. Vac. Sci. & Technol. 16 (1979) 1244.
[95] F. Stucki, G.J. Lapyere, R.S. Bauer, P. Zurcher and J.C. Mikkelsen, J. Vac. Sci. & Technol. 1 (1983) 865.
[96] R.S. Bauer, Thin Solid Films 89 (1982) 419, and references therein.
[97] B. Mrstik, Surf. Sci. 124 (1982) 253.
[98] J.E. Northrop, private communication.
[99] E.A. Kraut, J. Vac. Sci. & Technol. B 1 (1983) 645.
[100] B.A. Joyce and C.T. Foxon, J. Cryst. Growth 31 (1975) 122.
[101] W. Monch and H. Gant, Phys. Rev. Lett. 48 (1982) 512.
[102] A.D. Katnani, P. Chiaradia, H.W. Sang Jr and R.S. Bauer, J. Electron. Mater. 14 (1985) 25.
[103] S.P. Kowalczyk, R.W. Grant, J.R. Waldrop and E.A. Kraut, J. Vac. Sci. & Technol. B 1 (1983) 684.
[104] S.P. Sevensson, J. Kanski, T.G. Andersson and P.O. Nelson, Surf. Sci. 124 (1983) L31.
[105] A.D. Katnani, P. Chiaradia, Y. Cho, P. Mahowald, P. Pianetta and R.S. Bauer, Phys. Rev. B 32 (1985) 4071.
[106] G. Margaritondo, N. Stoffel, A.D. Katnani and L. Brillson, Appl. Phys. Lett. 37 (1980) 917.

CHAPTER 4

INTERFACE CONTRIBUTIONS TO HETEROJUNCTION BAND DISCONTINUITIES: X-RAY PHOTOEMISSION SPECTROSCOPY INVESTIGATIONS

R.W. GRANT, E.A. KRAUT, J.R. WALDROP and S.P. KOWALCZYK *

Rockwell International Corporation
Science Center
Thousand Oaks, CA 91360, USA

———
* Permanent address: IBM Thomas J. Watson Research Center, Yorktown Heights, NY, USA.

Heterojunction Band Discontinuities: Physics and Device Applications
Edited by F. Capasso and G. Margaritondo
© Elsevier Science Publishers B.V., 1987

Contents

1. Introduction 169
2. The XPS technique for observing interface dipoles and determining ΔE_v 170
   2.1. Measurement of interface contributions to ΔE_v by XPS 170
   2.2. Determination of ΔE_v 176
3. Experimental observations of interface contributions to ΔE_v 181
   3.1. Dependence of ΔE_v on crystallographic orientation 181
      3.1.1. Orientation dependence of Ge/GaAs interface dipoles 181
      3.1.2. Polar versus nonpolar heterojunction interfaces 185
   3.2. Growth-sequence dependence of ΔE_v 190
      3.2.1. Interface chemistry 191
      3.2.2. Compound–elemental (110) semiconductor heterojunctions 192
         3.2.2.1. Compound semiconductor–Ge(110) ΔE_v systematics 193
         3.2.2.2. Instability of the GaAs/Ge(110) interface 194
         3.2.2.3. Possible role of antiphase disorder 198
      3.2.3. Isocolumnar (110) heterojunctions 200
   3.3. Transitivity of ΔE_v 201
4. Summary 204
References 205

1. Introduction

Semiconductor heterojunctions are used in electronic devices to provide selective control over carrier transport by means of energy band discontinuities. The band gap discontinuity, ΔE_g, at the interface between two dissimilar semiconductors is the sum of the valence-band discontinuity, ΔE_v, and the conduction-band discontinuity, ΔE_c. The distribution of ΔE_g between the valence and conduction bands is thus of fundamental interest and importance, both to semiconductor device design and to the interpretation of device characteristics. Reproducible heterojunction fabrication techniques and an understanding of factors that affect heterojunction band discontinuities will be necessary before semiconductor heterojunctions are optimally utilized in device applications.

To be most useful in semiconductor device design and fabrication, heterojunction band discontinuities must be known and controlled to approximately the thermal energy, ~ 0.025 eV at room temperature. Recent progress in fabricating abrupt heterojunction interfaces by molecular beam epitaxy (MBE) and chemical vapor deposition has renewed interest in measuring and predicting heterojunction band discontinuities. Band discontinuities measured on such abrupt heterojunction interfaces are considered to be more reproducible and appropriate for comparison with theoretical models of ideal interfaces. There are several theories to predict heterojunction band discontinuities [1–4]. These theories are based on linear models which estimate band discontinuities from differences in absolute energies associated with the two semiconductors that form the heterojunction. Comparison of Harrison's LCAO model [1] with selected experimental data has suggested [5–8] that the accuracy of prediction is at present ~ ±0.2 eV; similar predictive accuracy has been indicated [9] for Tersoff's heterojunction model [4]. Linear heterojunction models do not account for band discontinuity contributions that are due to the presence of interface dipoles caused by interface microscopic structure. Although the contribution of these interface dipoles may in most cases be only a few tenths of an eV, they are significant and may account in part for the differences observed between theory and experiment.

Because interface dipoles can alter heterojunction band discontinuities by substantially more than the thermal energy, it is important to understand the origin of these dipoles and to provide results for critical tests and refinements of theoretical models. For the investigation of interface contributions to heterojunction band discontinuities, it is essential to utilize a measurement technique that will detect changes in band discontinuities with an experimental uncertainty considerably smaller than the observed effect. Various techniques for measuring heterojunction band discontinuities have recently been critically reviewed [6]; in most cases measurement accuracy better than ±0.1 eV is difficult to achieve in other than specialized cases. Although important results have been obtained at this level of accuracy, an improvement would clearly be beneficial. A technique based on the use of X-ray photoemission spectroscopy (XPS) has been developed [10,11] that is capable of measuring changes in ΔE_v with an uncertainty of ±0.01 eV. This XPS technique is thus particularly well suited to study interface dipole contributions to ΔE_v for a given heterojunction formed between two semiconductors as a function of interface preparation variables. The XPS technique is also capable of measuring the absolute value of ΔE_v with sufficient accuracy for comparison with theory [11–13]. As will be discussed, this measurement requires the determination of core-level to valence-band maximum binding energy differences and the consequent introduction of additional uncertainty into the measurement. However, in favorable cases absolute ΔE_v values have been determined to ±0.04 eV, which compares well with most other measurement techniques.

This chapter will provide a description of the XPS technique for measuring interface contributions to heterojunction band discontinuities and the extension of this method used to obtain absolute values of ΔE_v. Several studies that directly observe interface dipoles at specific heterojunction interfaces (by using the XPS method) will be discussed. Possible origins of these interface contributions to ΔE_v will be noted.

2. The XPS technique for observing interface dipoles and determining ΔE_v

This section discusses the XPS method for measuring interface contributions to heterojunction band discontinuities and the extension of the method which enables the measurement of absolute ΔE_v values.

2.1. Measurement of interface contributions to ΔE_v by XPS

A key point made in this section is that it is not necessary to measure the absolute magnitude of ΔE_v with precision in order to measure changes (interface contributions) in ΔE_v with precision.

The most common application of XPS is the analysis of interface elemental and chemical composition [14]. However, it is also well established [15,16] that the kinetic energy, E_k, of electrons emitted from a semiconductor depends on the position of the Fermi level, E_F, within the semiconductor band gap. This latter aspect of XPS makes it possible to determine E_F relative to the valence-band maximum, E_v, in the region of the semiconductor from which the photoelectrons originate. XPS can therefore be used as a contactless nondestructive technique to measure interface-potential related quantities such as heterojunction band discontinuities.

The escape depth, λ, for photoelectrons produced within a solid by incident radiation increases monotonically from ~ 5 to ~ 25 Å as the electron E_k increases from about 100 to 1500 eV [17]. A monochromatic AlKα ($h\nu = 1486.6$ eV) X-ray source is convenient for XPS measurements. Most semiconductors will contain an element that has an outer core level with binding energy, E_B, less than 100 eV. The escape depth for photoelectrons excited from an outer core level will therefore be ~ 25 Å and thus the photoelectron signal is averaged over many atomic layers. It is desirable to use outer core levels in XPS heterojunction band discontinuity studies because photoelectrons originating from these levels will in general have the narrowest linewidths (as determined by final-state lifetimes); in addition, when measuring absolute values of ΔE_v (see sect. 2.2), there is an advantage to keeping the core level to E_v binding energy difference relatively small.

A schematic diagram of a typical heterojunction sample that is suitable for XPS band discontinuity measurements is illustrated in fig. 1. The overlayer thickness of semiconductor B must be comparable to λ and the interface width must be a small fraction of λ.

The measurement of interface contributions to ΔE_v is shown in fig. 2 by a schematic energy band diagram appropriate for the sample shown in fig. 1 (with an abrupt interface). The quantities E_{CL}^A, E_v^A, and E_c^A are, respectively, a core-level binding energy, the valence-band maximum, and the conduction-band minimum associated with semiconductor A; similar energies are defined for semiconductor B. The core-level binding energy difference across the interface is $\Delta E_{CL} = E_{CL}^A - E_{CL}^B$. If a change in an interface dipole occurs, all energy levels in semiconductor A will be shifted by an equal amount relative to the energy levels in semiconductor B (as indicated by dotted lines in fig. 2). Thus a change in the heterojunction

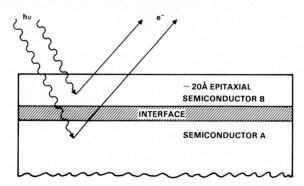

Fig. 1. Schematic illustration of a heterojunction sample that is suitable for XPS band discontinuity measurements.

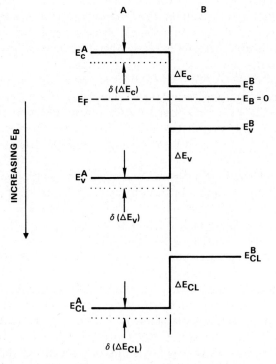

Fig. 2. Schematic energy-band diagram that illustrates the XPS measurement of interface contributions to ΔE_v.

valence-band discontinuity, $\delta(\Delta E_v)$, caused by an interface dipole, will be accompanied by a change in the heterojunction conduction-band discontinuity, $\delta(\Delta E_c)$, and a corresponding change in the core-level binding energy difference, $\delta(\Delta E_{CL})$. A change in an interface contribution to a heterojunction band discontinuity will therefore be observed as $|\delta(\Delta E_v)|$ = $|\delta(\Delta E_c)|$ = $|\delta(\Delta E_{CL})|$, with the signs of these energy differences determined by the relative positions of the corresponding energy levels in semiconductors A and B. Interface contributions to heterojunction band discontinuities are measured by using XPS to determine $\delta(\Delta E_{CL})$. As shown in fig. 1, photoelectrons which originate in semiconductor A pass through any dipole layer at the interface in order to be emitted from the surface and detected while photoelectrons that originate in semiconductor B do not pass through this layer. For example, a photoelectron from semiconductor A passing through the dipole layer from lower-to-higher electron density will experience a deceleration. The resulting decrease in relative E_k is proportional to the dipole moment per unit area at the interface. This E_k decrease will directly be measured as a relative increase in E_{CL}^A. For the energy band diagram shown in fig. 2, the resulting increase in the core-level binding energy difference, $\delta(\Delta E_{CL})$, is a direct measure of the equivalent increase in the heterojunction valence-band discontinuity, $\delta(\Delta E_v)$. For well-resolved outer core levels, it is possible to measure ΔE_{CL} to ± 0.01 eV [10,11]. As an example, table 1 provides ΔE_{CL} measurements on two different heterojunction samples of Ge grown on GaAs(110). Five independent measurements of $\Delta E_{CL} = E_{Ga3d}^{GaAs} - E_{Ge3d}^{Ge}$ were obtained for sample A

Table 1
Measurement of Ga3d–Ge3d core-level binding energy differences for two different Ge/GaAs(110) heterojunctions.

Sample	Ge layer thickness (Å)	$E_{Ga3d}^{GaAs} - E_{Ge3d}^{Ge}$ * (eV)
A	14	−10.194
		−10.200
		−10.195
		−10.202
		−10.201
B	17	−10.210
		−10.208

* Uncertainty is ± 0.01 eV.

and two independent ΔE_{CL} measurements were obtained for sample B. The scatter in values for a single sample is < 0.01 eV (however, consideration of spectrometer energy-scale calibration accuracy increases the measurement uncertainty to ±0.01 eV). Thus, as noted at the beginning of this section, it is possible to measure interface contributions to heterojunction band discontinuities with an uncertainty of ±0.01 eV without measuring the absolute value of ΔE_v.

The measurement of interface contributions to ΔE_v as described has been simplified in some important ways. One assumption is that the interface is atomically abrupt. It is well-known [18] that chemical reactions and interdiffusion can occur on a monolayer scale to broaden semiconductor interfaces even when they are formed at room temperature. These effects may alter the potential distribution due to the formation of an interfacial dipole layer of finite width in the immediate vicinity of the interface. In addition, interface chemical bonding [19] and the vacuum interface may contribute chemically shifted components to the core-level peaks; if not experimentally resolved, these chemical shifts could alter the apparent value of ΔE_{CL}. Even for an ideally abrupt interface, calculations have shown that potential variations occur within one or two atomic layers normal to the interface (see, e.g., refs. [20,21]). These considerations suggest that there is an advantage to collecting the photoelectron signal primarily from a region near, but not precisely at, an abrupt interface. An ideal situation would be to have sufficient energy resolution to resolve photoelectron signals which originate in the interface region from those that originate in the bulk semiconductor very near the interface. As a rough generality, the energy resolution associated with the various forms of photoelectron spectroscopy improves for decreasing E_k analysis. However, for low E_k electrons that originate from outer core levels, the fraction of the total electron signal that originates in the bulk semiconductor very near the interface may be small due to the small λ. Thus, in some cases it may be advantageous to sacrifice energy resolution in order to obtain a fairly large λ and thus minimize the fraction of photoelectron signal which originates from the monolayer or two interfacial region. In practice, the linewidths for the XPS photoelectron peaks and a dependence of results on overlayer thickness are monitored to assess the influence of interface potential variations or chemical shifts on the measured ΔE_{CL}.

Even without the possible complications in measuring ΔE_{CL} that are of microscopic origin (as discussed in the previous paragraph), the measurement of interface contributions to ΔE_v (as illustrated in fig. 2) has been simplified by assuming a flat-band condition on both sides of the heterojunction interface. The potential variation away from the interface region follows a solution of the Poisson equation. For a fixed interface E_F position, the depletion width for a given semiconductor will increase for decreased doping density. Thus, when using large λ to minimize the fraction of the photoelectron signal which originates from the monolayer or two interfacial region, it is beneficial to have moderately or lightly doped semiconductors to minimize the potential variation within the sampling depth. In general, a measurement error of < 0.01 eV is expected for a doping density $\leq 10^{17}$ cm^{-3} and $\lambda \sim 25$ Å. The practical lower limit on doping is that the sample does not become insulating and thus charges under the influence of the X-ray beam.

Sample preparation is, of course, an important aspect of applying the XPS method of measuring interface contributions to heterojunction band discontinuities. In most cases, it is necessary to fabricate samples of the type shown in fig. 1 within the ultrahigh vacuum system of the XPS spectrometer to avoid contamination and/or oxidation of the sample surface. A fairly general-purpose experimental apparatus for XPS interface studies is shown schematically in fig. 3. With this apparatus a clean substrate is prepared by ion sputtering and/or thermal annealing. The surface cleanliness is determined by XPS and surface order is assessed by low-energy electron diffraction (LEED). MBE is used to fabricate the ~ 20 Å thick epitaxial overlayer and again XPS and LEED are used to assess surface cleanliness and order.

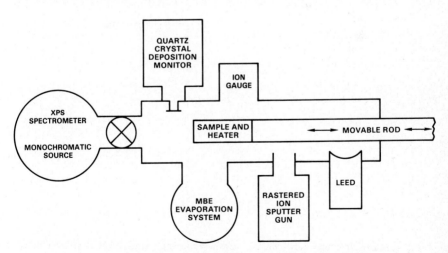

Fig. 3. Schematic diagram of an XPS system for ΔE_v measurements.

2.2. Determination of ΔE_v

The XPS method for measuring interface contributions to heterojunction band discontinuities can be extended to determine absolute values of ΔE_v. The approach is illustrated by the schematic band diagram shown in fig. 4. The assumptions made in drawing this idealized flat-band diagram were discussed in sect. 2.1. While the investigation of interface contributions to ΔE_v requires only the measurement of ΔE_{CL}, the determination of the absolute magnitude of ΔE_v requires measurement of two additional parameters, namely the core level to E_v binding energy difference in bulk semiconductors A and B, $(E_{CL}^A - E_v^A)$ and $(E_{CL}^B - E_v^B)$ respectively. By inspection of fig. 4 it is seen that

$$\Delta E_v(A - B) = (E_{CL}^B - E_v^B) - (E_{CL}^A - E_v^A) + \Delta E_{CL}(A - B). \tag{1}$$

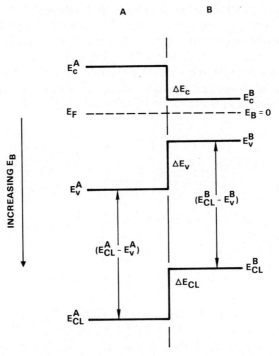

Fig. 4. Schematic energy-band diagram that illustrates the measurement of the absolute value of ΔE_v by XPS.

Thus, to apply XPS for ΔE_v measurements it is essential to determine the bulk semiconductor material parameters ($E_{CL} - E_v$) for those semiconductors which are to form the heterojunctions of interest. A method for determining these parameters with precision is outlined in this section.

Many ($E_{CL} - E_v$) measurements for semiconductors have been reported in the literature (see ref. [22] for a recent tabulation). In general, the precision of most measurements has been limited to about ± 0.1 eV. A primary difficulty with the measurement of ($E_{CL} - E_v$) is the accurate determination of the E_v position in photoemission spectra. The most frequently employed method involves extrapolation of a tangent line to the leading edge of the valence-band spectrum to the energy axis; this intercept is defined as E_v. It has been noted that this procedure can lead to substantial uncertainties [23]. In addition to the XPS method for measuring ΔE_v (described in this section), there are other photoemission methods for measuring ΔE_v that employ the linear extrapolation procedure of determining E_v in the photoemission spectrum (see, e.g., ref. [24]). Use of this extrapolation procedure accounts in large part for the typical ± 0.1 eV uncertainty in ΔE_v associated with these other photoemission measurements.

A method that largely overcomes the difficulty in determining the E_v position in XPS data has been reported [12,13]. In essence the approach involves a least-squares fitting of the XPS data in a limited region around the estimated position of E_v with an instrumentally broadened valence-band density of states (VBDOS) defined as

$$N_v(E) = \int_0^\infty n_v(E') \, g(E - E') \, dE', \qquad (2)$$

where $n_v(E')$ is the theoretical VBDOS and $g(E)$ is the instrumental response function. The XPS spectral intensity $I(E)$ is assumed to have the form

$$I(E) = SN_v(E - E_v) + B, \qquad (3)$$

where S is a scale factor, and B is a constant random noise background. As an example of determining the position of E_v, angle-resolved XPS data collected from a Ge(110) surface are shown in fig. 5. The solid curve is a least-squares fit of $N_v(E)$ to the experimental data where a nonlocal pseudopotential VBDOS [25] was used for $n_v(E')$. The position of E_v^{Ge} corresponds to $N_v(0)$.

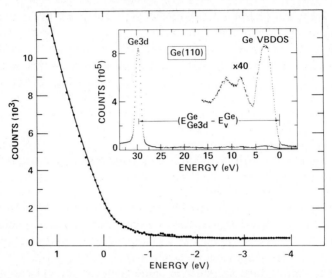

Fig. 5. Least-squares fit of the instrumentally broadened VBDOS (solid curve) to XPS data in the region of E_v. Inset shows the XPS spectrum which contains the VBDOS and the Ge3d core level. The energy scale is zero at E_v. Data were taken from ref. [13].

A primary difficulty with the determination of E_v in XPS spectra is the possibility of VBDOS distortion due to occupied surface states in the vicinity of E_v. To detect possible VBDOS distortion, sets of angle-resolved measurements are analyzed and compared. Because the XPS photoelectron cross section is expected to depend on the orbital character of filled surface states [26,27], it should be possible to detect the presence or absence of these states by studying the angular variation of the XPS valence-band spectrum in the vicinity of E_v. In fig. 6a, a convenient polar-coordinate system is defined to relate the photoemission direction e to the crystallographic axes of the Ge(110) surface. The polar angle for the measurements described here was 51.5° and only the azimuthal angle α was varied. Although the Ge(110) surface has been studied by LEED [28] little is known about the corresponding electronic structure. Figure 6b shows the results of analyzing two sets of Ge(110) data for $\alpha = 0°$ and 90°. The error bars represent a 95% confidence interval for the least-squares procedure [13]. The quantity E_{max} is the end point of the fitting interval. The relatively constant value of $(E_{Ge3d}^{Ge} - E_v^{Ge}) = 29.52$ eV independent of E_{max} and α suggests that any filled Ge(110) surface states below E_v^{Ge} are either very weakly localized near the surface or lie well outside the energy interval

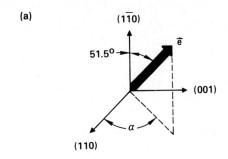

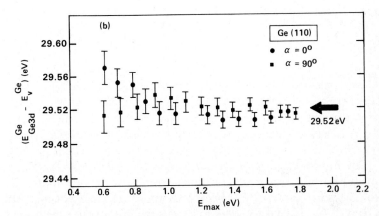

Fig. 6. (a) Polar-coordinate system relating photoelectron emission direction e to crystallographic axes for a (110) crystal surface. The azimuthal angle α is in the plane of the crystal surface. (b) Position of E_v^{Ge} measured relative to E_{Ge3d}^{Ge} as a function of the end point, E_{max}, of the fitting interval. Results are shown for data obtained at two different photoelectron emission directions. Data were taken from ref. [13].

analyzed. In either case, surface states do not appear to complicate the determination of $(E_{Ge3d}^{Ge} - E_v^{Ge})$ from Ge(110) XPS data; the situation is quite different for Ge(111) XPS data where the apparent value of $(E_{Ge3d}^{Ge} - E_v^{Ge})$ is found to depend on the fitting interval [13].

As can be seen from eq. (1), the uncertainty in the bulk semiconductor $(E_{CL} - E_v)$ binding energy difference directly influences the uncertainty in the measurement of ΔE_v by the XPS method. Factors that contribute to the uncertainty in determining $(E_{CL} - E_v)$ bulk binding-energy differences

Table 2
$E_{CL} - E_v$ binding energy differences in eV as measured by XPS.

Semiconductor surface	core level	$(E_{CL} - E_v)_{XPS}$	$(E_{CL} - E_v)_b$
Ge(110)	Ge3d	29.52 ± 0.03	29.52 ± 0.03
GaAs(110)	Ga3d	18.78 ± 0.03	18.75 ± 0.03
	As3d	40.70 ± 0.03	40.74 ± 0.03
ZnSe(110)	Zn3d	8.86 ± 0.03 *	–
InAs(100)	In4d	17.38 ± 0.03 *	–
	As3d	40.72 ± 0.03 *	–
CdTe($\bar{1}\bar{1}\bar{1}$)	Cd4d	10.24 ± 0.03	–
HgTe($\bar{1}\bar{1}\bar{1}$)	Hg5d$_{5/2}$	7.66 ± 0.04	–
AlAs(100)	As3d	40.16 ± 0.04	–
	Al2p	72.70 ± 0.04	–

* Error limits refined subsequent to initial publication.

have been considered in detail [13]. These factors include band bending, surface chemical shifts, background effects associated with inelastic scattering processes, instrumental line shape, and spectrometer calibration accuracy. It is concluded that $(E_{CL} - E_v)$ can be determined for the 3d levels of Ge and GaAs with an uncertainty of $\leqslant 0.026$ eV. As discussed in sect. 2.1, ΔE_{CL} can be measured with an uncertainty of ± 0.01 eV. If $(E_{CL} - E_v)$ parameters for bulk semiconductors are known with an uncertainty of 0.026 eV, eq. (1) indicates that XPS measurements of ΔE_v are possible with an uncertainty of $\sim \pm 0.04$ eV; this is a considerable improvement over the ± 0.1 eV measurement accuracy reported for many ΔE_v measurement techniques.

One final point is that surface chemical shifts can alter XPS measured $(E_{CL} - E_v)$ parameters by a few hundredths of an eV. If these surface chemical shifts are well known, as for the GaAs (110) surface [29], an accurate correction can be made to obtain the bulk semiconductor $(E_{CL} - E_v)$ value. The XPS measured and the surface chemical shift corrected bulk semiconductor values of $(E_{CL} - E_v)$ will be denoted as $(E_{CL} - E_v)_{XPS}$ and $(E_{CL} - E_v)_b$, respectively. Values of these parameters for Ge [13], GaAs [13], ZnSe [30], InAs [31], CdTe [32], HgTe [32] and AlAs [33], are collected in table 2 [34]. To increase the generality of the XPS method for measuring ΔE_v, additional $(E_{CL} - E_v)_b$ values will need to be determined for semiconductors of interest.

3. Experimental observations of interface contributions to ΔE_v

This section primarily discusses studies that use the XPS method (sect. 2.1) to determine interface contributions to ΔE_v. As previously noted, there is considerable evidence that interface dipoles can alter ΔE_v for some heterojunctions by at least a few tenths of an eV. A few tenths of an eV variation in ΔE_v is large compared to kT and thus can influence the characteristics of semiconductor heterostructure devices. Observations of a ΔE_v dependence on crystallographic orientation and on growth sequence will be discussed. Also, a test of ΔE_v transitivity for a series of three heterojunctions will be described; such tests may be useful to select values of ΔE_v least influenced by interface dipoles and therefore most suitable for comparison with linear predictive models.

3.1. Dependence of ΔE_v on crystallographic orientation

A study of Ge/GaAs heterojunctions has provided clear evidence that ΔE_v depends on crystallographic orientation for these heterojunctions. A fundamental difference between heterojunctions formed on polar and nonpolar crystallographic faces that offers an explanation of the ΔE_v crystallographic orientation dependence is discussed.

3.1.1. Orientation dependence of Ge/GaAs interface dipoles

Observation of a ΔE_v crystallographic orientation dependence for abrupt vapor-grown Ge/GaAs heterojunctions based on transport measurements has been reported [35]. Unfortunately, the experimental uncertainty in these measurements is about as large as the measured effects (a few tenths of an eV). A study of the ΔE_v crystallographic orientation dependence for abrupt MBE-grown Ge on GaAs heterojunctions that utilized the XPS method has also been reported [10,11,36]. The results of this latter study are discussed here.

Heterojunctions of Ge/GaAs were prepared on GaAs substrates with (110), (100) and (111) orientations that had been cut from the same boule of GaAs material. The substrates were cleaned by sputtering with 750 eV Ar^+ ions and annealed at $\simeq 460\,°C$ to remove sputter damage. Prior to growth of the epitaxial layers, the room temperature GaAs substrate LEED patterns were: (110) (1 × 1), (111)Ga (2 × 2), ($\overline{111}$)As (1 × 1) and (100)Ga c(8 × 2); the (100)As LEED pattern was either c(2 × 8) or (2 × 4). Very thin (~ 20 Å) epitaxial layers of Ge were grown at ~ 1 Å/s deposition rates under ultra-high vacuum (UHV) conditions on these substrates. A low-growth temperature (~ 340°C) was used to maximize interface abruptness. Following growth the samples were cooled to room temperature and LEED was used to confirm the epitaxy of the Ge overlayers.

The Ge3d and Ga3d core levels were studied to determine the crystallographic orientation dependence of ΔE_v; these core levels are well-resolved and are in a region of the XPS spectrum where the inelastically scattered electron background is smooth and featureless. A typical XPS spectrum obtained from a heterojunction that consisted of a thin epitaxial layer of Ge grown on a GaAs(110) (1 × 1) substrate is shown in fig. 7. To determine

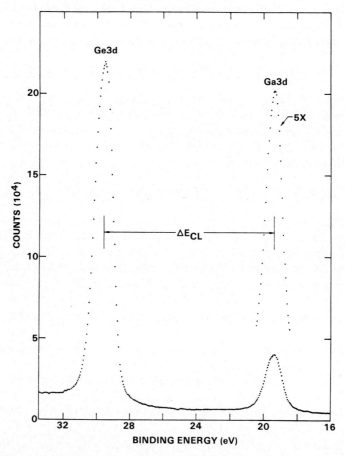

Fig. 7. XPS spectrum in the binding energy region of the Ga3d and Ge3d core levels obtained from a Ge/GaAs(110) heterojunction. Data were taken from ref. [10].

$\Delta E_{\text{CL}} = E_{\text{Ga3d}}^{\text{GaAs}} - E_{\text{Ge3d}}^{\text{Ge}}$, a background function that is proportional to the integrated photoelectron peak area was subtracted from the XPS data to correct for the effect of inelastic photoelectron scattering. The quantity ΔE_{CL} was measured between the centers of the peak widths at half of the peak heights. This procedure makes it unnecessary to resolve the spin–orbit splitting of the Ge3d and Ga3d levels (~ 0.5 eV) to obtain high precision peak positions.

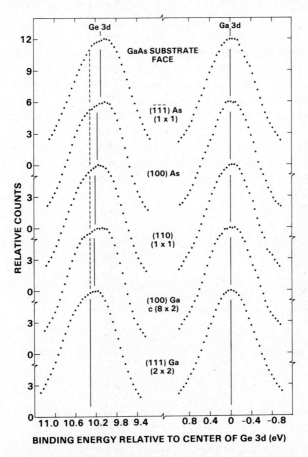

Fig. 8. XPS spectra in the binding energy region of the Ga3d and Ge3d core levels obtained for five different crystallographically oriented heterojunctions. The GaAs substrate crystal faces on which the thin epitaxial Ge overlayers were grown are indicated in the figure. The vertical lines indicate the centers of the various peaks as discussed in the text. Data were taken from ref. [11].

Thirty-three independent measurements were made on eight different Ge/GaAs heterojunctions. Figure 8 shows representative background subtracted XPS spectra from samples having each of the crystallographic faces studied. For easy comparison, each peak shown in this figure has been normalized to an equal height and the center (half-width at half-height) of each peak is indicated by a vertical solid line in the figure. The centers of the five Ga3d peaks have been aligned. The dashed vertical reference line that runs through the Ge3d peaks is the center of the Ge3d peak observed for the heterojunction that was grown on the GaAs(111)Ga(2×2) substrate. Inspection of fig. 8 directly indicates that ΔE_{CL} is dependent on the crystallographic orientation of the interface. As noted in sect. 2.1., the change in ΔE_{CL} is a direct measure of the change in ΔE_v and it is not necessary to know the actual magnitude of ΔE_v in order to detect the crystallographic orientation dependent interface contributions to ΔE_v.

Measurement results on eight different Ge/GaAs interfaces are given in table 3. In general, three to five independent measurements were made on each interface and the average ΔE_{CL} values are given in the table. In every case the measurement reproducibility for ΔE_{CL} was $< \pm 0.01$ eV and in most cases it was $< \pm 0.005$ eV. Spectrometer energy-scale calibration uncertainty increases the total error limit for ΔE_{CL} to ± 0.01 eV. The average linewidths, Γ, of the Ge3d and Ga3d photoelectron peaks that were measured at half of the peak height and the Ge epitaxial layer thickness for the eight Ge/GaAs interfaces are also given in table 3. If a

Table 3
The Ge epitaxial layer thickness, Ge3d and Ga3d photoelectron linewidths, Ge3d–Ga3d core-level binding energy differences, the average variation in ΔE_v relative to the (110) interface, and the ΔE_v value for eight different Ge–GaAs interfaces.

Substrate surface	Ge layer thickness (Å)	Γ(Ga3d) (eV)	Γ(Ge3d) (eV)	$E_{Ga3d}^{GaAs} - E_{Ge3d}^{Ge}$ [a] (eV)	$\delta(\Delta E_v)_{AVE}$ (eV)	(ΔE_v) [b] (eV)
(111)Ga (2×2)	13	1.17 ± 0.02	1.25 ± 0.01	-10.27	~ -0.085	0.50
	20	1.22 ± 0.02	1.26 ± 0.01	-10.31		0.46
(100)Ga c(8×2)	22	1.19 ± 0.02	1.25 ± 0.01	-10.22	-0.015	0.55
(110) (1×1)	14	1.13 ± 0.01	1.29 ± 0.01	-10.20	0	0.57
	17	1.16 ± 0.01	1.27 ± 0.01	-10.21		0.56
(100)As	14	1.15 ± 0.02	1.25 ± 0.01	-10.17	$+0.035$	0.60
($\bar{1}\bar{1}\bar{1}$)As (1×1)	13	1.21 ± 0.01	1.32 ± 0.01	-10.11	$+0.10$	0.66
	18	1.22 ± 0.01	1.28 ± 0.01	-10.10		0.67

[a] Uncertainty is ± 0.01 eV.
[b] Uncertainty is ± 0.04 eV.

sizeable potential variation occurred, either within the heterojunction area sampled or within the photoelectron escape depth, Γ would be broadened. A similar Γ broadening would occur if interface or surface chemical shifts were substantially affecting the measurements. Although the Γ values scatter somewhat, there is little significant systematic variation with crystallographic orientation or with Ge overlayer thickness. The ΔE_{CL} measurements for the two heterojunctions grown on each of the (110) (1 × 1) and ($\bar{1}\bar{1}\bar{1}$)As(1 × 1) substrates reproduce within 0.01 eV. This agreement indicates that while ΔE_{v} for the heterojunction formed on the ($\bar{1}\bar{1}\bar{1}$)As(1 × 1) substrate is ~ 0.10 eV larger than ΔE_{v} for the heterojunction formed on the (110) (1 × 1) substrate, the sample-to-sample reproducibility of ΔE_{v} for heterojunctions formed on these two substrates is 0.01 eV. The 0.04 eV variation in ΔE_{CL} for the two heterojunction samples formed on the (111)Ga(2 × 2) substrates appears to be outside of experimental error and most likely represents a real sample preparation-dependent ΔE_{v} difference between these two samples. In table 3, the column labeled $\delta(\Delta E_{\mathrm{v}})_{\mathrm{AVE}}$ is the difference in the average ΔE_{v} values for the different crystallographic orientations referred to the results for the (110) interface; as discussed in sect. 2.1, this was evaluated from $\delta(\Delta E_{\mathrm{v}}) = -\delta(\Delta E_{\mathrm{CL}})$. The $(E_{\mathrm{CL}} - E_{\mathrm{v}})$ values listed in table 2 for Ge3d and Ga3d have been used to evaluate the values of ΔE_{v} shown in the last column of table 3.

An explanation for the ΔE_{v} crystallographic orientation dependence of Ge/GaAs heterojunctions is offered in the next section.

3.1.2. Polar versus nonpolar heterojunction interfaces

A consideration of electrostatics leads to the conclusion that there is a fundamental difference between polar and nonpolar heterojunction interfaces [37] that offers a mechanism for obtaining a crystallographic orientation dependent interface dipole. To illustrate the concept, the electrostatics associated with a nonpolar (110) and a polar (100) Ge/GaAs interface will be compared.

Figure 9 shows a nonpolar Ge/GaAs (110) interface viewed along a [$\bar{1}$10] direction. To simplify the discussion an ideal tetrahedral bonding arrangement is assumed and bond polarization effects are ignored. In an ideal tetrahedral bonding arrangement all bonds contain two electrons. This is independent of whether the bonds are Ge–Ge, Ge–Ga, Ge–As, or Ga–As. The net charge on an interface plane is, therefore, determined by the relative nuclear charge on the plane. A Ga atom has one proton less than a Ge atom while an As atom has one more proton. For the (110)

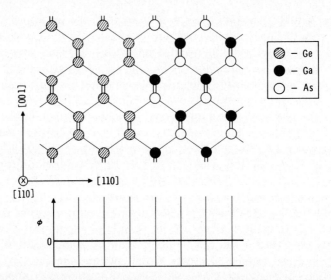

Fig. 9. A (110) heterojunction between Ge and GaAs. Every plane of atoms parallel to this nonpolar heterojunction is on average charge neutral. The symbols used to identify specific atoms are defined in this figure, and are the same for figs. 10–12. All atoms are tetrahedrally bonded; the "double" bonds schematically illustrated are two tetrahedral bonds projected onto the plane of the figure. If the potential, ϕ, is zero in the Ge, it will also be zero in the GaAs as shown at the bottom of the figure. This figure and figs. 10–12 were reproduced from ref. [37].

Ge/GaAs interface, the net charge on atom planes parallel to the interface is the same in Ge and GaAs as equal numbers of Ga and As atoms occupy the planes in GaAs. If the potential, ϕ, is zero in the Ge part of the heterojunction, integration of the Poisson equation from the Ge on the left through the interface into the GaAs on the right will cause no potential variation, as shown at the bottom of fig. 9. This corresponds to the absence of both a dipole layer and a charge accumulation at the interface.

The situation is quite different for a polar interface. Figure 10 illustrates a (001) Ge/GaAs heterojunction viewed along a [$\bar{1}$10] direction where the GaAs terminates with a Ga plane. The net charge on atom planes parallel to the interface is now different in GaAs and Ge. If it is again assumed that ϕ is zero in the Ge part of the heterojunction, integration of the Poisson equation from left to right yields the potential variation shown at the bottom of fig. 10. Upon crossing the first plane of negatively charged Ga atoms the potential gradient becomes positive and constant; the potential gradient returns to zero upon crossing the next plane of positively charged

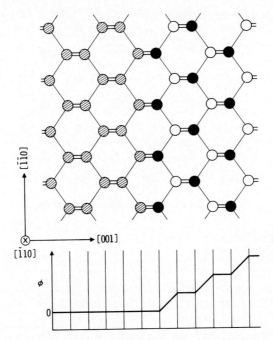

Fig. 10. A (001) heterojunction between Ge and GaAs. Note that the first atomic plane to the right of the junction consists entirely of Ga atoms that (without bond polarization) are negatively charged. The potential (assumed to be zero in Ge) averaged over planes parallel to the junction, is obtained by integrating the Poisson equation from left to right, as shown at the bottom of the figure.

As atoms. The result is a potential staircase function that has an average gradient and a fluctuating component. This average potential gradient is associated with interface charge accumulation [37] and produces a potential that cannot be sustained in a real heterojunction because the potential variation over a few atom distances would exceed the bandgap and result in spontaneous carrier generation.

The interface charge accumulation can be eliminated by adjusting the atomic composition of the interface region. One possibility is shown in fig. 11 where a transition plane with an equal number of Ga and Ge atoms is inserted between the bulk Ge and GaAs. Again the potential variation can be obtained (as shown at the bottom of fig. 11) by assuming $\phi = 0$ in the Ge and integrating the Poisson equation through the heterojunction. The single transition plane has eliminated the average potential gradient in the GaAs and produced an interface dipole shift, δ, which is estimated to be

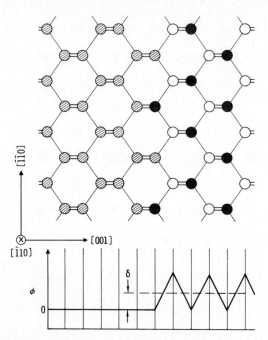

Fig. 11. A (001) heterojunction as in fig. 10, but with half of the Ga atoms in the junction plane replaced by Ge atoms. The potential obtained as previously described, is shown at the bottom of the figure. Although the average electric field in the GaAs has been eliminated, there remains a dipole shift, δ, which is much larger than is experimentally observed.

~ 0.37 eV [37]. The observed dipole shifts of polar Ge–GaAs interfaces relative to the nonpolar Ge–GaAs(110) interface (sect. 3.1.1.) are much smaller than 0.37 eV, which suggests that the simplest ideal single transition plane interface geometry needed to eliminate charge accumulation at a polar interface does not occur.

By adjusting the atomic composition of the interface region on two transition planes as shown in fig. 12, it is possible to eliminate both charge accumulation and the dipole shift. The potential variation for this interface is shown at the bottom of fig. 12. Thus, a polar interface modification that eliminates both charge accumulation and dipole shifts requires an interface region with at least two transition planes. It seems likely that the growth process produces a nonplanar polar heterojunction as an interface of lowest energy. To explain the observed ~ 0.1 eV dipole shifts between polar and nonpolar Ge–GaAs interfaces (sect. 3.1.1), a deviation from the idealized interface structure shown in fig. 12 must occur. The simplest kind of

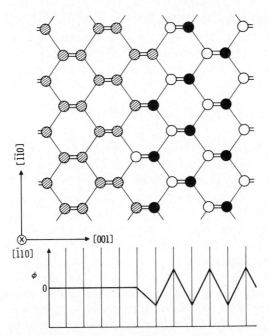

Fig. 12. A (001) heterojunction as in figs. 10 and 11 but with two compositionally adjusted junction transition planes. This is the simplest interface geometry that eliminates both charge accumulation and a dipole shift.

deviation would involve interchange of atom pairs with differing nuclear charge. An interchange of about one in fifteen interface atom pairs (which resulted in the transfer of one unit of nuclear charge between adjacent planes) would produce a dipole shift of ~ 0.1 eV [37].

Kroemer [6] has pointed out that it is unlikely for the polar interface atomic rearrangement to proceed far enough toward completion so as to completely eliminate interface charge accumulation and dipole shifts. He, therefore, argues that ΔE_v and interface charge for polar heterojunctions should be expected to be sample preparation-dependent and hence poorly reproducible.

As a final point, it has been shown that polar interfaces differ fundamentally from nonpolar interfaces by requiring at least two transition planes to eliminate charge accumulation and interface dipole contributions to ΔE_v that are substantially larger than experimental observations. It may appear that considerably nonabrupt interfaces are favored for polar heterojunctions. Recent tight-binding calculations of atomic substitution energies have shown that atomic exchange at many ideal abrupt interfaces (including the Ge–GaAs interface) is not favored [38]. Competing factors may thus favor an optimum width for abrupt polar heterojunction interfaces of about two transition planes.

3.2. Growth-sequence dependence of ΔE_v

The order in which a heterojunction is fabricated (i.e., the growth sequence) can result in a ΔE_v variation [30,39,40]. This observation of an interface contribution to ΔE_v due to growth sequence indicates that the atomic arrangement near an interface must depend on details of the growth process which at least in some cases are likely determined by chemical considerations. It has been noted [41] that in some cases thermodynamics may favor the formation of compounds more stable than either of the semiconductors that form the heterojunction.

As discussed in sect. 3.1, interface dipoles are observed and should be expected for polar heterojunction interfaces formed between semiconductors that contain elements from different columns of the periodic table. Any mechanism which causes atoms from different columns of the periodic table to transfer across a heterojunction interface may produce interface dipoles. To separate interface contributions to ΔE_v that are growth-sequence related in origin from those that depend on crystallographic orientation, heterojunctions formed on the nonpolar (110) interface can be studied. For this reason all experimental observations described in this section involve (110) interfaces.

The formation of a heterojunction interface by MBE involves a nonequilibrium growth process. Thus, it may not be surprising if in some cases interface chemistry is found to depend on growth sequence. An extreme example of this type is noted in sect. 3.2.1. The growth of a compound semiconductor on an elemental semiconductor involves an ambiguity in nucleation site which may cause antiphase domain disorder; the reverse growth sequence has no nucleation site ambiguity. Experimental observations of a growth sequence dependence for interfaces formed between compound semiconductors and Ge(110) are discussed in sect. 3.2.2 and a possible means by which antiphase domain disorder may contribute to interface dipoles is given. The fabrication of a nonpolar (110) heterojunction from semiconductors that contain elements from the same columns (isocolumnar) of the periodic table is perhaps a case where one would least expect to observe growth-sequence variations in ΔE_v. However, even in this case, as noted in sect. 3.2.3, interface contributions to ΔE_v of growth-sequence origin have been found.

3.2.1. Interface chemistry

If atoms from different columns of the periodic table are exchanged across a heterojunction, interface dipoles may be formed. A driving force for atom exchange at an interface may be chemical in origin. If interface chemical reactions depend on the order in which a heterojunction is formed, these reactions may contribute to a growth-sequence dependence of ΔE_v. An extreme example of interface chemistry growth-sequence dependence is given in this section.

The semiconductors Ge, GaAs, ZnSe and CuBr, which contain elements from row four of the periodic table, can be used to form many well-lattice-matched heterojunctions. It has been observed [42] that CuBr can be formed epitaxially on Ge(110) at $\sim 150\,°C$ with no evidence of interface chemical reaction. However, the situation is markedly different for the reverse growth sequence [43].

The growth-sequence dependence of interface chemical reactivity for the Ge–CuBr interface was investigated by XPS [43]. This phenomenon was studied by comparing core-level spectra from Ge on CuBr(110) [Ge/CuBr(110)] and CuBr on Ge(110) [CuBr/Ge(110)] interfaces that were formed at room temperature and then annealed for 1 min at $25\,°C$ temperature steps between 25 and $150\,°C$. Core-level spectra of Ge3d, Br3d, and Cu3p were obtained at room temperature after each anneal. The CuBr/Ge sample consisted of ~ 15 Å of CuBr deposited at room temperature on a Ge(110) substrate; the Ge/CuBr sample consisted of ~ 20 Å of Ge deposited onto a ~ 1000 Å thick film of CuBr that was grown epitaxially on a GaAs(110) substrate. Relative changes in the Br3d/Cu3p, Ge3d/Cu3p, and Ge3d/Br3d peak area ratios were used to monitor the interface chemistry for each heterojunction. Data for the Br3d/Cu3p peak area ratio are shown in fig. 13. For the CuBr/Ge(110) interface the Br3d/Cu3p peak area ratio is ~ 1.4. This ratio remains constant up to $\sim 125\,°C$ and is the same value as is observed for a thick stoichiometric CuBr film. Above $\sim 150\,°C$, the XPS peak area ratio data indicate that CuBr evaporates from the Ge(110) surface. For temperatures between 25 and $125\,°C$ the constant Br3d/Cu3p peak area ratio is consistent with the CuBr/Ge interface being essentially abrupt and chemically inert. The Br3d/Cu3p peak area ratio data for the Ge/CuBr(110) interface are also shown in fig. 13. These data suggest that a chemical reaction occurs

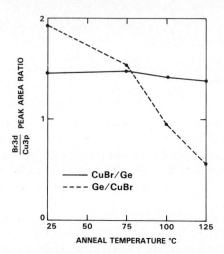

Fig. 13. Variation of the Br3d/Cu3p core-level peak area ratio for CuBr/Ge(110) and Ge/CuBr(110) interfaces as a function of anneal temperature. Data were taken from ref. [43].

immediately upon Ge deposition and continues at each anneal step. The peak area ratio changes are consistent with the reaction

$$Ge + 2 CuBr \rightarrow 2 Cu + GeBr_2 \uparrow,$$

in which there is a loss of Br and Ge from the sample surface and an accumulation of metallic Cu.

The interface between CuBr and Ge is therefore an extreme example of an interface chemistry growth-sequence dependence with the CuBr/Ge(110) interface being essentially inert and the Ge/CuBr(110) interface reacting to form many monolayers of chemically distinct products. The chemical reactivity for the Ge/CuBr(110) interface is so severe that it is not possible to prepare an epitaxial Ge/CuBr(110) heterojunction. This asymmetry in chemical reactivity is not explained by bulk thermodynamic considerations and may result from the nonequilibrium conditions under which these interfaces were formed. As chemical reactivity may be a driving force to cause atom exchange across an interface, a growth sequence dependent chemical reactivity may be a contributing factor to observed growth sequence variations in ΔE_v.

3.2.2. Compound–elemental (110) semiconductor heterojunctions

The growth of a compound semiconductor on an elemental semiconductor involves a site allocation ambiguity which is not present for the reverse growth sequence. The importance of solving this site allocation problem for producing device quality heterojunctions from compound on elemental semiconductor growth has been discussed [6,8]. It has been suggested that antiphase domain disorder may cause the formation of an interface dipole at a heterojunction formed by growing a compound on elemental semiconductor [44]. In this section some systematics of ΔE_v growth sequence variations for compound semiconductor–Ge(110) interfaces are reviewed, observations of a time dependent ΔE_v variation for the GaAs/Ge(110) heterojunction interface are discussed, and a mechanism by which antiphase domain disorder may produce an interface contribution to ΔE_v is considered.

3.2.2.1. Compound semiconductor–Ge(110) ΔE_v systematics

The observation of a growth sequence variation for a compound semiconductor–Ge(110) interface has been reported for the ZnSe–Ge semiconductor pair [30]. The valence-band discontinuity for Ge formed epitaxially on ZnSe (Ge/ZnSe) was 1.53 ± 0.04 eV, while ΔE_v for ZnSe formed epitaxially on Ge (ZnSe/Ge) was 1.30 ± 0.04 eV [45]. A growth-sequence variation has also been reported [39] for the GaAs–Ge(110) interface, where it is found that $\Delta E_v(\text{Ge/GaAs}) > \Delta E_v(\text{GaAs/Ge})$ by about 0.2 eV; independent measurements have confirmed this result [40].

A third compound semiconductor–Ge(110) interface for which a growth sequence can be inferred involves CuBr and Ge. It is observed [42,43] that the difference between $\Delta E_v[\text{Ge/GaAs}(110)] + \Delta E_v[\text{CuBr/GaAs}(110)]$ and $\Delta E_v[\text{CuBr/Ge}(110)]$ is $+0.70 \pm 0.05$ eV. Antiphase disorder at the CuBr/Ge(110) interface has been suggested [44] as a possible cause of this large nonzero result (a further discussion of the ΔE_v relationship between these semiconductors is given in sect. 3.3). Unfortunately, as discussed in sect. 3.2.1., the preparation of a heterojunction with the reverse growth sequence [i.e., Ge/CuBr(110)] is not possible due to interface chemical reactions that occur when Ge is deposited on CuBr(110). However, if the large positive deviation from zero for the difference $\{\Delta E_v[\text{Ge/GaAs}(110)] + \Delta E_v[\text{CuBr/GaAs}(110)]\} - \Delta E_v[\text{CuBr/Ge}(110)]$ is assumed to be associated primarily with the CuBr/Ge(110) interface, it follows that $\Delta E_v[\text{CuBr/Ge}(110)]$ would be less than $\Delta E_v[\text{Ge/CuBr}(110)]$.

The ΔE_v for three compound semiconductors (i.e., GaAs, ZnSe and CuBr) grown on Ge(110) is systematically found to be smaller than ΔE_v for heterojunctions formed by the reverse growth sequence. These heterojunctions involve band alignments in which the smaller Ge band gap is completely contained within the larger band gap of the compound semiconductor. Thus the systematically smaller ΔE_v for compound semiconductor/Ge(110) interfaces as compared to the reverse growth sequence is consistent with the formation of an interface dipole at the compound semiconductor/Ge(110) interface that has positive charge on the Ge side of the interface and negative charge in the compound semiconductor overlayer. The systematic variation of the compound semiconductor/Ge(110) growth-sequence effect suggests that a similar mechanism may be involved in each case.

3.2.2.2. *Instability of the GaAs/Ge(110) interface*

The interface dipole present at the GaAs/Ge(110) interface is found to vary with both time after interface formation and subsequent annealing conditions [40,46,47]. Thus, although as noted in sect. 3.2.2.1. the size of the interface dipole at the GaAs/Ge(110) interface causes $\Delta E_v[\text{Ge}/\text{GaAs}(110)]$ to be larger than $\Delta E_v[\text{GaAs}/\text{Ge}(110)]$, the magnitude of the difference depends on several factors (which will be discussed in this section) and indicates that the GaAs/Ge(110) interface is unstable.

A typical XPS spectrum for a GaAs/Ge(110) heterojunction is shown in fig. 14 where the binding energy region contains the Ga3d, Ge3d and As3d core levels. This heterojunction sample (C) and two others (D and E) were

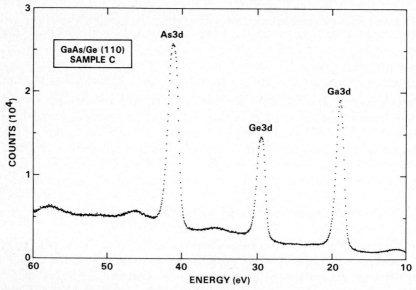

Fig. 14. Typical XPS spectrum of a GaAs/Ge(110) sample in the Ga3d, Ge3d, and As3d binding energy region. Data were taken from ref. [46].

prepared by sputtering and annealing a Ge(110) substrate, growing a few hundred Å thick epitaxial Ge layer (at ~ 450°C), and growing a thin (~ 22 Å) epitaxial GaAs layer by MBE at ~ 350°C. An additional sample (F) was prepared by growing the GaAs epitaxial layer directly on the sputtered and annealed Ge(110) substrate. Variations in the interface dipole present at the GaAs/Ge(110) interface were monitored by measuring $E_{\text{Ge3d}}^{\text{Ge}} - E_{\text{Ga3d}}^{\text{GaAs}}$.

Room-temperature spectra similar to that shown in fig. 14 were recorded at several times t following interface formation. Because data collection required several hours, t was taken as the midpoint of the data accumula-

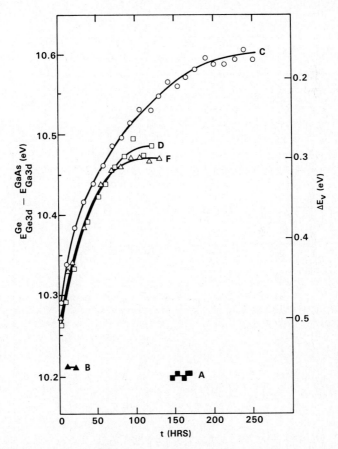

fig. 15. Variation of $E_{\text{Ge3d}}^{\text{Ge}} - E_{\text{Ga3d}}^{\text{GaAs}}$ (and of ΔE_v) with time after interface formation for three GaAs/Ge(110) heterojunctions (labelled C, D and F) and for two Ge/GaAs(110) heterojunctions (labeled A and B). Data were taken from refs. [40,46].

tion period. The core-level binding energy difference, $E_{Ge3d}^{Ge} - E_{Ga3d}^{GaAs}$ (and the corresponding value of ΔE_v) measured as a function of t for samples C, D and F is plotted in fig. 15. For comparison, data from table 1 for two Ge/GaAs(110) heterojunctions (labeled A and B) prepared as discussed in sect. 3.1.1 are also shown. As can be seen from this figure, there is a substantial variation of ΔE_v with time for GaAs/Ge(110) heterojunctions. This variation indicates interface dipole formation caused by GaAs/Ge(110) interface structural instability. No similar instability for heterojunctions formed by the reverse growth sequence [i.e., Ge/GaAs(110)] is observed.

The Ga3d, Ge3d and As3d linewidths, the Ga3d/As3d and Ge3d/Ga3d

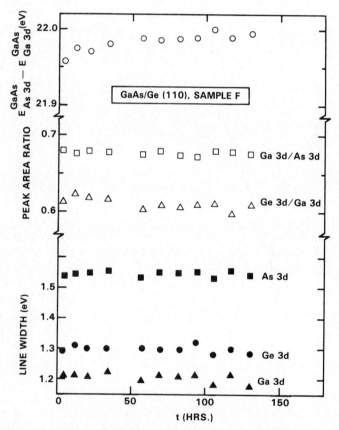

Fig. 16. XPS data for sample F obtained as a function of time after GaAs/Ge(110) interface formation. Top: $E_{As3d}^{GaAs} - E_{Ga3d}^{GaAs}$; middle: Ga3d/As3d and Ge3d/Ga3d peak area ratios; bottom: As3d, Ge3d, and Ga3d linewidths. Data were taken from ref. [40].

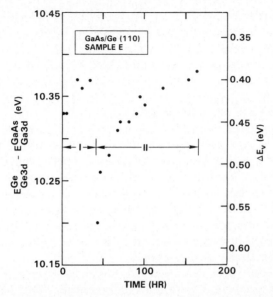

Fig. 17. Variation of $E^{Ge}_{Ge3d} - E^{GaAs}_{Ga3d}$ (and of ΔE_v) with time for a GaAs/Ge(110) heterojunction (sample E). Interval I follows the normal heterojunction growth procedure. Interval II follows a 10 min vacuum anneal at ~ 325°C. Data were taken from ref. [47].

peak area ratios and the As3d to Ga3d core-level binding energy difference, $E^{GaAs}_{As3d} - E^{GaAs}_{Ga3d}$, were also monitored as a function of t. As an example, these parameters derived from XPS data on sample F are shown in fig. 16. Except for a very small increase (~ 0.03 eV) in $E^{GaAs}_{As3d} - E^{GaAs}_{Ga3d}$ (which may be associated with a change in GaAs surface chemical shifts) there is no systematic variation in any of the parameters. The data in fig. 16 rule out the possibility that substantial interface chemical reactions are occurring to cause the large (> 0.2 eV) variations in $E^{Ge}_{Ge3d} - E^{GaAs}_{Ga3d}$ shown in fig. 15.

An additional observation of GaAs/Ge(110) interface instability is derived from XPS data [47] on sample E, as shown fig. 17. At ~ 42 h after interface formation, this sample was annealed at ~ 325°C in vacuum for 10 min. The value of $E^{Ge}_{Ge3d} - E^{GaAs}_{Ga3d}$ decreased to 10.20 eV, which is the same value as observed for the stable Ge/GaAs(110) interface (see table 1 and the data for samples A and B shown in fig. 15). This decrease indicates that the GaAs/Ge(110) interface dipole has been removed by the annealing treatment. Following this anneal the value of $E^{Ge}_{Ge3d} - E^{GaAs}_{Ga3d}$ again slowly increased as the interface dipole was reformed. Several annealing studies (both in vacuum and with an As overpressure) have been performed on

GaAs/Ge(110) and GaAs/Ge(100) samples [47]. Although there are some sample-to-sample variations (as also indicated by the data in fig. 15), results similar to those shown in fig. 17 are always obtained. A possible mechanism by which antiphase domain disorder may be related to the observed GaAs/Ge(110) interface instability is considered in the next section.

3.2.2.3. Possible role of antiphase disorder

The valence-band discontinuity for a heterojunction formed by growth of a lattice-matched compound semiconductor on Ge(110) is systematically smaller than for the reverse growth sequence. There is evidence (see sect. 3.3.) that interface contributions to ΔE_v may be small for Ge grown on compound semiconductor (110) heterojunctions. The relative magnitude of the compound semiconductor–Ge(110) growth-sequence effect can therefore be used to infer that at a compound semiconductor/Ge(110) interface, positive charge is transferred into the Ge and negative charge into the compound semiconductor overlayer.

The possible presence of antiphase disorder is a distinct growth-sequence difference for compound–elemental semiconductor (110) heterojunctions. The presence of this disorder would not necessarily affect ΔE_v unless it caused charge to be transferred across the interface to create an interface dipole. As discussed previously, atom transfer across a heterojunction interface that involves atoms from different columns of the periodic table can produce interface dipoles. If the compound semiconductor–Ge(110) ΔE_v growth-sequence effect is caused by atom transfer, as suggested by the long time constants associated with GaAs/Ge(110) interface instability, the electrostatic model of heterojunction interfaces (sect. 3.1.2) can be used to infer that anions (i.e., As, Se and Br) rather than cations preferentially exchange with Ge at a compound semiconductor/Ge(110) interface.

A large variety of antiphase domain walls can be imagined. However, the crystallographic planes associated with these walls may be characterized by whether one, two or three bonds between like atoms exist at the domain boundary. If attention is restricted to only those cases where the common atom bonds involve atoms in the compound semiconductor plane that is adjacent to the last Ge(110) plane, only three nearest-neighbor bonding arrangements are possible for the abrupt (110) interface as shown in fig. 18.

The energetics of atomic exchange across an interface are primarily associated with bonds formed or broken during the exchange. Energies of atomic substitution calculated by tight-binding theory [38] have shown that atomic exchange at an ideal abrupt compound semiconductor–Ge interface is not favored. However, at an antiphase domain boundary like those

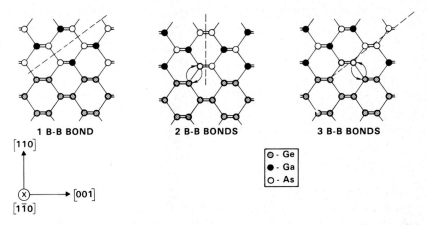

Fig. 18. Examples of arsenic antiphase domain boundaries (------) with one, two or three As–As bonds. The arrows suggest an atomic exchange across the GaAs/Ge(110) interface that may be related to the observed interface instability. Figure was taken from ref. [40].

shown in fig. 18, there are bonds between like atoms that are not present at the ideal interface. The atoms involved in these bonds may favor exchange. This possibility was investigated [40] by considering the sum of bond formation energies associated with a single atomic exchange as, for example, is shown by arrows in the center and at the right of fig. 18. The result was that a bonding arrangement that involved only one bond between like atoms (shown at the left of fig. 18) is stable while interfaces that involve two (middle of fig. 18) or three (right-hand side of fig. 18) bonds between like atoms favor the exchange indicated by the arrows. The present accuracy of the theory does not specify a preference for anion or cation exchange.

Some insight into the cause of the observed ΔE_v growth-sequence effect at compound–elemental semiconductor heterojunctions can be obtained although a detailed mechanism is not yet known. The slow time constant at room temperature for formation of the GaAs/Ge(110) interface dipole suggests that atomic motion is involved. The sign of the growth-sequence effect infers that anions are preferentially interchanged with Ge at a compound semiconductor/Ge(110) interface. Bond formation energy considerations favor atom exchange where certain antiphase domain boundaries intersect the Ge(110) interface. If the atom exchange mechanism has a constant probability per unit time and there are a fixed number of available sites at which the exchange can occur, an exponential variation of ΔE_v with time would be expected, as can be seen in fig. 15. For the GaAs/Ge(110) interface, the XPS linewidth data indicate that the dipole is localized over a very few atomic planes at the interface. The reversible variation of the GaAs/Ge(110) interface dipole with suitable sample annealing conditions (fig. 17) suggests that a mobile atom species (perhaps interstitial As) is involved in the dipole formation mechanism. Finally, variations in antiphase domain structure between samples could account for the observed sample-to-sample differences in the growth sequence effect.

3.2.3. Isocolumnar (110) heterojunctions

As noted previously an interchange of atoms from different columns of the periodic table can produce interface dipole contributions to ΔE_v. As atom interchange may be impossible to prevent, it might be expected that heterojunctions formed on nonpolar surfaces between semiconductors that contain elements from the same columns of the periodic table (isocolumnar) would be least likely to exhibit a growth-sequence effect. However, as noted in this section, a growth-sequence ΔE_v variation has been observed [33] even for isocolumnar (110) heterojunctions formed between AlAs and GaAs.

Heterojunction samples of AlAs–GaAs with a structure similar to that shown in fig. 1 were prepared by MBE. The substrates were (110) GaAs with a ~ 50 Å GaAs epitaxial buffer layer. The AlAs/GaAs heterojunctions were formed by growing ~ 30 Å of epitaxial AlAs onto the GaAs buffer layer. The GaAs/AlAs heterojunctions were formed by growing ~ 100 Å of AlAs followed by ~ 25 Å of GaAs onto the GaAs buffer layer. All of the epitaxial layers were grown at ~ 550°C.

The Al2p to Ga3d core-level binding energy difference was measured for four GaAs/AlAs(110) and three AlAs/GaAs(110) samples. The results are shown in table 4. As discussed in sect. 2.1, a change in the core-level binding energy difference, $\delta(\Delta E_{CL})$, is a direct measure of a change in the valence-band discontinuity, $\delta(\Delta E_v)$; in this case $\delta(\Delta E_{CL}) = \delta(\Delta E_v)$. It is therefore observed that $\Delta E_v[\text{GaAs/AlAs}(110)]$ is smaller than $\Delta E_v[\text{AlAs/GaAs}(110)]$ by ~ 0.13 eV; a 0.1 eV range of ΔE_v values is noted in the GaAs/AlAs(110) ΔE_v results. A similar ΔE_v growth-sequence effect with the same sign and approximately the same magnitude has been observed [33] for AlAs–GaAs(100) heterojunctions prepared by the same procedure described above. Corresponding values of ΔE_v are ~ 0.1 eV smaller for the (100) than for the (110) heterojunction interfaces.

The ΔE_v growth-sequence results for the AlAs–GaAs heterojunction

Table 4
Al2p to Ga3d core-level binding energy differences for several GaAs/AlAs(110) and AlAs/GaAs(110) heterojunctions.

Interface	Sample	ΔE_{CL} [a] (eV)	$(\Delta E_{CL})_{AVG}$ (eV)
GaAs/AlAs	1	54.31	
	2	54.41	54.37
	3	54.39	
	4	54.38	
AlAs/GaAs	5	54.50	
	6	54.51	54.50
	7	54.50	

[a] Uncertainty is ± 0.02 eV.

system indicate that the atomic arrangement in the interface region depends on growth sequence at least for the conditions used to prepare these samples. The growth of $Ga_{1-x}Al_xAs$ on GaAs(100) exhibits a smooth surface morphology, while growth on GaAs(110) can be complicated by relatively rough surface morphology and alloy clustering [48]. The exchange reaction between Al and Ga at the GaAs(110) surface [49] may be a cause of the alloy clustering phenomenon. For a given crystallographic orientation, an exchange reaction between Ga and Al might be expected to depend on growth sequence. As these elements are from the same column of the periodic table and have very similar electronegativities, an interface dipole formed directly by this exchange reaction would be expected to be extremely small. However, if antisite and/or antistructure defects were formed as a result of the exchange reaction, it is possible that a sizeable growth-sequence dependent interface dipole could result. Although further study of microscopic interface structure will be needed to understand the origin of interface contributions to ΔE_v for AlAs–GaAs heterojunctions, it is plausible that the effects mentioned here will be contributing factors.

3.3. Transitivity of ΔE_v

As mentioned previously, the several theories [1–4] which are currently in use to predict ΔE_v at abrupt interfaces are all based on linear models that estimate band discontinuities from the difference in an absolute energy associated with each of the semiconductors that form the heterojunction. These theories, therefore, specifically ignore the interface contributions to

ΔE_v that have been discussed in this chapter. If one considers ΔE_v for heterojunctions formed between three semiconductors (A, B and C), a fundamental property of all linear heterojunction models is that if $\Delta E_v(A-B)$ and $\Delta E_v(B-C)$ are known, $\Delta E_v(A-C)$ is also specified; this transitive property of ΔE_v for linear models has been previously noted [2]. It is, therefore, possible to test experimentally an underlying assumption of all linear models by measuring ΔE_v for three appropriate heterojunctions. Conversely, although not on a rigorous logical basis, it may be possible to use ΔE_v transitivity tests to select experimental ΔE_v values least affected by interface dipole contributions and, therefore, most suitable for comparison with the prediction of linear models.

If ΔE_v for a set of three heterojunctions is transitive, the following relation is, by definition, satisfied

$$\Delta E_v(A-B) + \Delta E_v(B-C) - \Delta E_v(A-C) = 0, \tag{4}$$

where in order to specify a unique sign convention, the semiconductor band gaps have been chosen such that $E_g^A > E_g^B > E_g^C$. Equation (1) expressed $\Delta E_v(A-B)$ as $(E_{CL}^B - E_v^B) - (E_{CL}^A - E_v^A) + \Delta E_{CL}(A-B)$. Substitution of this expression and similar expressions for $\Delta E_v(B-C)$ and $\Delta E_v(A-C)$ into eq. (4) results in a cancellation of all the $E_{CL} - E_v$ terms; in order for ΔE_v to be transitive, the quantity $T(\Delta E_{CL})$ must be zero where

$$T(\Delta E_{CL}) = \Delta E_{CL}(A-B) + \Delta E_{CL}(B-C) - \Delta E_{CL}(A-C). \tag{5}$$

Thus, the relation in eq. (5) can be used to provide a sensitive test of transitivity.

Values of ΔE_{CL} measured by XPS for abrupt (110) heterojunction interfaces are given in table 5. The core levels are identified and to specify a unique sign convention, the binding energy associated with the core level in the semiconductor with the largest band gap was subtracted from the binding energy of the core level in the semiconductor with the smallest bandgap. Also listed in table 5 are corresponding values of ΔE_v for each interface. These ΔE_v values were evaluated from eq. (1) by using the appropriate $E_{CL} - E_v$ values given in table 2. Error limits are not specified for ΔE_v associated with heterojunctions that involve CuBr because the value of $E_{Br3d}^{CuBr} - E_v^{CuBr}$ has only been approximately determined.

To test for ΔE_v transitivity, experimental ΔE_{CL} data for appropriate heterojunction sets formed between at least three semiconductors are needed. By using the data in table 5, five tests of ΔE_v transitivity can be obtained.

Table 5
Core-level binding energy differences and associated ΔE_v values at (110) heterojunction interfaces as measured by XPS.

Interface	Core levels	ΔE_{CL} (eV)	ΔE_v (eV)	Ref.
Ge/GaAs	Ga3d–Ge3d	−10.21 ± 0.01	0.56 ± 0.04	[10,42]
	As3d–Ge3d	11.78 ± 0.02		
ZnSe/GaAs	Zn3d–Ga3d	−8.95 ± 0.02	0.94 ± 0.04	[30]
CuBr/GaAs	Br3d–As3d	28.77 ± 0.03	0.8	[42,43]
InAs/GaAs	Ga3d–In4d	1.55 ± 0.06	0.18 ± 0.07	[31]
AlAs/GaAs	Al2p–Ga3d	54.50 ± 0.02	0.55 ± 0.05	[33]
GaAs/AlAs	Al2p–Ga3d	54.37 ± 0.06 [a]	0.42 ± 0.08 [a]	[33]
Ge/AlAs	Al2p–Ge3d	44.22 ± 0.02	1.04 ± 0.05	[43]
Ge/ZnSe	Zn3d–Ge3d	−19.13 ± 0.02	1.53 ± 0.04	[30]
ZnSe/Ge	Zn3d–Ge3d	−19.36 ± 0.02	1.30 ± 0.04	[30]
CuBr/Ge	Br3d–Ge3d	39.85 ± 0.03	0.65	[42,43]

[a] Because of the range of values given in table 4, a maximum variation is quoted rather than a measurement uncertainty.

The five possible combinations of three heterojunction interfaces are listed in table 6 along with the experimental value of $T(\Delta E_{CL})$. As can be seen from table 6, the first two heterojunction sets provide a positive test of transitivity; sets 3 and 4 are markedly nontransitive while set 5 is less so.

Some systematic trends are apparent from the data in table 6. The first two heterojunction sets where ΔE_v transitivity is satisfied involve only growth of interfaces where site allocation is not an issue. In contrast, the nontransitive ΔE_v observation for sets 3 and 4 involve interfaces formed by growing a compound on an elemental semiconductor where site allocation ambiguity may be present. These observations are consistent with the suggested role of antiphase disorder in contributing to interface dipole formation.

Table 6
Transitivity tests for five (110) heterojunction sets.

Set	Interfaces			$T(\Delta E_{CL})$ (eV)
	A	B	C	
1	ZnSe/GaAs	Ge/GaAs	Ge/ZnSe	−0.03 ± 0.03
2	GaAs/AlAs	Ge/GaAs	Ge/AlAs	−0.06 ± 0.06
3	CuBr/GaAs	Ge/GaAs	CuBr/Ge	+0.70 ± 0.05
4	ZnSe/GaAs	Ge/GaAs	ZnSe/Ge	+0.20 ± 0.05
5	AlAs/GaAs	Ge/GaAs	Ge/AlAs	+0.07 ± 0.03

The only difference between heterojunction sets 2 and 5 involves the AlAs–GaAs heterojunction growth sequence. As it was noted in sect. 3.2.3 that this growth sequence leads to an ~ 0.13 eV difference in ΔE_v, it is not surprising that if set 2 is transitive, set 5 exhibits a small deviation from transitivity. Some possible causes for the growth-sequence effect on ΔE_v(AlAs–GaAs) were previously given.

The last point considered here is the possible use of transitivity tests to select data least affected by interface dipole contributions to ΔE_v for comparison with theories that ignore interface effects. The obvious flaw in this logic is that interface contributions to ΔE_v for different members of a given heterojunction set could have opposite signs thus producing a cancellation and a $T(\Delta E_{CL})$ of ~ 0. However, in lieu of a better selection criterion, experimental ΔE_v values for the five interfaces involved in the first two heterojunction sets of table 6 would appear to be good candidates for comparison with predictions of linear models.

4. Summary

Interface contributions to heterojunction band discontinuities can be substantially larger than the thermal energy and can therefore influence semiconductor devices that incorporate heterojunction interfaces. To investigate these interface contributions to ΔE_v it is essential to utilize a measurement technique that has an experimental uncertainty considerably smaller than the observed effects. A technique based on X-ray photoemission spectroscopy has been described that is capable of measuring changes in interface contributions to ΔE_v with an uncertainty of ± 0.01 eV. In favorable cases, the technique can be extended to yield absolute ΔE_v values with an uncertainty of ± 0.04 eV.

Several experimental observations of interface contributions to ΔE_v have been described that include crystallographic orientation and growth-sequence effects. A consideration of electrostatics leads to the conclusion that when interfaces are formed on polar crystallographic planes between semiconductors that contain elements from different columns of the periodic table, interface dipole contributions to ΔE_v may be expected. Several observations of heterojunction growth-sequence effects were described and possible mechanisms that can contribute to charge transfer and interface dipole formation were reviewed. It is possible to test ΔE_v transitivity from data derived solely by the XPS technique. These tests suggest a means to select ΔE_v data least likely affected by interface dipoles for comparison with existing ΔE_v theories based on linear models and to identify interfaces with large interface dipole contributions to ΔE_v for systematic study. A better understanding of interface contributions to heterojunction band discontinuities is clearly needed if semiconductor heterojunctions are to be used most advantageously in device applications.

Acknowledgments

The authors acknowledge the collaboration of Prof. W.A. Harrison (Stanford University) on many of the subjects reviewed in this chapter and several useful discussions with Prof. H. Kroemer (University of California, Santa Barbara). Much of the original work which this chapter reviewed was supported by the Office of Naval Research.

References

[1] W.A. Harrison, Electronic Structure and the Properties of Solids (Freeman, San Francisco, 1980) p. 252.
[2] W.R. Frensley and H. Kroemer, Phys. Rev. B 16 (1977) 2642.
[3] R.L. Anderson, Solid State Electron. 5 (1962) 341.
[4] J. Tersoff, Phys. Rev. B 30 (1984) 4874.
[5] G. Margaritondo, Surf. Sci. 132 (1983) 469.
[6] H. Kroemer, Surf. Sci. 132 (1983) 543.
[7] H. Kroemer, J. Vac. Sci. & Technol. B 2 (1984) 433.
[8] H. Kroemer, in: Proc. NATO Advanced Study Institute on Molecular Beam Epitaxy and Heterostructures, Erice, Sicily, 1983, eds L.L. Chang and K. Ploog (Martinus Nijhoff, The Hague, 1984) p. 331.
[9] G. Margaritondo, Surf. Sci. 168 (1986) 439.
[10] R.W. Grant, J.R. Waldrop and E.A. Kraut, Phys. Rev. Lett. 40 (1978) 656.
[11] R.W. Grant, J.R. Waldrop and E.A. Kraut, J. Vac. Sci. & Technol. 15 (1978) 1451.
[12] E.A. Kraut, R.W. Grant, J.R. Waldrop and S.P. Kowalczyk, Phys. Rev. Lett. 44 (1980) 1620.
[13] E.A. Kraut, R.W. Grant, J.R. Waldrop and S.P. Kowalczyk, Phys. Rev. B 28 (1983) 1965.
[14] K. Siegbahn, C. Nordling, A. Fahlman, R. Nordberg, K. Hamrin, J. Hedman, G. Johansson, T. Bergmark, S.-E. Karlsson, I. Lindgren and B. Lindberg, Nova Acta Regiae Soc. Sci. Ups. 20 (1967).
[15] J. Auleytner and O. Hörnfeldt, Ark. Fys. 23 (1963) 165.
[16] J. Hedman, Y. Baer, A. Berndtsson, M. Klasson, G. Leonhardt, R. Nilsson and C. Nordling, J. Electron Spectrosc. & Relat. Phenom. 1 (1972/3) 101.
[17] M.P. Seah and W.P. Dench, Surf. & Interface Anal. 1 (1979) 2.
[18] L.J. Brillson, Surf. Sci. Rep. 2 (1982) 123.
[19] G. Margaritondo, A.D. Katnani, N.G. Stoffel, R.R. Daniels and Te-Xiu Zhao, Solid State Commun. 43 (1982) 163.

[20] G.A. Baraff, J.A. Applebaum and D.R. Hamann, J. Vac. Sci. & Technol. 14 (1977) 999.
[21] W.E. Pickett and M.L. Cohen, Phys. Rev. B 18 (1978) 939.
[22] R.W. Grant, E.A. Kraut, S.P. Kowalczyk and J.R. Waldrop, J. Vac. Sci. & Technol. B 1 (1983) 320.
[23] J. Olivier and R. Poirier, Surf. Sci. 105 (1981) 347.
[24] A.D. Katnani and G. Margaritondo, Phys. Rev. B 28 (1983) 1944.
[25] J.R. Chelikowsky and M.L. Cohen, Phys. Rev. B 14 (1976) 556.
[26] See, e.g., V.G. Aleshin and Yu.N. Kucherenko, J. Electron Spectrosc. 9 (1976) 1.
[27] U. Gelius, in: Electron Spectroscopy, ed. D.A. Shirley (North-Holland, Amsterdam, 1972) p. 311.
[28] B.Z. Olshanetsky, S.M. Repinsky and A.A. Shklyaev, Surf. Sci. 64 (1977) 224.
[29] D.E. Eastman, T.-C. Chiang, P. Heimann and F.J. Himpsel, Phys. Rev. Lett. 45 (1980) 656.
[30] S.P. Kowalczyk, E.A. Kraut, J.R. Waldrop and R.W. Grant, J. Vac. Sci. & Technol. 21 (1982) 482.
[31] S.P. Kowalczyk, W.J. Schaffer, E.A. Kraut and R.W. Grant, J. Vac. Sci. & Technol. 20 (1982) 705.
[32] S.P. Kowalczyk, J.T. Cheung, E.A. Kraut and R.W. Grant, Phys. Rev. Lett. 56 (1986) 1605.
[33] J.R. Waldrop, R.W. Grant and E.A. Kraut, J. Vac. Sci. & Technol., (1987) in press.
[34] Some of these $E_{CL} - E_v$ values have been adjusted by either -0.04 eV or -0.05 eV to correct for a computer program error that was present in the original analyses. In this chapter those ΔE_v values affected by this adjustment have been corrected.
[35] F.F. Fang and W.E. Howard, J. Appl. Phys. 35 (1964) 612.
[36] J.R. Waldrop, E.A. Kraut, S.P. Kowalczyk and R.W. Grant, Surf. Sci. 132 (1983) 513.
[37] W.A. Harrison, E.A. Kraut, J.R. Waldrop and R.W. Grant, Phys. Rev. B 18 (1978) 4402.
[38] E.A. Kraut and W.A. Harrison, J. Vac. Sci. & Technol. B 3 (1985) 1267.
[39] P. Zurcher and R.S. Bauer, J. Vac. Sci. & Technol. A 1 (1983) 695.
[40] R.W. Grant, J.R. Waldrop, S.P. Kowalczyk and E.A. Kraut, J. Vac. Sci. & Technol. B 3 (1985) 1295.
[41] D.-W. Tu and A. Kahn, J. Vac. Sci. & Technol. A 3 (1985) 922.
[42] J.R. Waldrop and R.W. Grant, Phys. Rev. Lett. 43 (1979) 1686.
[43] J.R. Waldrop, R.W. Grant, S.P. Kowalczyk and E.A. Kraut, J. Vac. Sci. & Technol. A 3 (1985) 835.
[44] J.C. Phillips, J. Vac. Sci. & Technol. 19 (1981) 545.
[45] The uncertainty analysis of these results was refined after initial publication, see ref. [43].
[46] R.W. Grant, J.R. Waldrop, S.P. Kowalczyk and E.A. Kraut, Surf. Sci. 168 (1986) 498.
[47] J.R. Waldrop, R.W. Grant and E.A. Kraut, J. Vac. Sci. & Technol. B 4 (1986) 1060.
[48] P.M. Petroff, A.Y. Cho, F.K. Reinhart, A.C. Gossard and W. Wiegmann, Phys. Rev. Lett. 48 (1982) 170.
[49] C.B. Duke, A. Paton, R.J. Meyer, L.J. Brillson, A. Kahn, D. Kanani, J. Carelli, J.L. Yeh, G. Margaritondo and A.D. Katnani, Phys. Rev. Lett. 46 (1981) 440.

CHAPTER 5

MEASUREMENTS OF BAND DISCONTINUITIES USING OPTICAL TECHNIQUES

Geoffrey DUGGAN

Philips Research Laboratories
Redhill, Surrey, UK

Heterojunction Band Discontinuities: Physics and Device Applications
Edited by F. Capasso and G. Margaritondo
© *Elsevier Science Publishers B.V., 1987*

Contents

1. Introduction 209
2. Theoretical background 210
3. Optical absorption and photoluminescence excitation spectroscopy 217
    3.1. The GaAs–(AlGa)As system 217
    3.2. Other heterojunction systems 233
4. Photoluminescence 237
5. Light scattering 246
6. Internal photoemission 254
7. Concluding remarks 257
References 259

1. Introduction

This chapter is an attempt to evaluate the reliability of a variety of optical techniques in measuring the conduction and valence band energy steps which characterise any semiconductor heterojunction. The magnitude of these "band offsets" is of central concern to the applied physicist wishing to utilise heterojunction technology in any new device concept, fully comprehend the performance of existing heterojunction devices or possibly extend the operation of devices to the only interesting operating temperature, i.e. room temperature. In addition, the magnitude of these offsets is fundamentally important to the solid-state theorist wishing to develop a satisfactory microscopic model of heterojunction formation. Reliable and accurate measurement of the offsets are needed of those semiconductor heterojunction pairs that have been afforded care and technological investment in their fabrication so that the theorists have a suitable test-bed for their models and that a discriminating audience can have some faith in the ability of these microscopic models to predict the offsets which are expected at other, perhaps more, speculative heterojunction combinations.

The heterojunction which has seen the largest investment in its growth, characterization and technological exploitation is that formed between the closely lattice matched III–V compound GaAs and the random, ternary alloy (AlGa)As. This system, therefore, forms the natural test-bed for the various theoretical models of heterojunction formation and provides the most suitable vehicle for illustrating the variety, degree of sophistication and accuracy of the optical techniques employed to discover the conduction of valence band offset existing at this heterojunction. Whilst this system provides the most convenient vehicle for illustrating the limitations of the optical techniques in band offset determination I will not limit the results and critique to this system alone.

Almost without exception the optical techniques described here, whether they be absorption spectroscopy, photoluminescence or light scattering, are all indirect methods of determining the heterojunction band offsets. They all rely on a preconceived model of what the heterojunction looks like which generally means the heterojunction is abrupt, the band offsets are commutative and that the interface is formed free from technological or extrinsic factors.

The semiconductor heterojunction is the fundamental building block of the potentially technologically important and fascinating structure of the semiconductor quantum well. Almost all of the optical techniques described and analysed in this chapter utilise the unique electronic properties of the quantum well heterostructure and the interpretation of the results of the optical experiments relies to a greater or lesser extent on the understanding of the electronic states in terms of some theoretical model of those states. It seems appropriate therefore to organise this chapter in such a way that any discussion of the optical techniques be prefaced by a brief discussion of the calculation of electronic states in quantum well heterostructures. Pertinent to the present review will be the energetic dependence of those states upon measurable quantities such as quantum well width, residual or intentional dopant levels and their sensitivity to the quantities of interest, i.e., the fraction of energy gap discontinuity that occurs either in the valence or in the conduction band and the functional dependence of those discontinuities on, for instance, molar percentage of the cation or anion elements in the binary or ternary confining, barrier layers.

2. Theoretical background

A semiconductor quantum well is fabricated by sandwiching a layer of a smaller band gap material, e.g. GaAs, between two layers of a larger band gap material, e.g. (AlGa)As. A schematic diagram of the band alignment at such a heterojunction is shown in fig. 1. If the thickness of the GaAs is reduced so that the layer thickness L_z is smaller than the de Broglie wavelength of the particles, i.e. electron or light and heavy holes, then quantum size effects manifest themselves in the direction perpendicular to the layer planes. A series of confined particle states appears in both the conduction and valence bands of the lower band gap material; the number and energetic position of these states being dependent on the size of the conduction and valence band offsets, ΔE_c and ΔE_v, and on the well width L_z. The calculation of the energy levels of a particle confined to a one-dimensional potential well of infinite or finite depth is one of the more elementary quantum mechanical problems [1] which points to the appeal of experiments that rely on the use of such a simple model. In the case that the

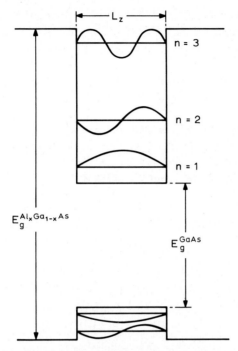

Fig. 1. Schematic of the eigenvalues, eigenfunctions and conduction- and valence-band edges at a (AlGa)As–GaAs quantum well heterostructure. (From ref. [3].)

confining potential is infinite then the one-dimensional Schroedinger equation reads

$$-\frac{\hbar^2}{2m}\frac{\partial^2 \psi}{\partial z^2} = E\psi, \tag{1}$$

where m is the mass of the particle in the direction of confinement. The solutions E_n are given by

$$E_n = \frac{\hbar^2}{2m}\left(\frac{n\pi}{L_z}\right)^2 \quad (n = 1, 2, \ldots). \tag{2}$$

The confined particle wavefunctions are alternately sine and cosine functions whose amplitude is zero at the well boundary. In the plane of the quantum wells, the x–y plane for defitiveness, then the particle is still free to move and the wavevectors in the plane remain good quantum numbers. In the simplest approximation that these are still free-particle-like states, then the total energy of a particle in this three-dimensional system becomes

$$E = E_n + \frac{\hbar^2}{2m}(k_x^2 + k_y^2),$$

where k_x and k_y are the wavevectors in the plane. This x–y plane is often referred to as the $k_\parallel$ direction. Equation (2) illustrates the strong quadratic dependence of the confined states on the well width L_z. For wells of finite depth, V, then the Schroedinger equation is

$$-\frac{\hbar^2}{2m}\frac{\partial^2 \psi_w}{\partial z^2} - E\psi_w = 0 \tag{3}$$

in the wells and

$$-\frac{\hbar^2}{2m}\frac{\partial^2 \psi_b}{\partial z^2} + (V - E)\psi_b = 0 \tag{4}$$

in the barriers. Assuming that the particle has the same mass in the wells as in the barriers then matching the wave functions and their first derivatives with respect to z at both boundaries results in the transcendental equation

$$k_w \tan(\tfrac{1}{2} k_w L_z) = k_b, \tag{5}$$

where k_w and k_b are the wavevectors in the wells and barriers, respectively, and are given by

$$k_w = \frac{(2mE)^{1/2}}{\hbar}; \quad k_b = \frac{[2m(V-E)]^{1/2}}{\hbar}. \tag{6}$$

The dependence of the confined states on well depth V enters only through the dependence of the k-vector in the well on the square root of the barrier height. The uppermost levels in a rectangular, wide quantum well will be the ones which have the greatest sensitivity to the barrier height. Alternatively, if the wells are narrow then the few confined particle states are again pushed nearer to the top of the quantum well, increasing the sensitivity to the value of V.

The above "particle in a box" description of the electronic states is the simplest description of the quantum well heterojunction (QWH) system. It is physically appealing; the model is simple and the role of the various physical parameters transparent. It has thus proved very popular in the analysis of optical spectra observed from QWHs, being used in the pioneering work of Dingle and co-workers [2,3] and despite it shortcomings it has continued to be used almost up to the present day [4]. What is not included in the above is any note that we are dealing with real crystals; crystals whose band structure deviates from ideal parabolic bands implicity assumed above and crystals whose effective masses are different on either side of the heterojunction.

When it is applied to the problem of calculating the confined states in quantum wells the "particle in a box' description falls into that class of models formulated in the effective mass or envelope function approximation [5]. Assuming that k_x and k_y are both zero then the crystal wavefunction is assumed to be of the form

$$\psi = \sum_j F_j(z) u_j(r). \tag{7}$$

where $u_j(r)$ is the cell periodic Bloch part and $F_j(z)$ describes the envelope part of the wavefunction which is slowly varying on the scale of the lattice constant a. The index j runs over the appropriate number of bands. In practice, this sum is model-dependent and for example in a simple two-band isotropic model of a direct semiconductor it would simply extend over the electron and hole levels at the Γ point. Let us continue with the simple two-band model as it illustrates that the matching of the envelope function and its derivative at the boundaries is incorrect.

If we consider the problem of a quantum well formed between two constituent semiconductors A and B where E_A, the energy gap of material A, is smaller than E_B, then identical Schroedinger equations can be written down for each of the two bands involved, these are

$$i\hbar \frac{\partial}{\partial z}\left[\frac{1}{m_{e,h}(z)} i\hbar \frac{\partial}{\partial z}\right] F_{e,h} + (V_s(z) - E) F_{e,h} = 0, \tag{8}$$

where the labels e and h refer to electron and hole states and V_s is the step in the conduction band at the interface. Note that the spatial dependence of the effective mass has been explicitly accounted for. Integration of this

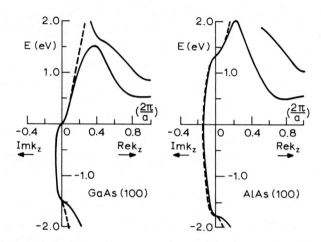

Fig. 2. Comparison of the complex band structure of GaAs and AlAs calculated using the two-band Kane model (dashed lines) and a tight-binding scheme (solid lines). (After ref. [12].)

equation across the interface provides us with the boundary condition that

$$\frac{1}{m_{e,h}} \frac{\partial F_{e,h}}{\partial z} \qquad (9)$$

be continuous and not just the derivative of F at the well edges. That eq. (9) is the appropriate boundary condition for quantum wells was realized by White and Sham [6] and derived earlier by Ben Daniel and Duke [7] to ensure current continuity at a single heterointerface.

The two-band isotropic model is too simple to describe what happens in real III–V crystals. A somewhat more complicated model to describe quantum well confinement is due to Bastard [8]. Bastard's model uses the Kane $\mathbf{k} \cdot \mathbf{p}$ description [9] of the band structure of the bulk semiconductors. A comparison between the Kane model and a tight-binding calculation of the band structure of GaAs and AlAs near to $\mathbf{k} = 0$ is given in fig. 2. A clear advantage in using the Kane model is that it accounts in a reasonably accurate manner for the electron non-parabolicity which becomes important when discussing confinement in narrow quantum wells. Furthermore, the Kane three-band model allows one to account for the effects of the spin–orbit split-off band. In the Kane model the heavy hole band is completely decoupled from the other bands and confinement effects for this band have to be considered in some other way. This is usually done by assuming the heavy hole band to be parabolic and using matching condition (9) above. Bastard's implementation of the Kane model has a number of disadvantages:

(a) the coupling of the electron and light hole bands to higher bands is neglected; and

(b) only one adjustable parameter describes the dispersion of the light hole and electron bands.

This parameter is the Kane momentum matrix element P^2. To correctly account for the different dispersion of the electron and light hole bands this parameter has to be chosen differently for each band; being given a value consistent with the observed electron and light-hole effective masses. An unified description of these bands has been achieved by Altarelli [10] who uses more adjustable parameters associated with the coupling to higher bands and also by Schuurmans and 't Hooft [11]. The latter authors make use of a Kane-type model which includes the electron, light hole, heavy hole and spin–orbit split bands explicitly. The coupling to higher bands is included perturbatively. The model for each constituent material contains four parameters which are fixed by the effective masses of the electron, light hole, heavy hole and spin–orbit split effective mass at the Γ point. Thus an unified picture is achieved. Comparison of the results of this model compare very favourable with calculations made using the computationally complex tight-binding method of Schulman and Chang [12] as we see from fig. 3. Despite its many incarnations the proliferation of envelope function type calculations has continued. The most recent publication of yet another

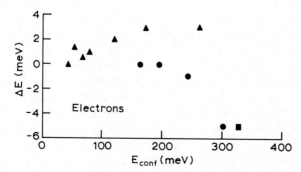

Fig. 3. Difference ΔE in the electron confinement energy obtained from the extensive Chang–Schulman tight-binding analysis and the confinement energy calculated using the $k \cdot p$ approach of Schuurmans and 't Hooft. (After ref. [11].)

variant is from Potz and co-workers [13]. Again the model for the constituent semiconductors is based on the Kane $k \cdot p$ approach this time accounting for remote-band effects in second-order perturbation theory.

The feature that remains central to all of these versions of the envelope function approach is the assumption that the cell periodic part of wavefunction, i.e. $u(r)$, remains invariant across the interface. As pointed out by Schuurmans and 't Hooft this assumption cannot be correct exactly. However, the excellent comparison between their results and those from the Schulman and Chang tight-binding scheme, which does not embrace that assumption, goes some way to demonstrate the usefulness of the assumption at least for the case of (AlGa)As–GaAs quantum wells. Whether this remains true for other widely different heterojunction combinations has yet to be proven. The other situations where the envelope function approach is clearly inappropriate will be when considering short-period superlattices or when the band edges involved on either side of the heterojunction are derived from different conduction-band minima. In this latter case then the pseudopotential approach proposed by Marsh and Inkson [14] seems more appropriate.

There are, of course, more sophisticated methods of calculating the electronic structure of quantum wells and electronic superlattices. For example, Schulman and Chang [12,15] have presented an elaborate tight-binding scheme to calculate the confinement in quantum wells (or more correctly very large period superlattices). The complex band structure of the materials is calculated using a s,p and s* nearest-neighbour, empirical tight-binding description with the tight-binding parameters being adjusted to reproduce the bulk band structures of the constituent materials. The tight-binding scheme is computationally time consuming but is regarded as providing an accurate description of the confined states in QWHs [15]. However, as is pointed out above there is at least one version of the envelope function approach which can do just as well in certain circumstances [11]. In addition a variety of authors [16–19] have adopted various pseudopotential schemes to determine the electronic states of superlattices. These schemes are all computationally time consuming and the accuracy of the pseudopotential schemes and their sensitivity to input parameters remains largely undetermined. Because of its simplicity and the clear role of material parameters the envelope function approach is firmly established as being the method used by most authors to analyse their optical data.

3. Optical absorption and photoluminescence excitation spectroscopy

The careful analysis of absorption spectra from single and uncoupled multiple quantum well heterojunctions by Dingle and co-workers [2,3] of (AlGa)As–GaAs pioneered this particular technique for the measurement of heterojunction discontinuities.

This spectroscopic method is indirect. It assumes that the quantum wells are symmetric, the interfaces are atomically abrupt and that there is no difference in band offset when material A is deposited on material B or when material B is deposited on material A. In other words the heterojunction offsets are commutative. Observed features in the absorption or excitation spectra have to be correctly assigned and a knowledge of the binding energy of quasi-two-dimensional excitons as a function of quantum well width has to be known.

The method relies heavily on some chosen theoretical model to calculate the positions of the confined particle states and therefore for narrow wells, this method is very sensitive to a knowledge of the quantum well width. If the quantum wells are thin enough then the method is not affected by any electrostatic potentials which may result from charge transfer from the barriers to the wells.

3.1. The GaAs–(AlGa)As system

The investigation, in 1974, by Dingle et al. [2] was made possible thanks to the emergence and refinement of molecular beam epitaxy (MBE) as a growth technique for the fabrication of III–V semiconductors. Although nowadays an alternative technology of metal-organic chemical vapour deposition (MOCVD) has emerged as a challenger to MBE in the production of thin films, MBE remains the enabling technology for most fundamental researchers in the quantum well field with MOCVD coming into its own for the production of, for example, quantum well lasers [20]. A further ingredient of vital importance to the early work by Dingle et al. was the development of selective chemical etches which were needed to remove the absorbing GaAs substrate without damaging the thin epitaxial layers of the quantum well heterostructures. Subsequent investigators have circumvented the need for the removal of the substrate by employing the technique of photoluminescence excitation (PLE) spectroscopy [21].

Dingle and co-workers [2,3] measured the 2 K infra-red absorption spectra seen from a number of single, double and multiple quantum well heterostructures. In all the samples studied the wells were GaAs and the

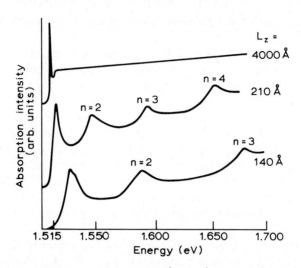

Fig. 4. Optical absorption spectra (2 K) of 4000 Å, 210 Å and 140 Å thick GaAs layers sandwiched between $Al_{0.2}Ga_{0.8}As$ barriers. Peaks at higher energies are associated with excitonic absorption between excited sub-bands. (From ref. [3].)

barrier material was (AlGa)As, where the molar fraction of Al was close to 0.2. As the well width was reduced, quasi-two-dimensional exciton peaks associated with each confined particle state were clearly seen emerging in the spectra (see fig. 4). In the widest sample of 400 Å, heavy hole excitons were seen up to $n = 7$, where n labels the confined particle states. The energetic positions of the observed excitonic transitions were derived using the simple "particle in a box" analysis described above. Parabolic bands were assumed and continuity of wave function and its spatial derivative was used as the matching conditions. The effective mass of carriers in the direction perpendicular to the layers was assumed to be the correct one to use in the calculation and further assumed to be identical in well and barrier materials. The effective masses for GaAs used by Dingle et al. are shown in table 1. An example of the sort of fits Dingle et al. were able to get to their spectra is shown in fig. 5. Here an impressive fit is achieved up to $n = 6$ with a conduction to valence band offset ratio of 88 : 12. Analysis of many more spectra (see fig. 6) allowed these authors [2] to arrive at two conclusions:

(a) the closeness of experiment and calculation for a wide range of energetic positions meant that the wells were symmetric and the heterojunctions abrupt; and

Table 1
Summary of the effective masses used by various groups in their absorption or PLE studies of (AlGa)As–GaAs quantum wells. All masses m are given in units of m_0.

Research group	m_{hh}	m_{lh}	m_e	Al (%)	Q_c
Dingle et al. [2]	0.45	0.08	0.0665	0.21	0.85
Miller et al. [29]	0.45	0.08	0.0665	0.30	0.5
Miller et al. [23]	0.34	0.094	0.0665	0.30	0.57
Dawson et al. [22]	0.403	0.08	0.0665	0.26–0.36	~0.75
Duggan et al. [37]	0.34	0.094	0.0665	0.36	0.65
Duggan et al. [96]	0.38	0.094	0.0665	0.35	0.65–0.67
Meynadier et al. [25]	0.45	0.08	0.067	0.14	0.6

(b) 0.85 ± 0.03 of the direct energy gap difference between the well and barrier material appears in the conduction band.

The fraction of the offset which appears in the conduction and valence band we will denote by Q_c and Q_v, respectively.

To make a connection between theory and experiment Dingle [3] had to make some postulation about the binding energy of the quasi-two-dimensional excitons of the system. The exciton binding energy was assumed constant for a given series, e.g. all the heavy hole excitons. From an extrapolation Dingle deduced that the binding energy was increased to about 9 meV for well widths smaller than 100 Å; about a factor of 2 larger than the three-dimensional value in GaAs. We will return to the importance of the value of the exciton binding energy later in this section. The values of well width, which is the single most important parameter in determining the confined particle states, were assumed to be known from prior growth rate calibrations on thicker samples.

There seems no doubt that at the time Dingle was well justified [3] in his assertion that this was "the most detailed and the best documented account of the quantum size effect in a solid". Despite results only being presented at one temperature and one alloy composition the 85:15 ratio became almost universally accepted even into the indirect (AlGa)As composition range. This 85:15 split has since become known as the Dingle "rule" although, in fairness, nowhere does Dingle claim that his 85:15 split is applicable to any situation other than for the set of samples he reports results for, it is others who have made the extrapolation of this result to the whole of the (AlGa)As composition range. However, more recent optical results from a number of groups using more sophisticated analysis [22,23]

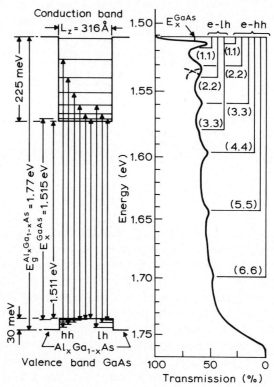

Fig. 5. Effective mass analysis of a 316 Å thick GaAs–Al$_{0.21}$Ga$_{0.79}$As quantum well. The predicted energy levels and transition energies are shown in the left-hand panel. A comparison between the predicted transition energies, corrected for exciton effects, and the measured absorption spectrum is shown in the right-hand panel. The offset ratio for the calculation was 88:12. (From ref. [3].)

and/or more sophisticated potential well profiles [23–25] have challenged this offset ratio even in the direct composition range.

The "optical" challenge to the 85:15 ratio did not start until a decade after the publication of the original paper by Dingle and co-workers [2]. In 1984, Miller and co-workers [23] published low-temperature PLE spectra from a number of synthesized "parabolic" quantum wells, an example of which is reproduced in fig. 7. Such spectra contain a wealth of transitions and care has to be taken in correctly assigning the spectral features. Miller et al. used circular polarization techniques [26] to aid in the identification of the lower energy peaks. Once the splitting between any pair of states

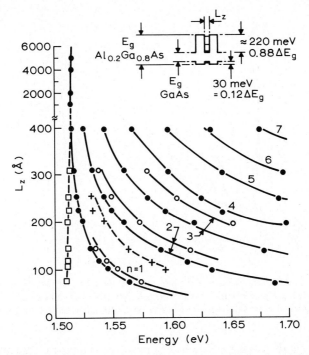

Fig. 6. Comparison of the effective mass predictions of energy transition in GaAs quantum wells with widths ranging from 70 Å to 400 Å. Solid points involve heavy-hole transitions whilst the open circles involve light holes. The crosses are transitions that violate the $\Delta n = 0$ selection rule and probably involve absorption from a $n = 3$ heavy hole to a $n = 1$ electron state. (From ref. [3].)

involving, for example, heavy holes is established then this indicates which of the higher transitions is due to heavy holes since for a parabolic well the confined particle states should be equally spaced in energy. This is somewhat idealized though since the spectral features observed are excitonic and the exciton binding energy will vary as the confined particle energy increases. Additionally evenly spaced levels are only expected if the well is infinitely deep.

Using parabolic wells increases the sensitivity to the offset fraction Q_c as the electron energy levels in a parabolic well are given approximately by

$$E_n = 2(n - \tfrac{1}{2}) \frac{\hbar}{L_z} \left[\frac{2Q_c \Delta E_g}{m_e} \right]^{1/2}, \tag{10}$$

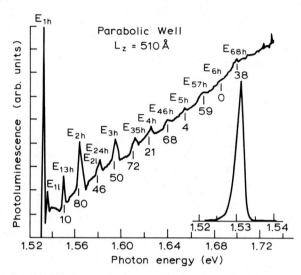

Fig. 7. The 5 K photoluminescence excitation spectrum from a quasi-parabolic GaAs quantum well. The various assignments to exciton transitions are shown in the figure. The inset shows the corresponding 5 K photoluminescence emission spectrum. (From ref. [23].)

where ΔE_g is the total band gap difference. A similar expression holds for the hole levels. Unlike the square well case [eq. (5)] the offset fraction enters the expression for E directly as it is determines the curvature of the parabolic potential.

The sharp excitonic peaks in the spectra were assigned to ladders of $\Delta n = 0$ and $\Delta n \neq 0$ transitions. Miller et al. [23] tried to fit the spacing between the individual components of the ladder, e.g. the electron shifts, using effective masses for holes and electrons given in table 1. They were unable to describe the energy separations successfully using 85:15 as the offset ratio and concluded that an approximately even split between the valence- and conduction-band discontinuities was necessary to explain their data. The value of the heavy hole effective mass of $0.45m_0$ was consistent with that used by Dingle [3] and in accord with an early cyclotron resonance value [27].

A further clue that the offset ratio at the GaAs–(AlGa)As heterointerface was not 85:15 can be found in hindsight from the reported PLE spectra on a double quantum well structure by Bastard et al. [28]. Here the authors found two peaks involving absorption by hole states which had to be energetically positioned above the top of the valence-band potential well should the offset ratio of 85:15 hold. The authors quite reasonably used the 85:15 ratio and explained their observation in terms of resonant hole states above the hole potential well. An alternative explanation which is now clear is that the absorption involves confined hole states and that the hole well must be deeper than predicted by the 85:15 "rule".

Following their work on the synthesized parabolic wells Miller et al. [29] returned to examine many spectra obtained from rectangular quantum wells. The authors claimed that they could arrive at an unified description of both parabolic and square wells by proposing a new offset ratio and suggesting alternative hole masses for both light and heavy holes. The offset ratio was 57:43 at an Al fraction of 0.3. The light hole mass was adjusted to a value of $0.094m_0$ whilst the preferred heavy hole mass was $0.34m_0$. Examples of the fits to observed exciton creation peaks in the PLE spectra are shown on a double logarithmic plot in fig. 8. In some cases the fits are poor but this is probably due to uncertainty in the well widths. Generally the agreement between calculation and experiment is good. The confinement energies were calculated using the "current continuity" matching condition given by eq. (9). The use of mainly wide wells led the authors to deduce that non-parabolicity of the electron (or light hole) band would not affect their conclusions. If one only considers the $\Delta n = 0$ transitions and sticks to the effective masses of ref. [3] (table 1) then satisfactory fits can be obtained to the data using an offset ratio of 85:15. It is the assignment and energetic position of the parity allowed $\Delta n = 2$ transitions, E_{13h}, that are essential to the revision of the offset ratio to 57:43 and the need for a smaller heavy hole mass. The confirmation that the observed $\Delta n = 2$ transition involved a heavy hole has come from PLE measurements using circularly polarized excitation [26].

The use of parabolic wells should increases the sensitivity to the band offset Q_c. However, the lack of control of the Al flux during MBE growth means that such structures have to be synthesized by precise control of alloy and binary layer thicknesses during deposition. This is technologically challenging and as the sensitivity to L_z for these wells is still stronger than Q_c [see eq. (10)], an accurate knowledge of the well width is still vital. The use of wide rectangular wells decreases the sensitivity to Q_c often making it possible to describe the lowest states accurately in the "infinite well" approximation [eq. (2)].

The use of narrow rectangular wells which will push confined particle states closer to the top of the potential wells should increase the sensitivity to the band offset. However, the use of narrow wells introduces additional complications:

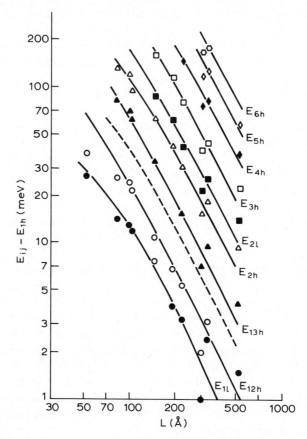

Fig. 8. Logarithm of the observed exciton transitions E_{ij} measured from the $n=1$ heavy hole exciton position, E_{1h}, versus the logarithm of the well width L. The solid lines are calculated using $Q_c = 0.57$ and $m_{hh} = 0.34 m_0$. The dashed curve is the $E_{13h} - E_{1h}$ splitting calculated with $Q_c = 0.85$ and $m_{hh} = 0.45$. (From ref. [29].)

(i) the electron states are pushed up into a region where the dispersion is non-parabolic; and

(ii) the well width must be accurately known as single monolayer fluctuations in wells of around 50 Å width will significantly change the confined states [30].

Mindful of these two considerations recent spectroscopic work from Dawson et al. [22] has concentrated on increasing the sensitivity to Q_c by using narrow GaAs wells (~ 55 Å). Particular attention was paid to the accurate determination of well widths. They were either determined from calibrated RHEED oscillation measurements on a stationary substrate prior to epitaxial growth on a rotating substrate or directly from real-time observation of the periodic intensity variation of the specularly reflected RHEED beam as GaAs and (AlGa)As were deposited on a 2.5×2.5 mm^2 stationary substrate. The authors made calculations of the confined subband energy positions using their implementation of Bastard's version of the Kane model referred to above. The calculation includes non-parabolicity of the electron band. Different P^2 values were used for the electron and light hole bands corresponding to effective masses of $0.0665m_0$ for the electron band and $0.087m_0$ for the light hole band. The preferred value of the heavy hole mass was $0.403m_0$ which is the value determined from possibly the most reliable cyclotron measurements of Skolnick et al. [31] on bulk p-type GaAs. The effective masses in the barrier region were determined from a linear interpolation between the GaAs values quoted above and those of AlAs [32].

The PLE spectra measures the position of the exciton creation peaks associated with the various subbands whilst the theory calculates the position of those subbands. To facilitate a comparison between the two, Dawson et al. used exciton binding energies calculated by Greene and co-workers [33]. All the heavy hole excitons were assumed to have the same binding energy as were all the light hole excitons. Rather than use a measured value of L_z, Dawson et al. chose to let the well width and band offset be adjustable parameters. Confidence in the theory used would be gained if it were able to predict the well width of the sample whose thickness was precisely known from counting the RHEED oscillations.

An example of the 5 K PLE spectrum from one of the multiple quantum well samples is shown in fig. 9. The results of the theoretical calculations were displayed as clearly as possible by plotting contours such as those shown in fig. 10. The allowed $\Delta n = 0$ electron–heavy hole and electron–light hole transitions were calculated as a function of L_z and Q_c the result of which is a three-dimensional energy surface for each of the transitions. The experimentally determined transition energy is then identified as a constant-energy contour on the appropriate surface. If the theory is correct then the projection of all the contours onto a two-dimensional plane should result in an unique crossing point which simultaneously determines both L_z and Q_c. The results of this procedure for the sample of fig. 9 are shown in fig. 10. There are two contours drawn for each transition corresponding to the measurement uncertainty in locating the exciton peaks. The contours for the $n = 2$ electron–light hole transition terminate abruptly at about 61

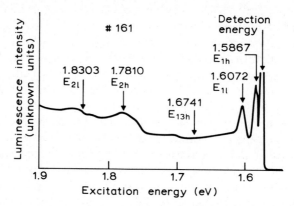

Fig. 9. 5 K PLE spectrum from a GaAs–Al$_{0.36}$Ga$_{0.64}$As multiquantum well sample. (From ref. [22].)

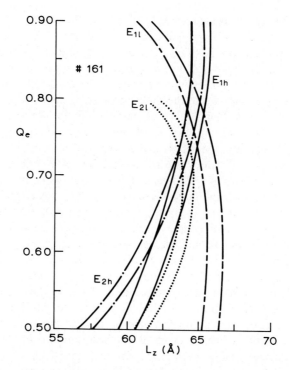

Fig. 10. Contours corresponding to the $\Delta n = 0$ transitions seen in the MQW sample of fig. 9. Calculations are made using Bastard's implementation of the Kane model. (From ref. [22].)

Å indicating that one of the states is no longer confined in the GaAs potential well. Aside from the $n=2$ e–lh transition all the contours intersect at a common point whose (Q_c, L_z) coordinates are (0.79, 64.5 Å). From the calibrated RHEED growth rate measurements the best estimate one can make for the well width is 62 ± 3 Å which embraces the value obtained from the contour plot. The value of offset ratio is closer to the original Dingle value of 0.85 than that proposed by Miller et al. [29]. However, what is clear from the shape of the contours in this diagram is that the offset is not particularly well-determined using the procedure of fitting only the $\Delta n = 0$ transitions. Whilst the curvature for the higher lying levels is increased somewhat, no contour bends sufficiently to pin-point the value of Q_c accurately. The technique appears much more suitable to the evaluation of well width! That this is the case is borne out by studying the contour plot of fig. 11. This is the contour plot of the $\Delta n = 0$ transitions from a single-well sample of GaAs clad with (AlGa)As with an Al fraction

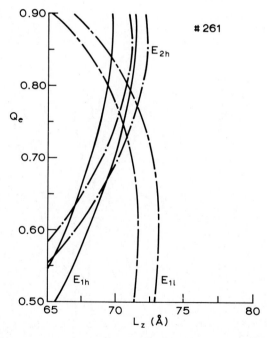

Fig. 11. Same as fig. 10 but for a single quantum well sample whose width was independently measured by real time in-situ counting of the RHEED oscillations seen during growth. (From ref. [22].)

of 0.26. Substrate rotation was not employed for this sample and the well width was determined from real-time RHEED oscillations. Under steady-state conditions the RHEED oscillations precisely reflect monolayer growth [34] and the error in well width determination, of about one monolayer, is introduced at each interface due to the finite response time of the Al shutter and the unknown Al flux profile as the shutter is closed. For this sample the well width is determined to be 73.5 ± 2.85 Å from the RHEED oscillations. The contour plots yield $L_z = 71$ Å and $Q_c \sim 0.75$ for this sample. The well width is comfortably within the bounds of the RHEED measurement and is a strong indication that we are making the calculation properly and that we have a good optical technique for measuring film thickness. Further contours on other multiple quantum well samples reinforced the conclusion that the offset was not very well determined by this method but lay in the range 0.75 ± 0.05 for Al fractions in the range 0.25–0.36.

Another fact clearly emerges from this analysis and that is that the inclusion of non-parabolicity is essential to the description of the states in

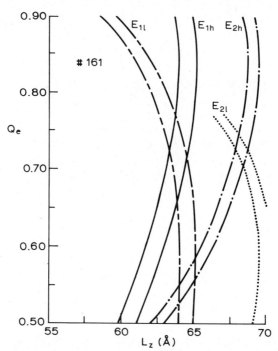

Fig. 12. Same as fig. 10 but on this occasion the contours were calculated using the calculational scheme described by Miller et al. [29]. (From ref. [22].)

these narrow wells. If we make calculations of the confined states using the calculational scheme described by Miller et al. [29] and assume parabolic bands then the result is contour plots like that of fig. 12. This is again for the sample shown in fig. 9. No common crossing point occurs, allowing no determination of either the well width or band offset from such plots.

So far we have considered the application of the optical method of absorption spectroscopy or PLE spectroscopy from three different sources, using three slightly different versions of the envelope function approach, using three different sets of material parameters and have arrived at three different values for the offset ratio. There appears to be significant uncertainty in determining Q_c by this method. Can we reconcile these sets of data or is any particular set of data supported by further PLE measurements on other structures where the sensitivity to Q_c may be increased?

In comparing the apparently contradictory results of Miller et al. [23,29] an Dawson and co-workers [22,35] the following points need to be noted. Miller et al. [23,29] based much of the faith in their result both for offset and new material parameters in the ability to describe the energetic positions of the $\Delta n = 2$ parity allowed transition E_{13h} involving the transition between a $n = 1$ electron and a $n = 3$ heavy hole. Dawson et al. only considered $\Delta n = 0$ transitions. Clearly visible in the excitation spectrum of fig. 9 is a feature between the pairs of $n = 1$ and $n = 2$ excitons. This is most likely the E_{13h} transition. Using $Q_c = 0.75$, their preferred hole masses, exciton binding energy of 9 meV and derived well width fails to locate this feature within 20 meV. This is a clear failure of the parameters. The confirmation that this transition involves a heavy hole has come from PLE on samples of similar width using circularly polarized excitation [36]. If the effective masses preferred by Miller et al. [29] are used in combination with our calculational scheme then a contour plot for this sample yields $Q_c = 0.66$ and a well width of 64 Å [37]. The position of the peak is then located to within 4 meV. Further illustration of the effect of the mass change and including the E_{13h} transition is provided by the contour plot of fig. 13. The shape of the E_{13h} contour is most sensitive to the offset and pinpoints its value far more accurately than the $\Delta n = 0$ transitions alone. The offset ratio of 66 : 34 is consistent with $I-V$ and $C-V$ measurements [38–40] at this alloy composition and closer to the value derived by Miller et al.

The importance of $\Delta n \neq 0$ transitions and changes in well shape have been illustrated in the above as aiding the sensitivity to Q_c. Changing the well shape to be triangular has been suggested [41] as being the potential profile most sensitive to Q_c, as far as I am aware no one has yet attempted

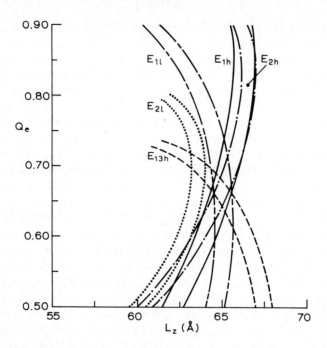

Fig. 13. Same as fig. 10 but now including the $n = 3$ heavy hole to $n = 1$ electron excitonic absorption and a heavy hole mass of $0.34 m_0$ (From ref. [37].)

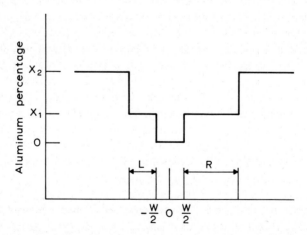

Fig. 14. Aluminium percentage profile of a quantum well separate confinement heterostructure (After ref. [25].)

the fabrication of such a structure. Meynadier et al. [41] have sought to exploit the dependence on $\Delta n \neq 0$ transitions by studying the PLE spectra from symmetric and asymmetric separate confinement heterostructures (SCH). The structure consists of a small well situated either at the centre or off-centre in a larger quantum well (see fig. 14). A typical well width W was 45 Å and $L = R = 100$ Å for the symmetric case and $L = 50$ Å and $R = 100$ Å for one of the asymmetric examples. X1 was close to 14% with X2 near 34%. Analysis of the intersubband structure including transitions between electron states in the narrow well and hole states in the wide well persuaded these authors that Q_c was 0.59 ± 0.03 for an Al fraction of 0.14. The effective masses used by these authors were given in table 1. Later work by Miller et al. [24] using half parabolic wells and a symmetric SCH structure leant weight to their earlier assertion that Q_c was 0.60 ± 0.03 for Al fractions in the range 0.2–0.3.

Before we consider the few measurements on other heterojunction systems we will spend some time discussing the importance of some of the material and physical parameters used in the analysis of the spectra.

One input parameter essential to the comparison between experiment and theory is the quasi-two-dimensional exciton binding energy of heavy and light hole excitons. In the true two-dimensional limit the binding energy of an exciton attains a value of 4 times its three-dimensional one [42]. Variational calculations by Bastard [43] assuming infinite potential barriers were extended by Greene et al. [33] using a slightly different vibrational function and including the effect of a finite barrier. The results of Greene et al. are shown in fig. 15. The results are clearly physically reasonable with the binding energy increasing as confinement becomes important until eventually enough of the exciton wavefunction "leaks" into the barrier regions so that the system becomes more three-dimensional with the exciton binding energy finally reaching the value one would expect for the barrier regions. Optical absorption measurements by Miller et al. [44] of the energy difference between the ground and excited states of the exciton in 90 Å wells tended to support the above theoretical work. The results of Greene et al. [33] were used by both Miller et al. [29] and Dawson and co-workers [22] to interpret their optical spectra. Recent magneto-optical measurements reported independently by two groups [45,46] point to the exciton binding being larger than previously thought. The most serious discrepancy occurring at a well width of about 50 Å; the variational calculations predict a heavy hole exciton binding energy of about 9 meV whilst the magneto-optic measurement is almost a factor of two larger. The experimentalists [45,46] speculated that the discrepancy was due to the

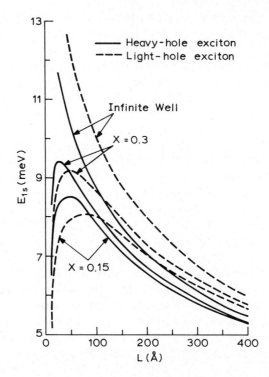

Fig. 15. Variation of the ground-state binding energy of a heavy hole exciton (solid lines) and a light hole exciton (broken lines) as a function of the GaAs well width L for Al fractions of 0.15 and 0.3 and for an infinite potential well. (From ref. [33].)

neglect in the theory of the light and heavy hole mixing in the plane of the quantum wells [47]. Strong mixing of light and heavy hole states might be expected when the heavy and light hole confined states are close in energy. This situation could readily occur for states where $n > 1$ and for the $n = 1$ states in wide quantum wells. However, in a 50 Å GaAs quantum well confined by $Al_{0.35}Ga_{0.65}As$ barriers the separation between the lowest heavy and light hole confined states is about 25 meV, so that we would not expect much perturbation of the heavy hole states due to the mixing. This is precisely the case in which one gets the largest discrepancy between experimental and theoretical values of the heavy hole exciton binding energy. Recent theoretical work [48] has included the effect of light and heavy hole mixing for just this case, with the results that the exciton binding energy is increased by only about 1–2 meV above that deduced by

Greene et al. The resolution of the dilemma lies in the need to extrapolate Landau level positions to zero applied magnetic field from a position quite far from the origin. Support for this explanation has come from very recent magneto-optic work by Rogers et al. [49] on GaAs–$Al_{0.35}Ga_{0.65}As$ multiple quantum wells. Landau levels were observed at fields at low as 2.5 T making extrapolation to zero field less hazardous. Their results are in quite good agreement with calculation [33,48]. Continued use of the exciton binding energies of Greene et al. appears to be quite justified.

The importance of $\Delta n \neq 0$ transitions has been emphasized many times. What has emerged from the work of Miller et al. [23,29] is that the change in heavy hole effective mass is essential for a correct description of these transitions. In their early papers Miller et al. [23] and Kleinman [50] point to theoretical work by Lawaetz [51] and Baldereschi and Lipari [52] as lending support for this value. Later work [24] quotes the review article of Bimberg [53] as the source of the value for the reduced mass. Dawson et al. [22] drew attention to the this large change in mass from the $0.403m_0$ calculated from the band parameters measured by Skolnick et al. [31]. However, choosing the opposite extrema of uncertainty in Skolnick's results would allow one to embrace the proposed value of $0.34m_0$. A theoretical analysis of magnetoreflection data by Hess et al. [54] yields band parameters which combine to give a heavy-hole mass of $0.377m_0$. This value depends as much on the correctness of the theory used as on the measurement uncertainty so its true accuracy is difficult to estimate.

What is clear so far is that if one assumes the same exciton binding energy for the whole series of heavy hole excitons then the empirical use of $m_{hh} = 0.34m_0$ reproduces more spectral features and depending on structure and calculational scheme yields a value for Q_c between 0.6 and 0.66 for Al fractions in the range of 0.14 to 0.36. What is also clear is the need for a good measurement of m_{hh} to determine whether continued use of the small value of $0.34m_0$ is justified.

3.2. Other heterojunction systems

Following (AlGa)As, the heterojunction combination that has seen the most investment in its growth and characterisation has been $In_{0.53}Ga_{0.47}As$ which can be grown lattice-matched on InP. Quantum wells of this material have been produced with either InP as the barrier material [55] or alternatively $Al_{0.52}In_{0.48}As$ as the higher band gap material [56]. Using this materials combination it is possible to produce optical components, such as lasers and photodetectors, in the important telecommunications wavelength

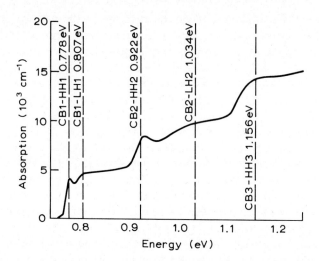

Fig. 16. Room-temperature absorption spectrum of an (InGa)As–(AlIn)As multiple quantum well heterostructure with a well width of 110 Å. The dashed lines represent the excitonic transition energies calculated using a finite square well model. (From ref. [56].)

range of 1.3–1.55 μm. Reports of intrinsic excitonic absorptions in this system are scarce however. Prior to the work of Weiner et al. [56] the only report was of a small bump in the absorption spectrum of bulk material at low temperature [57]. It is all the more remarkable then that the results of Weiner et al. are at room temperature from 110 Å wide quantum wells of (InGa)As sandwiched between barriers of (AlIn)As. Analysis of the spectrum (shown in fig. 16) followed that of Dingle and co-workers [2,3] for the (AlGa)As system. The (AlIn)As band gap was fixed at 1.47 eV [58] and adjustment of the band discontinuities yielded $\Delta E_c = 0.44$ eV and $\Delta E_v = 0.29$ eV. The relative discontinuity in the conduction band is slightly smaller than that determined earlier by People et al. [59] by a C–V method. Weiner et al. [56] were clearly mindful of the scrutiny applied to the first absorption measurements in the (AlGa)As system and point to a variety of places where their analysis could be improved, e.g., the effective masses used, the exciton binding energy used and improvement in sensitivity to the offset through special potential profile structures. The uncertainty associated with the measurements of offsets in this system is demonstrated by comparing the above work with low-temperature PLE measurements by Wagner et al. [60] on nominally the same heterojunction system. Here comparison of measured, intrinsic excitation peaks with a square well model of the transition energies (this time accounting for the different well and barrier masses) was effected with offsets of 0.5 and 0.2 eV in the conduction and valence bands. These are the values found from C–V measurements [59]. Satisfactory agreement was obtained for well widths between 10 nm and 100 nm. It is not possible to distinguish which of these two values is "correct". What they do demonstrate is the difficulty of using this method with much accuracy; the accuracy being limited not only by the theoretical models and the use of excessively large wells, but in this case by the lack of knowledge of key materials and sample parameters. For example, the alloy composition in the wells and barriers has to be known to determine the fundamental gaps and partition the offset ratio correctly. Key material parameters such as effective masses, particularly those of the holes, have not received much investment in their determination. Their importance in the offset determination at the GaAs–(AlGa)As heterojunction has been adequately covered above and it is likely that their precise values will be no less important here. Finally, if one makes the attempt to increase sensitivity to the offsets through using narrower wells the effects of non-parabolicity on both the electron and light-hole bands for this system cannot be ignored [61].

The heterojunction combination of (InGa)As–InP is most conveniently grown by MOCVD and 5 K excitation spectra have recently been reported by Razeghi et al. [55] from samples with nominal well widths of 35 Å and 75 Å. After tentative identification of electron to heavy and electron to light hole transitions the authors turned attention to the possibility of using the measured transition energies to determine the band offsets. With the limited sample and materials data available the authors wisely chose not to speculate on the value of the offset ratio, because for instance, they were unsure of the precise alloy composition and hence fundamental band gap, the well width was uncertain and the hole masses are not too well-determined. Their reasons for not speculating on the offset ratio serve to underline the problems with this sort of determination when control over the epitaxial growth is not as precise as one would like and the basic materials informations may be unreliable or unavailable.

Whilst there appears to be a not insignificant uncertainty associated with the accuracy of the optical absorption determinations of the offset fraction there is no doubting its value in revealing the electronic structure of the particular quantum well heterojunction system of interest. Consider, for example the strained layer quantum well system of $In_xGa_{1-x}As$ ($x = 0.15$) grown on a GaAs substrate with GaAs as the barrier material. Low-temperature excitation spectroscopy measurements by Marzin et al. [62] on this

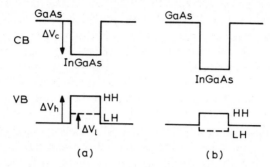

Fig. 17. Possible energy band configurations for a strained (InGa)As–GaAs quantum well heterostructure. CB and VB refer to the conduction and valence bands of the host materials. In the (InGa)As the valence band is split into heavy hole (HH) and light hole (LH) bands as a results of elastic strain from the lattice mismatch between the two materials. In (b) the system is simultaneously type-I for heavy holes and type-II for light holes. (From ref. [62].)

system indicate that the electrons and heavy holes are confined in the lower band gap (InGa)As yet the light holes are confined in the GaAs layers. In other words, a combination of confinement and bi-axial compression of the (InGa)As has led to a system that is simultaneously type-I for electrons and heavy holes but is type-II for electron and light hole recombination (see fig. 17). Confirmation of the assignment of the transitions seen in the various samples studied is achieved rather cleverly using the polarization properties of exciting light incident through the edge of the transparent substrate.

A further illustration of the power of absorption spectroscopy in understanding the fundamental electronic structure of heterojunction systems is evidenced by work of Voisin et al. [63] on the strained GaSb–AlSb system. Absorption spectra from this system exhibit well-defined free-exciton peaks and the shrinkage of the GaSb band gap is simultaneously accompanied by a reversal of the heavy and light hole subbands. Again, like the strained (InGa)As system, there are two valence-band offsets, one for each type of hole. This is because the strain reduces the overall symmetry of the constituent bulk semiconductors causing a lifting of the degeneracy of the light and heavy holes at the valence band extremum at the Γ point, which is a quantum mechanical feature seen in all III–V semiconductors. Voisin and co-workers were able to understand the spectral features using a simple envelope function calculation [63], unlike the strained (InGa)As system the system is type-I for both types of hole with band offsets of > 40 meV and > 70 meV for heavy and light holes, respectively.

4. Photoluminescence

Photoluminescence (PL), usually at cryogenic temperatures, has been employed by a number of groups to determine the band offsets not only in the (AlGa)As system. There are a number of variants of this technique but without exception they rely on being able to assign transitions in the recombination spectra which are intrinsic. The assignment of peaks in absorption is generally much more straightforward than in emission where the competing processes of recombination of excitons or carriers bound at impurities may play a significant part in the emission spectrum. A further effect which needs to be taken into account is the so-called "Stokes shift" which occurs between the exciton creation peaks seen in the PLE or absorption spectra and the principal peak seen in the PL spectrum at the same temperature [64]. We will return to this point in a little while.

In 1978, Dupuis and others [65,66] inferred the valence-band offset at the heterojunction between GaAs and $Al_{0.36}Ga_{0.64}As$ by examining the stimulated emission spectrum seen from photopumped platelets. The active region of the MOCVD fabricated lasers consisted of a single QW whose width was estimated from growth parameters to be 80 Å. The 77 K spectra are shown in fig. 18. The sample geometry was such that the thickness of the cladding regions were asymmetric being 1 μm and 0.3 μm thick, respectively. Pumping from the 1 μm side results in the spectrum labeled (a). Lasing occurs close to the band edge of the (AlGa)As. Pumping the 0.3 μm results in spectrum (b). Here we again seen lasing near to the (AlGa)As band edge and some low-energy features labeled A and B. Peak A is attributed to the recombination of electrons at the (AlGa)As band edge with holes that are confined in the GaAs hole well. Dupuis et al. [65] argue that an 80 Å quantum well acts selectively in the capture of injected electrons and holes. It is claimed that electrons do not scatter effectively over distances of the order of 80 Å and are not captured efficiently by the GaAs QW so that the optically generated electrons remain "hot" with respect to the GaAs quantum well band edge. Assuming this to be true then the position of peak A relative to the (AlGa)As band edge is

$$E_g - E_A \sim \Delta E_v, \tag{11}$$

making Q_v about 0.15, in agreement with the first absorption data of Dingle [3]. Peak B is then assumed to be due to the recombination of "free" electrons and a $n=2$ hole confined in the GaAs. Some uncertainty is expressed by the authors as to whether A labels the GaAs valence-band

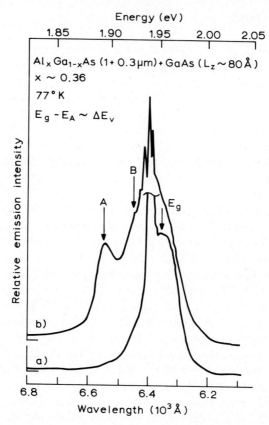

Fig. 18. 77 K photoluminescence spectra from a GaAs quantum well nominally 80 Å wide. Peak A is attributed to the recombination of "hot" electrons in the GaAs with confined holes. Peak B corresponds to recombination at the second confined hole level. See the text for further details of the optical pumping configuration. (From ref. [65].)

edge or is really the $n = 1$ heavy hole state, either way the uncertainty introduced is about 8 meV, which is the confined hole energy. The same experimental technique has been recently by Brunemeier et al. [67] to evaluate the offset ratio for those composition combinations of (InGa)(AsP) that are lattice matched to InP. For all of these strain-free combinations, the "hot" electron to confined hole luminescence permits the determination of the fractional discontinuity residing in the valence band to be between 0.31 and 0.36.

The dependence of this determination on recombination via the lowest confined particle states means that the results are not sensitive to the confined particle states and really only rely on theory to a very small degree in providing a refinement in the calculation of the hole confinement energy. As the lowest hole confined particle state is not shifted appreciably from the GaAs valence-band edge then an accurate value of the well width is not necessary either. The determination is limited therefore by a knowledge of the Al fraction which is presumably rather well known and could be measured from the low excitation PL spectrum. What is uncertain is that the recombination is due to the assignment given by the authors. It does seem likely that this may be recombination due to electron–acceptor related transitions in the (AlGa)As cladding regions. No real evidence that this transition is intrinsic is offered other than that it appears energetically in a position which supports a valence-band offset fraction of 0.15. An essential point in the argument is that electrons participating in the recombination do not readily thermalize into the 80 Å GaAs QW and therefore remain "hot" with respect to the GaAs band edge. Whilst it is an experimental fact that the threshold current of unmodified single quantum well lasers does increase with decreasing well width it is also clear that laser action does occur from 80 Å QWs. Therefore, sufficient carrier capture must occur to allow lasing from the lowest confined particle states which seems at variance with the observation above of weak recombination from the GaAs quantum well and significant recombination from the "hot" electrons and bound holes.

A popular method [4,68,69] for the determination of valence- or conduction-band offsets is by studying the PL emission, usually at liquid-He temperatures, from a series of quantum wells of different width, L_z, usually in the same sample; the idea being to fit the observed dependence of emission wavelength on the quantum well width using the valence- or conduction-band offset as the adjustable parameter. As far as I am aware, the first experiment of this sort was not used to determine the offset ratio but rather used, by Frijlink and Maluenda [70], to show that their control over quantum well width and interface abruptness by MOCVD growth was excellent, i.e. the interface grading extended over less than one unit cell. The authors compared experiment and calculation choosing to ignore exciton effects. Their conclusion is a little misleading as the potential profile chosen to mimic the grading was exponential. What their analysis determines is the characteristic length that appears in the exponent describing the graded potential. This means that the potential is actually graded over much more than 5 Å if their potential profile is realistic. If the grading is due to diffusion for instance then a more suitable profile would be a

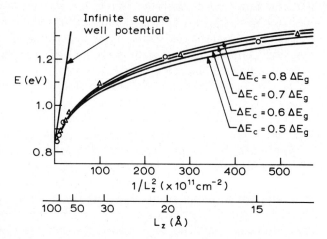

Fig. 19. Calculated $n=1$ electron to heavy hole transition energy as a function of well width L_z for (InGa)As–(AlIn)As quantum wells. The percentage of band gap difference in the conduction band is a parameter for each of the calculated curves. Experimental points from two separate growth runs are superimposed on the theoretical plots. (From ref. [4].)

complimentary error function or even a linear dependence as a first approximation. This experiment was repeated by Kawai et al. [71] again to demonstrate the abruptness of their MOCVD grown QWs. Their analysis of the data included the same potential profile as Frijlink and Maluenda, the same band offsets and recourse to exciton emission was made to explain some discrepancies between calculation and observation.

Welch et al. [4] measured the 4 K steady-state photoluminescence from a number of single QWs of the lattice matched combination of (InGa)As–(AlIn)As grown by MBE and deposited on an InP substrate. The peaks of the PL positions were plotted as a function of well width and theoretical curves of appropriate electron to heavy hole transitions were generated and compared to the experimental points using the conduction band offset fraction Q_c as a parameter. Their results are shown in fig. 19. The theoretical points were generated using the simple "particle in a box" model [1]. The effective masses of the (AlIn)As were assumed to be no different from the well material. Allowance was made for the non-parabolicity of both electrons and heavy holes through an energy dependent effective mass. The transition seen in the spectra was assumed to be due to the recombination of free electrons with free holes. No supporting evidence was offered. The slope of the E versus L_z^{-2} near the origin was used to fix

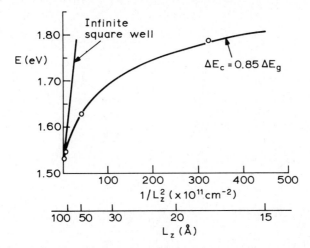

Fig. 20. As for fig. 19 but for GaAs–(AlGa)As quantum wells. (From ref. [4].)

the growth rate and this was assumed to be a constant over the time range of interest. From the two samples studied the fraction of the band gap difference occurring in the conduction band was about 0.7. This value is rather imprecise as experimental points span the range > 0.8 to 0.6 (see fig. 19). The reader is then offered a determination of the offset ratio in GaAs–(AlGa)As as an example that the technique will reproduce the "well known" Q_c of 0.85. This plot and the appropriate materials parameters are reproduced in fig. 20. The authors say that the points agree favourably with the value of 0.85. No comment is made about the nature of the recombination mechanism in this sample of several SQWs. If one assumes that the 4 K PL is intrinsic and due to free excitons, then this data can be equally well represented with an offset ratio of 77 : 23 for example [72]. I would suggest that other combinations of offset ratio and effective masses would equally fit the data with the same adequacy. I would further conclude that this is not a very reliable and sensitive measurement of the conduction- to valence-band offset ratio.

Other examples exist in the literature. Razeghi and co-workers [68] attempted to determine the band offsets at the $In_{0.47}Ga_{0.53}As$–InP interface from a set of SQWs fabricated in a single sample. Well widths were nominally 25, 50, 100 and 200 Å with barriers of 500 Å. PL at low temperature was consistent with the existence of four wells but there were enough significant deviations between theory and experiment not to allow

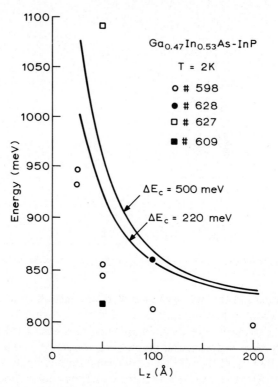

Fig. 21. Plot of measured photoluminescence peak energy versus well width for a number of single and multiple quantum well samples. Solid lines are calculations made in the envelope function approximation. (From ref. [68].)

any determination of the offset in the conduction band (see fig. 21). Possibilities for the discrepancies included incorrect well widths, variation in well composition and impurity participation in the recombination process. Similar work by Marsh and co-workers [73] in the same system but this time grown by MBE underlined the unreliability and inappropriateness of the technique for such a determination.

A clear limiting factor in using this technique for offset determination is the need to assign the recombination correctly. Even if the well width is known accurately then often "intrinsic" PL does not occur at the predicted or expected energy. This is because of the so-called "Stokes" shift between excitons created in an absorption or PLE experiment and the peak position of the transition seen in the corresponding PL. An example of such a shift

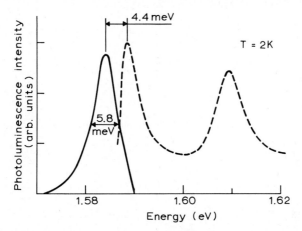

Fig. 22. Photoluminescence spectrum (solid line) and PLE spectrum (dashed line) from a 70 Å thick GaAs quantum well. (From ref. [64]).

is shown in fig. 22. For this 70 Å well a shift to lower energy of the PL by 4.4 meV is clearly seen at low temperatures. The origin of the shift is debatable and impurity participation and well width fluctuations have been suggested as being the origin. If the shift is due to well width fluctuations then the shift is explicable in terms of photocreated excitons relaxing or tunneling to wider parts of the layers where they reside at a slightly lower energy which is observed in recombination. What one needs in the comparison between theory and experiment is the position of the exciton creation peak for the known well width and not the recombination energy of these "localized" excitons.

A much more direct photoluminescence measurement of the valence-band offset at the $Al_{0.37}Ga_{0.63}As-AlAs$ interface has recently been reported by Dawson et al. [74]. The experiment is made possible by the large valence-band offset fraction that exists between (AlGa)As and AlAs.

The more reliable of the recent electrical and optical determinations place Q_v for the (AlGa)As system at a value between 0.3 and 0.4. If this fraction is extended into the indirect-composition range, as suggested by the electrical measurements [75] then a change in the band alignments for this system from type-I to type-II will occur at some critical Al fraction. This is illustrated in fig. 23. Here we plot the variation of the direct energy gap E_I in the lower band gap $Al_yGa_{1-y}As$ as the molar fraction of Al is increased. The barrier material is assumed to be the binary AlAs and E_{II} is the

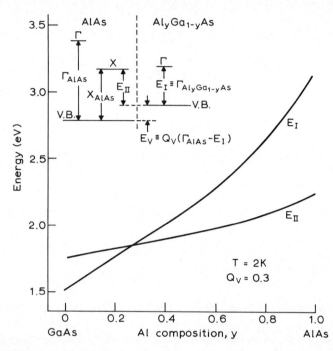

Fig. 23. Calculated energy transitions at the $Al_yGa_{1-y}As$–AlAs heterojunction as a function of alloy composition for a fixed valence-band offset fraction Q_v of 0.3. The meanings of E_I and E_{II} are evident from the inset.

energetic difference between the lowest conduction band state, X, in the AlAs and the top of the valence-band in the $Al_yGa_{1-y}As$. The fraction of the direct band gap difference that appears in the valence band is taken at 0.3. At a critical Al fraction the lowest conduction-band state in the system is in the AlAs and the band alignment has become a staggered one. In this case the Al concentration at the type-I to type-II crossover is at 0.26.

In their experiment, Dawson et al. [74] looked for this crossover to type-II behaviour in MQW samples of (AlGa)As–AlAs. The energy level scheme of fig. 23 is then slightly modified by the need to include confinement effects which will alter the crossover composition slightly. Comparison of the sense of the shift of the PL emission from (AlGa)As–AlAs quantum well structures whose well composition y lies well to either side of the crossover should test whether the hypothesis is correct. For a sample with a small value of y then a shift to higher energy due to confinement effect should occur over a bulk sample of the same Al fraction. The authors

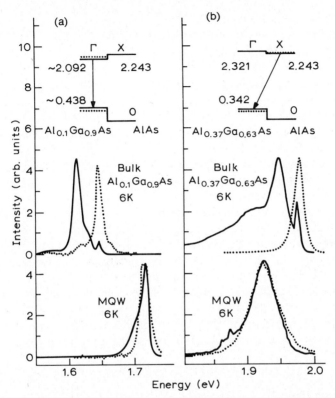

Fig. 24. Photoluminescence spectra of 6 K for bulk (AlGa)As alloys (super traces) and for MQW samples with these allos as well material and AlAs as the barriers. Solid lines are QW spectra and dotted are time resolved. The energy level scheme for each case is shown in the inset. (From ref. [74].)

observe such a shift as shown in fig. 24a. A shift to higher energy in the 60 Å QW compared to the bulk alloy being quite clear. Keeping the well width approximately constant and moving to a larger Al fraction (above the crossover) results in a downshift in energy of the PL from the QW system as opposed to the bulk alloy as shown in fig. 24b. At least two possibilities for the origin of the PL shift to lower energies; one, of course, is that we have passed the type-II crossover, and the second is that what is observed is impurity luminescence. That the PL involves a free carrier was inferred from the high-energy side of the peak which is approximately exponential with a slope corresponding to a free-carrier temperature of 10 K, close to the lattice temperature. Additional measurements of the movement of the peak position as a function of lattice temperature [76] were in good agreement with the temperature dependence of the X minima. In addition the recombination PL is weak in intensity, typical of indirect transitions and lends further weight to one thinking the assignment of the peak is correct. The PL line seen is a no-phonon line due to the mixing of the states caused by the proximity of the heterobarrier.

Taking the confinement energy of the hole (~ 16 meV) and the smaller confinement energy of the electron (~ 1 meV) [74] into account together with the direct bulk and gaps gives a valence-band offset of $\Delta E_v = 342 \pm 4$ meV. This is equivalent to an offset fraction in the valence band of 0.3. The small uncertainty of 4 meV arises not only from location of the peaks in the spectrum but also from a consideration of the effect of the various hole effective masses on its confinement energy.

Given that the assignment of the transition is correct then this is clearly a far more direct and reliable method than the rather unsophisticated attempts above. The actual value of offset fraction is in reasonable agreement with the PLE result from Dawson and co-workers [74] and somewhat smaller than that suggested by Miller et al. provided that the offset fraction is a constant across the whole (AlGa)As composition range. Supporting evidence for the value found by this method has recently come from light scattering measurements [77]. We will discuss them further below.

5. Light scattering

Since the observation of light scattering by intersubband excitations between the discrete confined particle states at a single GaAs–(AlGa)As heterojunction [78] the technique has been extensively applied to both nominally undoped [79] and n- and p-type modulation doped QWs [80,81]. Although until very recently [77] the experiment has not been undertaken with the express purpose of determining the conduction- or valence-band offsets. Most authors usually refer to the subband splitting seen spectroscopically as showing good or satisfactory agreement with calculation which often meant a simple "particle in a box" calculation using the original offset ratio of 85:15 proposed by Dingle et al. [2,3]. Below we examine some of the light scattering data and discuss its significance only in relation to the determination of the band offsets, in practice this means concentrating on single-particle excitations seen from rather wide, undoped and modulation doped QWs. It will emerge that careful sample choice is critical

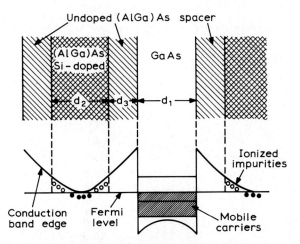

Fig. 25. Sequence of layers and structure of the conduction band edge of n-type modulation doped MQW GaAs–(AlGa)As heterostructures. (From ref. [80].)

for this technique to be sensitive. The promise of a more sensitive determination using light scattering from the hole bands is also addressed.

A lot of light scattering data exists on n-type modulation doped multiple quantum wells [80], whose conduction-band edge structure is illustrated schematically in fig. 25. A doped (AlGa)As layer, of thickness d_2 is separated from the lower band gap GaAs layer, thickness d_1, by an undoped spacer layer of (AlGa)As of width d_3. The GaAs layer is nominally undoped. The discontinuity at the heterobarrier makes it energetically favourable for the ionized electrons (supplied by the Si donors) to reside at a lower potential in the GaAs layers. Indeed it is the transfer of charge into this lower band gap region which forms the basis of the method used by Wang and co-workers [82,83] to further challenge the 85:15 offset ratio. The Fermi level resides above the top of the GaAs band edge and a quasi-two-dimensional electron gas is formed in each quantum well of the stack. The transfer of the charge introduces electrostatic potentials over and above the energy gap difference intrinsic to the heterojunction. The subsequent band-bending modifies the confined particle energies away from those of an ideal square well potential as is shown in the inset of fig. 26. Strictly, the quantum confined states should be calculated in a self-consistent fashion [84]. In addition if sufficient charge is transferred into the quantum well then it is no longer sufficient to ignore many-body effects [85,86].

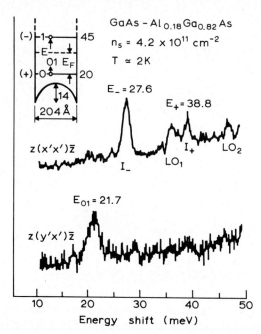

Fig. 26. Light scattering spectra of a modulation doped MQW GaAs–Al$_{0.18}$Ga$_{0.82}$As heterostructure. E_{01} labels the single-particle excitation from between the lowest and first excited electron states. The inset shows calculated confined particle energies, band bending and the Fermi energy. (From ref. [80].)

It is beyond the scope of this chapter to discuss in detail the selection rules involved in the physics of light scattering and the reader is referred to the review articles by, for example, Pinczuk [87] or Abstreiter [88] for further insight and appropriate references. We are interested principally in the single-particle excitations that occur between the subbands in the quantum well. It is usual to perform the light scattering close to a resonance condition, then excitations by free carriers show a strong enhancement close to the optical gaps where occupied states are involved in the transitions. In the case of n-type modulation doped or residual n-type GaAs then extensive use is made of laser excitation lines close to the energy gap of the conduction band and the spin–orbit split-off band. Figure 27 illustrates the transitions that are important in the single-particle excitations in a quantum well. Excitation of an electron out of the split-off subband to an energy above the Fermi level is followed by a downward transition of an electron in the occupied lowest electron subband into the vacant hole in the valence

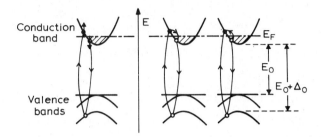

Fig. 27. Schematic of the transitions involved in resonant electronic Raman scattering. (From ref. [88].)

band. The energetic difference between the exciting radiation and the emitted radiation is then a measure of the energetic separation between subbands. An example of the signal observed is shown in the lower trace of fig. 26, where a distinct peak seen in the backscattering geometry is interpreted as a single-particle excitation between the lowest and first excited sub-bands. Comparison of the measured separation with that of the calculations (see inset) shows quite good agreement. A further example showing more observed single-particle excitations is shown in fig. 28 for another modulation doped MQW of 250 Å wide. Although it is not explicitly stated by the authors [80] just how these calculations are made it is not unreasonable to assume that an offset ratio of 85:15 was employed for this heterobarrier between GaAs and $Al_{0.12}Ga_{0.88}As$. The agreement for the E_{01} transition is good. This is not unexpected since the lowest particle states are the least sensitive to the conduction-band barrier height. The agreement becomes progressively worse with increasing energy separation. This could be due to a combination of factors:
(1) an incorrect well width (probably known to no better than 10% from growth rate estimates);
(2) neglect of the electron non-parabolicity;
(3) the neglect of many-body effects;
(4) an incorrect estimate of the band bending, and
(5) not using a self-consistent calculational scheme.
It is difficult to be precise about the magnitude of each of these factors but what is clear is that this is not the most suitable structure to which to apply a conceptually simple experimental method. The reliance on theoretical estimates and unknown or uncertain experimental parameters is too great.

The introduction of modulation doping as far as the offset determination is concerned is clearly only of nuisance value. Far better would be to

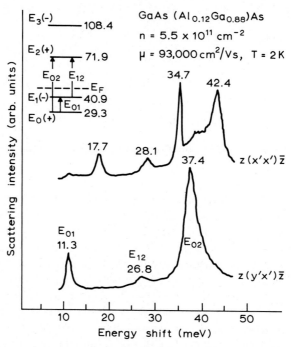

Fig. 28. A further example of light scattering spectra from an n-type modulation doped MQW whose nominal well width is 250 Å. (From ref. [80].)

determine single-particle, intersubband transitions on nominally undoped quantum wells where many-body effects could be neglected and where electrostatic effects were minimized. Such experiments were performed on MBE grown $GaAs$–$Al_{0.2}Ga_{0.8}As$ MQWs by Pinczuk et al. [79]. Light scattering spectra from one of the samples studied is shown in fig. 29. The $z(yx)\bar{z}$ spectra correspond to spin density intersubband excitations. The fact that the bands do not shift in energy as a function of injected carrier density is taken as confirmation that these are indeed single-particle excitations. The calculated energies of the three lowest conduction-band levels in the GaAs quantum wells are shown in the inset of the figure. E_{01} and E_{02} are predicted at 20 meV and 52 meV, respectively. Agreement with the simple square well, "particle in a box" calculation is achieved to within 2 meV. The source of the calculation is quoted as the early work of Dingle and co-workers [2] so the offset ratio used is 85:15 and an electron effective mass of $0.067m_0$. The low residual impurity concentration means

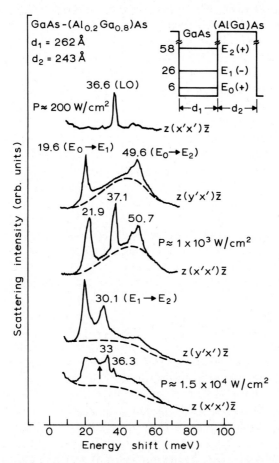

Fig. 29. Light scattering from a nominally undoped MQW heterostructure as a function of excitation density. Thickness d_1 and d_2 were defined in fig. 25. (From ref. [79].)

that band bending would occur over many μm and that the assumption of a flat conduction-band profile in the QWs is a good one. It could be argued that the quite good agreement between calculation and observation was supporting evidence for the 85:15 split, however, calculations in the same spirit as those above but now using 67:33 as the offset ratio and an electron effective mass of $0.0665m_0$ yields E_{01} and E_{02} values of 19.2 meV and 50.7 meV, respectively, in excellent agreement with the observations. No doubt other combinations of offsets, effective mass, well width and theoretical model would again reproduce the transitions satisfactorily. The problem here is that the electron well is relatively deep and the accessible experimental energies limit investigations to those bands that lie lowest in the quantum well and show the least sensitivity to the offsets.

To my knowledge the only attempt to use light scattering with the single objective of determining the conduction-band offset has been the very recent work by Menendez and co-workers [77] again using an undoped GaAs–(AlGa)As MQW sample. The additional feature of this investigation was the determination of the total energy gap difference by experiment, using resonant Raman scattering by the (AlGa)As phonons in the same sample. The sample studied consisted of ten periods of alternating layers of GaAs, 334 Å thick, with (AlGa)As of 59 Å thickness. The thicknesses were measured by TEM and the error in their determination was estimated to be 3 Å. The electronic light scattering spectra from the sample are shown in fig. 30. Intersubband transitions are seen involving the four lowest lying levels in the electron well. The single-particle energies are taken to be independent of the injected photoexcited carrier density because the photoexcitation simultaneously creates an electron and hole plasma so that the total charge density tends to cancel out. The energetic positions of the transitions labeled in fig. 30 were found to be independent of incident laser power, tending to confirm this hypothesis and is consistent with the authors' earlier work [79]. The measured transition energies are given in table 2. Comparison between theory and experiment was made using the envelope function approach, this time with the variation proposed by Potz et al. being employed [13]. This version uses $k \cdot p$ theory to calculate the band structure of the constituents and accounts for the effects of remote bands via second-order perturbation theory. Excellent agreement between theory and experiment is achieved with a conduction-band step of 55.9 meV and a well width of 335.1 Å. The thickness is within the errors set by the TEM determination. The energy gap difference is measured at 81 ± 3 meV. This makes the fraction of discontinuity appearing in the conduction band, Q_c, to be 0.69 ± 0.03. Although not presented in detail in their paper [77] Menendez et al. indicate that complimentary studies of parabolic wells give further support for this offset fraction.

This determination of Q_c at 0.69 ± 0.03 at $x \sim 0.06$ is in good agreement with the value determined from PL by Dawson et al. [74], embraces the PLE value of Duggan and co-workers [37] and is outside the range found by Miller and co-workers [23,29].

Recently, Pinczuk et al. [81] reported the first observation of light scattering from holes in p-type modulation doped quantum wells. A rela-

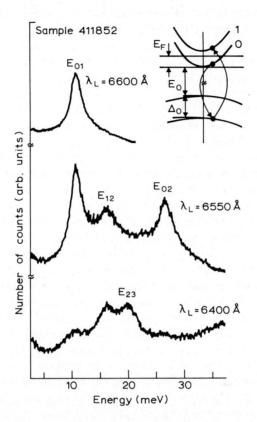

Fig. 30. Depolarized light scattering spectra for three different laser wavelengths. Sample details are given in the text and the inset shows schematically the two-step process responsible for the light scattering. (From ref. [77].)

Table 2
Measured and calculated single-particle transitions (in meV) in a single GaAs quantum well 335.1 Å wide with barriers of (AlGa)As whose Al fraction was about 0.06.

	Q_c	E_{01}	E_{12}	E_{23}	E_{02}
Light scattering [77]	0.69	10.42±0.2	16.03±0.3	19.92±0.3	26.59±0.2
Calculation after ref. [13] [77]	0.69	10.29	16.25	19.92	26.55
Four-band Kane model [11]	0.69	10.30	16.3	20.1	26.6
Three-band Kane model [8]	0.85	10.1	16.3	21.0	26.4

tively sharp spectral feature was identified as a single-particle excitation between the two lowest confined heavy hole states, whilst a much broader feature was assigned to a transition between the lowest heavy hole state and the lowest confined light hole state (see fig. 30). The measured h_0–h_1 transition was claimed to be in good agreement with the heavy hole interlevel spacing predicted when using the offset ratio and heavy hole mass preferred by Miller et al. [29]. The calculated h_0–ℓ_0 splitting using this offset and a light hole mass of $0.094m_0$ lie about 20% higher than the measurements, although the broad feature associated with this transition makes the precise assignment of its spectral position difficult. The broad nature of this feature is an indication that the two valence subbands h_0 and ℓ_0 are not parallel in the plane of the quantum wells. This point is reinforced by the detailed calculations of the valence-band structures of quantum wells by Altarelli and co-workers [89]. The sharp nature of the h_0–h_1 transition argues that these two subbands are parallel over k-vectors smaller than the Fermi wavevector. In principle, the hole dispersion in the plane of quantum wells is quite sensitive to the band offset Q_v and calculations could be made to model the energetic positions and shape of the features seen in the Raman spectra. However, uncertainty about materials parameters (the hole bands in III–Vs are highly anisotropic) and the need to do self-consistent calculations on the modulation doped samples will surely limit the usefulness of this particular technique.

In common with most of the optical techniques the available experimental evidence from light scattering does not conclusively determine the offset ratio at the GaAs–(AlGa)As interface, but the combination of the impressive agreement between experiment and theory for the undoped rectangular well sample [77], the inability to correctly describe the p-type sample and the quoted supplementary support from studies of parabolic wells persuades one that the offset ratio for an Al fraction of 0.06 is likely to be close to the value of 69:31 found by Menendez et al. [77].

6. Internal photoemission

Abstreiter and co-workers [90] have determined the band offsets at a single (AlGa)As–GaAs heterointerface using the method of internal photoemission. The method makes use of the photoexcitation of carriers at the interface from either the conduction or valence band over a barrier and into the conduction band of the wider gap material. The internal photoemission signal is detected as an induced photocurrent normal to the interface. The

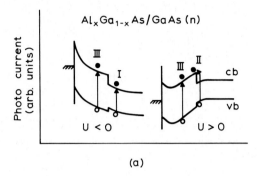

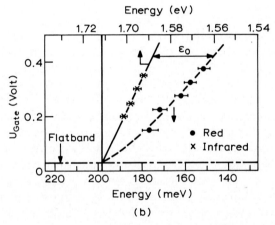

Fig. 31. (a) Schematic diagram of the band edges of the GaAs–(AlGa)As heterojunction under forward and reverse bias conditions; (b) Experimentally determined onsets of the photocurrent versus gate voltage. Crosses relate to excitations from the valence band whilst circles refer to excitations out of the conduction band. (After ref. [90]).

principle is illustrated in fig. 31. The method should require no theory to measure the conduction-band offset. The onset of the measured photocurrent is a function of the applied gate voltage and extrapolation to the flat-band voltage yields the conduction-band offset. Figure 31 shows experimental points for the photocurrent onset as a function of applied gate voltage for a GaAs–$Al_{0.2}Ga_{0.8}As$ single heterojunction when illuminated with either red light or infra-red radiation. Extrapolation to the flat-band condition yields a barrier height of 197 meV for electrons or a Q_c of 0.8.

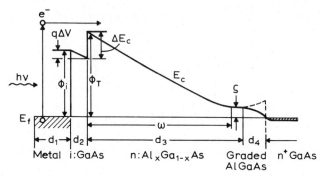

Fig. 32. Conductor-band edge arrangement at the heterojunction studied by the IBM group (From ref. [92].)

The estimated error on the offset fraction is estimated to be ±0.03. This value is more in line with the older value of Dingle [3] rather than the newer values of between 0.6 and 0.7. In a later discussion of the same experimental results Abstreiter et al. [91] point out that they have assumed *k*-conservation not to be important. If the process were such that first one excites an electron–hole pair in GaAs high enough in energy so that the electron can overcome the barrier then *k*-conservation may be required. In which case the dispersion of the heavy hole band becomes important and should be included in the analysis. Such a consideration reduces Q_c to about 0.71. Abstreiter et al. [90] argue that such an interpretation seems unlikely since the same considerations are not necessary for direct excitation from the conduction band when using infra-red radiation and the experiment returns the same result for Q_c of 0.78. The reason for the discrepancy between 0.8 and the most recent determinations of between 0.6 and 0.7 is left open.

The situation is confused further by recent results from the IBM group [92] using photoemission from a metal barrier through a GaAs layer and over the conduction-band step between GaAs and (AlGa)As of various compositions. The band diagram for the measurement is shown in fig. 32. The total barrier height, Φ_T, for the photoexcited electrons is

$$\Phi_T = \Phi_1 - \eta + \Delta E_c - q\Delta V, \tag{12}$$

where Φ_1 is the metal–GaAs barrier height, ΔV is the voltage drop across the thin GaAs layer; η includes image force and tunneling current corrections. In the direct Al composition range the authors find that the offset fraction is 0.62 ± 0.02 of the total band gap difference.

The appeal of the internal photoemission experiments is clear – they should provide a direct measurement of the barrier height. However, it also clear from the spread of results from just these two reports that the understanding of the experimental situation is far from complete if the results from both of these groups are to be reconciled. In particular, one must question the role of interface dipoles in controlling the apparent barrier height measured experimentally; Capasso and co-workers [93] having already shown that such interface dipoles can radically alter or even be made to tailor the apparent barrier height for electrons and holes.

7. Concluding remarks

The evaluation of the band offsets at the (AlGa)As–GaAs heterojunction provides an object lesson for those wishing to determine the energy discontinuities by an optical measurement. The time evolution of the offset ratio clearly indicates the need for careful examination of not only the measurement technique and the theoretical analysis to be used, but also to ask whether the sample structure chosen is suitable and sensitive enough to yield a trustworthy value for the band edge discontinuity.

Optical absorption spectroscopy or photoluminescence excitation spectroscopy should provide a reliable answer for the discontinuities. Yet this method alone when applied to the (AlGa)As system has produced offset ratios ranging from 85 : 15 [2,3] through 75 : 25 [22] and 65 : 35 [37] down to 53 : 47 [29]. This spread of values is, in part, attributable to the use of insensitive quantum well widths and to simplification of boundary conditions in the analysis, alongside the neglect of non-parabolicity and uncertainty in material parameters, in particular, the hole effective masses. Using the correct version of the envelope function approximation, accounting for non-parabolicity and reliable estimates of the exciton binding energies allows one to use this method to arrive at a very accurate value for the quantum well width, which is the most critical parameter when exploring the confined states of rectangular wells. The sensitivity to the offsets can be increased by tailoring the potential profile in some other way, e.g. by making the wells parabolic [23]. However, this at the expense of a degree of control over the technology. Continuous parabolic grading of the Al concentration is extremely difficult and synthesising the profile using the binary compounds alone calls for sophisticated controls over cell shutter speeds and growth rates. In all measurements of this sort one has to be sure that the structure is very well characterised, independent measurements of well width being essential for confidence in the results. Here the power of in situ RHEED monitoring is clear.

Where calculations are an integral part of the analysis then reliable values of material parameters of both well and cladding regions are of paramount importance. The debate over the possible values of heavy hole effective mass in GaAs illustrates the need for further investment in careful measurement of fundamental bulk materials parameters. The situation is worse when considering the multi-ternary heterojunction system of (InGa)As–(AlIn)As where accurate compositional determination is also important for band gap determinations and the knowledge of effective mass values is even more scarce.

In terms of knowledge of materials parameter, electronic light scattering has an advantage over absorption spectroscopy. In the former case, the electron effective masses need to be known whilst the uncertainty of the hole masses is clearly important for the latter method. Furthermore, concentration on single-particle excitations avoids the necessity for excitonic corrections. The light scattering determination on modulation doped structures are complicated by electrostatic effects, many-body considerations and the complex nature of the valence bands. The best chance of a reliable measurement seems to be on judiciously chosen undoped samples, again narrow wells and/or low Al content barriers will increase the sensitivity of the offsets.

Luminescence results that depend on fitting to "intrinsic" transitions as a function of well width I would regard as unreliable. In addition to suffering the same flaws as the optical absorption method one has to be sure that the recombination has no extrinsic origin. Most protagonists of this method make no attempt to measure the well width independently but infer them from growth rates which is probably only accurate to about 10%.

If the size of valence-band offset means that a transition from a straddled to a staggered alignment is possible, as happens in the (AlGa)As system, then the luminescence technique adopted by Dawson et al. [74] should yield values for the offsets with meV accuracy. That such a transition to type-II behaviour occurs qualitatively confirms the large valence-band offset in this system. If no such large offset occurs it may still prove possible to induce such a transition in other systems by judicious choice of quantum well width and the application of hydrostatic pressure [94].

The preceding discussion, I hope, indicates that optical measurements are limited in their ability to provide offset determination on the mV scale. A combination of inadequate theory, materials, growth and technological uncertainties limit their usefulness to the tens of mV scale.

Given these constraints can we reach some positive conclusion about the value of the GaAs–(AlGa)As heterojunction offsets as measured by optical techniques? In my opinion the variety of values reported and the limited range of Al fraction, x, studied makes it impossible to say that the valence band offset varies linearly with x over the entire range with a value equal to $0.54x$ (meV) as claimed by Batey and Wright [75] from the combination of I–V and C–V measurements. In a recent critique [95] of newest electrical experimental results Kroemer selects an offset ratio of 62:38 as the "best" value – a value that is in good agreement with the upper bound of the work of Miller et al. on half parabolic and SCH quantum wells [29], but one which is outside the range of the most recent optical determinations of 69:31 by light scattering [77] and luminescence [74]. The most recent PLE analysis [96] yielding a value of Q_c of 0.67 ± 0.03.

It is outside the scope of this chapter to say whether the optical techniques or electrical measurements give the "right" answer for the offset ratio. Suffice it to say that the electrical measurements of Batey and Wright [75] provide quite good evidence for the offset ratio to be a constant over the entire Al range, taking the Γ point gap as the appropriate gap to consider, and that the optical measurements by three different techniques of PL, PLE and light scattering are converging toward a value closer to 70:30 than 60:40.

Acknowledgements

I am grateful to Barbara Wilson, Milan Jaros, J. Menendez and G. Abstreiter for preprints of their work and to J. Menendez and A. Pinczuk for their comments on parts of the manuscript. It is a pleasure to acknowledge the contributions of my colleagues at Philips: Phil Dawson, Gert 't Hooft, Karen Moore, Karl Woodbridge and especially Hugh Ralph.

References

[1] L.D. Landau and E.M. Lifshitz, Quantum Mechanics, translated by J.B. Sykes and J.S. Bell (Pergamon, New York, 1959).
[2] R. Dingle, W. Weigmann and C.H. Henry, Phys. Rev. Lett. 33 (1974) 837.
[3] R. Dingle, Festkörperprobleme 15 (1975) 21.
[4] D.F. Welch, G.W. Wicks and L.F. Eastman, J. Appl. Phys. 55 (1984) 3176.
[5] G. Bastard, in: Molecular Beam Epitaxy and Heterostructures, eds. L.L. Chang and K. Ploog (Martinus Nijhoff, The Hague, 1985) p. 381.

[6] S.R. White and L.J. Sham, Phys. Rev. Lett. 47 (1981) 879.
[7] D.J. Ben Daniel and C.B. Duke, Phys. Rev. 152 (1966) 683.
[8] G. Bastard, Phys. Rev. B 12 (1982) 7584.
[9] E.O. Kane, J. Phys. & Chem. Solids 1 (1957) 249.
[10] M. Altarelli, Phys. Rev. B 28 (1983) 8842.
[11] M.F.H. Schuurmans and G.W. 't Hooft, Phys. Rev. B 31 (1985) 8041.
[12] J.N. Schulman and Y.C. Chang, Phys. Rev. B 24 (1981) 4445.
[13] W. Potz, W. Porod and D.K. Ferry, Phys. Rev. B 32 (1985) 3868.
[14] A.C. Marsh and J.C. Inkson, J. Phys. C 17 (1984) 6561.
[15] J.N. Schulman and Y.C. Chang. Phys. Rev. B 31 (1985) 2056.
[16] W. Andreoni, A. Baldereschi and R. Car, Solid State Commun. 17 (1978) 821.
[17] E. Caruthers and P.J. Lin-Chung, Phys. Rev. Lett. 38 (1977) 1543.
[18] W. Pickett, S.G. Louie and M.L. Cohen, Phys. Rev. B 17 (1978) 815.
[19] M. Jaros, K.B. Wong and M.A. Gell, Phys. Rev. B 31 (1985) 1205.
[20] N. Holonyak Jr and K. Hess, in: Synthetic Modulated Structures, eds. L.L. Chang and B.C. Giessen (Academic Press, New York, 1985) and references therein.
[21] C. Weisbuch, R.C. Miller, R. Dingle, A.C. Gossard and W. Wiegmann, Solid State Commun. 37 (1981) 219.
[22] P. Dawson, G. Duggan, H.I. Ralph, K. Woodbridge and G.W. 't Hooft, Superlattices and Microstruct. 1 (1985) 231.
[23] R.C. Miller, A.C. Gossard, D.A. Kleinman and O. Munteanu, Phys. Rev. B 29 (1984) 3470.
[24] R.C. Miller, A.C. Gossard and D.A. Kleinman, Phys. Rev. B 32 (1985) 5443.
[25] M.H. Meynadier, C. Delalande, G. Bastard, M. Voos, F. Alexandre and J.L. Lievin, Phys. Rev. B 31 (1985) 5539.
[26] R.C. Miller and D.A. Kleinman, J. Lumin. 30 (1985) 520.
[27] A.L. Mears and R.A. Stradling, J. Phys. C 3 (1970) L21.
[28] G. Bastard, U.O. Ziemelis, C. Delalande, M. Voos, A.C. Gossard and W. Weigmann, Solid State Commun. 49 (1984) 671.
[29] R.C. Miller, D.A. Kleinman and A.C. Gossard, Phys. Rev. B 29 (1984) 7085.
[30] An envelope function approximation calculation shows that the $n = 1$ electron level increases by about 6 meV when the well width is reduced by one monolayer (2.85 Å) whilst the $n = 2$ level increases by about 10 meV. (The calculation assumes that $x = 0.35$ in the barriers and an offset ratio of 65:35.)
[31] M.S. Skolnick, A.K. Jain, R.A. Stradling, J. Leotin, J.C. Ousset and S. Askenasy, J. Phys. C 9 (1976) 2809.
[32] B.A. Vojak, W.D. Laidig, N. Holonyak Jr, M.D. Camras, J.J. Coleman and P.D. Dapkus, J. Appl. Phys. 52 (1981) 621.
[33] R.L. Greene, K.K. Bajaj and D.E. Phelps, Phys. Rev. B 29 (1984) 1807.
[34] J.H. Neave, B.A. Joyce, P.J. Dobson and N. Norton, Appl. Phys. A 31 (1983) 1.
[35] P. Dawson, G. Duggan, H.I. Ralph and K. Woodbridge, Inst. Phys. Conf. Ser. 74 (1984) 391.
[36] P. Dawson and K.J. Moore, unpublished results.
[37] G. Duggan, H.I. Ralph and K.J. Moore, Phys. Rev. B 31 (1985) 8395.
[38] C.M. Wu and E.S. Yang, J. Appl. Phys. 51 (1980) 2261.
[39] J. Batey, S.L. Wright and D.J. Maria, J. Appl. Phys. 52 (1985) 484.
[40] T.W. Hickmott, P.W. Solomon, R. Fischer and H. Morkoc, J. Appl. Phys. 57 (1985) 2844.
[41] W. Potz and D.K. Ferry, Phys. Rev. B 32 (1985) 3863.

[42] F. Bassani and G. Pastori Parravicini, Electronic States and Optical Transitions in Solids (Pergamon Press, Oxford, 1985).
[43] G. Bastard, E.E. Mendez, L.L. Chang and L. Esaki, Phys. Rev. B 26 (1982) 1974.
[44] R.C. Miller, D.A. Kleinman, W.T. Tsang and O. Munteanu, Phys. Rev. B 24 (1981) 1134.
[45] S. Tarucha, H. Okamoto, Y. Iwasa and N. Miura, Solid State Commun. 52 (1984) 815.
[46] J.C. Maan, G. Belle, A. Fasolino, M. Altarelli and K. Ploog, Phys. Rev. B 30 (1984) 2253.
[47] A. Fasolino and M. Altarelli, in: Two Dimensional Systems, Heterojunctions and Superlattices, eds. G. Bauer, F. Kuchar and H. Heinrich (Springer, Berlin, 1984).
[48] G. Duggan, H.I. Ralph, K.S. Chan and R.J. Elliott, Semiconductor Quantum Well Structures and Superlattices VI, in: Proc. Mater. Res. Soc., Strasbourg, 1985, ed. K. Ploog (Les Editions de Physique, Paris, 1986) p. 47.
[49] D.C. Rogers, J. Singleton R.J. Nicholas, C.T. Foxon and K. Woodbridge, Phys. Rev. B 34 (1986) 4002.
[50] D.A. Kleinman, Phys. Rev. B 28 (1983) 871.
[51] P. Lawaetz, Phys. Rev. B 4 (1971) 3460.
[52] A. Baldereschi and N.O. Lipari, Phys. Rev. B 3 (1971) 349.
[53] D. Bimberg, Festkörperprobleme 17 (1977) 195.
[54] K. Hess, D. Bimberg, N.O. Lipari, J.U. Fischbach and M. Altarelli, in: Proc. 13th Int. Conf. on the Physics of Semiconductors, Rome, Italy 1976, ed. F.G. Fumi (North-Holland, Amsterdam, 1977) p. 142.
[55] M. Razeghi, J. Nagle and C. Weisbuch, Inst. Phys. Conf. Ser. 74 (1984) 379.
[56] J.S. Weiner, D.S. Chemla, D.A.B. Miller, T.H. Wood, D. Sivco and A.Y. Cho, Appl. Phys. Lett. 46 (1985) 619.
[57] Y.S. Chen and O.K. Kim, J. Appl. Phys. 52 (1981) 7392.
[58] A.Y. Cho, Thin Solid Films 100 (1984) 291.
[59] R. People, K.W. Wecht, K. Alavi and A.Y. Cho, Appl. Phys. Lett. 43 (1983) 118.
[60] J. Wagner, W. Stolz and K. Ploog, Phys. Rev. B 32 (1985) 4214.
[61] S. Yamada, A. Taguchi and A. Sugimura, Appl. Phys. Lett. 46 (1985) 675.
[62] J.Y. Marzin, M.N. Charasse and B. Sermage, Phys. Rev. B 31 (1985) 8298.
[63] P. Voisin, C. Delalande, M. Voos, L.L. Chang, A. Segmuller, C.A. Chang and L. Esaki, Phys. Rev. B 30 (1984) 2276.
[64] G. Bastard, C. Delalande, M.H. Meynadier, P.M. Frijlink and M. Voos, Phys. Rev. B 29 (1984) 7042.
[65] R.M. Kolbas, Ph.D. thesis (University of Illinois, Urbana-Champaign, IL, 1979).
[66] R.D. Dupuis, P.D. Dapkus, R.M. Kolbas, N. Holonyak Jr and H. Shichijo, Appl. Phys. Lett. 33 (1978) 596.
[67] P.E. Brunemeier, D.G. Deppe and N. Holonyak Jr, Appl. Phys. Lett. 46 (1985) 755.
[68] M. Razeghi, J.P. Hirtz, U.O. Ziemelis, C. Delalande, B. Etienne and M. Voos, Appl. Phys. Lett. 43 (1983) 585.
[69] H. Kawai, J. Kaneko and N. Watanabe, J. Appl. Phys. 58 (1985) 1263.
[70] P.M. Frijlink and J. Maluenda, Jpn. J. Appl. Phys. Lett. 21 (1982) L574.
[71] H. Kawai, J. Kaneko and N. Watanabe, J. Appl. Phys. 56 (1984) 463.
[72] G. Duggan, unpublished results.
[73] J.H. Marsh, J.S. Roberts and P.A. Claxton, Appl. Phys. Lett. 46 (1985) 1161.
[74] P. Dawson, B.A. Wilson, C.W. Tu and R.C. Miller, Appl. Phys. Lett. 48 (1986) 541.
[75] J. Batey and S.L Wright, J. Appl. Phys. 59 (1986) 200.
[76] P. Dawson, unpublished results.

[77] J. Menendez, A. Pinczuk, A.C. Gossard, J.H. English, D.J. Werder and G. Lamont, in: Proc. 13th Conf. on the Physics and Chemistry of Semiconductor Interfaces, Pasedena, CA, 1986, J. Vac. Sci. & Technol. B 4 (1986) 1041.
[78] G. Abstreiter and K. Ploog, Phys. Rev. Lett. 42 (1979) 1308.
[79] A. Pinczuk, J. Shah, A.C. Gossard and W. Weigmann, Phys. Rev. Lett. 46 (1981) 1341.
[80] A. Pinczuk and J.M. Worlock, Surf. Sci. 113 (1982) 69.
[81] A. Pinczuk, H.L. Stormer, A.C. Gossard and W. Wiegmann, in: Proc. 17th Int. Conf. on the Physics of Semiconductors, San Francisco, CA, 1984, eds. J.D. Chadi and W.A. Harrison (Springer, Berlin, 1984) p. 329.
[82] W.I. Wang, E.E. Mendez and F. Stern, Appl. Phys. Lett. 45 (1984) 639.
[83] W.I. Wang and F. Stern, J. Vac. Sci. & Technol. B 3 (1985) 1280.
[84] G. Fishman, Phys. Rev. B 27 (1983) 7611.
[85] T. Ando, J. Phys. Soc. Jpn. 54 (1985) 1528.
[86] T. Ando, J. Phys. Soc. Jpn. 51 (1982) 3893.
[87] A. Pinczuk, J. Phys. Coll. (France) C 5 (1984) 477.
[88] G. Abstreiter, in: Molecular Beam Epitaxy and Heterostructures, eds. L.L. Chang and K. Ploog (Martinus Nijhoff, The Hague, 1985).
[89] M. Altarelli, U. Ekenberg and A. Fasolino, Phys. Rev. B 32 (1985) 5138.
[90] G. Abstreiter, U. Prechtel, G. Weimann and W. Schlapp, Physica B 134 (1985) 433.
[91] G. Abstreiter, U. Prechtel, G. Weimann and W. Schlapp, Surf. Sci. 174 (1986) 312.
[92] M. Heiblum, M.I. Nathan and M. Eizenberg, Surf. Sci. 174 (1986) 318.
[93] F. Capasso, K. Mohammed and A.Y. Cho, J. Vac. Sci. & Technol. B 3 (1985) 1245.
[94] D.J. Wolford, T.F. Keuch, J.A. Bradley, M.A. Gell, D. Ninno and M. Jaros, Proc. 13th Conf. on the Physics and Chemistry of Semiconductor Interfaces, Pasedena, CA, 1986, J. Vac. Sci. & Technol., to be published.
[95] H. Kroemer, Surf. Sci. 174 (1986) 299.
[96] G. Duggan, H.J. Ralph, K.J. Moore, P. Dawson, P.J. Dobson, C.T. Foxon and K. Woodbridge, unpublished results.

CHAPTER 6

THE DIRECT OPTICAL DETERMINATION OF GaAs/Al_xGa_{1-x}As VALENCE-BAND OFFSETS

D.J. WOLFORD and T.F. KUECH

IBM Thomas J. Watson Research Center
Yorktown Heights, NY 10598, USA

and

M. JAROS

The University of Newcastle upon Tyne
Newcastle on Tyne, UK

Heterojunction Band Discontinuities: Physics and Device Applications
Edited by F. Capasso and G. Margaritondo
© *Elsevier Science Publishers B.V., 1987*

Contents

1. Introduction 265
2. Methods 266
3. Results 268
4. Discussion 274
5. Summary 279
References 280

1. Introduction

The electronic structure of GaAs/Al$_x$Ga$_{1-x}$As quantum wells and superlattices has been frequently studied experimentally and theoretically [1]. However, with the exception of ultrathin layers, all efforts have concentrated on determining the properties of quantum states lying deep in the GaAs wells. Such states can be modelled in terms of the so-called effective mass approximation in which the rapidly varying part of the wavefunction is ignored [1,2]. The solution of the Schrödinger equation then depends only on the slowly varying envelope functions whose energy and localization reflect – often sensitively – the well depth and width [2]. Although progress has recently been made in determining well depths, there is still considerable uncertainty concerning the precise value of the barrier heights for both electrons and holes [3]. This is partly because the experimental techniques used to measure barrier heights have been rather indirect [3].

Early work by Dingle and co-workers [2,4] established a consensus of opinion which prevailed for nearly a decade and which identified the fraction of the energy gap difference (between GaAs and Al$_x$Ga$_{1-x}$As) appearing at the conduction-band edge as 0.85. This was achieved by fitting the barrier height (band offsets) to account for exciton recombination energies. However, closer examination showed that other fractions may fit the data better and, more significantly, that the fit is quite insensitive to the choice of band offset. Fractions ranging from 0.85 to 0.50 have been quoted by authors using the optical fitting approach [3,5–7]. More recently, charge-transfer studies interpreted through model calculations [8,10], internal photoemission studies [11], thermal transport over barriers [12,13], C–V profiling techniques [14], and Hall measurement in modulation-doped AlAs/Al$_x$Ga$_{1-x}$As superlattices [15] have all been used to deduce the conduction- or valence-band offsets. The errors quoted by most authors are generally $\geq \pm 50$ meV for GaAs/Al$_x$Ga$_{1-x}$As interfaces (e.g. $\Delta E_c \sim 0.62 \pm 0.05$ eV). This is a composite error involving assessment of both experimental and model-dependent variables and its value is difficult to establish accurately. In general, however, results involving electrical measurement lead to estimated offsets of $\sim 0.6:0.4$ distributed between the conduction and valence bands for the GaAs/Al$_x$Ga$_{1-x}$As system [3,8–15].

In this study we present a new optical method which possesses the spectroscopic resolution and ultimate simplicity to potentially resolve this issue of band offsets. We show that a direct and more accurate measurement of the barrier height is possible if high hydrostatic pressure is used, at low temperatures, to drive the lowest confined states associated with the Γ minimum of the conduction-band edge of GaAs up in energy, to the crossing point with the indirect (X) edge of the $Al_xGa_{1-x}As$ barriers. This crossing occurs in a narrow range of energies in which the confined states, pinned to the *secondary* X minima of the $Al_xGa_{1-x}As$, hydridize with the zone-center (Γ) states of the GaAs, and, as a result, yield observable optical spectra. Our theory predicts that these new X-like states are well confined and dispersionless in the quantum wells and superlattices in question, and their spectra can be resolved – with meV accuracy – with respect to bulk GaAs and $Al_xGa_{1-x}As$ band edges. This enables us to deduce the barrier heights, or the so-called band offsets, *directly* without introducing model-dependent parameters.

Our study also provides the first experimental evidence concerning physical properties of confined states having an X character, e.g. their energies, wavefunctions, and transition probabilities. Although the existence of some such states has been hinted at before [16–21], none of the data or theory available in the literature can be used to determine their physical properties such as, e.g., their relationship to the states of different momenta with which they are degenerate in energy. This relationship determines the shape and intensity of observable spectra associated with these states. The experimental results presented here are in good agreement with predictions based on full-scale pseudopotential calculations [20–22]. Analogous studies can be made in other microstructures.

2. Methods

Samples were prepared epitaxially by MOCVD at 700–750°C in high (>50) AsH_3/MO ratios on (100) GaAs substrates [23–25]. Figure 1 compares photoluminescence spectra at 2 K for a single ~ 5 μm GaAs layer and a similarly prepared $GaAs/Al_xGa_{1-x}As$ multi-quantum well (MQW) [24,25]. The bulk spectrum illustrates the high crystal quality attained, showing sharp near-edge exciton and carbon-acceptor structure indicative of low doping level ($\leq 5 \times 10^{14}$ cm^{-3}, n-type) [26]. Such undoped, n-type material makes up the wells of the MQW and the superlattices (SLs) used here. The $Al_xGa_{1-x}As$ barrier materials were also

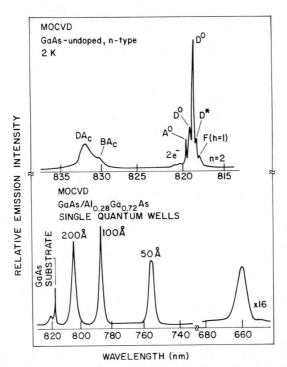

Fig. 1. Photoluminescence spectra (2 K) of a three-well GaAs/Al$_{0.28}$Ga$_{0.72}$As MQW prepared by MOCVD. Well widths are approximately 200 Å, 100 Å and 50 Å; barriers are 500 Å. MQW transitions are $\Gamma_{1e}-\Gamma_{1hh}$, found together with GaAs substrate and barrier emission. Shown for comparison is photoluminescence (2 K) of undoped, n-type ($\leq 3 \times 10^{14}$ cm^{-3}) GaAs prepared under similar MOCVD growth conditions, and showing sharp excitonic and near-edge direct-gap structure indicative of high crystal quality.

nominally undoped, and p-type ($10^{16}-10^{17}$ cm^{-3}) [23], with compositions x estimated by PL [27] and X-ray microprobe analysis. The MQW in fig. 1 is composed of single 200, 100 and 50 Å wells bounded by 500 Å Al$_x$Ga$_{1-x}$As ($x \approx 0.28$) barriers, and prepared on an undoped GaAs buffer and GaAs substrate. Emission is seen from each sample region, with the quantum wells showing strong Γ-confined transitions ($\Gamma_{1e}-\Gamma_{1hh}$) for each well, of 4–6 meV halfwidth. Approximate well widths were deduced from growth rates, TEM micrographs, and electron diffraction analysis in similarly prepared SLs. Superlattices reported here were composed of 40 periods of GaAs/Al$_x$Ga$_{1-x}$As ($x \approx 0.28$ and 0.70), with ~ 70 Å well widths, and prepared on undoped GaAs buffer layers and GaAs substrates. Relatively narrow well widths have been chosen for the SLs to eliminate possible influence of band bending.

For pressure application, samples were thinned to ≤ 50 μm and cleaved into squares ~ 80 μm on a side [28,29]. Single samples were loaded into steel-gasketed diamond anvil cells, together with an ~ 10 μm ruby crystal and a transparent pressure-transmitting fluid [30]. For measurement, cells were held in an optical cryostat at 8 K in flowing He. Pressures were hydrostatic to better than 1 part in 600 [29] and were determined by monitoring the shift in ruby R_1 emission [30]. Photoluminescence was excited by the 4579 Å line of an Ar^+ laser at power densities incident on the cells of 10^3 W cm^{-2}. Emission was collected in back-scattering and analyzed with a double-grating spectrometer and photon counting.

In the theoretical analysis we employ a pseudopotential method which has been used successfully to model the electronic structure of quantum wells and graded structures at zero pressure [20,22]. The Hamiltonian is $H = H_0 + V$ where H_0 represents bulk GaAs. The potential V is the difference between the pseudopotentials associated with the atoms of $Al_xGa_{1-x}As$ and GaAs in the alloy layer. We ignore the difference between lattice constants of the constituent materials and employ the virtual crystal approximation to model the alloy (bulk) band structure. The input potential is fitted to reproduce accurately the band structure of GaAs and $Al_xGa_{1-x}As$, and the pressure dependence of GaAs (using recent experimental results) [28,29]. The relative position of the two bulk band structures is shifted rigidly in order to adjust the band offset. No other adjustments are involved. We expand the wavefunction ψ in terms of the eigenfunctions of H_0 and solve the Schrödinger equation $(H - E)\psi = 0$ by direct diagonalization.

3. Results

Figure 2 presents high-pressure PL results for the three-well MQW of fig. 1. As at atmospheric pressure, each of the Γ-confined transitions ($\Gamma_{1e}-\Gamma_{1hh}$) appear at the lowest pressure of 8.7 kbar as strong, narrow PL lines. They are, however, shifted up in energy (together), compared to room pressure, due to the pressure-induced gap shift of the bulk GaAs [18,31–33]. This continues at higher pressures, until above ~ 26 kbar PL intensity rapidly diminishes, as evidenced by the scale factors and comparison with the substrate emission [31]. Importantly, this intensity decrease occurs for the narrowest (50 Å), highest-energy well first, with the next narrower well

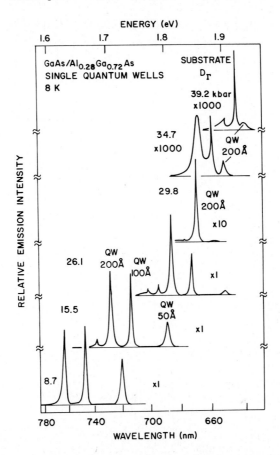

Fig. 2. Photoluminescence spectra (8 K) of a three-well GaAs/Al$_{0.28}$Ga$_{0.72}$As MQW prepared by MOCVD and held at indicated pressures in a diamond anvil cell. Well widths are approximately 200 Å, 100 Å and 50 Å; barriers are 500 Å. Energies of the Γ-confined transitions (Γ_{1e}–Γ_{1hh}) increase with pressure, following the shift (~10.7 meV/kbar) of the direct GaAs band edge. Note the sharp intensity decrease of well emissions (over 10^3) between 26 and 35 kbar.

following in succession. Above 34 kbar only a weak remnant of the 200 Å well remains – here reduced by a factor of $> 10^3$ compared to its initial low-pressure intensity – and spectra are dominated by direct near-edge (donor-bound-exciton and band-to-acceptor) substrate emission. This continues through 39.2 kbar, a pressure which is still somewhat below the Γ–X crossover of bulk GaAs occurring at 41.3 kbar [28,29].

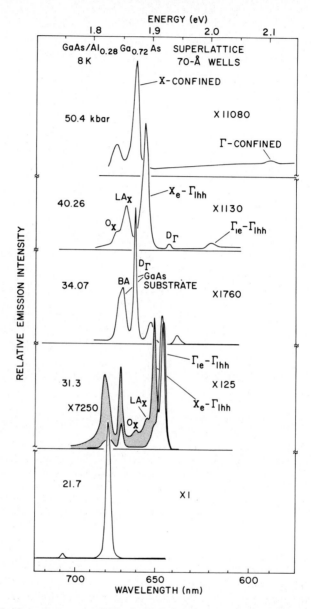

Fig. 3. Photoluminescence spectra (8 K) of a 40-period GaAs/Al$_{0.28}$Ga$_{0.72}$As SL with ~ 70 Å wells prepared by MOCVD and held at indicated pressures in a diamond anvil cell. Energies of the Γ-confined transitions ($\Gamma_{1e}-\Gamma_{1hh}$) increase with pressure, following the shift (~10.7 meV/kbar) of the direct GaAs band edge (note GaAs substrate emission labeled D$_\Gamma$ and BA), while intensities rapidly diminish near and above 31 kbar. A staggered-band, X-confined transition (X$_e-\Gamma_{1hh}$) appears here, and shifts down in energy and weakens with increasing pressure; phonon replicas follow. Stipled spectrum represents a 10× reduction in incident power density.

Figure 3 shows corresponding pressure results for a 70 Å well GaAs/Al$_{0.28}$Ga$_{0.72}$As superlattice for pressures from 21.7 kbar to well beyond the bulk GaAs Γ–X crossover. The narrow (7.4 meV) line at 21.7 kbar results from the Γ_{1e}–Γ_{1hh} transition, while the weak accompanying line results from substrate emission. For this sample, SL intensity remains strong and virtually independent of pressure from room pressure to ~ 30 kbar. Above 30 kbar, however, intensity decreases sharply with pressure (note scale factor of ×125 at 31.3 kbar), and a set of new PL lines appears, together with relative strengthening of the substrate emission. Occurring first as a shoulder on the main line at 31.3 kbar, this new emission becomes more prominent as excitation power density is reduced (stipled). Note that at this pressure these new processes lie *above* the bulk GaAs band edge, but below the band edge of the Γ-confined quantum states. As pressure is increased still further these lines shift toward *lower* energies, eventually passing through the GaAs band edge, while the Γ_{1e}–Γ_{1hh} line continues to follow the direct gap to higher energies.

Superlattice emission shifting under pressure counter to, or independently of, the Γ_{1e}–Γ_{1hh} lines has been recently reported [18,32]. One report identifies this as X-edge emission in the GaAs [18]. The other identifies it as sub-gap interface-impurity recombination [32]. The new lines shown in fig. 3 cannot, however, be explained by either of these interpretations. As we will demonstrate, the transitions we have observed originate from spatially separated, X-confined subband electrons located within the Al$_x$Ga$_{1-x}$As barriers, recombining with Γ-confined subband heavy holes in the GaAs wells.

Figure 4 shows pressure results proving this new PL may also be seen in other related structures, in this case a 70 Å well superlattice composed of GaAs wells with Al$_{0.70}$Ga$_{0.30}$As barriers. Here we directly compare the usual Γ-confined (Γ_{1e}–Γ_{1hh}) transition and the new emission (at different pressures), with their peaks aligned vertically. The principal lineshapes are surprisingly similar, both being nearly symmetric with half-widths of ~ 9.5 meV. They differ, however, in both intensity and low-energy structure. The Γ_{1e}–Γ_{1hh} line is some 650 × stronger for these indicated pressures, under identical laser excitation conditions. And unlike the Γ_{1e}–Γ_{1hh}-line, the new line labeled X$_e$–Γ_{1hh} is always found in our data accompanied by phonon replicas spaced by the energies appropriate to Al$_x$Ga$_{1-x}$As near-zone-boundary LA and optical (unidentified) phonons [34,35]. These same replicas may be seen in the sample of fig. 3 as well, and depending on pressure, have relative intensities of 0.1–0.4. Such accompanying phonon structures have not been reported in other recent studies of quantum well

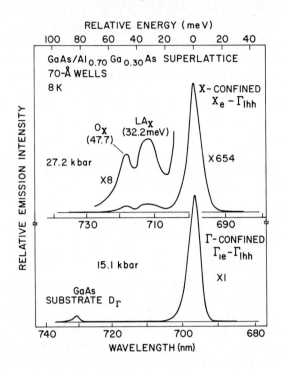

Fig. 4. Photoluminescence spectra (8 K) of a 40-period GaAs/Al$_{0.70}$Ga$_{0.30}$As SL with an ~ 70 Å well prepared by MOCVD and held at indicated pressures in a diamond anvil cell. Lineshape of the usual Γ-confined transition ($\Gamma_{1e}-\Gamma_{1hh}$) and the weaker staggered-band, X-confined transition (X$_e-\Gamma_{1hh}$) are similar (8–9 meV half-widths). Phonon replicas of the X$_e-\Gamma_{1hh}$ transition are from near zone boundary.

emission under pressure [18,32], perhaps suggesting we have observed different processes.

Figure 5 compares the collected pressure results from the two SLs of differing barrier composition ($x \approx 0.28$ and 0.70). For reference the Γ and X band-gap dependences on pressure of bulk GaAs are shown, together with the bulk edge dependences of the two barrier materials, all referenced to the top of the bulk GaAs valence band [28,29,36]. Shown also for comparison is the region, in both pressure and energy (stipled), over which $\Gamma_{1e}-\Gamma_{1hh}$ intensities quench by a factor of more than 10^3 in the three-well MQW ($x \approx 0.28$) of figs. 1 and 2. Immediately apparent is that, as might be expected based on effective-mass treatment of quantum wells [1,2], the $\Gamma_{1e}-\Gamma_{1hh}$ transitions closely track the Γ-gaps of bulk GaAs (and

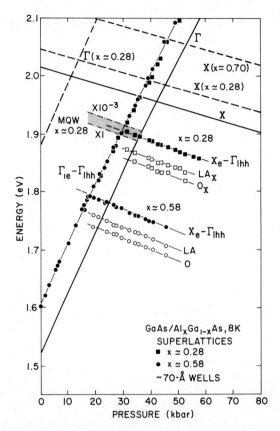

Fig. 5. Photoluminescence transition energies (8 K) of 40-period GaAs/Al$_x$Ga$_{1-x}$As SLs with ~ 70 Å wells versus pressure in diamond anvil cells. Shown also are the assumed Γ and X conduction-band edges of GaAs and the Al$_x$Ga$_{1-x}$As barriers versus pressure, *all* referenced to the top of the GaAs valence band. Γ-confined transitions ($\Gamma_{1e}-\Gamma_{1hh}$) follow the shift (~ 10.7 meV/kbar) of the direct band edges. Intensities decrease sharply above 18 kbar for $x \approx 0.70$ and above 30 kbar for $x \approx 0.28$. Above these thresholds the staggered-band, X-confined transitions ($X_e-\Gamma_{1hh}$) appear and follow the indirect band edges; phonon replicas follow. Threshold region for quenching of MQW ($x \approx 0.28$) transitions of fig. 3 is shown (stipled). Valence-band offsets are derived *directly* from these data according to fig. 6.

Al$_x$Ga$_{1-x}$As). As we show elsewhere, however, detectible deviations do occur as confinement increases and "non-effective-mass" intervalley mixing increases [31,37]. In contrast to the Γ-confined states, the $X_e-\Gamma_{1hh}$ transitions in both superlattices follow closely the pressure dependence of the X conduction-band edges. This result, together with the accompanying near-zone-boundary replicas, identify the bound electron as being formed from X-conduction states.

An additional striking result in fig. 5 is that the crossings between these new X-confined states and the Γ_{1e}-states occur at very different threshold pressures in the two SL structures. For $x \approx 0.28$ this crossing occurs near 29.5 kbar, while for $x \approx 0.70$ it occurs near 18 kbar. These crossings also correspond precisely in pressure and energy to the sharp intensity quenching of the $\Gamma_{1e}-\Gamma_{1hh}$ transitions in both SLs. Further, the level crossing in the SL with $x \approx 0.28$ barriers agrees in energy and pressure with the intensity threshold for the three wells of the MQW of similar barrier composition. These similarities and differences are thus directly related to the Al-mole-fraction in the *barriers*. Indeed, as we will show, they are the key to understanding the quantum-well band-structure and, therefore, the spectroscopic determination of the band offsets.

4. Discussion

Our results may be readily explained by the one-electron GaAs/Al$_x$Ga$_{1-x}$As QW and SL band diagrams of fig. 6. These illustrations allow us not only to interpret and correlate MQW and SL results, but also provide the first *simple, direct* method for obtaining valence-band offsets, ΔE_v, for a hetero-interface. To best illustrate the simplicity of our method, we have labeled in fig. 6 only those energies necessary for obtaining ΔE_v from our data. Here, $E_{gB}^X(P)$ represents the X band gap of the Al$_x$Ga$_{1-x}$As barriers versus pressure; $\Sigma_{1e-1hh}^{\Gamma-\Gamma}(P)$ and $\Sigma_{1e-1hh}^{X-\Gamma}(P)$ represent the energy separations ($\Gamma_{1e}-\Gamma_{1hh}$ and $X_e-\Gamma_{1hh}$) of the one-electron ground-state quantum levels, respectively; Δ_{1hh}^Γ equals the ground-state Γ-heavy-hole confinement energy; and Δ_{1e}^X equals the ground-state X-electron confinement energy. We have neglected in this one-electron diagram only:

(1) the internal exciton binding energies E_x which amount to only ~ 10 meV for the $\Gamma_{1e}-\Gamma_{1hh}$ exciton [38], and even less for the $X_e-\Gamma_{1hh}$ exciton (diminished by the electron–hole spatial separation), and

(2) the small (≤ 5 meV) "Stokes shifts" which we detect between peak PL emission and peak PL-excitation (i.e. absorption) [24]. In practice, such exciton corrections act to reduce, or even cancel, the already small effect of the localization energies, Δ_{1hh}^Γ and Δ_{1e}^X, in deducing ΔE_v from our data.

In both MQWs and SLs the $\Gamma_{1e}-\Gamma_{1hh}$ transitions increase in energy with pressure according to the Γ-gap increase of the bulk GaAs (and bulk Γ-gap of the Al$_x$Ga$_{1-x}$As barriers). At a critical pressure, however, the electron

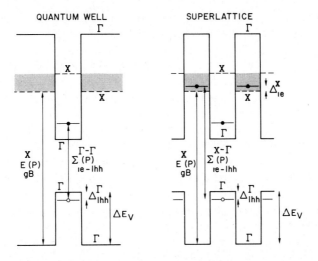

Fig. 6. Schematic band diagram showing for the GaAs/Al$_x$Ga$_{1-x}$As heterostructure those energies necessary for obtaining valence-band offsets, ΔE_v, from pressure experiments (symbols are explained in text). Minima formed in X states by valence-band offset are shown stipled. Only Γ-confined bound states exist for isolated quantum wells. For superlattices X-confined electron bound states may exist within the Al$_x$Ga$_{1-x}$As barriers.

bound state (Γ_{1e}) becomes degenerate with the X conduction edge of the barrier material in MQWs, or an electron-bound state confined to the X conduction-band "wells" formed by the barriers. This occurs at pressures much lower than the bulk GaAs Γ–X band crossing (41.3 kbar), but well-above any possible band crossings of the barriers (fig. 5). It may also occur at pressures well-below the point at which the Γ_{1e} state becomes degenerate with the X states of the *GaAs wells*, because of the existence of a valence-band offset ΔE_v. When ΔE_v is small enough, such as would be the case for the originally accepted $\Delta E_v \approx 0.15 \Delta E_g$ offsets, an energy-well for electrons would form for X states in the GaAs, for barriers of composition $x \approx 0.28$. Indeed, for $x \approx 0.28$, this occurs for all valence-band offsets of $\lesssim 0.20 \Delta E_g$. For larger offsets, however, the absolute position of the X band-edge of the Al$_{0.28}$Ga$_{0.72}$As may be depressed sufficiently (relative to GaAs) to create, instead, energy-wells within the Al$_x$Ga$_{1-x}$As "barriers". This causes a "staggered" band alignment and can lead to spatial separation of nonequilibrium (photo-excited) electrons and holes, especially when these X edges represent the lowest energy of the heterostructure conduction band. This is the example represented in fig. 6 for the case of a

0.7:0.3 offset. These subtleties relating to the possible band alignments were misunderstood in other recent pressure studies of quantum-well structures [18,32]. Thus, it led these authors to conclude incorrectly (based on an *assumed* 0.6:0.4 offset) that all existing energy wells for both electrons and holes must lie within the GaAs, and that the bulk GaAs X-band edges must therefore be 45–50 meV lower in energy than we show in fig. 6 and have reported in ref. [29].

Therefore, we conclude from fig. 6 that quenching observed under pressure for Γ_{1e}–Γ_{1hh} transitions in both MQWs and SLs is due to transfer of Γ-confined electrons into the barriers when degeneracy with these free-electron X states is reached. Note this results in *both* spatial and *k*-space separation of the photo-excited electrons and holes. Although Γ-confined electron states continue to exist above these crossings as conduction-band resonances (note the weak, but persistent Γ_{1e}–Γ_{1hh} lines at high pressures in figs. 3 and 5), oscillator strength for their transition weakens with energy separation between the Γ_{1e} and X_{1e} states, as a result of reduced inter-valley and inter-layer mixing [37]. In addition, the Γ_{1e} electron states become naturely depopulated as a result of rapid thermalization of electrons from these resonances toward the existing accessible lower energy states provided by the barriers. Such electron transfer from the GaAs to the $Al_xGa_{1-x}As$ occurs across the hetero-interface as a consequence of quantum mechanical tunnelling, or "leakage", of the wavefunction into the adjacent layer. Indeed, it is this leakage, or the finite amplitude of the Γ component of the electron wavefunction outside the alloy layers, which makes observable the optical transition between the confined X-like electron state near the indirect $Al_xGa_{1-x}As$ band gap and the Γ-like heavy hole state at the top of the GaAs valence band. This is demonstrated in fig. 7 where we plot the charge density $|\psi|^2$ associated with an indirect-edge state near the Γ_{1e}–X_{1e} crossover pressure. Here we present a result for a 50 Å GaAs well bounded by 450 Å $Al_{0.28}Ga_{0.72}As$ barriers, to illustrate that the Γ component is present, even in the case of wide barriers. These Γ-related spatial tails of the wavefunction, seen most clearly in fig. 7b, are responsible for the relatively strong zero-phonon lines found in figs. 3–5. Without these $k = 0$ components of the electron bound state (shown dashed in fig. 7b with $k = 0$ component removed) the near-zone-boundary momentum-conserving phonons found in figs. 3–5 would dominate the recombination, as in bulk studies of indirect-gap excitons [39].

Considering now the issue of band offsets from our data and the band diagrams of fig. 6, we conclude the following. From the photoluminescence energies $PL^{\Gamma-\Gamma}_{1e-1hh}(P)^* = \Sigma^{\Gamma-\Gamma}_{1e-1hh}(P)^* - E^{\Gamma-\Gamma}_x - \Delta^{\Gamma-\Gamma}_{ss}$ at which the Γ_{1e}–Γ_{1hh}

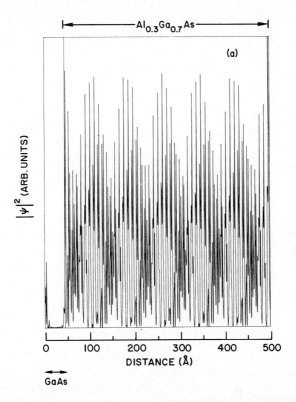

Fig. 7a. Spatial wavefunction $|\psi|^2$ of the lowest staggered-band, X-confined electron state formed in 450 Å-wide $Al_{0.3}Ga_{0.7}As$ "barriers", surrounding a 50 Å GaAs "well," assuming a valence-band offset of $0.3\Delta E_g$. Contributions from all sampling points across the first Brillouin zone are included. Nodal structure arises from the complex, dominantly X-like character of the electron bound state.

exciton transitions quench for each well width in an MQW, the valence-band offset ΔE_v may be directly deduced using the simple expression

$$\Delta E_v = E_{gB}^X(P) - \Sigma_{1e-1hh}^{\Gamma-\Gamma}(P)^* + \Delta_{1hh}^{\Gamma}, \qquad (1)$$

where $E_x^{\Gamma-\Gamma}$ is the internal exciton binding energy of 9 meV [38], and $\Delta_{ss}^{\Gamma-\Gamma}$ is the "Stokes shift" of ~4 meV found between peak PL and peak PL-excitation [24]. Knowing the barrier indirect-gap dependence on pressure $E_{gB}^X(P)$ from fig. 5, the only uncertainty is the hole localization energy Δ_{1hh}^{Γ} which may be easily computed for each well width. Because of X-well

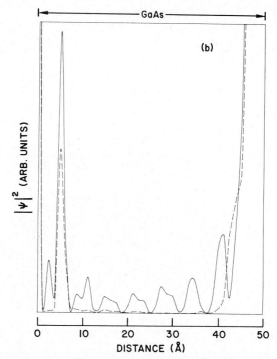

Fig. 7b. Spatial wavefunction $|\psi|^2$ of fig. 7a largely confined to the $Al_{0.3}Ga_{0.7}As$ "barriers," but tunnelling into the 50 Å GaAs "well" (solid), compared to result with $\Gamma(k=0)$ contributions removed from the wavefunction (dashed). This result indicates that the strong zero-phonon line (X_e–Γ_{1hh}) found in figs. 3–5 arises primarily from $k=0$ electron–hole recombination from within the GaAs layers.

formation in SLs (fig. 6), this threshold point must also precisely correspond to appearance of X_e–Γ_{1hh} recombination from across the interface. This is indeed confirmed in fig. 5 for the MQW and SL of similar barrier composition, $x \approx 0.28$. Using fig. 6 it is correspondingly obvious that for SLs the valence-band offset may be directly deduced from the X_e–Γ_{1hh} transition energies $PL^{X-\Gamma}_{1e-1hh}(P) = \Sigma^{X-\Gamma}_{1e-1hh}(P) - E^{X-\Gamma}_x - \Delta^{X-\Gamma}_{ss}$ using the expression

$$\Delta E_v = E^X_{gB}(P) - \Sigma^{X-\Gamma}_{1e-1hh}(P) + \Delta^{\Gamma}_{1hh} + \Delta^X_{1e}. \qquad (2)$$

Here, $E^{X-\Gamma}_x$ is the internal binding energy of the indirect-gap exciton formed between the electron and hole in adjacent superlattice layers which

Table 1
Valence-band offsets ΔE_v and fractional direct-gap discontinuity appearing in the valence band, as derived from pressure-dependent quantum-well PL and eqs. (1) and (2).

x, Al-mole-fraction	ΔE_v (eV)	Fractional ΔE_g
0.28 ± 0.01	0.110 ± 0.008	0.31 ± 0.03
0.70 ± 0.01	0.320 ± 0.010	0.34 ± 0.02

– because of minimal electron–hole spatial overlap – we shall neglect [40], and $\Delta_{ss}^{X-\Gamma}$ is an assumed 4 meV Stokes shift between peak emission and excitation.

As Δ_{1e}^{X} and Δ_{1hh}^{Γ} are small for 70 Å wells, and because these terms are effectively cancelled by the small exciton terms of opposite sign, the offset ΔE_v may be estimated from the SL data using *only* $E_{gB}^{X}(P)$ and $PL_{1e-1hh}^{X-\Gamma}(P)$. However, from our calculations for 70 Å wells [20–22], we obtain $\Delta_{1e}^{X} \approx 4$ meV and $\Delta_{1hh}^{\Gamma} \approx 10$ meV (effective-mass theory provides similar results). Thus, using the data of fig. 5 surrounding the X_{1e}–Γ_{1e} crossings and the $E_{gB}^{X}(P)$ at the same pressures, we obtain the valence-band offsets displayed in table 1. As shown, we find an approximate 0.68 : 0.32 division of the direct gap for both the direct(0.28)- and indirect(0.70)-gap compositions studied. This differs considerably from the original 0.85 : 0.15 rule [2,4] and somewhat from the recent ~ 0.6 : 0.4 consensus [3,5–15]. Complete composition-dependence for the GaAs–Al$_x$Ga$_{1-x}$As system will be reported elsewhere [37].

The principal uncertainties in determining band offsets from our data and model arise from accurately determinating $E_{gB}^{X}(P)$, and from the assumption that, with respect to GaAs, the conduction and valence bands of the Al$_x$Ga$_{1-x}$As barriers move together with pressure. These considerations are easily settled, with meV resolution, however, as will be shown [37]. We thus conclude that by the methods described here, band offsets may be reliably obtained, for the first time, with spectroscopic precision.

5. Summary

In this chapter we have discussed experiment and theory of the pressure dependence of quantum-well bound states formed in the GaAs/Al$_x$Ga$_{1-x}$As heterostructure system. Using MQWs and SLs of various barrier compositions, x, we studied in photoluminescence spectra

(8 K), and in full scale pseudopotential calculation, the pressure-induced evolution of the lowest spatially confined states within the wells. With increasing pressure, Γ-confined states follow the shift to higher energies of the direct GaAs band gap. At critical pressures a crossing occurs between these Γ bound states and the *barrier* indirect X states. Here, Γ intensities plunge and new emission tracking the X-edges appears. Confirmed in wavefunction calculation, these new transitions occur across the hetero-interface, between X-confined electrons within the $Al_xGa_{1-x}As$ and Γ-confined holes within the GaAs. Arising from valence-band offset-induced staggered band alignment, critical pressures for observation of these states decrease with increasing Al-mole-fraction. We thus obtain, with meV resolution, the first *direct* measure of valence-band offsets. For $x \approx 0.28$ and 0.70 we find $\Delta E_v \approx (0.32 \pm 0.02) \Delta E_g^\Gamma$. Taken together, these results comprise one of the most precise and spectroscopically detailed accounts of staggered band alignment yet reported for a semiconductor interface. The optical method reported on here should apply to other interface systems. Our results also represent the first spectroscopic evidence concerning the physical properties of confined states associated with secondary (X) minima. An excellent agreement between theory and experiment is achieved.

Acknowledgements

This work was performed in colaboration with J.A. Bradley of IBM, and M.A. Gell and D. Ninno of The University of Newcastle Upon Tyne. We thank K.K. Bajaj and D. Bimberg for helpful discussions, and the US Office of Naval Research (under contracts N00014-80-C-0376 and N00014-85-C-0868) and the Croucher Foundation, RSRE, Malvern, SERC (UK) for support. Reprinted from the Journal of Vacuum Science and Technology [B 4 (1986) 1043] with permission of the American Institute of Physics, New York, NY, USA.

References

[1] T. Ando, A.B. Fowler and F. Stern, Rev. Mod. Phys. 54 (1982) 437.
[2] R. Dingle, Festkörperprobleme 15 (1975) 21.
[3] G. Duggan, J. Vac. Sci. & Tech. B 3 (1985) 1224.
[4] R. Dingle, W. Wiegmann and H.C. Casey Jr, Phys. Rev. Lett. 33 (1974) 827.
[5] R.C. Miller, D.A. Kleinman and A.C. Gossard, Phys. Rev. B 29 (1984) 7085.

[6] R.C. Miller, A.C. Gossard, D.A. Kleinman and O. Munteanu, Phys. Rev. B 29 (1984) 3470.
[7] P. Dawson, G. Duggan, H.I. Ralph, K. Woodbridge and G.W. 't Hooft, J. Superlattices and Microstructures 1 (1985) 231.
[8] W.I. Wang, E. Mendez and F. Stern, Appl. Phys. Lett. 45 (1984) 639.
[9] W.I. Wang and F. Stern, J. Vac. Sci. & Tech. B 3 (1985) 1280.
[10] W.I. Wang, Solid-State Electron. 29 (1986) 133.
[11] M. Heiblum, M.I. Nathan and M. Eizenberg, Appl. Phys. Lett. 47 (1985) 503.
[12] J. Batey, S.L. Wright and D.J. DiMaria, J. Appl. Phys. 52 (1985) 484.
[13] J. Batey and S.L. Wright, J. Appl. Phys. 59 (1986) 200.
[14] M.O. Watanabe, J. Yoshida, M. Mashita, T. Nakanisi and A. Hojo, J. Appl. Phys. 57 (1985) 5340.
[15] T.J. Drummond and I.J. Fritz, Appl. Phys. Lett. 47 (1985) 284.
[16] E.E. Mendez, L.L. Chang, G. Landgren, R. Ludeke, L. Esaki and F.H. Pollak, Phys. Rev. Lett. 46 (1981) 1230.
[17] P.L. Gourley, R.M. Biefeld, G.C. Osbourn and I.J. Fritz, Inst. Phys. Conf. Ser. 65 (1982) 249.
[18] U. Venkateswaran, M. Chandrasekhar, H.R. Chandrasekhar, T. Wolfram, R. Fischer, W.T. Masselink and H. Morkoc, Phys. Rev. B 31 (1985) 4106;
U. Venkateswaran, M. Chandrasekhar, H.R. Chandrasekhar, T. Wolfram, R. Fischer and H. Morkoc, Bull. Am. Phys. Soc. 30 (1985) 453.
[19] J.N. Schulman and T.C. McGill, Phys. Rev. B 19 (1979) 6341.
[20] M. Jaros and K.B. Wong, J. Phys. C 17 (1984) L765.
[21] M. Jaros, K.B. Wong and M.A. Gell, Phys. Rev. B 31 (1985) 1205.
[22] M. Jaros, K.B. Wong, M.A. Gell and D.J. Wolford, J. Vac. Sci. & Tech. B 3 (1985) 1051.
[23] T.F. Kuech, E. Veuhoff, D.J. Wolford and J.A. Bradley, Inst. Phys. Conf. Ser. 74 (1984) 181.
[24] R.D. Dupuis, R.C. Miller and P.M. Petroff, J. Cryst. Growth 68 (1984) 398;
R.C. Miller, R.D. Dupuis and P.M. Petroff, Appl. Phys. Lett. 44 (1984) 508.
[25] S.D. Hersee, M. Krakowski, R. Blondau, M. Baldy, B. De Cremoux and J.P. Duchemin, J. Cryst. Growth 68 (1984) 38.
[26] U. Heim and P. Heisinger, Phys. Status Solidi 66 (1974) 461.
[27] R. Dingle, R.A. Logan and J.R. Arthur, Inst. Phys. COnf. Ser. 33a (1977) 216.
[28] D.J. Wolford, J.A. Bradley, K. Fry, J. Thompson and H.E. King, Inst. Phys. Conf. Ser. 65 (1982) 477.
[29] D.J. Wolford and J.A. Bradley, Solid State Commun. 53 (1985) 1069.
[30] A. Jayaraman, Rev. Mod. Phys. 55 (1983) 65.
[31] D.J. Wolford, J.A. Bradley, T. Kuech and W. Wang, Bull. Am. Phys. Soc. 30 (1985) 454.
[32] S.K. Hark, B.A. Weinstein and R.D. Burnham, Bull. Am. Phys. Soc. 30 (1985) 453;
B.A. Weinstein, S.K. Hark and R.D. Burnham, J. Appl. Phys. 58 (1985) 4662.
[33] S.W. Kirchoefer, N. Holonyak Jr, K. Hess, D.A. Gulino, H.G. Drickamer, J.J. Coleman and P.D. Dapkus, Solid State Commun. 42 (1982) 633;
S.W. Kirchoefer, N. Holonyak Jr, K. Hess, D.A. Gulino, H.G. Drickamer, J.J. Coleman and P.D. Dapkus, Appl. Phys. Lett. 40 (1982) 821.
[34] G. Dollinger and J.L.T. Waugh, in: Proc. Int. Conf. on Lattice Dynamics, Copenhagen, ed. R.F. Wallis (Pergamon Press, New York, 1964) p. 19.
[35] R. Trommer, E. Anastassakis and M. Cardona, in: Light Scattering in Solids, eds. M. Balkanski, R.C.C. Leite and S.P.S. Porto (Flammarion, Paris, 1976) p. 396.

[36] D.J. Wolford and J.A. Bradley, unpublished.
[37] D.J. Wolford, T.F. Kuech, J.A. Bradley, M.A. Gell, D. Ninno and M. Jaros, to be published.
[38] R.L. Greene and K.K. Bajaj, Phys. Rev. B 31 (1985) 6498.
[39] R.G. Humphreys, U. Rossler and M. Cardona, Phys. Rev. 18 (1978) 5590.
[40] Although an indirect-gap exciton generally has a large binding energy [39] which will increase with exciton confinement in quantum wells [38], we neglect it here because the X-electrons and Γ-holes are confined to separate layers and thus have minimal spatial overlap. That this Coulombic correction may be safely neglected in SLs is confirmed by the precise agreement in energy (for identical barrier composition, $x = 0.28$) in fig. 5 between the sharp intensity thresholds (stipled) for the Γ-confined states in the MQW and the appearance of the $X_{1e} - \Gamma_{1hh}$ transitions in the SL.

CHAPTER 7

BAND DISCONTINUITIES IN HgTe–CdTe SUPERLATTICES

J.P. FAURIE

Department of Physics
University of Illinois, Chicago
P.O. Box 4348, Chicago, IL 60680, USA

and

Y. GULDNER

Groupe de Physique des Solides
l'Ecole Normale Superieure
24, Rue Lhomond, 75231 Paris Cedex 05, France

Heterojunction Band Discontinuities: Physics and Device Applications
Edited by F. Capasso and G. Margaritondo
© Elsevier Science Publishers B.V., 1987

Contents

Introduction	285
1. HgTe–CdTe SL band-structure calculations	286
2. Growth of HgTe–CdTe SLs	291
3. Magneto-optical measurements	292
4. SL infrared transmission at 300 K	302
5. Photoluminescence and resonant Raman scattering	305
6. Conclusion	307
References	309

Introduction

HgTe–CdTe superlattices (SL) are new and important materials which present a great technical and fundamental interest. They have been, for instance, proposed as a novel infrared material [1] for wavelengths around 10 μm. The specific characters of this system arise mainly from the zero-gap configuration of HgTe. More generally, II–VI superlattices involving a zero-gap mercury compound and an open-gap semiconductor, such as HgTe–CdTe SLs, form a new class of heterostructures which are called Type-III SLs. The band structure of these heterostructures can be calculated by using the LCAO [2] or the envelope function [3] models which give very similar results. An important parameter, which determines most of the HgTe–CdTe SLs properties, is the valence-band discontinuity Λ between HgTe and CdTe and the value of Λ is presently disputed. From the phenomenological common-anion rule [4] and from the LCAO approach of Harrison [5], one can deduce that Λ is small ($\Lambda \leq 0.1$ eV) because the valence-band energy depends essentially on the anion and because HgTe and CdTe are closely matched in lattice constant (within 0.3%). Nevertheless, recent theoretical results, based on the role of interface dipoles and the analogy of an heterojunction with a metal–metal junction, do not support the common-anion rule and predict a much larger value $\Lambda \sim 0.5$ eV [6]. It is important to point out that these theoretical values for Λ are obtained from an energy difference between two large quantities. This means that Λ cannot be known from theoretical calculations to better than a fraction of an eV, which is clearly insufficient to predict most of the electronic properties of these materials. A more accurate determination can be obtained experimentally, for example from optical measurements. High-quality HgTe–CdTe SLs have been grown recently by molecular beam epitaxy [7] and experimental studies of the electronic properties of these heterostructures have been undertaken by different groups [8–13]. In the present review, we shall present a survey of the optical data obtained on HgTe–CdTe SLs in the temperature range 2–300 K. The band-structure calculations in the envelope function formalism are presented in sect. 1 and the crystal growth in sect. 2. Section 3 will deal with the magneto-absorption measurements at low temperatures, and in sect. 4 we shall present infrared transmission experiments at 300 K. Finally, photoluminescence and resonant Raman scattering results will be described in sect. 5. All the experimental data are interpreted in the envelope function formalism and we shall show that these data are consistent with a small positive offset between the HgTe and CdTe valence bands, in agreement with the common-anion rule for lattice-matched heterostructures. However, X-ray photoelectron spectroscopy (XPS) experiments recently carried out at 300 K are consistent with a larger valence-band offset of 0.35 eV [43,44]. XPS results are presented by R.W. Grant et al. in chapter 4 of the present volume.

1. HgTe–CdTe SL band-structure calculations

The bulk band structure of HgTe and CdTe near the Γ point and the band line-up of these two materials are shown in fig. 1. CdTe is an open-gap semiconductor with a direct gap at the Brillouin zone center. At $k = 0$, the conduction band Γ_6 has a s-type symmetry whereas the upper valence band is degenerate and has a p-type symmetry ($J = \frac{3}{2}$). The spin–orbit split-off Γ_7 band (p-type symmetry, $J = \frac{1}{2}$) is located below the Γ_8 states with $\Delta = E_{\Gamma_8} - E_{\Gamma_7} \sim 0.93$ eV. HgTe is a zero-gap semiconductor (or a semi-

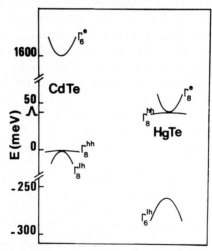

Fig. 1. Band structure of bulk HgTe and CdTe at 4 K. The lh, hh and e indices refer to light holes, heavy holes and electrons, respectively.

metal) due to the inversion of the relative positions of the Γ_6 and Γ_8 edges. What was the Γ_8 light-hole band in CdTe forms the conduction band in HgTe and the Γ_6 conduction band in CdTe becomes a light-hole band in HgTe. The ground valence band is the Γ_8 heavy-hole band so that the Γ_8 states represent both the top of the valence band and the bottom of the conduction band yielding a zero-gap configuration. The spin–orbit separation Δ is ~ 1.05 eV in HgTe. The evidence for the inverted structure of HgTe was mainly provided by magneto-optical measurements.

The band structure of HgTe–CdTe SLs depends on the offset Λ between the Γ_8 band edges of HgTe and CdTe, this parameter being measured from the top of the CdTe valence band (fig. 1). It has been show that most of the HgTe–CdTe heterostructures grown by molecular beam epitaxy present a p-type conduction at low temperature [14]. From that observation, one can conclude that Λ must be positive, otherwise electron transfer would occur between the CdTe valence band and the HgTe conduction band yielding a n-type conduction which would not be compatible with the experiments. As will be shown later on from the analysis of the experimental data, Λ is found to be positive. The first experimental determination of Λ was obtained from far-infrared magneto-optical techniques and yielded $\Lambda \sim 40$ meV [8]. This positive value implies that the HgTe layers are potential wells for heavy holes while the situation for light particles (electrons or light holes) is more complicated because the bands which contribute most significantly to the light-particle SL states are the Γ_8 conduction band in HgTe and the Γ_8 light-particle valence band in CdTe. These two bands have opposite curvatures and the same Γ_8 symmetry. This mass-reversal for the light particles at each of the HgTe–CdTe interfaces is a unique property of the type-III SLs, in particular of HgTe–CdTe SLs. An important consequence of these very unusual features is the existence of interface states [15,16] in the energy region $(0, \Lambda)$ which are evanescent in both the HgTe and CdTe layers with a wavefunction peaking at the interfaces. This special situation met in type-III SLs contrasts with the more common one corresponding, for instance, to GaAs–AlGaAs SLs (type-I) where the SL states arise mainly from bands in GaAs and AlGaAs displaying the same curvature along with the same symmetry.

The simplest description of the SL band structure is obtained in the framework of the envelope function approximation [3,17]. The band structure of both HgTe and CdTe near the Γ point is described in this approach by the Kane model which takes into account the non-parabolicity of the Γ_6 and the light-particle Γ_8 bands. This non-parabolicity is important in HgTe where the separation ϵ_0 between the Γ_6 and Γ_8 edges is small. The

Table 1a
Band parameters of HgTe and CdTe at 4, 77 and 300 K, respectively. ϵ_0 is the interaction energy gap between the Γ_6 and Γ_8 edges. E_p is related to the square of the Kane matrix element.

	ϵ_0 (meV)			E_p (eV)
	300 K	77 K	4 K	
HgTe	−122	−261	−302	18
CdTe	1425	1550	1600	18

interaction with the higher bands is included up to the second order and is described by the Luttinger parameters γ_1, $\gamma_2 = \gamma_3 = \gamma$ (spherical approximation) and κ. The band parameters of HgTe and CdTe at 300, 77 and 4 K used throughout this chapter are given in table 1. At 4 K, the Luttinger parameters are well-known and are taken from refs. [18,19] for HgTe and CdTe, respectively. The temperature variation of γ_1, γ and κ between 4 K and 300 K is assumed to arise essentially from the variation of the interaction gap ϵ_0 between the Γ_6 and Γ_8 band edges. For a HgTe–CdTe heterostructure, the envelope function is a six component spinor [17] in each kind of layer, if one considers only the Γ_6 and Γ_8 bands. A system of six differential equations for the six-component envelope function is established from the 6×6 Kane Hamiltonian and the boundary conditions are obtained by writing the continuity of the wavefunction at the interfaces and by integrating the coupled differential equations across an interface [17]. This is compatible with the continuity of the probability current [3]. Taking into account the SL periodicity d (Bloch theorem), the dispersion relations of the SL bands are obtained along the growth axis, which is usually the [111] direction, and in the plane of the layers. The model depends on a single unknown parameter, the valence-band offset Λ, the others being well-established HgTe and CdTe bulk parameters as well as the HgTe and CdTe layer thicknesses d_1 and d_2, respectively.

Table 1b
The Luttinger parameters γ_1, γ and κ, of the Γ_6 band at 300 and 4 K.

	γ_1		γ		κ	
	300 K	4 K	300 K	4 k	300 K	4 K
HgTe	−44.8	−15.5	−23.55	−8.9	−25.50	−10.85
CdTe	5.75	5.29	2.12	1.89	1.50	1.27

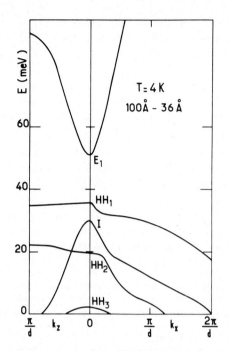

Fig. 2. Calculated band structure along k_z ([111] axis) and k_x (in the (111) plane) of a (100 Å)HgTe–(36 Å)CdTe superlattice. The zero of energy corresponds to the CdTe valence-band edge, d is the superlattice period and $\Lambda = 40$ meV.

Figure 2 presents the calculated band structure of a (100 Å)HgTe–(36 Å)CdTe SL along k_z (z being the [111] SL growth axis) and k_x [x being a direction of the (111) plane of the layers]. The zero of energy corresponds to the CdTe valence-band edge and the calculations are done for $\Lambda = 40$ meV and $T = 4$ K. The lowest conduction band, E_1, the ground light-particle band, I, and the heavy hole bands, HH_1, HH_2 and HH_3, are shown.

For $k_x = 0$, the light-particle and the heavy-particle bands are completely decoupled. The I band lies in the forbidden energy region $(0, \Lambda)$ for the light particles and corresponds to an interface state at $k = 0$ with an envelope function peaking at the interfaces [15,16]. This state results from the mass-reversal occurring for the Γ_8 light-particle band at each interface. The SL band gap E_g is defined as the separation between E_1 and HH_1 at $k = 0$ and is ~ 17 meV for the SL presented in fig. 2. The calculated bandgap E_g is found to decrease when Λ is increased and can even become negative for large Λ. The width of the E_1 band along k_z is small and, as a consequence, the calculated electron effective mass along k_z is found to be much larger than the very small value occurring in the $Hg_{1-x}Cd_xTe$ alloys with a similar band gap. This might be an important advantage of the HgTe–CdTe SLs as infrared detector materials [1] compared to the corresponding $Hg_{1-x}Cd_xTe$ alloys, because of the reduction of the tunneling effects which are usually important in small-gap materials. For $k_x \neq 0$, there is a hybridization between the I and the heavy-hole subbands which results in a complicated valence-band structure. In particular, it can be seen in fig. 2 that the in-plane mass of HH_1 is rather light for small k_x compared to the heavy-hole mass in bulk HgTe ($\sim 0.4 m_0$). This could explain the high hole mobility obtained in p-type SLs from Hall measurements [14].

In this analysis, the strain effects, due to the small lattice mismatch between HgTe and CdTe ($\sim 0.3\%$), are assumed to be negligible. The effects of strain were calculated by different groups [20–22]. They found

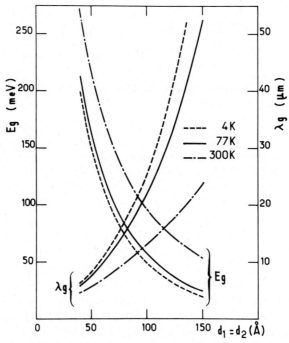

Fig. 3. Energy gap E_g and cut-off wavelength λ_g as a function of layer thickness for HgTe–CdTe SLs with equally thick HgTe and CdTe layers ($d_1 = d_2$).

that strains change the band energies only by a few meV and they have shown that the band structure of semiconducting SLs grown along the [111] direction is not significantly influenced by strain. The conduction band E_1 is nearly unaffected whereas the order of the light (I) and heavy-hole (HH_1) band can be reversed at $k = 0$ but the resulting valence-band structure along k_x is nearly unaffected because of the strong hybridization between the I and HH_1 bands [21,22]. In no case, the strain effects can strongly influence the experimental determination of the valence-band offset Λ.

Figure 3 shows the SL bandgap E_g and the corresponding cut-off wavelength λ_g calculated [23] at 300 K, 77 K and 4 K using $\Lambda = 40$ meV for SLs with equally thick layers of HgTe and CdTe (d_1 and d_2, respectively). For each temperature, a narrowing of the SL bandgap is predicted when the layer thickness increases. More generally, when $d_1 \neq d_2$, it is found that d_1 controls essentially the SL band gap while d_2 governs the width of the bands along k_z and therefore, the effective masses along the SL axis. Another important feature is that E_g increases when the temperature is raised as obversed in bulk $Hg_{1-x}Cd_xTe$ alloys with a similar energy gap [18]. Nevertheless, the temperature variation is calculated to be smaller for an SL than for the ternary alloy. In fig. 3, it can be noted that the interesting cut-off wavelengths λ_g for infrared detectors (8–12 μm) should be obtained at 77 K for layer thicknesses in the range 50–70 Å.

2. Growth of HgTe–CdTe SLs

HgTe–CdTe SLs were grown for the first time in 1982 by molecular beam epitaxy (MBE) on a CdTe($\overline{111}$)B substrate [7]. The MBE growth experiments are carried out using three different effusion cells containing CdTe, for the growth of CdTe, Te and Hg for the growth of HgTe. The CdTe substrate temperature must be above 180°C in order to grow high-quality SL crystals. At this temperature, the condensation coefficient for mercury is close to 10^{-3} [24]. This requires a high mercury flux during the growth of HgTe. Nevertheless, with a suitable pumping facility, the background pressure during the growth stays in the high 10^{-7} torr range. Most of the time the Hg cell is left open during the growth of the CdTe layers. Thus, a competition occurs between Hg and Cd. As a result it was found that thick CdTe layers grown under the same conditions contain a few percent of Hg (up to 5%) and are in fact (Cd, Hg)Te layers. This is not supposed to affect even slightly the calculations. Thus we will neglect this effect here. HgTe–CdTe superlattices have also been grown on $Cd_{0.96}Zn_{0.04}Te$ ($\overline{111}$)Te substrates and on GaAs(100) substrates [25]. On GaAs(100), both

(100)SL∥(100)GaAs and (111)SL∥(100)GaAs epitaxial relationships have been obtained. The orientation can be controlled by the preheating temperature as previously reported [26]. When the preheating temperature is 480°C or less, CdTe grows in the (100)∥(100) orientation. But when it is 580°C, CdTe grows in the (111)∥(100) orientation. Despite the large lattice mismatch between GaAs and HgTe–CdTe SL (14.5%), high-quality superlattices can be grown on GaAs if a 2 μm CdTe buffer layer is deposited prior to the growth of the superlattice. For CdTe(111) grown on GaAs(100), the orientation of the CdTe film is the ($\overline{111}$)B face according to selective etching, X-ray photoelectron spectroscopy and electron diffraction investigations [27]. SL have been grown on both CdTe($\overline{111}$)∥GaAs(100) and CdTe(100)∥GaAs(100) substrates and it turns out that growing on a (100) orientation requires about 4 times more mercury than growing on a ($\overline{111}$)B orientation [28].

In order to obtain high-quality superlattices, typical growth rates are 3–5 Å s^{-1} for HgTe and 1 Å s^{-1} for CdTe. This represents the best compromise between the low growth rate required for high crystal quality, especially for CdTe which should be grown at a higher temperature than 180°C, and the duration of the growth, which should be as short as possible in order to save mercury and to limit the interdiffusion process. A very important question for the application of HgTe–CdTe SLs to opto-electronic devices is the thermal stability of the HgTe–CdTe interface. Because of the lower temperature used in MBE compared to other epitaxial techniques, such as LPE of MOCVD, the diffusion processes are more limited in MBE, but the magnitude of this interdiffusion has not yet been fully determined. The interdiffusion constant were recently measured between 110 and 185°C [29]. It turns out that for a two hour growth run at 185°C, corresponding to a 2 μm thick superlattice, one can expect an interdiffused interface of 10 Å or less near the substrate [30]. The extent of the interdiffusion obviously changes in the sample along the growth axis.

The period of the SLs are measured using both a mechanical step profiler and the position of the SL satellite peaks as determined by X-ray. The values of the HgTe and CdTe layer thicknesses (d_1 and d_2, respectively) are then calculated using the Cd and Hg concentrations measured by energy dispersive X-ray analysis (EDAX).

3. Magneto-optical measurements

Interesting informations on the SL band structure can be obtained from far-infrared (FIR) magneto-absorption experiments. When a strong mag-

Table 2
Characteristics of HgTe–CdTe superlattices used in the magneto-optical investigations. d_1 = HgTe layer thickness and d_2 = CdTe layer thickness. The samples were grown in the (111)B orientation.

Sample	d_1 (Å)	d_2 (Å)	n	Substrate
S_1	180	44	100	CdTe
S_2	100	36	100	CdTe
S_3	77	38	70	GaAs
S_4	38	20	250	$Cd_{0.95}Zn_{0.05}Te$

netic field B is applied perpendicular to the layers, the SL bands are split into Landau levels. At low temperature, the FIR transmission signal being recorded at fixed photon energies as a function of B, presents pronounced minima which corresponds to resonant optical transitions between the different Landau levels. Intraband (namely cyclotron resonance) and interband magneto-optical transitions can be observed, depending on the Fermi level position at low temperature. From the theoretical analysis of the data, the SL band structure, in particular the valence-band offset Λ, is deduced [8,12,31].

The four samples (S_1, S_2, S_3 and S_4) used in the magneto-optical investigations reported here, were grown by molecular beam epitaxy at low temperature ($\sim 185\,°C$) on (111)CdTe, $Cd_{0.95}Zn_{0.05}Te$ or (100)GaAs substrates (see sect. 2). The HgTe and CdTe layer thicknesses (d_1 and d_2, respectively), the number of periods and the type of substrate for each sample are listed in table 2. Samples S_1, S_3, S_4 are p-type at liquid-helium temperatures and undergo a p- to n-type transition when the temperature is raised. Sample S_2 is n-type in the whole temperature range investigated (2 – 300 K) with a maximum Hall mobility at 77 K $\mu \sim 40.000$ cm^2/(V s).

The infrared magneto-absorption experiments reported here [31] were done at liquid-helium temperatures using a grating monochromator (3 $\mu m \leq \lambda \leq 5\ \mu m$), a CO_2 laser (9 $\mu m \leq \lambda \leq 11\ \mu m$), a far-infrared laser (41 $\mu m \leq \lambda \leq 255\ \mu m$) and carcinotrons (600 $\mu m \leq \lambda \leq 1$ mm). The magnetic field B, which was provided by a superconducting coil, was applied perpendicularly to the plane of the SL layers.

Figure 4 shows typical transmission spectra obtained in sample S_1 for several infrared wavelengths and fig. 5 gives the energy positions of the transmission minima (i.e., absorption maxima) as a function of B for this sample. The observed transitions extrapolate to an energy ~ 0 at $B = 0$ but they cannot be due either to electron cyclotron resonance because S_1 is

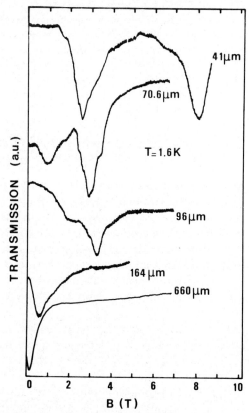

Fig. 4. Typical transmission spectra observed in sample S_1 as a function of the magnetic field B for different infrared wavelengths.

found to be p-type for $T < 20$ K, or to hole cyclotron resonance because they would lead to hole masses much too small. They are attributed to interband transitions from Landau levels of the top-most valence band HH_1 up to Landau levels of the ground conduction band E_1 arising in a zero-gap SL [8]. The band structure of sample S_1 along k_z (SL axis), calculated in the framework of the envelope function model using $\Lambda = 40$ meV, and $T = 4$ K, is shown in fig. 6, as well as the band structure of bulk HgTe and CdTe. For this particular value of Λ, S_1 presents a zero-gap configuration because E_1 and HH_1 are degenerate at $k_z = 0$ and this is qualitatively in agreement with the results presented in fig. 5. Note that similar results were also obtained from LCAO calculations [21].

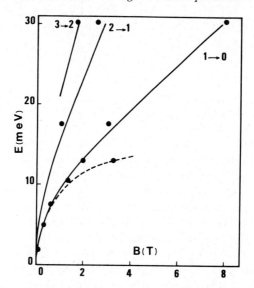

Fig. 5. Energy position of the transmission minima shown in fig. 4 as a function of B (solid dots). The solid lines are theoretical fits.

The energies $E_1(n)$ and $HH_1(n)$ of the Landau levels of index $n = 0, 1, 2, \ldots$, associated to E_1 and HH_1, were calculated using an approximate model where the influence of the higher bands are neglected [3]. The

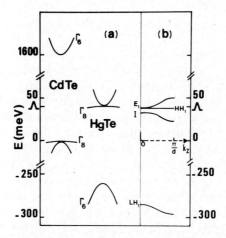

Fig. 6. (a) Band structure of bulk HgTe and CdTe and 4 K ($\Lambda = 40$ meV). (b) Calculated band structure of sample S_1 along k_z ([111] axis) at 4 K.

selection rule for the interband magneto-optical transitions $HH_1(n) \to E_1(n')$ are taken to be $n' - n = \pm 1$, as for the interband $\Gamma_8 \to \Gamma_8$ transitions in bulk HgTe [32]. The calculated transition energies using $n' - n = -1$ and $\Lambda = 40$ meV were shown in fig. 5 (solid lines).

For example, the curve labelled $1 \to 0$ corresponds to the transitions $HH_1(1) \to E_1(0)$. Note that the experimental data could be interpreted equally well with the selection rule $n' - n = +1$ except for the transition $1 \to 0$. The agreement between theory and experiment is fairly good for $\Lambda = 40$ meV. The deviation from the theoretical fit of the experimental data for the $1 \to 0$ transition around 2.5 T is not understood at the moment. The calculated band structure of sample S_1 (fig. 6b) is confirmed by the observation of interband transitions from Landau levels of LH_1, which is the top-most SL band arising from the Γ_6 HgTe states, up to the Landau levels of E_1, in the photon energy range 300–400 meV. Figure 7a shows a typical magneto-transmission spectrum observed in this energy range in the Faraday configuration. The position of the transmission minima are presented in fig. 7b and the solid lines correspond to the calculated transitions slopes using the approximate model neglecting the influence of the higher bands [3].

The selection rules are taken to be $\Delta n = \pm 1$, as those established for $\Gamma_6 \to \Gamma_8$ magneto-optical transitions in bulk HgTe [33] (Faraday configuration). The observed broad minima (fig. 7a) correspond to the two symmetric transitions $n \to n+1$ and $n+1 \to n$ which are not experimentally resolved. The agreement between theoretical and experimental slopes is rather good. The transitions converge to 344 meV at $B = 0$ while the energy separation between LH_1 and E_1 is calculated to be 330 meV at $k_z = 0$. The 14 meV difference can be explained by the approximations of the model. It might be also explained by the 0.3% lattice mismatch between HgTe and CdTe which results in an increase of the interaction gap $|\epsilon_0|$ in HgTe [21] and, therefore, in an increase of the separation between LH_1 and E_1. Note that these observations rule out any appreciable interdiffusion between HgTe and CdTe layers. Indeed, in the case of strongly interdiffused HgTe layers, the interaction gap $|\epsilon_0|$ of the resulting HgCdTe alloy would be significantly smaller than 302 meV, its value in pure HgTe at 4 K and, as a consequence, the energy separation between LH_1 and E_1 would be much smaller than 340 meV, which is not observed in the experiments.

Quite different results are expected for sample S_2 which is an open-gap SL (see the calculated band structure in fig. 2) with a n-type conduction at low temperature. Figure 8a shows typical transmission spectra obtained [12] for different FIR wavelength in sample S_2. A single broad minimum is

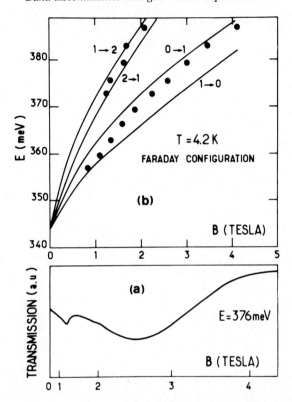

Fig. 7. (a) Transmission spectrum observed in sample S_1 for an infrared photon energy $E = 376$ meV. (b) Energy position of the observed transmission minima versus B (solid dots). The solid lines are the calculated $LH_1 \to E_1$ transitions slopes using $\Lambda = 40$ meV, as described in the text.

observed, whose energy position as a function of B is shown in fig. 8b. The transition extrapolates to an energy ~ 0 at $B = 0$ and is attributed to cyclotron resonance arising in the E_1 conduction band. The corresponding cyclotron mass at $B \sim 1$ T is $m = (0.017 \pm 0.003)m_0$. No transmission spectra are obtained around 20 meV which corresponds to the LO phonon energy in CdTe and to the restrahlen band of the substrate. When the magnetic field is tilted from the normal to the layers, the line becomes broader and the minimum is shifted to higher magnetic field because of the anisotropy of the E_1 band (fig. 2). To calculate the Landau level energies when a magnetic field is applied along the k_z direction, the model [12] is formally the same as that used at $B = 0$ and described in sect. 1, replacing $\boldsymbol{k}$

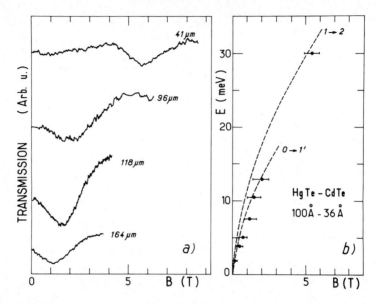

Fig. 8. (a) Typical transmission spectra obtained at 1.6 K in sample S_2 as a function of B for several infrared wavelengths. (b) Energy position of the transmission minima versus B (solid dots). The dashed lines correspond to theoretical fits of the E_1 cyclotron resonance.

by $k - (eA/c)$ in the Kane Hamiltonian and taking into account the direct coupling of the electron and hole spins to the field by introducing the additional Luttinger parameter κ [34]. The motion parallel to the layers is then described by a six-component vector [35].

$$\psi_n = (C_1\Phi_{n-1}, C_2\Phi_{n-2}, C_3\Phi_n, C_4\Phi_n, C_5\Phi_{n-1}, C_6\Phi_{n+1}),$$

where Φ_n is the n the harmonic oscillator function and $n = -1, 0, 1, 2, \ldots$. For $n \leq 1$, the coefficients C_i corresponding to the negative oscillator index vanish.

The calculated Landau levels associated with E_1, I, HH_1 and HH_2 are shown in fig. 9 using $\Lambda = 40$ meV. The situation is fairly complicated and the Landau levels are strongly mixed due to the coupling between the interface state I and the heavy-hole bands. The ground conduction level corresponds to $n = 1$ and the second level to $n = 0$. The first E_1 intraband transitions, fulfilling the selection rule $\Delta n = +1$ (cyclotron resonance), corresponds to $1 \to 2$ and $0 \to 1'$ (fig. 9). The dashed lines in fig. 8b are the calculated energies of those two transitions using $\Lambda = 40$ meV. At low photon energies ($E < 15$ meV), the dashed lines correspond fairly well to

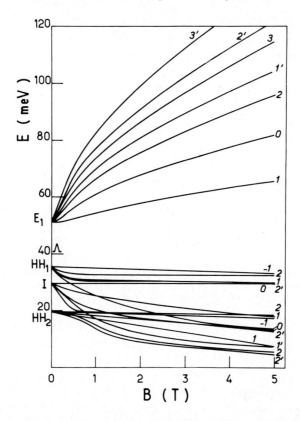

Fig. 9. Calculated Landau levels corresponding to the E_1, I, HH_1 and HH_2 bands (see fig. 2) in the case of sample S_2. The calculations are done for $T = 4$ K and $\Lambda = 40$ meV. For each band, the two Landau levels corresponding to the multi-component wavefunction ψ_n (see text) are labelled n and n'.

the observed broad FIR absorption showing that both the $n = 1$ and $n = 0$ levels are populated. For $E = 30$ meV, the calculated magnetic field separation between the two lines is larger than the observed absorption line. Only one transition, i.e., $1 \rightarrow 2$, is observed indicating that the $n = 1$ Landau level is populated at $B \sim 5$ T. The interband transitions between valence and conduction Landau levels are not observable in the investigated FIR region (0–30 meV) because of the population of the ground conduction levels and of the value of the superlattice bandgap. Such transitions have been investigated in the CO_2 laser energy region [22], as shown in fig. 10. Three transitions are observed in the energy region 110–130 meV and

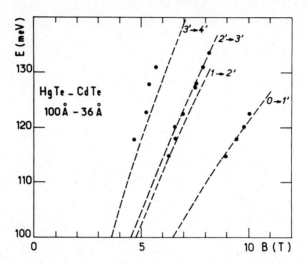

Fig. 10. Energy position of the transmission minima (solid dots) as a function of B corresponding to the interband $HH_1 \to E_1$ transitions observed in sample S_2 at 1.6 K. The dashed lines correspond to the theoretical fits.

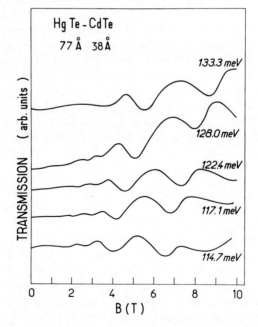

Fig. 11. Magneto-transmission spectra at 2 K associated to $HH_1 \to E_1$ transitions in sample S_3.

extrapolate to ~ 20 meV at $B = 0$ (fig. 10). They are interpreted as being due to $HH_1 \rightarrow E_1$ magneto-optical transitions obeying the selection rule $\Delta n = \pm 1$. The dashed lines in fig. 10 correspond to the calculated transitions using $\Delta n = +1$ and $\Lambda = 40$ meV. The experimental data could be interpreted also with the selection rule $\Delta n = -1$ due to the width of the observed absorption lines but, for the sake of simplicity, only one type of transition has been presented in fig. 10. The results for sample S_2 are consistent with a valence-band offset $\Lambda = 40$ meV. The sensitivity of the fitting procedure to the value of Λ was studied and it turns out that an acceptable agreement between experiment and the calculated transitions could be obtained for Λ within the limits 0–100 meV, if one takes into account the uncertainties on the sample characteristics, on the data (broad absorption minima) and on the band parameters of HgTe and CdTe used in the model. In addition, the S_2 bandgap becomes nearly zero for $\Lambda > 100$ meV, and interband transitions should then be observed, in addition to cyclotron resonance, in the 0–30 meV FIR region.

Figure 11 presents the magneto-optical spectra obtained [36] in sample S_3. The observed transmission minima are again interpreted as interband magneto-optical transitions between HH_1 and E_1 Landau levels and a bandgap $E_g = 45 \pm 10$ meV is deduced by extrapolating the energy of the

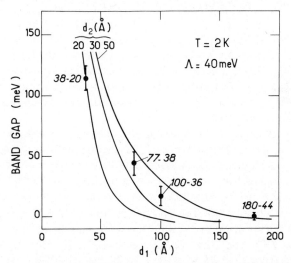

Fig. 12. Variation of the superlattice bandgap E_g as a function of the HgTe layer thickness d_1. The experimental data for samples S_1, S_2, S_3 and S_4 are given by the solids dots; for each sample, the first number corresponds to d_1 and the second one to d_2 (in Å). The solid lines are the theoretical variations $E_g(d_1)$ for three values of d_2.

observed transitions to $B = 0$. Finally, for each sample, a precise determination of the SL bandgap E_g at low temperature can be obtained from magneto-absorption experiments [30], and fig. 12 shows the value of E_g deduced from such experiments for the samples S_1, S_2, S_3 and S_4. The solid lines in fig. 12 are the theoretical variation of E_g as a function of d_1 calculated, as described in sect. 1, for $d_2 = 20$, 30 and 50 Å using $\Lambda = 40$ meV and $T = 4$ K. Experiments and theory show a very satisfying agreement. An acceptable agreement can in fact be obtained for Λ within the range 0–100 meV by taking into account the uncertainties on the sample characteristics, on the experimental data and on the band parameters of HgTe and CdTe.

4. SL infrared transmission at 300 K

In order to determine the SL bandgap E_g at 300 K, infrared transmission measurements were performed [13] between 50 and 600 meV on several SLs whose characteristics are reported in table 3. The absorption coefficient, α, was obtained by taking the negative of the natural logarithm of the transmission spectrum and then dividing by the thickness of the SL. The energy bandgap was defined to be the energy where α is equal to 1000 cm^{-1}. Even though the accuracy of this determination is questionable because the value of 1000 cm^{-1} for α is rather arbitrary, it is found that the values of the band gap determined in this way are in good agreement with those obtained from photoconductivity threshold at the same temperature [37]. It has been also shown that this method is more accurate for thick SLs ($e > 1.5$ μm) [38]. Figure 13 shows a typical infrared transmission curve for a HgTe–CdTe superlattice.

Figure 14 presents a comparison of the experimentally determined bandgap with the theoretical curves $E_g(d_1)$ calculated, as described in sect. 1, for $d_2 = 10$, 20, 30 and 100 Å using $\Lambda = 40$ meV and $T = 300$ K. There is a good agreement if one considers the uncertainties in the HgTe and CdTe parameters used in the theoretical calculation and in the experimental determination. The fact that the fit worsens for small d_2 (samples SL 1 and SL 3 for instance) is most probably due to the increased relative effect of interdiffusion [30]. The effect of interdiffusion is to shift E_g towards the smaller energies, due to the decreasing CdTe layer thickness. It can be seen in fig. 14, that the SL band gap is essentially governed by d_1 when $d_2 > 30$ Å.

Nevertheless, d_2 governs the width of the subbands along k_z, which

Table 3
Characteristics of HgTe–CdTe superlattices presented in fig. 13. d_1 = HgTe layer thickness, and d_2 = CdTe layer thickness. The superlattices were grown in the (111) orientation.

Sample	d_1, HgTe (Å)	d_2, CdTe (Å)
SL 1	38	20
SL 2	40	60
SL 3	45	17
SL 4	74	36
SL 5	97	60
SL 6	110	40
SL 7	100	36
SL 8	77	38
SL 9	58	35
SL 10	47	30
SL 11	85	45
SL 12	60	30
SL 14	74	32
SL 15	70	35
SL 16	61	25
SL 17	70	41
SL 18	37	61
SL 19	52	34

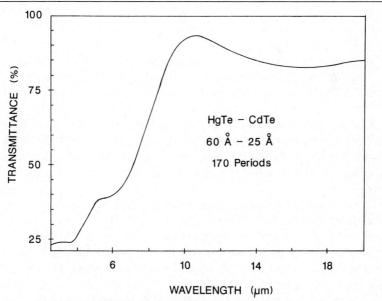

Fig. 13. Infrared transmission curve of SL 16.

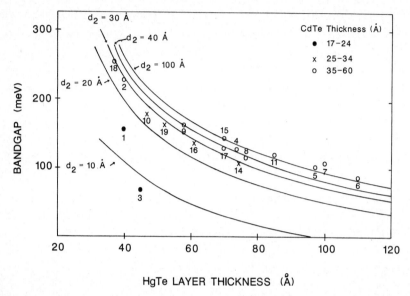

Fig. 14. Variation of bandgap E_g of different HgTe–CdTe superlattices at 300 K as a function of the HgTe layer thickness d_1. The sample characteristics are listed in table 3 and the experimental data are given by: solid circles for 17 Å $\leq d_2 \leq$ 24 Å, by crosses for 25 Å $\leq d_2 \leq$ 34 Å, and by open circles for 35 Å $\leq d_2 \leq$ 60 Å. The solid lines are theoretical variations $E_g(d_1)$ for different values of d_2 using $T = 300$ K and $\Lambda = 40$ meV.

strongly increases when d_2 is decreased. For $d_2 > 30$ Å, the band widths are small and E_g is nearly independent on d_2. When d_2 becomes less than 30 Å, the increase in the width of the E_1 and HH_1 bands strongly influences $E_g = E_1 - HH_1$ which is, therefore, no longer governed only by d_1. From the results of fig. 14, one can also deduce [13] that the SL cut-off wavelength is easier to control than that of the $Hg_{1-x}Cd_xTe$ alloy of the same band gap. This effect, predicted by Smith et al. [1], is another advantage of SLs as infrared detector materials, compared to the ternary alloy.

From the experimental SLs band gaps shown in fig. 14, we have reported an experimental curve of room temperature cut-off versus HgTe layer thickness d_1 along with an experimental equation relating the two when $d_2 \geq 35$ Å [39]. We have used this equation to show that indeed the cut-off wavelength of HgTe–CdTe SL is easier to control than that of HgCdTe alloy for $\lambda \geq 2.0$ μm.

Even though the band gap determination is rather arbitrary, it is clear that the absorption spectra could not be interpreted with a large value of Λ. For $\Lambda > 200$ meV, the calculated band gap energy is located below the onset of the infrared absorption for each sample and it is important to point out that the best agreement between experiments and theory is again obtained for Λ in the range 0–100 meV.

5. Photoluminescence and resonant Raman scattering

Resonant Raman scattering (RRS) was also applied recently to investigate electronic properties of HgTe–CdTe SLs [9]. RRS measurements were performed in back scattering geometry at 12 K using blue and green lines from Ar$^+$ and Kr$^+$ ion lasers. Resonant Raman scattering was carried out to investigate the SL valence states arising from the spin–orbit split-off Γ_7 bands of HgTe and CdTe. If the value of the valence-band offset Λ is between 0 and 100 meV, the confinement of Γ_7 holes is expected to be in the CdTe layers because the spin–orbit energy $\Delta = E_{\Gamma_8} - E_{\Gamma_7}$ is ~ 1.05 eV [18] and ~ 0.93 eV [9] in HgTe and CdTe, respectively. Figure 15 shows, for two SLs and for a CdTe layer, typical Stokes spectra excited with laser light in the neighborhood of the Γ_6–Γ_7 edge of CdTe (also called the $\epsilon_0 + \Delta$ gap). The spectra consist of sharp lines at energies $n \times 170$ cm^{-1} ($n =$ 1, 2, 3, 4) which correspond to Raman scattering by the zone-center LO phonons of CdTe, and of photoluminescence (PL) backgrounds indicated by arrows for the two superlattices SL 7 and SL 12 whose characteristics are described in table 3. No indication of such PL activity is detected for the CdTe layer. It appears that for different SLs the lineshapes shift to higher energies with decreasing d_2 as shown in the inset of fig. 15. We have also shown that a relationship exists between the PL and RRS spectra which indicates that the PL activity arises in the CdTe layers of the superlattices. It can be seen that a clear interplay takes place between the PL response and the resonant behavior of of the CdTe Raman phonons. Concomitant with the shift of the PL peaks to higher energies, the Raman lines with smaller Stokes shifts are enhanced. The 1- and 2-LO lines resonate more strongly in SL 12 than in SL 7.

In addition, we have recently investigated the temperature dependence of the PL spectral position in SL 12 [40]. With increasing temperature, the PL bands shift to lower energies. This dependence can be described empirically with the Varshni equation [41],

$$E_g(T) = E_g - \frac{\alpha T^2}{\beta + T},$$

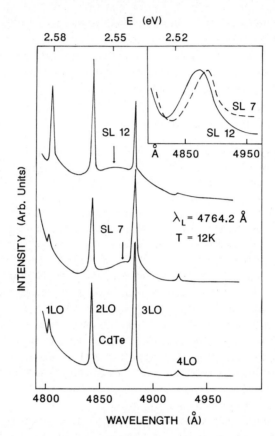

Fig. 15. Recombination spectra and resonant Raman scattering for a laser excitation λ_L in the neighborhood of the $\Gamma_6-\Gamma_7$ edge of CdTe. The spectra are obtained in a MBE-grown CdTe layer and in the two superlattices SL 7 and SL 12 whose characteristics are given in table 3.

using for α and β coefficients obtained for the ϵ_0 gap of CdTe and a value of $E_g = E_0 + \Delta(T=0) = 2.547$ eV. The calculated curve describes the measured behavior rather well. That re-inforces the interpretation of the origin of the PL bands as being transitions taking place in the CdTe layers. Thus, the PL signal has been attributed to radiative transitions across the $\Gamma_6-\Gamma_7$ gap of the CdTe layers in the SLs and the PL peak maxima were interpreted as a measure of this edge. An independent determination of the $\Gamma_6-\Gamma_7$ gap can also be obtained from the energy dependence of the RRS efficiencies [42]. We found that for the SLs the RRS results yield the same values as the PL peak maxima.

RRS measurements gave a Γ_6–Γ_7 separation value of 2.530 eV for CdTe and larger values for SLs. The larger value of the Γ_6–Γ_7 edge in the SLs can be explained by the confinement energy of the Γ_7 holes in the CdTe layers. This confinement results from a square well-like potential barrier formed for the Γ_7 holes of CdTe because $\Delta^{\mathrm{CdTe}} < \Delta^{\mathrm{HgTe}}$. If Λ has a value between 0 and 100 meV, a potential barrier $\Delta^{\mathrm{HgTe}} - \Delta^{\mathrm{CdTe}} - \Lambda$ of 120 to 20 meV is then expected for the Γ_7 holes in the CdTe layers. Assuming a value of 60 meV for the potential barrier and of $0.7m_0$ for the effective mass of the Γ_7 holes, one calculates a Γ_6–Γ_7 edge increase of 13 meV for SL 7 and of 19 meV for SL 12, in good agreement with measured energy shifts of 10 meV and 16 meV [9].

The observation of quantized Γ_7 valence states in CdTe constitutes, as well, a *direct* experimental verification that Λ is indeed small. An upper limit for its magnitude can be evaluated as $\Delta^{\mathrm{HgTe}} - \Delta^{\mathrm{CdTe}} \sim 120$ meV. The existence of the PL signals can be explained by invoking a large decrease in the lifetime of the Γ_8 holes of CdTe. The reduced quantum efficiency of the radiative Γ_6–Γ_8 transitions in the SLs compared to those in CdTe and their stronger temperature dependence are additional experimental evidence that the valence-band offset Λ is positive and that the Γ_8 holes are quantized in the HgTe layers. The photocreated Γ_8 holes in CdTe scatter to the energetically more favorable hole subbands in HgTe [9,40]. Finally, from these experiments, one can conclude that the Γ_7 holes are confined in the CdTe layers which implies an upper limit of 120 meV for Λ and that Λ is positive.

6. Conclusion

In this chapter, we have presented a survey of the optical measurements done in HgTe–CdTe SLs. These investigations show clearly that the band structure of these heterostructures is more complicated and subtle than that of common III–V compound systems (type-I SLs). All the experimental data reported here were interpreted in the envelope function approximation and are consistent with a small positive offset Λ with Λ in the range 0–120 meV between the HgTe and CdTe valence bands. This value is in good agreement with the predictions of the common-anion rule for lattice-matched heterostructures [4,5]. Recently, Λ was also measured by X-ray photoemission spectroscopy (XPS) [43] and a much larger value, $\Lambda \sim 0.35$ eV, was obtained, which supports the idea that lattice-matched heterojunctions with a common anion may present large valence-band discontinuities [6]. It is important to point out that the magneto-optical data at 2 K as well as the infrared transmission measurements at 300 K cannot be interpreted by using such a large valence-band offset in the envelope function model. Indeed, most of the investigated SLs are calculated to be semimetallic at 4 K for $\Lambda = 0.35$ eV, which is not compatible with the magneto-optical data. For example, no interband magneto-optical transitions are observed in the energy range 0–30 meV in the samples S_2, S_3, S_4, demonstrating the existence of a finite gap at 2 K in these samples. At 300 K, the calculated band gap energy E_g using $\Lambda = 0.35$ eV is found to be located far below the onset of the infrared absorption. Moreover, the resonant Raman scattering results show that the Γ_7 holes are confined in the CdTe layers and are interpreted with a positive valence-band offset Λ smaller than 120 meV. In order to understand the reason of such a discrepancy between the optical data and the XPS measurements, in situ XPS experiments are currently carried out at the University of Illinois at Chicago. From our XPS measurements carried out on CdTe $\|$ HgTe($\overline{111}$)B and HgTe $\|$ CdTe($\overline{111}$)B we found the valence-band discontinuity Λ to be insensitive to interface Fermi-level position, commutative and equal to 0.36 ± 0.05 eV [44], in excellent agreement with the value reported in ref. [43]. Thus it is important to point out that XPS measurements give definitively a larger value than 100 meV. The reason of such a discrepancy between optical and XPS experiments is not understood at the present time. XPS experiments are carried out at 300 K on single heterojunctions, whereas optical investigations are performed at low temperatures on superlattices. In this peculiar heterojunction between a semiconductor and a semimetal Λ might be temperature-dependent. XPS experiments have been performed on a single heterojunction. It is legitimate to wonder whether one would get the same value for Λ on a superlattice structure. The possible interdiffusion which can occur during the growth of the superlattice [30] could also contribute to a smaller value for Λ in superlattices compared to the one in single heterojunctions where no interdiffusion has been found. Experiments are currently carried out in order to answer this question. We wish also to point out that the type-III SLs formed from II–VI zero-gap compounds, widens the field of two-dimensional systems in a very interesting way, at least from the point of view of basic physics. For instance, $Hg_{1-x}Cd_xTe$–CdTe SLs have been grown recently [45,46] and a type-III → type-I transition is expected at 4 K in these heterostructures [46], corresponding to the semimetal → semiconductor transition which occurs in the alloys at ~ 0.16. A similar system, $Hg_{1-x}Mn_xTe$–CdTe SLs, should present attractive characteristics due to the magnetic properties of $Hg_{1-x}Mn_xTe$ alloys [47]. One may expect, for example, two-dimensional spin glasses to occur in such structures. The highly strained-layer ZnTe–HgTe Sls with a 6.5% difference between HgTe and ZnTe lattice parameters, were also grown recently [45,48] and are also promising type-III SLs for basic physics and as an infrared material.

Acknowledgements

The authors would like to thank G. Bastard, J.M. Berroir, J.P. Vieren and M. Voos from the École Normale Supérieure (Paris), together with J. Reno and I.K. Sou from the University of Illinois (Chicago), as well as D. Olego from Stauffer Chemical Company (Elmsford) and A. Million from the Laboratoire Infrarouge (Grenoble). The authors would like to acknowledge the support of the GRECO Expérimentation Numérique (Y.G.) and the support of DARPA under contract no. MDA 903-85K-0030 (J.P.F.).

References

[1] D.L. Smith, T.C. McGill and J.N. Schulman, Appl. Phys. Lett. 43 (1983) 180.
[2] J.N. Schulman and T.C. McGill, Appl. Phys. Lett. 34 (1979) 663, and Phys. Rev. B 23 (1981) 4149.
[3] G. Bastard, Phys. Rev. B 25 (1982) 7584.
[4] J.O. McCaldin, T.C. McGill and C.A. Mead, Phys. Rev. Lett. 36 (1976) 56.
[5] W. Harrison, J. Vac. Sci. & Techn. 14 (1977) 1016.
[6] J. Tersoff, Phys. Rev. B 30 (1984) 4874, and Phys. Rev. Lett. 56 (1986) 2755.
[7] J.P. Faurie, A. Million and J. Piaguet, Appl. Phys. Lett. 41 (1982) 713.
[8] Y. Guldner, G. Bastard, J.P. Vieren, M. Voos, J.P. Faurie and A. Million, Phys. Rev. Lett. 51 (1983) 907.
[9] D.J. Olego, J.P. Faurie and P.M. Raccah, Phys. Rev. Lett. 55 (1985) 328.
[10] S. Hetzler, J.P. Baukus, A.T. Hunter, J.P. Faurie, P.P. Chow and T.C. McGill, Appl. Phys. Lett. 47 (1985) 260.
[11] N.P. Ong, G. Kote and J.T. Cheung, Phys. Rev. B 28 (1983) 2289.
[12] J.M. Berroir, Y. Guldner, J.P. Vieren, M. Voos and J.P. Faurie, Phys. Rev. B 34 (1986) 891.
[13] J. Reno, I.K. Sou, J.P. Faurie, J.M. Berroir, Y. Guldner and J.P. Vieren, Appl. Phys. Lett. 49 (1986) 106.
[14] J.P. Faurie, M. Boukerche, S. Sivananthan, J. Reno and C. Hsu, Superlattices and Microstructures 1 (1985) 237.
[15] Y.C. Chang, J.N. Schulman, G. Bastard, Y. Guldner and M. Voos, Phys. Rev. B 31 (1985) 2557.
[16] Y.R. Lin Liu and L.J. Sham, Phys. Rev. B 32 (1985) 5561.
[17] M. Altarelli, Phys. Rev. B 28 (1983) 842.
[18] M.H. Weiler, in: Semiconductors and Semimetals, Vol. 16, eds R.K. Willardson and A.C. Beer (Academic Press, New York, 1981) p. 119.

[19] P. Lawaetz, Phys. Rev. B 4 (1971) 3460.
[20] G.Y. Wu and T.C. McGill, Appl. Phys. Lett. 47 (1985) 634.
[21] J.N. Schulman and Yia-Chung Chang, Phys. Rev. B 33 (1986) 2594.
[22] J.M. Berroir and J.A. Brum, in: Proc. 2nd. Int. Conf. on Superlattices and Microstructures, 1986, in press.
[23] Y. Guldner, G. Bastard and M. Voos, J. Appl. Phys. 57 (1985) 1403.
[24] J.P. Faurie, A. Million, R. Boch and J.L. Tissot, J. Vac. Sci. & Technol. A 1 (1983) 1593.
[25] J.P. Faurie, J. Reno and M. Boukerche, J. Cryst. Growth 72 (1985) 11.
[26] J.P. Faurie, C. Hsu, S. Sivananthan and X. Chu, Surf. Sci. 168 (1986) 473, and references therein.
[27] C. Hsu, S. Sivananthan, X. Chu and J.P. Faurie, Appl. Phys. Lett. 48 (1986) 908.
[28] S. Sivananthan, X. Chu, J. Reno and J.P. Faurie, J. Appl. Phys. 60 (1986) 1359.
[29] D.K. Arch, J.L. Staudenmann and J.P. Faurie, Appl. Phys. Lett. 48 (1986) 1588.
[30] J.L. Staudenmann, R.D. Horning, R.D. Knox, J. Reno, I.K. Sou, J.P. Faurie and D.K. Arch, Trans. Metallurg. Soc. AIME Semiconductor-based Heterostructures (1986) p. 41.
[31] J.M. Berroir, Y. Guldner and M. Voos, IEEE J. Quantum Electron. 22 (1986) 1793.
[32] S.H. Groves, R.N. Brown and C.R. Pidgeon, Phys. Rev. 161 (1967) 779;
J. Tuchendler, M. Grynberg, Y. Couder, H. Thome and R. Le Toullec, Phys. Rev. B 8 (1973) 3884.
[33] Y. Guldner, C. Rigaux, M. Grynberg and A. Mycielski, Phys. Rev. B 8 (1973) 3875.
[34] J.M. Luttinger, Phys. Rev. 102 (1956) 1030.
[35] A. Fasolino and M. Altarelli, Surf. Sci. 142 (1984) 322.
[36] Y. Guldner, SPIE Proc. 659 (1987) 24.
[37] M. DeSouza, M. Boukerche and J.P. Faurie, unpublished results.
[38] C.E. Jones, T.N. Casselman, J.P. Faurie, S. Perkowitz and J.N. Schulman, Appl. Phys. Lett. 47 (1985) 140.
[39] J. Reno and J.P. Faurie, Appl. Phys. Lett. 49 (1986) 409.
[40] D.J. Olego and J.P. Faurie, Phys. Rev. B 33 (1986) 7357.
[41] Y.P. Varshni, Physica 39 (1967) 149.
[42] J. Menendez and M. Cardona, Phys. Rev. B 31 (1985) 3696.
[43] S.P. Kowalczyk, J.T. Cheung, E.A. Kraut and R.W. Grant, Phys. Rev. Lett. 56 (1986) 1605.
[44] T.M. Duc, C. Hsu and J.P. Faurie, Phys. Rev. Lett. 58 (1987) 1127.
[45] J.P. Faurie, IEEE J. Quantum Electron. 22 (1986) 1656.
[46] J. Reno, I.K. Sou, P.S. Wijewarnasuriya and J.P. Faurie, Appl. Phys. Lett. 48 (1986) 1069.
[47] X. Chu, S. Sivananthan and J.P. Faurie, Appl. Phys. Lett. 50 (1987) 597.
[48] J.P. Faurie, Appl. Phys. Lett. 48 (1986) 785.

CHAPTER 8

MEASUREMENT OF ENERGY BAND OFFSETS USING CAPACITANCE AND CURRENT MEASUREMENT TECHNIQUES

S.R. FORREST

Departments of Electrical Engineering/Electrophysics and Materials Science, University of Southern California, Los Angeles, CA 90089-0241, USA

Heterojunction Band Discontinuities: Physics and Device Applications
Edited by F. Capasso and G. Margaritondo
© *Elsevier Science Publishers B.V., 1987*

Contents

1. Introduction 313
2. Capacitance–voltage intercept method 315
   2.1. Theory 315
   2.2. Error sources in the C–V intercept method 322
      2.2.1. Non-ohmic contacts 322
      2.2.2. Bulk traps 323
      2.2.3. Interfacial traps 325
      2.3.4. Non-uniform bulk free-carrier concentration 331
3. Capacitance–voltage measurement via the depletion method 332
   3.1. Theory 332
   3.2. Error sources in the C–V depletion technique 345
      3.2.1. Baseline error 345
      3.2.2. Fixed interfacial charge 348
      3.2.3. Interface grading effects 352
      3.2.4. Bulk trap effects 353
4. Current transport analysis techniques 354
   4.1. Current–voltage analysis 355
   4.2. Photocurrent analysis 365
   4.3. Error sources in the I–V method 369
5. Conclusions 372
References 373

1. Introduction

Differences in the electrostatic potentials of two dissimilar, contacting semiconductors give rise to a dipole of charge across the resulting heterojunction. For n–N isotype heterojunctions, for example, the potential difference between the two contacting materials appears as an offset in the energy bands at the conduction-band edge, with the rest of the bandgap energy difference appearing in the valence band. That is, if ΔE_g is the difference in band gap energies E_{g1} and E_{g2} between materials 1 and 2, respectively, then:

$$\Delta E_g = |E_{g2} - E_{g1}| = \Delta E_c + \Delta E_v, \tag{1}$$

where ΔE_c is the energy difference of the conduction bands (or conduction-band discontinuity), and ΔE_v is the valence-band discontinuity. The relative magnitude of ΔE_c and ΔE_v is of primary importance to our understanding of heterojunctions. In addition, knowledge of the magnitudes of these band offsets is essential if we are to utilize this property for the confinement of electrical and optical energy in practical devices fabricated from heterojunction systems.

It would appear that transport measurements commonly used for determining the size of potential barriers such as p–n junctions and Schottky barriers, are also an ideal means of determining the size of energy barriers existing at heterojunctions. Thus, for many years, analysis of the current–voltage (I–V), capacitance–voltage (C–V) and photocurrent spectroscopic characteristics of heterojunctions have proven to be important probes which have augmented absorption and transmission data obtained using complex multiple quantum well (MQW) structures, as well as photoemission measurements of high-energy radiation and absorption data arising from electronic excitations of core levels deep in the valence bands of the contacting species. Although the transport measurements have the advantage of being a relatively undemanding means of acquiring data using simple structures, the accuracy of these techniques has never been considered to be particularly high, basically due to the existence of "parasitic" phenomena giving rise to excess dark currents or stray capacitances, which confuse the measurement by introducing variables which cannot be easily treated in the overall analysis.

Recently, however, significant progress has been made in improving C–V analysis techniques such that they are now considered to be among the most reliable means of accurately measuring energy band offsets. In addition, the recent controversy surrounding measurements made on heterojunction discontinuites via MQW absorption data has also indicated that the other, heretofore "reliable" means of obtaining these critical data, are also subject to very large errors due to the necessity of making parametric fits to complicated data. Thus, to gain confidence in a value for the conduction- or valence-band discontinuites for a given heterojunction system, more than a single technique must be employed in the measurement, and differences between values obtained using the various techniques must be resolved prior to the acceptance of any particular physical picture. In the determination of band offsets, it has become apparent that transport measurements play a central role. In particular, recently modified C–V measurements of the heterojunction potential are the most reliable transport-related method for obtaining these vital data. As in all techniques, however, the C–V data are subject to large error if not interpreted or implemented correctly, and hence must be employed with a great deal of caution.

In this chapter we discuss details of the various transport measurements used in determining energy band offsets. Particular attention will be focussed on the implementation of C–V analysis, and the various sources of error commonly encountered in this method. Thus, in sect. 2 we will discuss C–V analysis of both isotype and anisotype heterojunctions using the voltage-intercept method, and in sect. 3 the discussion will be expanded to include the analysis of heterojunctions using the modified C–V technique whereby data is obtained using a sample with a p–n or metal–semiconductor junction placed in the vicinity of an isotype heterojunction. In sect. 4 we consider the I–V technique, which includes analysis of both dark current and photocurrent data of heterojunctions. In all of these discussions, the sources of errors and the ultimate accuracies of the various techniques will be considered in detail. In the several sections we compile data published in the literature concerning measurements made on several important heterojunction systems made using transport methods. Finally, in sect. 5 we present conclusions.

2. Capacitance–voltage intercept method

2.1. Theory

We begin our discussion of heterojunction measurement by examining the band diagram of an archetypal n–N isotype heterojunction shown in fig. 1a. The various quantities to be used in this and the analysis that follows are defined in the figure. It is apparent that the total diffusion potential due to the heterojunction dipolar charge layer is equal to the sum of the potential on the small bandgap side (V_{D1}) and on the large bandgap side (V_{D2}) of the junction. In this discussion, we assume that materials 1 and 2 refer to the small and large bandgap materials, respectively, and that variables referring to each of the two materials will have the corresponding subscript. The total diffusion potential is given by

$$V_D = V_{D1} + V_{D2}, \qquad (2)$$

and is the quantity that is actually measured by transport techniques. Of

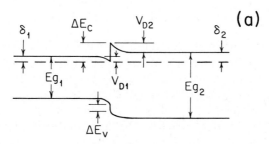

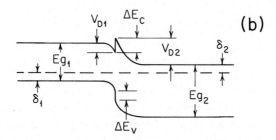

Fig. 1. Band diagram of: (a) n–N isotype, and (b) p–N anisotype heterojunction. Notation used in text is defined in the figure.

greater interest, however, is the conduction-band discontinuity energy, ΔE_c. This is related to the diffusion potential by

$$\Delta E_c = qV_D - \delta_1 + \delta_2, \tag{3}$$

where δ_1 and δ_2 refer to the position of the Fermi energies relative to the conduction-band minimum in the bulk of materials 1 and 2, respectively, and q is the electronic charge. That is:

$$\delta_1 = kT \log(N_1/N_{c1}), \tag{4}$$

with a similar expression for material 2. Here, kT is the Boltzmann energy at the temperature T, $N_1 = |N_{D1} - N_{A1}|$ is the net free-carrier concentration, and N_{c1} is the effective conduction-band density of states. Now, N_c is a function of the reduced effective mass of the electron (m^*) and of T. Therefore, the difference in the Fermi energies between materials 1 and 2 can be simplified to give

$$\delta_2 - \delta_1 = kT \left[\log(N_1/N_2) + \tfrac{3}{2}\log(m_2^*/m_1^*) \right]. \tag{5}$$

Thus once the diffusion potential is determined, it is relatively straightforward to obtain the conduction-band discontinuity. Indeed, as can be seen from the equations above, it is not necessary to have a highly precise measurement of any of the material parameters such as the bulk free carrier concentration or the effective density of states, since ΔE_c depends only logarithmically on these parameters. On the other hand, the dependence of ΔE_c on V_D is linear, and, therefore, it is important that the measurement of the diffusion potential be as accurate as possible.

For an anisotype p–N heterojunction (fig. 1b), eq. (3) is modified as

$$\Delta E_c = qV_D - E_{g1} + \delta_2 + \delta_1. \tag{6}$$

Here, material 1 is p-type, and hence δ_1 is the difference in energy between the Fermi level and the valence-band maximum in that material.

One method of obtaining V_D which is in principal very accurate is the technique of C–V measurement. To understand this technique, it is first necessary to calculate the capacitance of the anisotype heterojunction. Note that, as in the case of the p–n junction, the potential difference between materials 1 and 2 gives rise to a dipole of free charge at the edges of a depleted, or space-charge region. There is a capacitance associated with the space-charge region which can be obtained by solving Poisson's equation in one dimension. Using the boundary condition that electric displacement is continuous across the heterointerface region (i.e., $\epsilon_1 F_1 = \epsilon_2 F_2$, where ϵ is the permittivity and F the electric field), we obtain

$$W_1^2 = \frac{2}{q} \frac{N_2 \epsilon_1 \epsilon_2 (V_D - V)}{N_1 (\epsilon_1 N_1 + \epsilon_2 N_2)}, \quad \text{and} \tag{7}$$

$$W_2^2 = \frac{2}{q} \frac{N_1 \epsilon_1 \epsilon_2 (V_D - V)}{N_2 (\epsilon_1 N_1 + \epsilon_2 N_2)}, \tag{8}$$

for the depletion region widths on both sides of the heterojunction due to a total applied voltage, V.

Next, using the depletion approximation, and assuming that the capacitance of an anisotype heterojunction can be adequately modelled as being due to the series contribution of the capacitances on each side of the heterojunction, then the total capacitance per unit area is given by

$$C_D^2 = \frac{q N_1 N_2 \epsilon_1 \epsilon_2}{2(\epsilon_1 N_1 + \epsilon_2 N_2)(V_D - V)}. \tag{9}$$

Here, the parallel-plate approximation has been employed to calculate the total capacitance. In addition, it has been assumed that both materials 1 and 2 have a spatially uniform free-carrier concentration, or net doping density, in the heterojunction region and in the bulk of the sample. There are several other implicit assumptions made in calculating eq. (9) which occasionally have a significant influence on the measurement accuracy. These will be discussed in sect. 2.2.

It is readily apparent that a plot of $1/C_D^2$ versus V has a slope which is determined by the free-carrier concentrations and the permittivities of the various materials, and an intercept with the $-V$ axis equal to V_D. Thus, such a measurement of an ideal sample should directly yield the diffusion potential, and hence the conduction-band discontinuity via eq. (6). This technique is analogous to the method of determining barrier heights of metal–semiconductor junctions in Schottky barrier diodes, or of determining the built-in potential of a p–n junction.

Figure 2 shows the results of a $C-V$ measurement made on a n–P Ge–GaAs heterojunction made at two different temperatures [1]. The straight line intercepts of the $1/C_D^2$ data with the voltage axis yield a

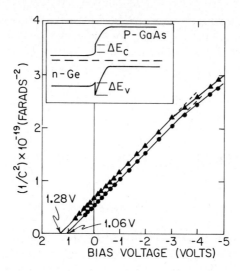

Fig. 2. Capacitance–voltage characteristics measured at 296 K (solid circles) and 77 K (solid triangles) of an n–P Ge–GaAs heterojunction. Voltage intercepts at 1.06 V and 1.28 V, respectivey, are also shown [1].

diffusion potential of 1.06 V at $T = 296$ K and 1.28 V at $T = 77$ K. The temperature dependence of the diffusion potential is expected from the variation of the Fermi level energy with temperature [eqs. (6) and (4)]. These values of diffusion potential then result in a valence-band offset for this material combination of $\Delta E_v = 0.60$ eV, and thus the conduction-band discontinuity energy is $\Delta E_c = 0.17$ eV. The energy band diagram inferred from this measurement is shown in the inset of fig. 2.

The situation is somewhat different for isotype heterojunctions due to the more complex potential distribution between the narrow and large bandgap semiconductors. Once again, using the condition of continuity of electric displacement across the heterojunction, the potential distribution is obtained from

$$\frac{\exp[q(V_{D1} - V_1)/kT] - q(V_{D1} - V_1)/kT - 1}{\exp[-q(V_{D2} - V_2)/kT] + q(V_{D2} - V_2)/kT - 1} = \frac{\epsilon_2 N_2}{\epsilon_1 N_1}. \quad (10)$$

Recognizing that the charge per unit area in the depletion layer of the large band gap material (material 2) is

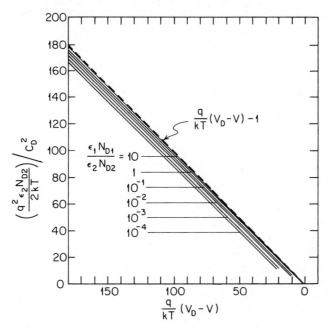

Fig. 3. Calculated dependence of capacitance on voltage for an isotype heterojunction, where the ratio of layer dopings is a parameter [2].

$$Q = \epsilon_2 \frac{dV_2}{dx}\bigg|_{x=0} = \{2kT\epsilon_2 N_2[q(V_{D2}-V_2)/kT - 1]\}^{1/2}, \quad (11)$$

then the capacitance is given by [2]

$$\frac{1}{C_D^2} = \frac{2kT}{q^2\epsilon_2 N_2} \frac{q(V_{D2}-V_2)/kT - 1}{1-(kT/q)[(V_D-V)-(V_{D1}-V_1)(1-\epsilon_1 N_1/\epsilon_2 N_2)]^{-1}}. \quad (12)$$

Clearly, from the expression given in eq. (12), a plot of $1/C_D^2$ versus V is linear only if $V_{D2} - V_2$ is also a linear function of V (in the limit of $V \gg kT/q$). Although this is not strictly the case, the relatively small voltage drop across the accumulated, narrow bandgap material makes this approximation reliable in most cases. In addition, the ratio $\epsilon_1 N_1/\epsilon_2 N_2$ must also be equal to one, if a linear dependence of $1/C_D^2$ versus V is to be a good approximation of the $C-V$ characteristics, as indicated by the plot in fig. 3.

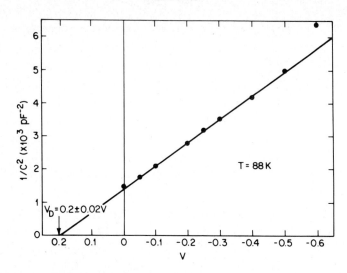

Fig. 4. Capacitance–voltage characteristics measured at 88 K of an n–N $In_{0.53}Ga_{0.47}As/InP$ heterojunction. A voltage intercept of 0.20 V is shown [3].

In spite of these apparent problems, isotype heterojunction band offsets can nevertheless be determined with reasonable accuracy using suitably prepared samples. For example, fig. 4 shows the results of such a measurement made on a n-$In_{0.53}Ga_{0.47}As$/N-InP heterojunction [3]. It is observed that the C–V data are well-behaved – i.e., they are linear over a significant range of voltages. From the voltage-axis intercept, the diffusion potential is found to be $V = 0.20$ V, from which a conduction-band discontinuity of $\Delta E_c = 0.21 \pm 0.02$ eV is inferred.

Note that the measurements of this heterojunction were done at $T = 88$ K to reduce the leakage current which ordinarily would give rise to significant error in the capacitance measurement. The I–V characteristics of the isotype heterojunction used in obtaining the C–V data of fig. 4 are shown in fig. 5. An energy band diagram of the n-$In_{0.53}Ga_{0.47}As$/N-InP heterojunction is shown for reference in the inset of fig. 5. From the I–V data, it is clear that extremely large leakage currents arise from thermionic emission of electrons at room temperature over the small energy barrier. On the other hand, by cooling the heterojunction to $T = 88$ K, the electron population is less energetic, and thus it is less likely that a carrier will be thermally emitted across the heterojunction barrier. For this reason, the low-temperature reverse-biased characteristics are reasonably free of large leakage currents particularly at low voltages, and reliable C–V data can

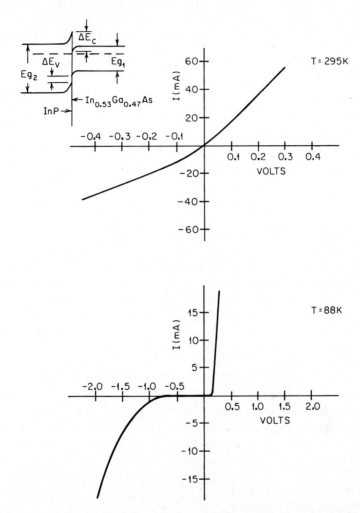

Fig. 5. Current–voltage characteristics of the heterojunction used in fig. 4 measured at 295 K and 88 K. Due to the small value of ΔE_c of 0.2 eV, it is necessary to cool the sample to low temperatures to obtain reliable capacitance data [3].

thus be obtained. An important qualitative result obtained from these I–V data is the fact that if the conduction-band discontinuity were absent in this device, there would be no barrier to current flow thus resulting in ohmic characteristics at all temperatures considered. As will be shown in sect. 4, the I–V characteristics can be analyzed to directly yield the conduction-band

Table 1
Representative band offset measurements using the $C-V$ intercept method.

Heterojunction		ΔE_c (eV)	ΔE_v (eV)	Ref.
Material 1	Material 2			
n-In$_{0.53}$Ga$_{0.47}$As	N-InP	0.21 ± 0.02		[3]
p-Ge	N,P-GaAs	0.15	0.55	[4]
n-Ge	P-GaAs	0.11	0.56	[1]
n-Ge	N-Si	0.15		[5]
p-InP	N-CdS	0.56		[6]

discontinuity, although the results are not as reliable as those obtained from $C-V$ data.

In table 1 we present data obtained for energy band measurements obtained on several heterojunction systems using the $C-V$ intercept method. In each case, it is stated if the measurements were made on isotype or anisotype junctions. Of the various techniques discussed in this chapter, the $C-V$ intercept method has been employed the least, although a wide range of heterojunction systems have been assessed by the technique.

2.2. Error sources in the $C-V$ intercept method

Although, in principle, the $C-V$ intercept method is a straightforward means for measuring energy band offsets, there are several potential sources of error which can systematically result in measured values of E which are in serious disagreement with the actual values. While some of the error sources are common to all transport techniques, some are peculiar to the $C-V$ intercept method. The main sources of error to be considered in this section are effects due to non-ohmic contacts, bulk traps, interfacial defects, and a non-uniform background free-carrier concentration in the material layers near the heterointerface.

2.2.1. Non-ohmic contacts

As discussed in sect. 2.1 the measurement of the energy band offset is simply a measurement of the diffusion potential arising at the heterojunction. However, such a measurement is easily obscured in samples where other potential barriers also exist. For example, measurements of isotype heterojunction discontinuities are typically made on samples consisting of two ohmic contacts made to the opposite surface of a sample where it is assumed that the only energy barrier present is the band edge discontinuity. For most semiconductor heterojunctions, the discontinuity is no more than a few tenths of an eV in magnitude. Thus, if there are small, non-ohmic barriers at the metal contacts or elsewhere in the sample, the diffusion potential from such a barrier will significantly contribute to the total value obtained for ΔE_c.

This error source can only be elimated by ensuring that truely ohmic contacts have been formed. One technique of accomplishing this is to measure the contact resistance of a homojunction sample at several temperatures, while monitoring the ohmic nature of the contact. In fact, careful measurements of this type are, in general, sufficient to eliminate this systematic error source, except possibly in samples where the band discontinuity is very small (less than 50 meV).

2.2.2. Bulk traps

Sah and Reddi [7] have shown that defect levels in the semiconductor bulk can influence both the shape and the intercept of the $1/C^2$ versus V data, and these effects can therefore lead to error in obtaining the energy band offsets using the C–V intercept method. The bulk traps can fall into two catagories: traps with fast thermal emission rates such that they are always in equilibrium with the AC capacitance measurement frequency, and traps with rates less than the excitation frequency.

Assuming that only one bulk trap exists, and is distributed uniformly throughout the semiconductor with a density N_T, it can be shown that the capacitance due to fast traps is given by

$$C_{FT}^2 = \frac{q\epsilon N_D}{2[(V_D - V) - (N_T\phi_T/N_D)]} \tag{13}$$

Here, $\phi_T = (E_F - E_T)/q$, where E_T is the energy of the trap. Although eq. (13) indicates a linear dependence of $1/C^2$ on V as in the trap-free case, it can be seen that the intercept is reduced by an amount equal to $N_T\phi_T/N_D$. Thus, as the trap density is increased to a value comparable to the background free-carrier concentration, significant errors are introduced into the value of the diffusion potential inferred from the voltage intercept.

The situation for slow traps is somewhat more complicated. Although the traps in this regime cannot respond to changes in the AC signal voltage, they can respond to external bias voltage changes, and, therefore, emission and capture of charges at the depletion region edge can occur, thereby influencing the background free-carrier concentration. In this case, it can be shown that the capacitance for slow traps is

$$C_{ST} = \frac{\left(\tfrac{1}{2} q \epsilon N_D\right)^{1/2}}{\left[N_D \phi_T/(N_D - N_T)\right]^{1/2} + \left[(V_D - V) - N_T \phi_T/N_D\right]^{1/2}}, \quad (14)$$

which is not a linear function of V. Furthermore, the intercept voltage is also shifted by a complicated function of the trap density and energy within the bandgap.

Figure 6 shows the capacitance–voltage characteristics measured at several frequencies for a Au-doped Si, p^+–n diode sample, where the deep level density due to the presence of Au is 7.3×10^{15} cm^{-3} [7]. It is clear from these data that the presence of the Au atoms strongly influences the shape and intercept of the data. Thus, large defect densities are expected to lead to error in obtaining V_D if not properly taken into account.

One means by which the data can be corrected for the presence of trapped space charge is to measure independently the densities and energy levels of the charge using a technique such as deep level transient spectroscopy [8], and then to apply the appropriate values to a model such as that proposed by Sah and Reddi to compensate the C–V data. However, the values of V_D thus obtained are to be used with extreme caution, since they are sensitive to the trap densities and energy levels measured in the ancillary experiments. Furthermore, the diffusion potential inferred will

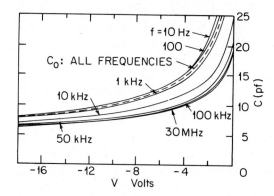

Fig. 6. Dependence of the capacitance–voltage (C–V) characteristics on measurement frequency due to the presence of deep trapping levels in a Au-doped, Si diode. Here, C_0 denotes the trap-free sample capacitance [7].

also be somewhat model-dependent. Thus, in general, the most reliable values can only be obtained from relatively bulk-trap-free samples.

2.2.3. Interfacial traps

In fig. 7 we show a model for a trap-dominated heterojunction assuming that the traps are acceptor-like (fig. 7a) or donor-like (fig. 7b). Here, "acceptor-like" refers to defect levels which are more negatively charged when ionized, whereas donor-like traps exhibit the opposite behavior. It can be seen that acceptor-like traps have the effect of bending the bands away from the Fermi-energy at the heterojunction since they increase the potential energy of the conduction-band electrons. On the other hand, donor-like defects raise the potential energy of holes. In either case, the bands are distorted from their equilibrium, trap-free values, and hence the diffusion potential is also changed by the amount of trap-related band bending. This trap charging is considered to be the most troublesome of all the error sources contributing to the $C-V$ intercept method since it is not easy to measure independently the trap-related contribution to the observed diffu-

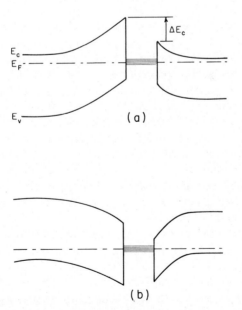

Fig. 7. Band diagrams of isotype heterojunctions with a high density of: (a) acceptor-like, and (b) donor-like deep levels residing at the heterointerface.

sion potential. Furthermore, one must have a priori evidence of the existence of such defects if appropriate corrections of the data are to be made.

We now estimate the magnitude of the interface charge density which is expected to influence significantly the measurement of ΔE_c obtained from C–V data [9]. The effect of this fixed charge is to create a built-in potential V_T which is superimposed on the dipole potential due to the heterojunction. The electric field due to a sheet of charge of density σ residing at the heterointerface is simply $F = q\sigma/\epsilon$, where the permittivity, ϵ, is chosen corresponding to its value on each side of the heterojunction. Furthermore, the maximum electric field due to the diffusion potential, V_D, is approximately

$$F = 2V_D/W = [2qN_DV_D/\epsilon]^{1/2}. \tag{15}$$

Setting these two values of electric field equal and solving for σ, we have

$$\sigma_M = [2\epsilon N_D V_D/q]^{1/2} \tag{16}$$

as the value of interface state density which results in a potential V_T, approximately equal to V_D. That is, if the density of interface states approaches σ_M in eq. (16), we can expect the measured value of V_D to be significantly different from the value which would be obtained for a trap-free, "intrinsic" heterobarrier. To estimate the magnitude of σ_M which might lead to such measurement errors, consider the case of an n–N heterojunction with $N_D = 1 \times 10^{16}$ cm^{-3} on both sides of the interface, and a conduction-band discontinuity of 0.2 eV which is typical of many common heterojunctions. Substituting these values into eq. (16) gives $\sigma_M = 2 \times 10^{12}$ cm^{-2} as the value for the interface charge density which would be expected to significantly distort measurements of the band offset values.

It can be seen that the band distortion due to interfacial traps should lead to a "double depletion" characteristic, i.e. the capacitance of the heterojunction decreases independent of the polarity of the applied potential [10]. Taking as an example the behavior of the n–N Ge–Si junction (fig. 7a), double depletion results since the bands are bent upward on both sides of the heterointerface. Since the total capacitance of the junction is equal to the series contributions to the depletion region width from both the narrow and wide bandgap layers, applying voltage of one polarity decreases the capacitance on the "reverse biased" side, and increases the capacitance on the opposite side of the interface. In the series-circuit approximation, the smallest capacitance (i.e. that contributed by the reverse-biased layer in symmetrically doped heterojunctions) must dominate, thereby reducing the total heterojunction capacitance. Due to the symmetry of the high defect-density junction, the same situation should be obtained when the polarity of the external potential is reversed. However, the magnitude of the effect is expected to be different since the total amount of band bending on a given side of the heterojunction is due to the superposition of the intrinsic heterojunction potential with that contributed by the interfacial trapped charge.

Experimental observation of such symmetrically depleted junctions was first made on n-N isotype Ge-Si heterojunctions [11]. In this case, the large lattice mismatch between these materials (approximately 4%) leads to a high density of dislocations, and hence a high density of defect centers which are localized at the heterointerface. Such a large mismatch is expected to result in a surface state density of greater than 10^{13} cm^{-2} – a value considerably larger than the criterion set in eq. (16) for interface measurements relatively free of trap-related effects. Thus, fig. 8a shows the bipolar C–V characteristics of such a heterojunction, where the double depletion is clearly observed by the increase in $1/C^2$ for both forward and reverse bias of the n-N heterojunction. Following the analysis of sect. 2.1, the capacitance–voltage characteristic can be approximated by

$$C_D^2 \simeq \frac{q}{2\epsilon_1 N_1(V_{D1} + V)}, \quad V > 0, \text{ and} \tag{17}$$

$$C_D^2 = \frac{q}{2\epsilon_2 N_2(V_{D2} - V)}, \quad V < 0. \tag{18}$$

Thus, the depletion potential on each side of the heterointerface can be obtained from intercepts of the data with both the positive and negative voltage axes. For the data in fig. 8, we therefore obtain $V_{D1} = 0.31$ V and $V_{D2} = 0.43$ V.

Furthermore, the total density of negative charge at the interface at equilibrium is simply

$$\sigma = \frac{1}{q}\left[(2q\epsilon_1 N_1 V_{D1})^{1/2} + (2q\epsilon_1 N_2 V_{D2})^{1/2}\right], \tag{19}$$

and was found for the Ge-Si interface to be 1.23×10^{12} cm^{-2}.

In principle, it is still possible to obtain the conduction-band discontinuity which is related to the difference of V_{D1} and V_{D2}. However, this technique is no longer reliable in the presence of such a high density of

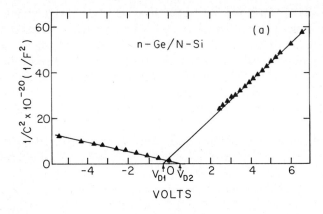

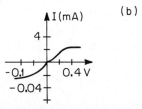

Fig. 8. (a) The capacitance–voltage characteristics of a double-depletion n–N Ge–Si heterojunction with a high density of defect states at the heterointerface. The voltage intercepts are denoted in the figure for both positive and negative applied potentials [11]. (b) Current–voltage characteristics of the diode shown in (a), showing double-saturation.

charge for several reasons. As pointed out by Van Ruyven and co-workers [12] such interfacial charge can have several effects, each contributing separately to the measured potential barrier. In particular, the interface charge contributes a sheet of charge which forms a Schottky-like barrier with each of the two contacting semiconductors. Additionally, the interfacial charge can be dipolar in nature due to the asymmetric environment created by the different potential energies of the atomic species adjoining both sides of the sheet of interface charge. Such an extrinsic dipolar layer of charge would be indistinguishable from the intrinsic dipole created by the defect-free contact of two different semiconductors.

The double depletion characteristic also gives rise to a distortion in the current–voltage ($I-V$) characteristics (fig. 8b). As will be shown in sect. 4, transport of charge over the heterojunction energy barrier occurs primarily by the process of thermionic emission. Assuming that a Schottky barrier-like

I–V characteristic would thus result from an ideal heterojunction, the characteristic would then increase exponentially under forward bias, whereas it would achieve a saturation current value when a small reverse voltage is applied. On the other hand, a high trap density, double depleted junction can be modelled as two back-to-back rectifiers. Thus, it would also be expected to exhibit current saturation independent of the polarity of the applied voltage, as is indeed observed for the Ge–Si heterojunction shown in fig. 8.

One further complication which arises in obtaining heterojunction offsets in the presence of a high density of interface charge, is due to thermal excitation of the trapped charge during the measurement. That is, in the case where the emission rate of charges from the interface traps is much smaller than both the AC capacitance measurement frequency, and the rate of change in the Fermi level induced by varying the external bias, the trap population will not be in equilibrium with the Fermi level. This leads to a time-dependent, or hysteretic C–V characteristic, which can induce large errors in measurements of the diffusion potentials V_{D1} and V_{D2}. Such effects can be identified by a non-linear $1/C^2$ versus V dependence near $V=0$.

A somewhat different situation exists when the AC measurement frequency is comparable to the emission rate. In this case, the differential AC current which is measured is phase-shifted by less than 90° from the AC voltage, indicating a pure reactive response of a capacitive heterojunction. That is, the trap emission contributes charge to the conduction band (assuming an n–N isotype heterojunction), thereby giving rise to a finite AC conductance signal [13]. The effect on the measured reactive signal can be understood if we consider the trap-dominated heterointerface to be appropriately modeled by a parallel RC network (fig. 9), where the conductances g_1 and g_2 are shunts corresponding to charge emission to the narrow and wide bandgap sides of the heterointerface. In this case, the admittance at an angular frequency ω can be shown to be

$$Y_T = G_T + j\omega C_T, \tag{20}$$

where

$$G_T = \frac{g_1 g_2 (g_1 + g_2) + \omega^2 (g_1 C_2^2 + g_2 C_1^2)}{(g_1 + g_2)^2 + \omega^2 (C_1 + C_2)^2}, \quad \text{and} \tag{21}$$

$$C_T = \frac{g_1^2 C_2 + g_2^2 C_1 + \omega^2 (C_1 + C_2) C_1 C_2}{(g_1 + g_2)^2 + \omega^2 (C_1 + C_2)^2}. \tag{22}$$

Here, C_1 (C_2) is the capacitance contributed from the large (small) bandgap side of the heterojunction. That is, the measured capacitance, C_T, is a complicated function of the conductance and capacitance contributions from both sides of the heterojunction. Although the voltage dependence of the diffusion potentials depends on C_1 and C_2 in the usual manner [eqs. (17) and (18)], the dependence of the conductance on voltage is complicated by the distribution and density of traps in the interfacial region. Thus, from eq. (22), it is seen that their values cannot be simply obtained from a measurement of capacitance except in the high frequency limit where $g_1 + g_2 \ll \omega(C_1 + C_2)$. In this case, the total capacitance reduces to its simple, series capacitance approximation.

One final complication is introduced by large saturation leakage currents shunting the two junction capacitances in the doubly saturated heterojunction case. This phenomenon can also be modeled according to the series-parallel network shown in fig. 9. In this case, however, the AC conductances are replaced by the effective DC (voltage-dependent) shunt resistances arising from the large saturation currents associated with the small heterointerface height. It has been shown [5] that the leakage current across high trap density n–N Ge–Si heterojunctions gives rise to a sharp minimum in the C–V characteristic when the Ge layer is reverse-biased with respect to the Si side of the junction (fig. 10). Note that this minimum disappears as the measurement frequency is increased. This is consistent with the model, which predicts that the shunt of the AC current through the reactive half of the circuit must increase with increasing frequency.

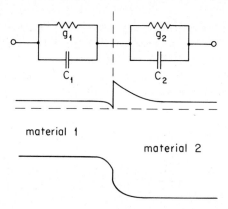

Fig. 9. Equivalent circuit representation of a heterojunction sample showing the conductances and capacitances corresponding to each side of the junction.

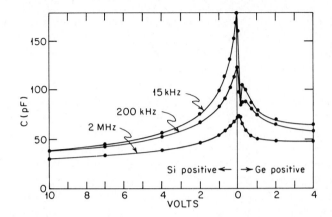

Fig. 10. Capacitance–voltage characteristic as a function of frequency of an n–N Ge–Si heterojunction. The complicated dependence of the low-frequency capacitance near $V = 0$ is due to the comparable magnitudes of the shunt conductances and reactances shown in the equivalent circuit in fig. 9 [5].

It is clear that, in this case, a simple plot of $1/C_D^2$ versus V will not be monotonically increasing with increasing voltage when the Ge is reverse-biased. In effect, the large trap-induced leakage current can result in significant distortions in the C–V data, leading to a concomitant error in the values obtained for the band edge discontinuities.

2.2.4. Non-uniform bulk free-carrier concentration

It has been shown that the accuracy of the C–V intercept method depends critically on the ability to extrapolate the data to the $1/C_D^2 = 0$ axis with high precision. Thus, any effect which lends to difficulty in performing this extrapolation reduces the overall accuracy of the technique. One common phenomenon which has this effect is due to doping non-uniformities in the material layers in the near-interface region. That is, the linear equation relating $1/C_D^2$ to voltage [eq. (9)] is derived assuming a spatially uniform free-carrier concentration. Deviations from this uniformity result in curvature (i.e. a non-vanishing second derivative of the $1/C_D^2$–V data) from which a reliable measurement of V_D cannot be obtained.

It is of course possible, in principle, to compensate for any non-uniformity in the doping by understanding the functional form of the free-carrier concentration profile, and then analyzing the data in the appropriate manner. For example, in the case of linearly graded free-carrier concentration profiles, a plot of $1/C_D^3$ versus V gives a linear plot from whose intercept V_D can be determined. Unfortunately, the dependence of the free-carrier concentration is not always known a priori, and thus only uniformly doped samples are considered to be reliable for these measurements.

Non-uniformities of the free-carrier concentration very near to the heterointerface can also lead to a non-linear $1/C_D^2-V$ dependence, and can induce even larger errors than non-uniformities existing in the bulk of the adjacent layers. Such non-uniformities can arise due to interdiffusion of dopant atoms across the heterointerface, and can be a large effect due to the large diffusion coefficients for many atomic species at the high temperatures used in growing the material layers. This problem can largely be eliminated by preparing heterojunctions with the same level of doping on both sides of the interface, and with the uniform doping extending deep into the bulk of the adjoining semiconductor layers.

3. Capacitance–voltage measurement via the depletion method

3.1. Theory

It has long been recognized that the free-carrier concentration in a non-uniformly doped semiconductor can be obtained via analysis of the capacitance–voltage data obtained by reverse-biasing a p–n junction or other rectifying contacts placed near to the region of interest. Thus, by measuring the capacitance of the depletion region as a function of a change in applied voltage, the change, dQ, in charge, Q, at the edge of the depletion region, and therefore the free carrier concentration, is obtained via

$$C_D = dQ/dV. \tag{23}$$

Now, if $N(x) = |N_D - N_A|$ is the net free-carrier concentration due to a number N_D of ionized donors and N_A of ionized acceptors at a distance x from the junction, then the total space charge contained in the depletion region of width W is

$$Q = q \int_0^W N(x) \, dx. \tag{24}$$

The main assumption employed in the depletion C–V technique is that the charge density at the edge of the depletion region (i.e. at W) abruptly vanishes. That is, in the depletion region, the free-carrier concentration is zero, whereas outside of this region, $N(x)$ assumes its equilibrium bulk value, and the transition from one region to the other occurs over a negligibly small distance. This is known as the depletion approximation, and it leads to negligible error when the background free-carrier concentration varies slowly with distance. However, for samples where the carrier concentration is rapidly varying with distance, such as at heterojunctions, the depletion approximation can lead to significant errors, as will be discussed more fully below. The second assumption usually employed is that a depleted region of a semiconductor can be treated as a parallel-plate capacitor for many device configurations. In this case, the capacitance (per unit area) is related to the depletion region width via the relationship

$$C = \epsilon/W. \tag{25}$$

Combining eqs. (23)–(25) leads to

$$N(x) = \frac{2}{q\epsilon} \frac{dV}{d(1/C^2)}, \tag{26}$$

where we have used the relationship that $dC/C^3 = -\frac{1}{2}d(1/C^2)$. Equation (23) states that the free-carrier concentration at the edge of the depletion region a distance x away from the rectifying contact can be obtained by measuring the incremental change in capacitance with respect to voltage. Here, the distance $x = W$ at which the measurement is made is obtained from eq. (25).

Thus, we would expect that such a measurement would be a simple means of determining the free-carrier concentration profile in the vicinity of a heterojunction. Analysis of the shape of this profile as compared with what would be expected for a sample without a heterojunction should then provide the magnitude of the diffusion potential which exists as a result of the heterojunction dipole. Unfortunately, as has been suggested earlier, the spatial resolution – i.e. the ability to resolve changes in N with distance x – is limited in all C–V techniques to the finite width of the depletion region edge. In effect, the resolution of the C–V technique is not limited if we assume that the depletion approximation applies, although, in fact, use of the depletion approximation can lead to significant errors if it is applied to samples with a free-carrier concentration which is a strong function of position within the sample bulk.

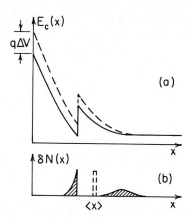

Fig. 11. (a) Band displacement due to a small increase in applied potential of an isotype heterojunction placed near a Schottky barrier contact at $x = 0$. (b) Change in charge corresponding to the change in potential shown in (a). Here $\langle x \rangle$ represents the mean position of the depleted charge [14].

To calculate the ultimate resolution of C–V techniques, we refer to the diagram in fig. 11 which shows the conduction band edge of an n–N isotype heterojunction where it is seen that the free-carrier concentration in the region of the heterojunction is varying over very short distances. When an incremental voltage, dV, is applied at the top contact (at $x = 0$), the free-carrier concentration on *both* sides of the heterojunction must change by an amount δN due to the non-abrupt nature of the charge density distribution at the edge of the depletion region. In effect, although an incremental voltage perturbs the actual charge distribution over an extensive distance, the C–V technique assumes that all of the charge is depleted from a single, average location, $\langle x \rangle$ given by

$$\langle x \rangle = \frac{\int_0^\infty x\, \delta N\, dx}{\int_0^\infty \delta N\, dx}. \tag{27}$$

The total charge density thus determined via a measurement of the sample capacitance does not have a one-to-one correspondence with the charge at a specific location x defined as the depletion region width, and therefore this analysis can lead to large errors under certain circumstances [15]. The distance over which the depletion-region edge extends is calculated from the gradient of the electric field resulting from a change of δN charges at the edge of the depletion region. From Gauss' law, therefore,

$$\frac{dE}{dx} = \frac{q\,\delta N}{\epsilon}. \tag{28}$$

Also, the current density must be equal to the sum of the drift and diffusion contributions. In equilibrium, the current density for electrons, J_n, and holes, J_p, must both be independently equal to zero such that no net build-up of either carrier occurs with time across the junction. Thus, taking the case for electrons only, we have

$$J_n = q\left(\mu_n |N_A - N_D| E - D_n \frac{d\,\delta N}{dx}\right) = 0, \tag{29}$$

where D_n is the diffusion constant, and μ_n is the mobility for electrons. Substituting into eq. (28) the first derivative of eq. (29), we then obtain for the spatial distribution of electrons

$$\delta N(x) = \delta N_0 \exp(-x/L_D), \tag{30}$$

where δN_0 is the amount of charge depleted at $x = 0$, and L_D is the Debye screening length given by

$$L_D = \left[kT\epsilon/q^2 |N_A - N_D|\right]^{1/2}. \tag{31}$$

From eq. (31), the Debye length is seen to depend on the permittivity of the semiconductor and on the temperature, and is inversely proportional to the square root of the free-carrier concentration of the bulk semiconductor. In fig. 12 we plot the room-temperature Debye length as a function of $|N_D - N_A|$ for GaAs and observe that it varies from approximately 100 Å at a doping level of 10^{17} cm^{-3}, to a few thousand Å for dopings in the range 10^{14}–10^{15} cm^{-3}. However, most practical heterojunctions are only 10–50 Å wide. Here, the width of the heterojunction is taken as the distance over which the semiconductor composition changes completely from material 1 to material 2. The heterointerface thus consists of an intermediate material consisting of a spatially varying ratio of the constituents from the two bulk semiconductors. It would appear that the C–V technique does not have sufficient resolution to determine the rapidly varying concentration of carriers at the heterojunction.

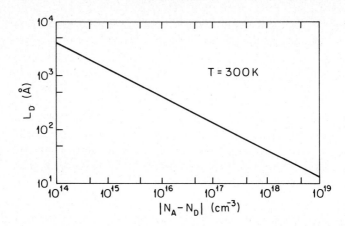

Fig. 12. Effective Debye length at room temperature versus the net free-carrier concentration for GaAs.

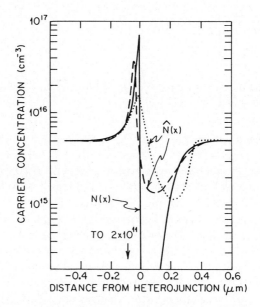

Fig. 13. Calculated free-carrier concentration profile (solid line) in the vicinity of an N–n $Al_{0.3}Ga_{0.7}As/GaAs$ heterojunction as a function of position in the region of the heterointerface. Also shown is the calculated apparent carrier concentration obtained from capacitance–voltage data where the nearby Schottky barrier is placed on either the low conduction band (dotted line) or the high conduction band (dashed line) side of the heterojunction [14].

To obtain an appreciation of the magnitude of the error in assuming the simple depletion approximation, fig. 13 shows the actual free-carrier concentration (solid line) and the apparent free-carrier concentration (dashed and dotted lines) obtained for an N–n $Al_{0.3}Ga_{0.7}As/GaAs$ heterojunction with a conduction-band discontinuity of $\Delta E_c = 0.317$ eV. In this case, all of the various concentrations are calculated solutions to Poisson's equation. Although the general shape of all of the curves reflects the trends of depletion and accumulation of carriers on the two sides of the heterojunction, it is immediately apparent that the measured concentrations underestimate both the maximum and minimum concentrations by significant amounts. This is to be expected if we assume that the effect of the Debye screening is to "spread out" the actual concentration over an extended region. Furthermore, the apparent concentration, $\hat{N}(x)$, is also influenced by the direction from which carrier depletion occurs. For example, if the rectifying contact is assumed to be on the low conduction band material (dotted line), the apparent free-carrier concentration is shifted to larger distances into the semiconductor bulk than if the contact were on the high conduction band side of the junction (dashed line).

Although there cannot be a simple correspondence between the measured and the actual free-carrier concentrations in the vicinity of a heterojunction, Kroemer and coworkers [14,16] have shown that by analyzing the perturbed apparent carrier concentrations over long distances on both sides of the heterojunction, the conduction-band discontinuity can still be accurately determined from simple C–V analysis. In this technique, it was recognized that the total charge density contained in the interface dipole is

$$\sigma = q \int_0^\infty \left[N_D(x) - \hat{N}(x) \right] dx, \tag{32}$$

where $N_D(x)$ is the background carrier concentration (assuming n-type material) as a function of position on both sides of the heterojunction. Note that since the heterojunction gives rise to a charge dipole, the integral in eq. (32) must vanish except in the presence of fixed charge in the heterointerface region. The fixed charge can exist due to several extrinsic sources such as lattice mismatch, atomic vacancies, etc.. However, unless it is present in high densities, it will not affect the accuracy of the measurement. Further discussions of the effects of fixed charge will be given in the section on error analysis (sect. 3.2.2).

Furthermore, the first moment of the charge distribution is also unperturbed as long as the integration is carried out sufficiently far from the region where the carrier concentration is rapidly varying with position. Thus, we can write

$$\int_0^\infty N_D(x) \, x \, dx = \int_0^\infty \hat{N}(x) \, x \, dx. \tag{33}$$

Integrating Poisson's equation by parts then yields the diffusion potential due to the heterojunction dipole via

$$V_D = \frac{q}{\epsilon} \int_0^\infty [N_D(x) - \hat{N}(x)] \, [x - x_I] \, dx, \tag{34}$$

where x_I is the position of the metallurgical heterojunction. From fig. 13 it is apparent that x_I is accurately determined as the position of the peak in the apparent free-carrier concentration profile. Once V_D is determined from eq. (34), the conduction-band discontinuity is obtained in the standard manner using eq. (3). Note that this same technique can be applied to p-type heterojunctions, where the valence-band discontinuities are thus obtained.

In practice, the technique for heterojunction measurement is accomplished as follows:

(i) The bulk carrier concentrations, $N_{D1}(x)$ and $N_{D2}(x)$, on each side of the heterojunction are determined from standard C–V analysis of the sample at positions x such that $|x - x_I| \gg 0$.

The accuracy of this technique depends on an accurate determination of $N_D(x)$ in the region of the heterointerface. However, since the free-carrier concentration in this region is strongly influenced by the presence of the interfacial dipole, values of $N_{D1}(x)$ and $N_{D2}(x)$ obtained in step (i) must be extrapolated into the region where $x \simeq x_I$. For this reason, it is important that $N_D(x)$ be a simple and well-behaved function of x. The highest accuracy can practically be achieved if N_D is uniform throughout the sample.

(ii) The position of the heterointerface, x_I, is determined from the peak of the apparent free-carrier concentration.

(iii) V_D is obtained from a numerical integration of the data using eq. (34).

Alternatively, a more accurate measurement of V_D can be achieved by fitting the measured apparent free-carrier concentration to a calculated curve as shown in fig. 14 which assumes a value of V_D which fits best the measurement. This "back-fitting" technique also allows for the most accurate determination of the fixed charge density, as given in eq. (32).

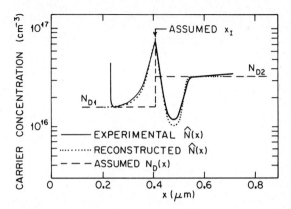

Fig. 14. Experimental (solid line) and reconstructed (dotted line) carrier concentration profiles of an N–n $Al_{0.3}Ga_{0.7}As/GaAs$ heterojunction sample. The dashed line indicates the background doping "baselines" assumed for the reconstructed data. From these data, a conduction-band discontinuity of 0.248 eV is obtained [14].

An example of such an analysis is shown in fig. 15 for an n–N heterojuntion consisting of $In_{0.53}Ga_{0.47}As$ grown lattice matched to an InP buffer layer [17]. From these data, it was found that the conduction-band discontinuity was $\Delta E_c = 0.23 \pm 0.02$ eV, and the fixed charge at the hetero-

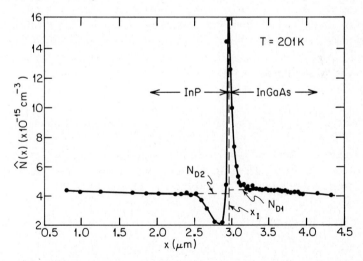

Fig. 15. Apparent free-carrier concentration profile of an n–N $In_{0.53}Ga_{0.47}As/InP$ heterojunction sample indicating the assumed baseline background dopings and the heterointerface position. These data give a conduction-band discontinuity of 0.23 eV [17].

interface was $(3.0 \pm 0.5) \times 10^{10}$ cm^{-2} at room temperature. In sect. 3.2 it will be shown that the fixed charge for these particular samples was found to be strongly dependent on temperature. Nevertheless, it is apparent that the sample used is nearly ideal for the purposes of determining the band offsets for this material system. For example, the free-carrier concentrations on both sides of the heterojunction are accurately determined by the ability to deplete deep into the bulk of the various material layers. Furthermore, the heterointerface is narrow on the scale of a Debye length, resulting in a sharp peak from which x_I is easily and accurately determined. Although it can be shown that the technique is, in principle, not subject to error when applied to samples with a heterointerface extending over long distances, in practice, such an effect can indeed introduce systematic errors. This will be discussed further below.

Note that the requirement for accurately determining the bulk free-carrier concentrations (or "baselines") has several implications relating to the desired magnitude of these carrier concentrations as well as on the placement of the heterojunction with respect to the rectifying contact. It is desirable to sweep the depletion region between positions which are far away from the heterojunction on both sides of the interface (i.e. into regions where $|x - x_I| \gg 0$) such that the baselines can be determined without error introduced by distortions in $N(x)$ arising from the heterojunction dipole itself. Thus, the heterojunction must be positioned relative to the rectifying contact such that the depletion region is swept through the heterojunction at some voltage midway to the breakdown of the contact. Now the breakdown voltage of an ideal, abrupt, uniformly doped p$^+$-n junction is given approximately by [18]

$$V_{BD} \simeq 60 (E_g/1.1)^{3/2} (N_D/10^{16})^{-3/4}, \tag{35}$$

and is seen to decrease monotonically with increasing free-carrier concentration, N_D. Here, E_g is in eV and N_D in cm^{-3}. Furthermore, the dependence of depletion region width on voltage is given by

$$W^2 = \frac{2\epsilon(V_{bi} - V)}{qN_D}, \tag{36}$$

for a uniformly doped sample. For the purpose of this argument, we will only consider heterojunctions with uniform and equal doping densities in both materials. Here V_{bi} is the built-in voltage of the rectifying contact. Combining eqs. (35) and (36) gives

$$W_{\text{BD}} \simeq \left[\frac{E_g^{3/2}\epsilon}{qN_D^{7/4}}\right]^{1/2} \times 10^7 \quad (\text{cm}) \tag{37}$$

for the maximum possible depletion layer width obtained in ideal p–n junction devices. It is therefore desirable that the heterojunction be located at a distance $\simeq \frac{1}{2}W_{\text{BD}}$.

In addition, it is important that the heterojunction be sufficiently far away from the rectifying contact such that the baseline in material 1 can be accurately determined. That is, the baseline for material 1 cannot be accurately determined in very low doped samples where the built-in potential depletes the interface region. Now, the width of the built-in depletion region W_0 is given by eq. (36) by setting the applied voltage, V, equal to zero. The theoretically "accessible" region of a sample is then between W_0 and W_{BD}, and is seen to become increasingly narrow with increasing background doping. This region is shown in fig. 16 for Si and GaAs-like samples. The center lines denote "x_1" indicating the best placement of the heterojunction with respect to the rectifying contact for the purpose of measuring the conduction-band discontinuity for n–N heterojunctions.

We note that the above calculation is done for ideal p^+–n junctions with no defects or other features which might reduce the breakdown voltage from that predicted in eq. (35). However, in general, such high voltages are not routinely obtained, placing severe constraints on the optimum positioning of the rectifying contact. For example, it is well-known that Schottky barrier diodes with no guard rings are subject to very low voltage breakdown due to concentration of the electric field at the edges of the metal contact electrode [19]. Thus, the use of Schottky barriers, while desirable due to their ease of fabrication, places stringent demands on the doping and layer thickness of the heterojunction sample. In addition, Schottky barrier heights for some materials are unacceptably small, thus giving rise to large reverse leakage currents which degrade the accuracy of the capacitance measurement at room temperature. For example, for InP and related compounds, the Schottky barrier height is less than $\simeq 0.5$ eV [20]. Thus, measurements on InP/InGaAsP heterojunctions have been enhanced by replacing the Schottky barrier contact with an organic-on-inorganic semiconductor contact barrier which more closely resembles an ideal abrupt p–n junction in its reverse-biased characteristics [21,22].

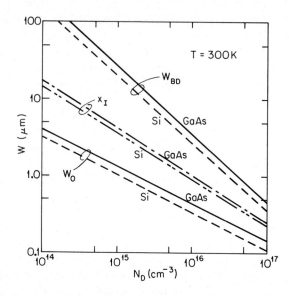

Fig. 16. Dependence of the depletion region width at breakdown (W_{BD}) and at zero voltage (W_0) of ideal, diffused-junction Si and GaAs diodes. Also shown is the optimum placement of the heterojunction (x_I) for band-offset measurements using the C–V depletion method.

The depletion technique has recently become the most generally accepted and accurate means of determining energy band offsets using transport techniques. Table 2 indicates results reported for such offset measurements made on several III–V material systems. In addition, comprehensive measurements made on the functional dependence of the conduction-band discontinuity on bandgap difference for the GaAs/AlGaAs and InP/InGaAsP material systems have also bee reported. The results of these two studies are shown in fig. 17a and fig. 17b, respectively. The values for the GaAs-based system [23], given by $\Delta E_c = 0.67 \Delta E_g$ for an aluminum concentration $x_{Al} < 0.5$, are consistent with measurements made on this same material system using photoabsorption and luminescence techniques [27]. Thus, at least in the case of GaAs/AlGaAs heterojunctions, several different techniques have been used to determine the conduction-band offsets as a function of alloy compositon. All reliable measurements made on this system indicate that the conduction-band discontinuity is roughly between 60%–70% of the bandgap difference for Al concentrations as high as 50%. Thus, due to the agreement between these measurements, these values are considered to be reasonably reliable, and the depletion technique affords

Table 2
Representative band offset measurements using the $C-V$ depletion method.

Heterojunction		ΔE_c (eV)	ΔE_v (eV)	Ref.
Material 1	Material 2			
n-GaAs	N-Al$_{0.3}$Ga$_{0.7}$As	0.248 [a]		[14]
n-GaAs	N-Al$_x$Ga$_{1-x}$As ($x < 0.42$)	0.67ΔE_g	0.33 ΔE_g	[23]
n-GaAs	N-Al$_{0.3}$Ga$_{0.7}$As	0.33eV [b,c]		[24]
n-In$_{0.53}$Ga$_{0.47}$As	N-InP	0.23 ± 0.02		[17]
n-In$_x$Ga$_{1-x}$As$_y$P$_{1-y}$ ($0 < y < 1.0$)	N-InP	0.39ΔE_g	0.61ΔE_g	[21]
n-In$_{0.53}$Ga$_{0.47}$As	N-InP	0.21		[25]
n-In$_{0.53}$Ga$_{0.47}$As	N-InP	0–0.4 [c,d]		[26]
n-In$_{0.53}$Ga$_{0.47}$As	N-In$_{0.52}$Al$_{0.43}$As	0.50 ± 0.05 [c]		[24]

[a] $\Delta E_c = 0.66 \Delta E_g$.
[b] $\Delta E_c = 0.88 \Delta E_g$.
[c] Poor baseline doping data.
[d] Value depends on lattice-mismatch.

the most accurate means of determining this important parameter via transport-related measurements.

The situation is somewhat more complicated for InGaAsP/InP heterojunctions (fig. 17b) where it has been found that $\Delta E_c = (0.39 \pm 0.01)\Delta E_g$. Here, reliable independent determination of these values has not yet been achieved using techniques other than transport measurements. This is due largely to the difficulty in growing multiple quantum well (MQW) structures consisting of these compounds, and hence performing absorption or other optical measurements useful in determining band offsets. In general, such MQW structures are grown by molecular beam epitaxy (MBE). However, until recently this technique has not been useful due to the low sticking coefficient of phosphorus. To date, the most reliable measurements made on this important material system have utilized depletion $C-V$ analysis on single heterojunction samples as shown in fig. 17.

It is interesting to note that the difference in the magnitude of the band offsets relative to the bandgap difference in these two material systems suggests that for AlGaAs/GaAs most of the bandgap difference appears in the conduction band, whereas for InGaAsP/InP heterojunctions the larger fraction of the difference appears in the valence band. This situation is consistent with the predictions of the "common anion rule" [28,29] which states that the electron affinity is determined by the chemical potential of the anion species of a compound semiconductor. Thus, the valence bands of

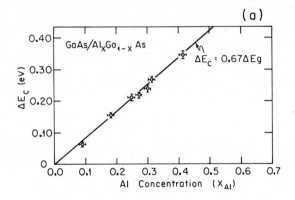

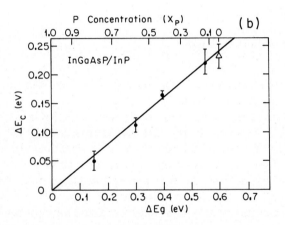

Fig. 17. Dependence of the conduction-band offset energy on: (a) Al concentration for AlGaAs/GaAs heterojunctions, where the Al concentration is varied from 0 to 0.5 [23], and (b) P concentration for InGaAsP/InP heterojunctions, where the P concentration is varied from 0 to 1.0 [21]. The data in this figure were obtained using the $C-V$ depletion technique.

materials with common anions (e.g. GaAs and AlGaAs) will tend to lie at the same energy, resulting in the difference in the bandgaps of the contacting materials to be taken up in the conduction band. Materials with dissimilar anions (e.g. InP and $In_{0.53}Ga_{0.47}As$ which is the ternary end-point of the InGaAsP system lattice-matched to InP [30]) however, are expected to have a conduction-band offset which is somewhat smaller since much of the valence-band offset is determined by the difference in anion potential.

3.2. Error sources in the C–V depletion technique

There are several sources of systematic error which can be introduced in the C–V depletion technique. These include errors in determining the background carrier concentrations in the bulk on each side of the heterointerface, fixed charge at the interface and in the layers themselves, compositional grading over long distances, and dipolar interfacial defects. Each of these potential sources of error will now be treated in detail.

3.2.1. Baseline error

Uncertainties in determining the baseline, or free-carrier concentration in the bulk of each layer is perhaps the greatest potential source of measurement error in the C–V depletion technique. This error is incurred when the depletion region is not swept sufficiently far away from both sides of the heterointerface (i.e. to $|x - x_\mathrm{I}| \gg 0$) such that the perturbation of the free-carrier concentration in the vicinity of the interface due to the heterojunction dipole makes it difficult or impossible to accurately determine either N_1 or N_2, or both. To understand the magnitude of the error thus incurred, we calculate the uncertainty in V_D due to an error, $\delta N_\mathrm{D}(x)$, made in estimating the baseline doping. Thus, from eq. (34),

$$\mathrm{d}V_\mathrm{D} = \frac{q}{\epsilon} \int_0^\infty \delta N_\mathrm{D}(x) \, [x - x_\mathrm{I}] \, \mathrm{d}x, \tag{38}$$

where we recognize that, although both x and x_I also depend sublinearly on $N_\mathrm{D}(x)$, the difference $\mathrm{d}x - \mathrm{d}x_\mathrm{I}$ is generally negligible. From eq. (38), it is apparent that the error induced in the diffusion potential can be very large since it is cumulative. That is, $|\mathrm{d}V_\mathrm{D}|$ increases as the integral is carried out over increasingly larger distances from the heterojunction. Such an error is common for uniformly doped samples where the difference $N_\mathrm{D} - \hat{N}(x)$ is constant over long distances, at least on one side of the junction. Note also that eq. (38) indicates that $\mathrm{d}V_\mathrm{D}$ is proportional to the first moment of $\delta N_\mathrm{D}(x)$. Thus, larger errors are incurred by misjudging N_D further away from the heterojunction than if the value of $N_\mathrm{D}(x)$ is uncertain near the interface where $|x - x_\mathrm{I}|$ is small.

An example of data which can lead to systematic ambiguities in estimates of the baseline free-carrier concentration is shown in fig. 18a. Here, the apparent free-carrier concentration profile taken for an N–n $In_{0.52}Al_{0.48}As/In_{0.53}Ga_{0.47}As$ heterointerface is obtained by C–V analysis using a Schottky barrier contact made on the large bandgap ($In_{0.52}Al_{0.48}As$) side of the junction [24]. Due to the proximity of the heterojunction to the

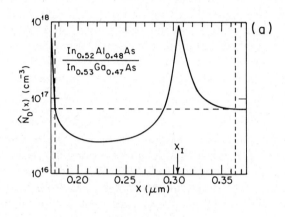

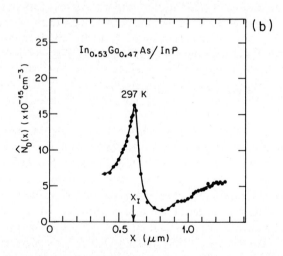

Fig. 18. Apparent free-carrier concentration profile of an: (a) InAlAs/InGaAs [24], and (b) InGaAs/InP heterojunction [26]. Both samples have ambiguous "baseline" dopings which might lead to error in determining the band offsets.

wafer surface, depletion on the large bandgap side of the heterointerface extends all the way to the top contact. For this sample, therefore, it is not possible to make a precise determination of the value for $N_D(x)$. To enable some estimate for ΔE_c for this sample, it was assumed that the free-carrier concentration was the same on both sides of the heterojunction, and therefore $N_D = 7.4 \times 10^{16}$ cm^{-3} was used throughout the integration of eq. (34). However, from eq. (38) we expect that such an assumption can, in fact, lead to substantial errors in determining V_D. A second example which shows data which can also lead to large errors due to a misjudgement of the background doping level is shown in fig. 18b where the $N_D(x)$ profile in the vicinity of an In$_{0.53}$Ga$_{0.47}$As/InP heterojunction is shown [26]. Once again, for these data, it is clear that the samples do not allow for sufficient depth of depletion away from the heterojunction such that $N_D(x)$ can be unambiguously determined both above and below x_I.

To estimate the potential for error due to misjudging the baseline value, we return to the data in fig. 15 obtained for an n–N InGaAs/InP heterojunction. From this figure, we observe that a uniform In$_{0.53}$Ga$_{0.47}$As free carrier concentration of $N_{D1} = 4.5 \times 10^{15}$ cm^{-2} was used, and that the integration was carried out from $x_I = 3.0$ μm to $x = 4.0$ μm. Assuming that the background carrier concentration was chosen to be 2% smaller than the value used, we obtain an underestimate of the diffusion potential of approximately 25%! On the other hand, if instead the integration extends only between x_I and $x = 3.5$ μm, the underestimate in V_D is reduced to only 6%. Thus, it is readily apparent that small errors in estimating $N_D(x)$ can have a particularly strong effect on the accuracy of measurement of V_D, particularly if the integration must be carried out over extensive distances.

One primary assumption of the C–V depletion technique is that since it is not possible to independently determine $N_D(x)$ at the heterojunction, values used in this region must be extrapolated from measurements made deep in the bulk of the two contacting material layers. Thus, for uniformly doped layers with carrier concentrations N_1 and N_2, it is assumed that these values remain uniform up to the heterointerface where they change abruptly along with the composition of the layers. Clearly, this assumption is only reliable for samples fulfilling two conditions:

(1) The carrier concentration is highly uniform in the layer bulks, and
(2) The carrier concentrations N_1 and N_2 are nearly equal.

This last requirement avoids the possibility that, during the growth and subsequent high-temperature processing of the sample, the shallow donors (or acceptors) from the more highly doped side of the junction diffuse into the adjacent layer, thus changing the doping in the interface region in an undetermined manner.

It thus appears that although the accurate determination of baselines is of great importance in assuring the accuracy of measuring V_D, it can be readily achieved in samples where the free-carrier concentration on both sides of the heterojunction are nearly equal and uniform, Furthermore, the magnitude of the carrier concentration and the placement of the heterojunction relative to the rectifying contact are optimized when the sample can be depleted over sufficiently large distances away from the heterojunction such that distortions induced in $\hat{N}(x)$ due to the existence of the interface charge dipole do not result in an unreliable estimate for the background doping density.

3.2.2. Fixed interfacial charge

As in the case of the $C-V$ intercept method, the $C-V$ depletion method is also sensitive to fixed charge at the heterointerface. Since both of these techniques measure the diffusion potential due to the barrier arising from the energy band offsets, any fixed charge which might also give rise to a potential in the interface region will be superimposed on the discontinuity potential, and will introduce error into the values thus obtained. The most troublesome type of fixed charges (sect. 2.2) are those which are dipolar, since they are accompanied by an electrostatic potential which is indistinguishable from the dipole originating from the heterojunction itself. Indeed, as discussed in previous sections, most defect charge residing at heterojunctions is *expected* to be dipolar due to the asymmetric potential environment which is intrinsic to semiconductor–semiconductor contacts between unlike atomic species. Therefore, all of the models and arguments used for the $C-V$ intercept method apply to the results obtained using the depletion technique. Nevertheless, several interesting observations of fixed charge have been made, using the depletion technique, which are of interest in understanding the effects of interface charge on the shape of the band discontinuities.

It was first observed [14] using the depletion technique that a significant density of fixed charge ($\sim 2.7 \times 10^{10}$ cm^{-2}) is found at the lattice-matched AlGaAs/GaAs heterojunction. Care must be taken, however, in assigning undo importance to such an observation due to the resolution of the measurement. For example, let us assume that the background carrier concentration of a sample is $N = 1 \times 10^{16}$ cm^{-3}, and that the integrals in eqs. (32) and (34) are calculated over a distance of 1.0 μm in order to obtain the various heterointerface parameters. Now, the relative uncertainty in measuring the background free-carrier concentration is usually greater than 1%. From eq. (32), this leads to an uncertainty in the fixed charge density of

$$d\sigma = \int_0^\infty \delta N_D(x)\,dx. \tag{39}$$

Using eq. (39) for $d\sigma$, we obtain $d\sigma = 1.0 \times 10^{10}$ cm^{-2}. Clearly, this level of uncertainty increases as both the limits of integration and the background carrier concentration increase as well.

It has recently been shown [9] that the fixed charge at the heterointerface does not significantly influence the values obtained for the conduction-band offsets measured for InGaAsP/InP heterointerfaces as long as σ remains below the value determined from eq. (16). Thus, in fig. 19 is shown a plot of V_D versus σ obtained for a sample where the bandgap of the InGaAsP was 0.95 eV. These data were measured for a single wafer, where each value of σ was obtained at a different location on the sample. Here, the fixed charge density varies between 0.5 and 8×10^{10} cm^{-2} with no apparent systematic effect on values obtained for V_D. However, there appears to be a larger randomness to the values obtained for V_D as σ increases. This effect, however, is not a direct result of the larger fixed charge density, but is probably due to uncertainties in determining the free-carrier concentration "baselines" in some samples, which by the above analysis also results in an anomalously large value for σ.

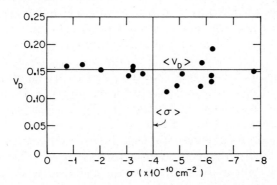

Fig. 19. Dependence of the measured diffusion potential on fixed interface charge density of n-N InGaAsP ($E_g = 0.95$ eV)/InP isotype heterojunctions obtained from capacitance–voltage data. Over the range of interface charge density shown, there is no systematic dependence of potential on charge [9].

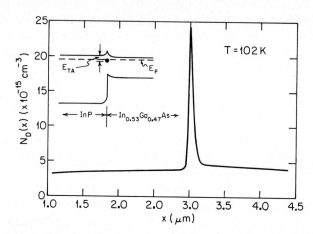

Fig. 20. Apparent free-carrier concentration profile at 102 K of the InGaAs/InP heterojunction sample of fig. 15. The loss of the "spike and notch" profile is attributed to the filling of an acceptor-like trap localized at the heterojunction (inset) with decreasing temperature [17].

Although in most reports the fixed charge at the heterointerface has been found to be small (i.e. in the range of 10^{10} cm^{-2}), in some cases the values obtained for σ are significantly greater than can be attributed solely to experimental uncertainties. Indeed, when this is the case, values obtained for V_D are often significantly influenced by the large fixed charge density. Such results were first reported by Forrest and Kim [17] for $\text{In}_{0.53}\text{Ga}_{0.47}\text{As}/\text{InP}$ heterointerfaces, where it was found that a dramatic reduction in V_D was observed as σ increased with decreasing sample temperature. More recently, Okumura and co-workers [31] have found that AlGaAs/GaAs band offsets are also affected by a high density of interface charge.

Figure 20 shows the apparent free-carrier concentration measured at low temperature for the $\text{In}_{0.53}\text{Ga}_{0.47}\text{As}/\text{InP}$ heterointerface. High-temperature data for this same sample was shown in fig. 15. The low-temperature measurement indicates a large distortion of the expected "spike and notch" profile obtained at room temperature. Qualitative examination of these data indicates that the heterointerface dipole has been dwarfed by the superposition of a large potential barrier due to a high-density deep level at an energy E_{TA} residing in the interfacial region. Quantitative analysis of σ and V_D as a function of temperature, shown in fig. 21, tends to support these qualitative observations. Here, it can be seen that the fixed charge density increases rapidly from a "background" high-temperature value of 3×10^{10}

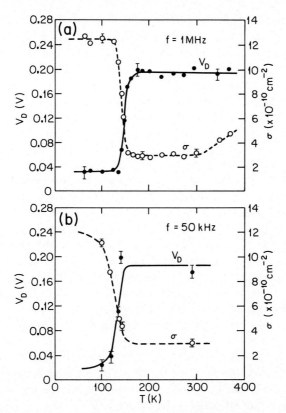

Fig. 21. Dependence of diffusion potential and fixed interface charge as a function of temperature for the InGaAs/InP sample of figs. 15 and 20. The data are obtained at a measurement frequency of: (a) 1 MHz, and (b) 50 kHz. The resulting shift in the curves with frequency is used to obtain the energy of the interface trap [17].

cm^{-2} to a low-temperature value of 1.3×10^{11} cm^{-2} – a nearly five-fold increase. Concomitant with this charge density increase is a decrease in the diffusion potential of 0.17 eV. Thus, it is readily apparent that the superposition of a high density of fixed charge on the background heterointerface dipole has the effect of screening the effects of the latter, more fundamental potential. Indeed, it would appear that the presence of the fixed charge distribution can lead to extremely large errors in the values obtained for the conduction-band discontinuity from these data.

It was further observed that the transition from small to large fixed charge densities is shifted to lower temperatures as the capacitance measurement frequency is reduced (fig. 21b). Analysis of the activation energy of this temperature shift yields the emission rate from the heterointerface trap, which was found to be $E_{TA} = 0.2$ eV below the conduction-band minimum. Studies of both these $C-V$ data and admittance spectroscopic measurements indicate that the change in occupancy with temperature of a high density deep level localized at the $In_{0.53}Ga_{0.47}As/InP$ heterojunction adequately explains the phenomena observed. The particular deep level is neutral at room temperature since it lies above the Fermi energy. As temperature is reduced, the Fermi-level moves toward the conduction band to an energy greater than the trap energy (inset of fig. 20). This results in the filling of the acceptor-like trap, thereby creating a strong band bending at the heterojunction, resulting in a small, double-depletion potential barrier characteristic of Ge-Si heterojunctions (sect. 2.2).

It is interesting to note that these traps have been variously attributed to lattice-mismatch between the $In_{0.53}Ga_{0.47}As$ and InP layers [26], or to atomic (phosphorus) vacancies in the InP lattice [17]. To date, however, no consistent picture accounting for the physical origin of these traps has yet been presented. Nevertheless, such a picture needs to be developed if we are to gain a complete understanding of the various factors which influence the measurement and the origin of the potential barriers observed at semiconductor heterojunctions.

It was also found that AlGaAs/GaAs heterointerfaces can contain a significant density of deep levels, although such high densities as those leading to band bending at $In_{0.53}Ga_{0.47}As/InP$ heterointerfaces have yet to be observed in the AlGaAs/GaAs system [31,32].

3.2.3. Interface grading effects

One of the strengths of the $C-V$ depletion technique is its insensitivity to compositional grading in the heterointerface region. Here a graded junction is one where the interface consists of a finite region where the semiconductor composition is a mixture of atomic constituents of both materials 1 and 2. As pointed out by Kroemer [33], the $C-V$ integrals [eqs. (32) and (34)] are calculated over the entire interface region, and therefore the diffusion potential measured is equal to the *total* electrostatic potential difference between the bulk values of both contacting materials. This total potential difference is thus the actual conduction-band discontinuity (in the absence of interface charge), and is not dependent on the thickness of the compositionally graded region.

Nevertheless, compositional grading over distances greater than a few

Debye lengths can limit the practical resolution of the depletion technique. The first, and perhaps less important source of error arises from difficulties in accurately obtaining the baselines in a compositionally graded junction [9]. That is, if the depletion and accumulation regions reflected in the apparent carrier concentration profile extend over several Debye lengths, it is not possible to deplete to sufficient distances on both sides of the interface such that the background doping density can be obtained in regions not influenced by carriers responding to the dipole potential.

A second, more serious effect of interface grading arises from difficulties in determining the heterojunction location, x_I, from a broad, very graded free-carrier concentration profile. From eq. (34), we see that the uncertainty in the diffusion potential V_D induced by an uncertainty in the heterointerface position x_I is given by

$$dV_D = \frac{q}{\epsilon} \int_0^\infty \left[N_D(x) - \hat{N}(x) \right] \delta x_I \, dx. \tag{40}$$

As in the case of baseline uncertanties, the effect of δx_I is cumulative – i.e. it increases as the limits of integration of the data are expanded to obtain an accurate value for the potential integral. Furthermore, in analogy to baseline errors, the errors due to δx_I are largest when the difference between the apparent and background carrier concentrations, $N_D(x) - \hat{N}(x)$, are largest – i.e. in the region of the heterojunction itself. Thus, due to the cumulative nature of the error, and due to its dependence on the rapidly varying charge distribution in the interface region, uncertainties in determining x_I due to interface grading can lead to significant errors in the measurement of V_D.

3.2.4. Bulk trap effects

In sect. 2.2 we have shown that trapped charges in the bulk of the contacting layers can distort the reverse-biased $C-V$ characteristics from their trap-free values. These deviations in the $C-V$ characteristics must also lead to errors in obtaining the free-carrier concentration, as it is inferred from the former, more basic $C-V$ measurement [eq. (26)]. Thus, we expect that a high density of bulk space charge which leads to departures of either the low- or high-frequency capacitance from their trap-free values [c.f. eqs. (13) and (14)] will also result in errors in determining $N_D(x)$, and hence V_D. The magnitude of the errors in determining the apparent carrier concentrations are further increased since it is derived from the derivative of the $1/C_D^2$ versus V characteristic.

Thus, as in the case of the intercept method, although the effects of bulk trapped charge can be reduced in the depletion method using direct measurements of the trap distributions and energies along with the appropriate capacitance model, more reliable results can be achieved using relatively bulk-trap-free samples.

4. Current transport analysis techniques

Perhaps the oldest means for measuring band edge offsets is via the analysis of the current–voltage (I–V) characteristics of both anisotype and isotype heterojunctions. The basic principle of this measurement is that the potential barrier will reduce the magnitude of the current flowing across the barrier by an amount proportional to the exponent of the barrier energy. A quantitative measurement of the junction current thereby yields the barrier height, and hence the energy band offsets.

Although the principle is quite simple, its implementation can be complicated by the existence of parasitic current paths which shunt the heterobarrier. For example, sources of current such as thermal generation of carriers in the semiconductor bulk or via deep levels at the heterointerface, tunneling across narrow heterojunction barriers, and shunt currents along the edge of the sample can all contribute to the measured current, thereby making it difficult to ascertain the value of the barrier-limited current. For this reason, the I–V method is not considered to be highly reliable, and whenever possible researchers have utilized C–V analysis for determining band offsets.

A technique related to the I–V method is the photocurrent technique, whereby the energy of an incident light beam is varied while monitoring the photocurrent generated in an external circuit connected to the heterojunction sample. By determining the light energy at which low energy "thresholds" are induced in the photocurrent, one can determine the electronic transitions which correspond to free carriers being generated by excitation over the heterojunction band discontinuity. This technique would also appear to be relatively straighforward and free of ambiguities. Unfortunately, as in the case of I–V analysis, interpretation of the photocurrent spectrum can be complicated by carrier tunneling, photoexcitation from bulk and interface deep levels, and energy loss of carriers generated more than a few mean free paths away from the heterointerface. Furthermore, the photocurrent technique relies on very abrupt heterojunctions if accurate barrier energies are to be obtained.

Although these techniques tend to suffer from several problems, they are nevertheless of great importance since they often provide data which is needed to understand $C-V$ or photoluminescence measurements. Furthermore, heterojunctions are often used to control the flow of current in many III–V semiconductor devices such as lasers and high electron mobility transistors (HEMT). Before such a device can be properly designed, it is essential that a basic understanding of current transport across the heterointerface be obtained by direct $I-V$ analysis. In this section we will consider both $I-V$ and photocurrent–voltage analysis, and the major sources of error inherent in these methods.

4.1. Current–voltage analysis

Current transport in ideal anisotype heterojunctions is limited by diffusion of minority carriers over the potential barrier formed by the energy band offsets at the interface. In this case, the current density is [4]

$$J = J_1 \exp(-qV_1^0/kT) - J_2 \exp(-qV_2^0/kT), \tag{41}$$

where J_1 and J_2 are the saturation currents associated with emission over the barriers in materials 1 and 2, and V_1^0 and V_2^0 are their respective barrier heights in the presence of an applied voltage. Current is limited by transport over the two potential barriers. At $V = 0$, the barrier in the wide bandgap material (fig. 1b) is V_{D2}, and the barrier in the narrow bandgap (p-type) material is $\Delta E_v - qV_{D1}$. Since at equilibrium these two current components are equal, we have the condition that

$$J_1 \exp[-(\Delta E_v - qV_{D1})/kT] - J_2 \exp(-qV_{D2}/kT) = 0. \tag{42}$$

Using this condition gives

$$J = J_2 \exp(-qV_{D2}/kT) \left[\exp(qV_2/kT) - \exp(-qV_1/kT)\right] \tag{43}$$

for anisotype heterojunctions. To determine the magnitudes of V_1 and V_2 relative to the applied potential, we refer to the results of sect. 2.1. The condition that the electric displacement is continuous across the heterointerface gives $(V_{D1} - V_1)N_1\epsilon_1 = (V_{D2} - V_2)N_2\epsilon_2$. Substituting this relationship into the above expression then implies

$$J = J_2 \exp(-qV_{D2}/kT) \{\exp(qV/mkT) - \exp[q(m-1)V/mkT]\}, \tag{44}$$

where

$$m = (1 + N_2\epsilon_2/N_1\epsilon_1). \tag{45}$$

Since the current is diffusion-limited, the saturation current J_2 in eq. (44) is given by [4]

$$J_2 = XqN_2(D_p/L_p), \tag{46}$$

where X is the probability that a sufficiently energetic hole crosses the junction energy barrier, N_2 is the acceptor density on the wide bandgap side of the P–n junction, and D_p and L_p are the hole diffusion constant and diffusion length, respectively.

From eqs. (44) and (45), it is apparent that the current does not saturate for either a positive or negative applied potential. In fact, the current varies approximately exponentially, independent of the direction of the applied voltage. Nevertheless, from the treatment above, using eq. (44) which assumes that the only source of current is due to diffusion of carriers over the P–n junction barrier, then the intercept of the current density with the $V = 0$ axis should give the diffusion potential on the large bandgap side of the heterojunction. One this is known, the diffusion potential on the narrow bandgap side is easily calculated using:

$$V_1 = (1 - 1/m)V, \tag{47}$$

from which the band discontinuity energy is easily inferred using eq. (6).

The analysis is somewhat different for isotype heterojunctions, largely due to the fact that the current is dominated by thermionic emission, and also since the division of voltage between the large and small bandgap materials is more complicated [eq. (10)]. Now, substituting this expression into eq. (43), we obtain for the current density in n–N heterojunctions [34]

$$J = J_0(1 - V/V_D)\left[\exp(qV/kT) - 1\right], \tag{48}$$

where the saturation current is given by

$$J_0 = q^2 N_{D2} V_D (2\pi m^* kT)^{-1/2} \exp(-qV_D/kT), \tag{49}$$

where we have taken the thermionic emission pre-factor to be equal to $qN_{D2}(kT/2\pi m^*)^{-1/2}$. Here, N_{D2} is the concentration of donors in the large bandgap material, and m^* the electron effective mass.

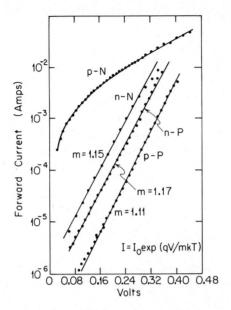

Fig. 22. Forward current–voltage characteristics of several isotype and anisotype Ge–GaAs heterojunction samples [4].

With the exception of a somewhat different pre-factor, it can be seen that the current transport across n–N heterojunction barriers is similar to that found in metal–semiconductor barriers. The two cases differ as a result of the differences in the density of states available in semiconductor–semiconductor junctions as opposed to that of metal–semiconductor contacts. Furthermore, it is easily seen that the current is limited by emission over a barrier of height V_D, and therefore the diffusion potential can once more be inferred from the intercept of the current–voltage characteristics with the $V = 0$ axis.

Figure 22 shows the forward I–V characteristics of p–N, n–N, n–P and p–P Ge–GaAs heterojunctions [4], where the value of m is obtained from fitting the data to the expressions for current given in the equations above. These data, taken at room temperature, suggest that the above analysis is indeed valid in some cases, and that the heterojunction discontinuity can be taken directly from the I–V characteristics as long as the bias voltage remains small. From these data, values of 0.15 eV and 0.55 eV are obtained for ΔE_c and ΔE_v, respectively.

A more direct means of plotting these same I–V data is to explicitly

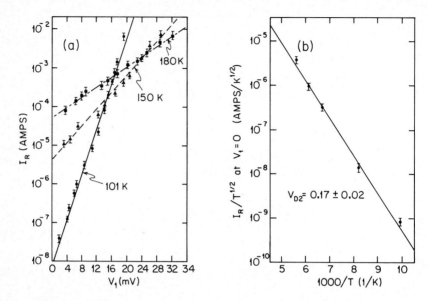

Fig. 23. (a) Reverse current characteristics at several temperatures of an n–N $In_{0.53}Ga_{0.47}As/InP$ heterojunction plotted as a function of voltage drop across the narrow bandgap side of the heterojunction. (b) Activation energy plot of the intercepts of the current–voltage characteristics with the $V_1 = 0$ axis of the data represented in (a). The activation energy data give a diffusion potential of 0.17 V [17].

calculate the division of voltage across the two isotype material layers, and then to plot the data as a function of V_1 or V_2 as opposed to the applied voltage. When this is done, the data lie on straight lines when plotted on a semi-logarithmic I–V graph, as shown in fig. 23a [17] for $In_{0.53}Ga_{0.47}As/InP$ heterojunctions. In this manner, the assumption that m is independent of voltage implied in the above treatment is not required. Then, by measuring the thermal activation energy of the reverse-bias characteristic intercept with the $V = 0$ axis, the diffusion potential is obtained (fig. 23b). From these data, a total barrier potential of 0.20 ± 0.02 V is inferred, giving a value for ΔE_c of 0.21 ± 0.02 eV for the $In_{0.53}Ga_{0.47}As/InP$ conduction-band discontinuity energy.

For samples with a higher level of doping in the various layers, or when high voltages are applied, significant deviations from the simple theory given above are often observed. For example, in fig. 24, we show the forward I–V characteristics of n–P Ge–Si heterojunctions [35], and find that there are clearly two regimes of forward current transport – one at low

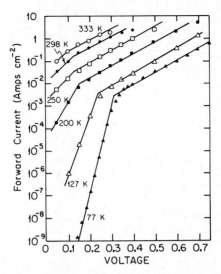

Fig. 24. Forward current–voltage characteristics of an n–P Ge–Si heterojunction as a function of temperature. The low-voltage data are consistent with thermionic emission over the heterojunction barrier, whereas the high-voltage data indicate the presence of tunneling [35].

voltage governed by thermionic emission, and a high-voltage regime due to carrier tunneling. Tunneling through narrow barriers at the heterointerface requires some modification of the simple thermionic emission theory. Tunneling of hot carriers incident on the nearly triangular barrier results in obtaining an effectively lower barrier than actually exists (fig. 25). Thus, it

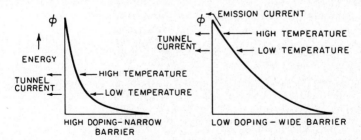

Fig. 25. Plot of the heterojunction energy barrier at the conduction-band edge between highly doped (left) and low doped (right) material layers. The energy, ϕ, is plotted as a function of distance from the heterojunction. Also shown is the mean electron energy at high and low temperatures for the two cases. Note that the effective width of the highly doped barrier is smaller than for low doped barriers, thus allowing for the tunneling of electrons at lower energies. This has the net effect of presenting a "lower" barrier to carrier transport [36].

is expected that the measured barrier height will be smaller for more highly doped samples, or when the measurement is made at high temperatures or under high forward bias.

A simplifying assumption in treating carrier tunneling is that the barrier is parabolic, and slowly varying on the scale of an electron's de Broglie wavelength (for n–N barriers) [36]. Using the WKB approximation, we obtain for the probability for quantum-mechanical tunneling for a particle with energy E

$$P(E) = \exp\left\{-\left(\frac{8m^*}{h^2}\right)^{1/2} \int_{x_1}^{x_2} [E - V(x)]^{1/2} \, dx\right\}, \quad (50)$$

where the limits of integration, x_1 and x_2, are the classical turning points of the carrier incident on the barrier. Solving eq. (50), we obtain

$$P(E) = \exp\left\{(4m^*\epsilon/Nqh^2)^{1/2} \phi \right. \\ \left. \times \left[(1-E/\phi)^{1/2} - (E/\phi) \ln\left\{\left[1 + (1-E/\phi)^{1/2}\right]/(E/\phi)^{1/2}\right\}\right]\right\}. \quad (51)$$

Here, ϕ is the barrier height as shown in fig. 25. Now the tunneling current generated by a charge of energy E is the product of the probability of barrier penetration [eq. (51)] and the total flux of charge incident on the barrier. Thus, the total current density is found from integrating this product over the total charge distribution extending from the conduction-band minimum in the semiconductor bulk to energy ϕ

$$J = (4\pi q m^*/h^3) \exp(-E_F/kT) \int_0^\phi \exp(-E/kT) \, P(E) \, dE. \quad (52)$$

It is found that the current density in eq. (52) is reasonably well described by

$$J \propto \exp(qV/m'kT), \quad (53)$$

where the "ideality factor", m' is obtained from numerical integration of the exact form of J given in eq. (52). Here, m' is dependent on the doping and the temperature of the heterojunction, since these two factors affect the barrier thickness (and hence its quantum-mechanical transparency) and the temperature of the free-carrier population, respectively. This model for thermally assisted tunneling has been found to give a reasonably accurate representation of the transport across moderate to highly doped isotype heterojunction samples. Unfortunately, it is apparent that applying such corrections to $I-V$ data used in heterojunction measurement can lead to significant errors due to the complexity of the tunneling phenomenon which requires the introduction of several new physical parameters (such as m') to which the data must be fit.

With the advent of high-quality molecular beam epitaxial (MBE) growth of multilayered structures, many of the difficulties associated with these techniques can be avoided using a modified $I-V$ method [37–39]. In this technique, the single isotype heterointerface is replaced by a potential barrier of a large bandgap material sandwiched between two narrow bandgap layers. For the experiments of Batey and Wright [38], a layer of undoped AlGaAs of thickness ranging between 50–100 Å was grown between two p-type GaAs layers forming a square potential barrier to the transport of holes (fig. 26a). As in other $I-V$ techniques, transport is assumed to be limited by thermionic emission. Note that, due to the uniform thickness of the barrier, effects of tunneling can largely be eliminated.

Provided that the applied bias is sufficiently small such that significant band bending (and concomitant barrier distortion) is not induced, the current density is given simply by

$$J = A^*T^2 \exp(-q\phi_B/kT), \tag{54}$$

where ϕ_B is the barrier height due to the valence-band discontinuity. A simple measurement of the thermal activation energy of the current density therefore yields ϕ_B. The results of such a procedure is shown in fig. 26c, where E is plotted against applied voltage to illustrate the effects of barrier distortion with increasing applied potential (fig. 26b). The valence-band offset of 0.19 eV is obtained from the data at 0 V for this sample containing 38% Al to 62% Ga. Making this measurement using p-type samples, Batey and co-workers were able to eliminate ambiguities of interpretation which arise when measuring conduction-band offsets for indirect bandgap materials. That is, AlGaAs becomes an indirect gap semiconductor at roughly a 50% (or larger) mole fraction of Al. As shown in fig. 27a, the data of these experiments are consistent with the rule

$$\Delta E_v = 0.55 x_{Al}, \tag{55}$$

where x_{Al} is the mole fraction of Al in the AlGaAs layer grown on GaAs.

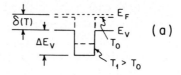

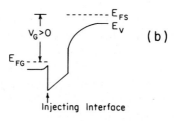

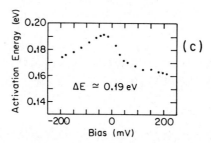

Fig. 26. (a) The energy barrier to hole transport at a p-GaAs/AlGaAs/p-GaAs double heterostructure at thermal equilibrium. The position of the barrier with respect to the Fermi energy is shown at two temperatures. (b) The lower energy barrier of (a) under applied bias. (c) Activation energy of the emission current over the energy barrier in (a) as a function of applied bias. The activation energy at 0 V corresponds to the valence-band discontinuity of a sample with a layer of $Al_{0.38}Ga_{0.62}As$ sandwiched between two GaAs layers [38].

This linear dependence on Al mole fraction then can be translated into a dependence on bandgap difference between the contacting materials, with the results compiled in fig. 27b. In addition to the valence-band offset data reported in the work of Batey and Wright, previously reported measurements of conduction band offsets [37] using similar sample geometries are also plotted, along with the photoluminescence measurements of Miller and co-workers [27]. An interesting result is that the dependence of the conduc-

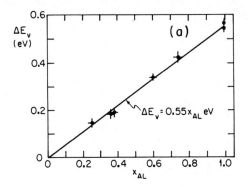

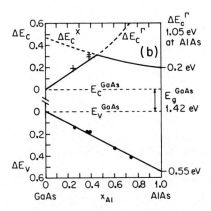

Fig. 27. (a) Dependence of the valence-band discontinuity of p-P GaAs/Al$_x$Ga$_{1-x}$As heterojunctions on x obtained via the double heterostructure I–V method. (b) Distribution of the energy gap difference between the valence and conduction bands as inferred from I–V and photoluminescence data for GaAs/Al$_x$Ga$_{1-x}$As heterojunctions. Note that the conduction-band discontinuity increases with bandgap until the Al mole fraction is increased to the point where Al$_x$Ga$_{1-x}$As becomes an indirect gap material. Thus, at $x_{Al} > 0.5$, the X minimum is less than the Γ minimum, and ΔE_c decreases with Al concentration while ΔE_v continues to increase linearly [38].

tion band offsets on Al mole fraction is only linearly increasing until x_{Al} is increased beyond the compositional point where AlGaAs becomes an indirect bandgap semiconductor. At larger values of x_{Al}, it is suggested that the conduction-band discontinuity actually decreases, following the energy of the X-minimum in the satellite valley of the AlGaAs band structure,

Table 3
Representative band offset measurements using the $I-V$ method

Heterojunction		ΔE_c (eV)	ΔE_v (eV)	Ref.
Material 1	Material 2			
p-GaAs	P-Al$_x$Ga$_{1-x}$As $0.25 < x < 1.0$	See text	$0.55 x_{Al}$	[38]
p-GaAs	P-Al$_x$Ga$_{1-x}$As $0.3 < x < 1.0$	$0.65 \Delta E_{g\Gamma}$	$0.35 \Delta E_{g\Gamma}$	[39]
n-GaAs	N-Al$_x$Ga$_{1-x}$As $0.2 < x < 0.65$	$0.85 \Delta E_g$		[40]
n-GaAs	N-Al$_{0.28}$Ga$_{0.72}$As	0.28 [a]		[41]
n-In$_{0.53}$Ga$_{0.47}$As	N-InP	0.21		[3]
p-Ge	N, P-GaAs	0.15	0.55	[4]
n-Ge	N-GaAs$_{1-x}$P$_x$			
	$x = 0.1$	0.62 [b], 0.53 [c]		[34]
	$x = 0$	0.50 [b], 0.47 [c]		[34]
n-Ge	N-Si	0.15		[5]
p-Si	P-GaP		1.4 [d]	[42]
n-In$_{1-x}$Ga$_x$As	P-GaSb$_{1-y}$As$_y$			
$x = 0.62$	$y = 0.64$	0.38		[43]
0.52	0.56	0.47		[43]
0.50	0.28	0.48		[43]
0.16	0.10	0.74 [e]		[43]

[a] $\Delta E_c = 0.8 \Delta E_g$.
[b] $\langle 111A \rangle$-GaAs$_{0.9}$P$_{0.1}$ substrates. Value of ΔE_c assumes $\delta_1 \cong \delta_2 \cong 0$.
[c] $\langle 111B \rangle$-GaAs substrates. Value of ΔE_c assumes $\delta_1 \cong \delta_2 \cong 0$.
[d] Value assumes $\delta_1 \cong \delta_2 \cong 0$.
[e] This is a type-II (misaligned) band lineup where $\Delta E_c > E_{g2}$ [46]. Thus E_{v2} is at a higher energy than E_{c1}.

while the remainder of the bandgap difference appears at the valence-band edge. These data represent the first observation of the effects of complex band structures – such as those found for indirect gap materials – on determining the energy barriers formed at the junction between two dissimilar semiconductors.

In table 3 the band offsets for several important heterojunction systems obtained using the current–voltage analysis technique are compiled. It can be seen that the $I-V$ method has been used extensively for a period of many years, and has been applied to almost all heterojunction systems of interest. The popularity of this technique is largely due to the ease with which it can be implemented, although as we will see in the following section, the $I-V$ technique is vulnerable to several serious sources of error.

Indeed, only the most reliable data have been included in this table. Nevertheless, considerable disagreement for such "canonical" heterojunction systems as AlGaAs/GaAs are reported using this technique. Note that the data of Batey and Wright [38] and of Arnold and co-workers [39] utilize the square-barrier double heterostructure, whereas the other data refer to single heterobarrier measurements.

4.2. Photocurrent analysis

One variation of the I–V analysis technique is to measure the spectral dependence of a current generated by light incident on an isotype heterojunction sample. As an example, fig. 28 shows an n–N heterojunction indicating five possible optically excited transitions due to light incident parallel to the plane of the junction. It is evident that the photocurrent will show abrupt increases in magnitude as the energy of the incident light is increased beyond the minimum necessary to induce one of the transitions labelled 1, 2, 3, 4 or 5. Of these transitions, only processes 1 and 2 yield information regarding the magnitude of the band offset energy, and are, therefore, to be considered in the following treatment. In addition, transition 2 requires the absorption of a very long wavelength photon such that an electron is given sufficient energy to surmount the conduction-band offset energy, ΔE_c. Although this is the most direct means of obtaining this parameter using photocurrent spectroscopic means, it is often difficult to implement. Thus transition 1 is employed most frequently to obtain ΔE_c. Transition 5 involves tunneling through the heterojunction energy barrier, and will be discussed in detail below.

The primary assumption of the photocurrent technique is that

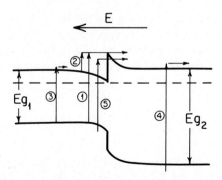

Fig. 28. Possible optically induced transitions in an n–N heterojunction.

electron–hole pairs are photogenerated in the presence of an applied electric field in a region sufficiently close to the heterojunction such that they do not lose energy prior to traversing the heterointerface barrier. In effect, the incident light raises the effective electron temperature such that the average kinetic energy of the charge population is roughly equal to the conduction-band discontinuity energy. Although the total photocurrent measured will consist of the sum of all the transitions allowed in fig. 28 corresponding to the energy of the incident light, it is expected that a steep increase in the photocurrent will occur at an energy equal to $E_{g1} + \Delta E_c$. This increase in photocurrent will continue as the photon energy is increased further, since the excess energy will compensate losses due to cooling of the carrier distribution generated at larger distances from the interface.

It is apparent that the least ambiguous data can be obtained from this type of experiment for samples with very abrupt heterointerfaces, and for carriers generated within a few mean free paths of the heterojunction. Although this places severe constraints on the samples used, this technique has been successfully employed for several heterojunctions of interest. Figure 29 shows the photocurrent spectrum of an n–N $In_{0.53}Ga_{0.47}As/InP$ heterojunction, with thresholds due to transitions 1, 3, and 4 clearly observed [17]. From the threshold energy corresponding to process 1, it was determined that the conduction-band discontinuity for the heterojunction was $\Delta E_c = 0.24 \pm 0.02$ eV, which was consistent with measurements obtained by other means.

Note that the voltage dependence of the photocurrent generated at energies intermediate between the small bandgap threshold (process 3 at $\lambda = 1.65$ μm) and the conduction-band threshold (process 1 at $\lambda = 1.25$ μm) has been attributed to the tunneling through the barrier by carriers with insufficient energy to be emitted over the barrier peak. Now, the probability of tunneling through a barrier depends on its width. Using the model of Korol'kov et al. [44], it can be shown that for uniformly doped semiconductors on both sides of the heterointerface, the tunneling probability is given by

$$P(V) = \exp\left[-\frac{8}{\pi h}\left(\frac{m^*\epsilon}{qN_D}\right)^{1/2} \Delta E_c^{3/2} (V_D - V)^{-1/2}\right]. \tag{56}$$

The photocurrent is proportional to $P(V)$. It has been shown that the voltage dependence of the photocurrent of the n–N $In_{0.53}Ga_{0.47}As/InP$ heterojunction follows eq. (56) in the energy region bracketed by processes

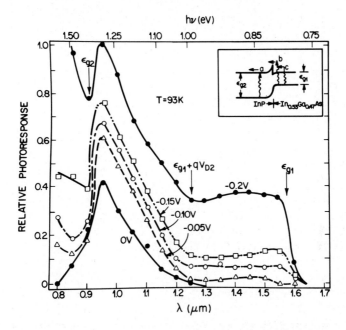

Fig. 29. The photocurrent spectrum of an n–N $In_{0.53}Ga_{0.47}As/InP$ heterojunction. The various transitions corresponding to those shown in fig. 28 are indicated [3].

1 and 3, indicating that tunneling is an important transport mechanism which can obscure both photocurrent and dark current measurements of ΔE_c.

The above analysis applies to heterojunctions operated in the applied potential mode, where the induced photocurrent flows in a direction determined by the applied electric field. However, considerably different results are often observed when the heterojunctions are operated in the open circuit, or photovoltaic mode. It has been observed [10,12] that interface states can play a significant role in determining both the direction of current flow in such a circuit, as well as the position of the energy thresholds. Indeed, although the photovoltaic mode experiment is not particularly useful in measuring band offsets, it has strongly established the back-to-back Schottky barrier model discussed in sect. 2 as a valid picture of some isotype heterojunctions with high interface state densities.

Returning to fig. 7, we notice that in addition to the transitions possible for a trap-free heterojunction, there also exists the possibility of generating photocurrent via the direct optical pumping of the interface states. Further-

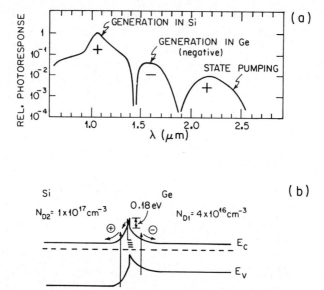

Fig. 30. (a) Relative photoresponse versus wavelength of an N–n Si/Ge heterojunction operated in the photovoltaic mode. Here, "+" and "−" refer to the relative direction of current flow in the external circuit. (b) Band diagram of the double-depletion heterojunction in (a) indicating the three transitions giving rise to the current observed in (a) [45].

more, since there is no externally applied potential, carriers can flow in both directions across the heterointerface. Indeed, the band bending resulting from the back-to-back depletion region in fact induces flow in both directions in the external circuit.

The results of illuminating a trap-dominated n–N Ge–Si heterojunction [45] is shown in fig. 30. Here, many of the processes discussed above are empirically observed. In the figure, "+" and "−" refer to opposite signs of the current flowing through the external circuit. For these experiments, the sample was illuminated in a direction perpendicular to the heterojunction and was incident via the large bandgap (Si) layer. Relatively high energy radiation (>1.1 eV) is absorbed in the Si layer. The photogenerated electrons are repelled by the interfacial energy barrier, resulting in current flow toward the heterojunction. On the other hand, photons of energy intermediate between the bandgaps of Si (1.1 eV) and Ge (0.67 eV) are absorbed in the Ge layer. Here, photogenerated electrons are also repelled by the interfacial energy barrier, and the current must therefore also flow toward the heterojunction, thereby resulting in a current sign reversal from that observed for light absorbed only in the Si layer. At even lower light energies (< 0.67 eV) the only photocarriers available are those directly pumped from the interface states themselves. For the n–N Ge–Si sample, it was observed that the photocurrent undergoes one additional sign reversal at these low excitation energies, indicating that the trapped carriers were preferentially pumped into the Si layer.

Thus, although the photocurrent method appears to reveal a multitude of physical phenomena which relate to the properties of the heterojunction band structure, the data are often difficult to interpret. For this reason, photocurrent analysis is rarely used as the primary means for determining band offsets. As has been shown above, capacitance and dark current analysis are generally more simple to implement and far more reliable, and are considered to be the preferred means of band offset measurement as opposed to photocurrent techniques.

4.3. Error sources in the I–V method

All current–voltage analysis techniques for offset measurement rely on the assumption that the magnitude of the observed current is a direct and understandable function of the barrier formed at the heterojunction between two semiconductors. Therefore, any physical property of the sample which either limits the current below that of the intrinsic heterojunction current, or acts as a source of excess current, will induce error into the barriers determined from $I-V$ analysis. In general, current-limiting processes arise from additional energy barriers in the sample, and are not as troublesome as are those phenomena giving rise to additional leakage. These latter phenomena can be due to bulk generation–recombination currents, surface shunt currents arising from a plethora of sources, and excess current generated via thermal excitation of trapped charges localized at imperfect heterojunctions. One additional excess current source – barrier tunneling – has already been discussed in sections 4.1 and 4.2, and will therefore not be discussed further here.

The effects that interface states have on the $I-V$ characteristics of isotype heterojunctions was first treated by Oldham and Milnes [10], and can be understood in terms of the band diagram in fig. 31 which depicts an n–N heterojunction with a high density of states at the interface. In the diagram, F_i refers to the bulk flux of majority carriers incident on the heterojunction from the ith (where i equals either 1 or 2) layer, R is the reflection coefficient from, and α the transmission coefficient through the

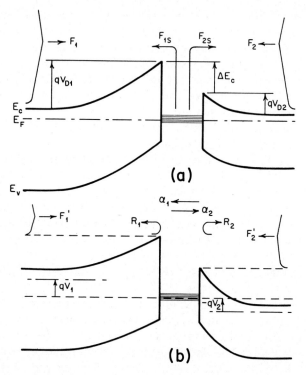

Fig. 31. Schematic diagram of a doubly-depleted heterojunction showing: (a) the current sources, and (b) the definitions of the current transport coefficients used in the text [10].

heterojunction barrier. The emission fluxes from the interface states themselves are given by F_{is}. Assuming Boltzmann statistics, the actual flux with sufficient energy to be transported across the barrier is given by

$$F_1' = F_1 \exp[-q(V_{D1} - V_1)/kT], \text{ and} \tag{57}$$

$$F_2' = F_2 \exp[-q(V_{D2} - V_2)/kT]. \tag{58}$$

Furthermore, the total flux to the right and left in the diagram is given, respectively, by

$$F_R = F_1'(1 - R_1) - F_{1s} - \alpha_2(1 - R_2)F_2', \text{ and} \tag{59}$$

$$F_L = -F_2'(1 - R_2) + F_{2s} + \alpha_1(1 - R_1)F_1'. \tag{60}$$

Using the equilibrium condition that $F_L = F_R = 0$, then

$$\frac{1 - R_2}{1 - R_1} = \frac{\alpha_1 F_1 \exp(-qV_{D1}/kT)}{\alpha_2 F_2 \exp(-qV_{D2}/kT)}. \tag{61}$$

Now, F_1/F_2 is simply proportional to the ratio of thermal velocities of the carriers incident from materials 1 and 2, which, in turn, is inversely proportional to the square root of the effective masses of electrons in these layers. Thus, eq. (61) simplifies to

$$\frac{1 - R_2}{1 - R_1} = \frac{\alpha_1 m_1' \exp(-qV_{D1}/kT)}{\alpha_2 m_2' \exp(-qV_{D2}/kT)}. \tag{62}$$

In effect, the heterobarrier behaves as an effective mass "filter" whereby carriers incident from the material layer with a higher effective mass have a lower probability for transport across the interface region.

Using the above relationships, we obtain two expressions for the total flux of charge across the heterointerface, at saturation, viz.:

$$F \simeq F_2 \exp(-qV_{D2}/kT)(1 - R_2)/(1 - \alpha_1), \quad \text{and} \tag{63}$$

$$F \simeq -F_2 \exp(-qV_{D2}/kT)(1 - R_1). \tag{64}$$

Equations (63) and (64) refer to the saturation fluxes with the junction potential positive and negative, respectively. These results therefore imply that the diode characteristic saturates for applied potentials of both polarities, with the relative magnitude of the two currents being $\alpha_1/(1 - \alpha_1)\alpha_2$. Note that such a double saturation result is observed only for heterojunctions which have approximately the same carrier concentration in both layers. If the heterojunction is strongly "one-sided", the highly doped side is unable to support a significant voltage drop, and hence the current will not saturate under these bias conditions.

Note that eqs. (63) and (64) are considerably more complicated than are the transport equations which refer to heterojunctions free of interface states, with several additional parameters (such as R and α) whose functional forms are still unspecified. This additional model dependence therefore makes values of the heterojunction offsets determined from such interfaces highly suspect as to their reliability for determining the diffusion potentials, V_{D1} and V_{D2}. For this reason, isotype heterojunctions which exhibit double saturation I–V characteristics (see fig. 8b for an example of an n–N Ge–Si heterojunction), cannot be reliably employed to determine band offsets.

In considering anisotype heterojunctions, the ability to determine the band offset is once again reduced in the presence of states which behave as recombination centers which introduce extraneous current sources. Thus, determination of the band offset barrier by simple diffusion theory is complicated by the need for additional models which accurately predict the temperature and voltage dependence of the magnitude of the recombination current. Since our understanding of such recombination processes is limited (it is not always possible to separate the various current contributions), the accuracy of measurement must correspondingly be reduced.

All of the above arguments can be applied equally well to other parasitic sources of current such as surface leakage, shunt currents through defects in the bulk of the semiconductor, and recombination via bulk deep levels. Since at least one of these extraneous current sources exists in most heterojunction samples, the $I-V$ technique is often subject to significant error.

5. Conclusions

In this chapter, various common means of band offset measurement have been described. These methods are implemented via capacitance and current measurement of samples containing either isotype or anisotype heterojunctions. Of all the techniques considered, the $C-V$ depletion technique first described by Kroemer and co-workers [14] appears to be the most reliable due to its relative insensitivity to interface compositional grading and low densities of interface states. Furthermore, analysis of $I-V$ data tend to give the least reliable measurements of the band offsets due to their strong dependence on poorly quantified shunt, or parasitic currents. On the other hand, a recent technique employing double heterojunction structures consisting of a single square well potential has been demonstrated [37] which appears to give considerably more reliable data than do conventional, single heterojunction samples when analyzed via the $I-V$ techniques discussed.

Independent of the relative strengths and weaknesses of the several techniques considered, all methods give important data and insight into the behavior of heterojunctions. Indeed, the most reliable values for the energy band offsets can only be obtained when several different techniques are employed such that a complete understanding of the details of the energy bands in the region of the heterointerface, along with the role played by interface states, can be indepently obtained.

References

[1] A.R. Riben and D.L. Feucht, nGe–pGaAs Heterojunctions, Solid-State Electron. 9 (1966) 1055.
[2] S.I. Cserveny, Potential distribution and capacitance of abrupt heterojunctions, Int. J. Electron. 25 (1968) 65.
[3] S.R. Forrest and O.K. Kim, An n-In$_{0.53}$Ga$_{0.47}$As/n-InP rectifier, J. Appl. Phys. 52 (1981) 5838.
[4] R.L. Anderson, Experiments on Ge–GaAs heterojunctions, Solid-State Electron. 5 (1962) 341.
[5] C. Van Opdorp and H.K.J. Kanerva, Current–voltage characteristics and capacitance of isotype heterojunctions, Solid-State Electron. 10 (1967) 401.
[6] J.L. Shay, S. Wagner and J.C. Phillips, Heterojunction band discontinuities, Appl. Phys. Lett. 28 (1976) 31.
[7] C.T. Sah and V.G.K. Reddi, Frequency dependence of the reverse-biased capacitance of gold-doped silicon p$^+$n step junctions, IEEE Trans. Electron. Devices ED-11 (1964) 345.
[8] D.V. Lang, Deep-level transient spectroscopy: A new method to characterize traps in semiconductors, J. Appl. Phys. 45 (1974) 3023.
[9] S.R. Forrest, P.H. Schmidt, R.B. Wilson and M.L. Kaplan, Measurement of the conduction band discontinuities of InGaAsP/InP heterojunctions using capacitance–voltage analysis, J. Vac. Sci. & Technol. B 4 (1986) 37.
[10] W.G. Oldham and A.G. Milnes, Interface states in abrupt semiconductor heterojunctions, Solid-State Electron. 7 (1964) 153.
[11] J.P. Donnelly and A.G. Milnes, The capacitance of double saturation nGe–nSi Heterojunctions, Proc. IEEE 53 (1965) 2109.
[12] L.J. Van Ruyven, J.M.P. Papenhuizen and A.C.J. Verhoeven, Optical Phenomena in Ge–GaP heterojunctions, Solid-State Electron. 8 (1965) 631.
[13] W.G. Oldham and S.S. Naik, Admittance of p–n junctions containing traps, Solid-State Electron. 15 (1972) 1085.
[14] H. Kroemer, Wu-Yi Chien, J.S. Harris and D.D. Edwall, Measurement of isotype heterojunction barriers by $C-V$ profiling, Appl. Phys. Lett. 36 (1980) 295.
[15] W.C. Johnson and P.T. Panousis, The influence of Debye length on the $C-V$ measurement of doping profiles, IEEE Trans. Electron Devices ED-18 (1971) 965.
[16] H. Kroemer and Wu-Yi Chien, On the theory of Debye averaging in the $C-V$ profiling of semiconductors, Solid-State Electron. 24 (1981) 665.
[17] S.R. Forrest and O.K. Kim, Deep levels in In$_{0.53}$Ga$_{0.47}$As/InP heterostructures, J. Appl. Phys. 53 (1982) 5738.
[18] S.M. Sze and G. Gibbons, Avalanche breakdown voltages of abrupt and linearly graded p–n junctions in Ge, Si, GaAs and GaP, Appl. Phys. Lett. 8 (1966) 111.
[19] P.A. Tove, Methods of avoiding edge effects on semiconductor diodes, J. Phys. D 15 (1982) 517.
[20] H. Morkoc, 1981, Schottky barriers and ohmic contacts on n-type InP based compound semiconductors for microwave FETs, IEEE Trans. Electron Devices ED-28 (1981) 1.

[21] S.R. Forrest, P.H. Schmidt, R.B. Wilson and M.L. Kaplan, Relationship between the conduction-band discontinuities and band-gap differences of InGaAsP/InP heterojunctions, Appl. Phys. Lett 45 (1984) 1199.
[22] S.R. Forrest, M.L. Kaplan, P.H. Schmidt and J.V. Gates, Evaluation of III–V semiconductor wafers using non-destructive organic-on-inorganic contact barriers, J. Appl. Phys. 57 (1985) 2892.
[23] H. Okumura, S. Misawa, S. Yoshida and S. Gonda, Determination of the conduction band discontinuities of GaAs/Al_xGa_{1-x}As interfaces by capacitance–voltage measurements, Appl. Phys. Lett. 46 (1985) 377.
[24] R. People, K.W. Wecht, K. Alavi and A.Y. Cho, Measurement of the conduction-band discontinuity of molecular beam epitaxial grown $In_{0.52}Al_{0.48}As/In_{0.53}Ga_{0.47}As$, N–n heterojunction by C–V profiling, Appl. Phys. Lett. 43 (1983) 118.
[25] P. Philippe, P. Poulain, K. Kazmierski and B. deCremoux, Dark-current and capacitance analysis of InGaAs/InP photodiodes grown by metalorganic chemical vapor deposition, J. Appl. Phys. 59 (1986) 1771.
[26] M. Ogura, M. Mizuta, K. Onaka and H. Kukimoto, A capacitance investigation of InGaAs/InP isotype heterojunction, Jpn. J. Appl. Phys. 22 (1983) 1502.
[27] R.C. Miller, D.A. Kleinman and A.C. Gossard, Energy gap discontinuities and effective masses for GaAs–Al_xGa_{1-x}As quantum wells, Phys. Rev. B 29 (1984) 7085.
[28] W.R. Frensley and H. Kroemer, Prediction of semiconductor heterojunction discontinuities from bulk band structures, J. Vac. Sci. & Technol. 13 (1976) 810.
[29] J.O. McCaldin, T.C. McGill and C.A. Mead, Correlation for III–V and II–VI semiconductors of the Au Schottky barrier energy with anion electronegativity, Phys. Rev. Lett. 36 (1976) 56.
[30] R.L. Moon, G.A. Antypas, and L.W. James, Bandgap and lattice constant of GaInAsP as a function of alloy composition, J. Electron. Mater. 3 (1974) 635.
[31] H. Okumura, S. Misawa and S. Yoshida, Reliability of the band discontinuity determination by capacitance–voltage method: The relation of the interface charge density and trap concentration near the interface, Second Int. Conf. on Modulated Semiconductor Structures, Kyoto, Japan, Sept. 9–13 (1985) p. 202.
[32] S.R. McAfee, D.V. Lang and W.T. Tsang, Observation of deep levels associated with the GaAs/As_xGa_{1-x}As interface grown by molecular beam epitaxy, Appl. Phys. Lett. 40 (1982) 520.
[33] H. Kroemer, Determination of heterojunction band offsets by capacitance–voltage profiles through nonabrupt isotype heterojunctions, Appl. Phys. Lett. 46 (1985) 504.
[34] L.L. Chang, Conduction properties of Ge–$GaAs_{1-x}P_x$ n–n heterojunctions, Solid-State Electron. 8 (1965) 721.
[35] J.P. Donnelly and A.G. Milnes, Current–voltage characteristics of p–n Ge–Si and Ge–GaAs Heterojunctions, Proc. IEEE 113 (1966) 1468.
[36] S.J. Owen and T.L. Tansley, Thermally assisted tunneling in certain GaAs heterostructures, J. Vac. Sci. & Technol. 13 (1976) 954.
[37] J. Batey, S.L. Wright and D.J. DiMaria, Energy band-gap discontinuities in GaAs: (Al, Ga)As heterojunctions, J. Appl. Phys. 57 (1985) 484.
[38] J. Batey and S.L. Wright, 1986, Energy band alignment in GaAs: (Al, Ga)As heterostructures: The dependence on alloy composition, J. Appl. Phys. 59 (1986) 200.
[39] D. Arnold, A. Ketterson, T. Henderson, J. Klem and H. Morkoc, Determination of the valence-band discontinuity between GaAs and (Al, Ga)As by the use of p^+-GaAs–(Al, Ga)As–p^--GaAs capacitors, Appl. Phys. Lett. 45 (1984) 1238.

[40] A.C. Gossard, W. Brown, C.L. Allyn and W. Wiegmann, Molecular beam epitaxial growth and electrical transport of graded barriers for nonlinear current conduction, J. Vac. Sci. & Technol. 20 (1982) 694.
[41] A. Chandra, and L.F. Eastman, A study of the conduction properties of a rectifying n-GaAs–n(Ga, Al)As heterojunction, Solid-State Electron. 23 (1980) 599.
[42] G. Zeidenbergs and R.L. Anderson, Si–GaP heterojunctions, Solid-State Electron. 10 (1967) 113.
[43] H. Sakaki, L.L. Chang, R. Ludeke, C.-A. Chang, G.A. Sai-Halasz and L. Esaki, $In_{1-x}Ga_xAs$–$GaSb_{1-y}As_y$ heterojunctions by molecular beam epitaxy, Appl. Phys. Lett. 31 (1977) 211.
[44] V.I. Korol'kov, V.G. Nikitin and D.N. Tret'yakov, Tunneling of photocarriers in p-GaAs–n-Al_xGa_{1-x}As heterojunctions, Sov. Phys.-Semicond. 8 (1975) 1535.
[45] J.P. Donnelly and A.G. Milnes, The photovoltaic response of nGe–nSi heterodiodes, Solid-State Electron. 9 (1966) 174.
[46] L. Esaki, Semiconductor superlattices and quantum wells, Proc. 17th Int. Conf. Physics of Semiconductors (Springer, Berlin, 1984) p. 473.

CHAPTER 9

MEASUREMENT OF BAND OFFSETS BY SPACE CHARGE SPECTROSCOPY

D.V. LANG

AT&T Bell Laboratories
Murray Hill, NJ 07974, USA

Heterojunction Band Discontinuities: Physics and Device Applications
Edited by F. Capasso and G. Margaritondo
© *Elsevier Science Publishers B.V., 1987*

Contents

Introduction 379
1. Deep-level transient spectroscopy 381
2. Admittance spectroscopy 385
References 396

Introduction

The methods of space-charge spectroscopy [1], most notably deep-level transient spectroscopy (DLTS) [2] and admittance spectroscopy [3], have been widely used to study deep-level defects in semiconductors. Recently, these methods have also been applied to the study of heterojunction band offsets [4–6]. In this chapter we will discuss their strengths and weaknesses for this new area of application. As we shall see, these techniques are quite complementary for probing high- and low-field transport regimes and allow considerable insight into perpendicular transport in superlattices as well as the determination of band offsets.

Space-charge spectroscopy techniques [1] involve the use of a p–n junction or Schottky barrier, hence they might be superficially confused with other junction methods used to determines band offsets, such as Kroemer's C–V method [7]. In fact, these methods are fundamentally very different. Kroemer's method is, in essence, a determination of the dipole moment associated with a single heterojunction. One then determine the band offset from the well-known fact of electrostatics that a dipole space-charge layer is related to a potential discontinuity by

$$M/\varepsilon A = \Delta V,$$

where M is the dipole moment, A is the area, ε is the permittivity of the material, and ΔV is the potential discontinuity, which is related to the band offset. The method uses the well-known C–V method of profiling the carrier concentration within the space-charge layer of a p–n junction or Schottky barrier [8]. The space-charge spectroscopy methods, on the other hand, are dynamic techniques and determine band offsets from thermal activation of carriers over the potential barrier formed by the heterojunction band offset. In fact, in the DLTS studies we will describe one might consider a single quantum well to be a "giant trap" with a depth related to the band offset.

One might also confuse these dynamic methods with band offset determinations from DC transport measurements of thermionic emission over single heterojunctions [9]. In the methods we will discuss, the heterojunction is usually fabricated either as a single quantum well or as a superlattice. This makes the Fermi level position easier to determine than in the case of a single heterojunction. Second, by using a reverse-biased junction and measuring transient phenomena on time scales of μs to ms it is easier to discriminate between thermionic emission over the band offset and other extraneous sources of leakage current. Indeed, as we shall discuss, the necessity for proper control of leakage currents is a major problem with any determination of band offsets using transport techniques. InP/GaInAs seems to be less sensitive to surface leakage than is GaAs/AlGaAs, thus most of our examples will be drawn from our work on the former materials system.

There has recently been considerable interest in the question of how the band-gap discontinuity at InP/GaInAsP heterojunctions is distributed between the conduction and valence bands. In the case of InP/$Ga_{0.47}In_{0.53}As$ heterojunctions, Forrest and co-workers [10,11] used capacitance–voltage (C–V) measurements to show that 39% of the discontinuity was in the conduction band. Ogura et al. [12] used the same measurements and obtained a fractional conduction-band offset of 50–70%. Both Forrest and Kim [10] and Ogura et al. [12] found that the apparent conduction-band offset approached zero for temperatures below about 170 K. On the other hand, Steiner et al. [13] reported C–V measurements showing that the valence-band offset is zero. Optical measurements have been similarly inconsistent. Brunemeier et al. [14] reported from luminescence measurements that 60% of the discontinuity is in the conduction band. Temkin et al. [15] recently analyzed photoluminescence and optical absorption data to conclude that the band-gap discontinuity is split roughly 50/50 between the conduction and valence bands. We will show how the recent admittance spectroscopy results of Lang et al. [6] resolve these inconsistencies by measuring for the first time both the valence-band and conduction-band offset in this material system using the same technique with the consistency check that the sum of these two values adds up to the well-known band-gap difference between InP and $Ga_{0.47}In_{0.53}As$.

In this chapter we will first discuss DLTS studies of single InP/$Ga_{0.47}In_{0.53}As$ quantum wells. Such measurements are strongly field-dependent, but are very useful in proving that the observed transient effects are due to true band offsets and not to interface traps. We will next discuss how admittance spectroscopy may be used in a somewhat unusual way to measure band offsets in heterojunction superlattices in the zero-field limit. This will lead us to a treatment of perpendicular transport in superlattices in order to relate the observed activation energies to band offsets. In the case of $InP/Ga_{0.47}In_{0.53}As$ the conduction- and valence-band offsets have been measured independently by this method and the sum agrees remarkably well with the well-known band-gap difference between these two materials. Finally, we will discuss how a proper understanding of the dynamic response of heterojunctions resolves some anomalous results in the band offset literature.

1. Deep-level transient spectroscopy

Deep-level transient spectroscopy (DLTS) [1,2] has been used previously in attempts to measure band offsets in p–n junctions containing heterojunctions [4,5]. The main conclusion from these studies is that such measurements are difficult to interpret. In both cases one saw "giant trap" signals which were clearly due to the presence of a heterojunction or quantum well in the space-charge layer of the diode. However, unlike typical defect-related DLTS signals, these band offset signals depended strongly on the experimental parameters of the measurement. The most dramatic effect was the strong dependence of the apparent activation energy on the junction bias voltage. This is an expected signature of carrier trapping at heterojunction band offsets, since there is a strong competition between thermionic emission over the offset barrier and phonon-assisted tunneling through the barrier. Since the shape of the carrier trapping barrier depends strongly on bias voltage, one expects strongly field-dependent effects much like the case of carrier emission over a Schottky barrier [8].

We will illustrate some of the problems encountered with DLTS measurements of this sort by considering the case of an $InP/Ga_{0.47}In_{0.53}As$ quantum well in an InP p^+–n junction. In previously reported work we were able to measure the thermal activation energy of the DLTS signal associated with the single quantum well [5]. Both types of DLTS excitations were tried: bias voltage (majority carrier) pulses and light (minority and majority carrier) pulses. No signals were seen for bias pulses, but it was possible to see a DLTS signal from the quantum well for light pulses. For these experiments we maintained the sample at a fixed reverse bias and used a strobe lamp (1 µs pulse length) through a silicon filter to inject a pulse of minority carriers into the $Ga_{0.47}In_{0.53}As$ layer. The DLTS spectra of the photo-induced minority-carrier (hole) emission transients in the 1 MHz capacitance at various bias voltages for a rate window of 200 µs are shown in fig. 1. The disappearance of the DLTS peak at low bias voltages

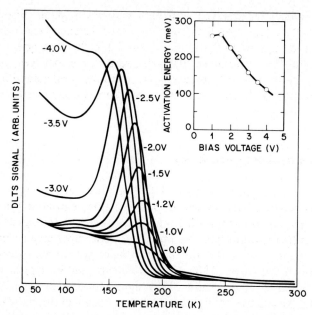

Fig. 1. DLTS spectra of emission of photo-induced carriers from a single 60 Å InP/Ga$_{0.47}$In$_{0.53}$As quantum well for various reverse-bias voltages at a rate window of 200 μs. The inset shows the measured DLTS activation energy versus voltage. (Ref. [5].)

corresponds to the quantum well no longer being within the depletion layer, i.e. the DLTS peak in fig. 1 is spatially correlated with the location of the quantum well. Note also the shift in the DLTS peak to lower temperatures and the eventual broadening of the peak as the bias voltage is increased. This corresponds to a transition from thermally activated emission (sharp DLTS peak) to a temperature-independent (tunneling) emission rate (broad DLTS peak) as the bias is increased.

By varying the rate window and measuring the activation energy of the DLTS peak as a function of voltage (inset in fig. 1) we can see that the shift in this peak is due to a very strong reduction in energy with increasing voltage. Our largest measured activation energy saturates at 260 meV as the signal approaches zero due to the quantum well leaving the depletion layer. If we add to this the heavy-hole quantum state energy of 16 meV from the bottom of a 60 Å InP/Ga$_{0.47}$In$_{0.53}$As quantum well and subtract the $2kT = 29$ meV correction due to the T^2-dependence of the thermionic emission prefactor, we have an energy of 247 meV. In our original work we pointed out that this DLTS energy is likely to be a lower limit for the valence-band offset ΔE_v in InP/Ga$_{0.47}$In$_{0.53}$As since one should expect thermal emission from a quantum well in the presence of a space-charge electric field to underestimate the true band edge discontinuity. In fact, our subsequent admittance spectroscopy measurements [9] on InP/Ga$_{0.47}$In$_{0.53}$As superlattices in zero field showed that $\Delta E_c = 250$ meV and $\Delta E_v = 346$ meV, much larger than the apparent DLTS hole emission energy of 247 meV. We will discuss the admittance spectroscopy results more fully below, however, to properly analyze the DLTS results we will need to make use of them here.

We can now shed some light on the apparent discrepancy between the DLTS and admittance spectroscopy results by considering the band bending diagram of the DLTS sample used in ref. [5] in more detail, as shown in fig. 2. This diagram has been constructed from the known position of the quantum well at 3800 Å from the p$^+$-layer and the apparent position on the quantum well (0.55 μm) in a light-induced C–V profile at low temperature (65 K). Note that unlike the C–V profile of a single InP/Ga$_{0.47}$In$_{0.53}$As interface [10], there is no obvious feature associated with the 60 Å single quantum well in C–V profiles measured in the dark at either low temperature or room temperature. The low-temperature, illuminated C–V profile shows a quantum well peak in $N(x)$ versus x at 0.55 μm, corresponding to 1–2 V bias. The case of 2 V bias is shown in fig. 2. The depletion layer width in the dark at 2 V is 0.65 μm; thus at this bias for the DLTS signal in fig. 1, the capacitance transient from the light-induced state to the dark equilibrium state corresponds to the depletion layer width increasing from 0.55 to 0.65 μm. The sign of such a transient is indicative of minority carrier (hole) emission since the amount of positive charge in the n-type space-charge layer is decreasing during the transient. However, from fig. 2 we see that in this case we must consider both electron and hole emission from the quantum well. Note that the Fermi level for electrons is well within the conduction-band quantum well for the light-induced state. This tendency of the strobe light to fill both the electron and hole quantum wells holds for the initial conditions for all DLTS bias voltages in fig. 1. Indeed, one needs to consider separate quasi-Fermi levels for both electrons and holes in the presence of light. Thus, while there is a net surplus of photo-induced holes in the Ga$_{0.47}$In$_{0.53}$As layer, the relaxation of the photo-induced initial state to the dark equilibrium state involves the emission of both holes and electrons.

Given the fact that both holes and electrons must be considered in the apparent "hole emission" DLTS signals in fig. 1, we may now relate the

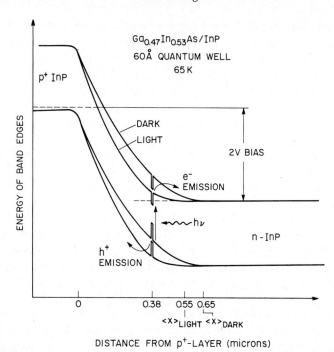

Fig. 2. Band bending diagram of p$^+$–n single quantum well sample used for DLTS measurements in fig. 1. The bias voltage corresponds to 2 V and the steady-state conditions are shown for the light-induced and dark states.

activation energy obtained by DLTS with the true band offsets obtained by admittance spectroscopy. From the admittance spectroscopy results, the hole emission energy in zero field is the band offset of 350 meV minus the heavy-hole energy of 16 meV for an activation energy of 334 meV. Similarly, for electrons the band offset of 250 meV is reduced by the bound-electron energy of 74 meV in the 60 Å quantum well to obtain a net emission energy of 176 meV in zero field. Note that these values are for zero occupancy of these levels and will be reduced a few tens of meV for finite occupancy by quasi-Fermi level shifts which we cannot easily calculate. Nevertheless, since the capacitance transient rate constant is the sum of the electron and hole emission rates, as an approximation the DLTS activation energy should be roughly the average of the hole and electron activation energy, or 255 meV (minus a Fermi level shift of 10–50 meV). This is to be compared with the measured value of 260 meV minus the

$2kT = 29$ meV correction, or 231 mev. Considering the uncertainties in this analysis, the agreement is satisfactory. In any event, the main point is to show that while one can see a quantum well "giant trap" by DLTS, in general this is not a very accurate way to determine band offsets. As we shall see, one needs a zero-field method such as series-resistance-limited admittance spectroscopy which can measure the activation energies of holes and electrons separately without the ambiguities of space-charge electric fields.

2. Admittance spectroscopy

In this section we discuss the use of admittance spectroscopy [3] to measure the band offsets in $InP/Ga_{0.47}In_{0.53}As$. Briefly, the method consists of analyzing the temperature dependence of the capacitance and AC conductance of a zero-biased p–n junction containing a superlattice of heterojunctions, thereby obtaining the activation energy of conduction perpendicular to the layers of the superlattice, which is simply related to the band offset. We have applied this technique to both n^+–p and p^+–n junctions containing $InP/Ga_{0.47}In_{0.53}As$ superlattices and can independently obtain both the conduction- and valence-band offsets [6]. These results indicate that 42% of the band-gap discontinuity is in the conduction band, in good agreement with the C–V measurements of Forrest and co-workers [10,11].

The admittance spectroscopy samples were grown by two approaches utilizing molecular beam epitaxy (MBE) growth chambers. Both used thermally decomposed AsH_3 and PH_3 for the As and P. One used MBE-like elemental-source effusion cells for the group-III elements (In and Ga) [16] while the other used MOCVD-like organo-metallic sources for In and Ga [17]. The layer sequence for the p^+–n superlattice sample started with a 1.0 μm buffer layer of n-InP (6×10^{17} cm^{-3}) on an n^+-InP substrate. This was followed by a 2.0 μm thick layer of n-type $Ga_{0.47}In_{0.53}As$ and a 10-period n-type superlattice grown with alternating layers of 300 Å of $Ga_{0.47}In_{0.53}As$ and 500 Å of InP. These layers had a uniform doping of 6×10^{16} cm^{-3}. Note, however, that the C–V profile of the free-carrier concentration in the superlattice was highly non-uniform, with the carriers from donors in the barriers spilling over into the wells, similar to modulation doping [18]. The p^+–n junction was formed by a 1.5 μm layer of p-InP (1×10^{18} cm^{-3}) followed by a contact layer of 1000 Å of p-InP (1×10^{19} cm^{-3}). The n^+–p samples were essentially the same with the n and p dopings reversed, i.e. grown on a p^+-InP substrate, etc.. Ohmic contacts were applied and mesa

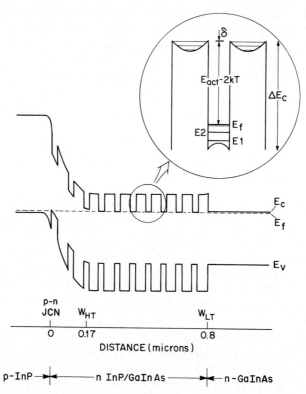

Fig. 3. Band bending diagram of p^+–n superlattice sample. The inset shows the effects of a uniform 6×10^{16} cm^{-3} doping and free-carrier redistribution on the small scale band bending of the $Ga_{0.47}In_{0.53}As$ quantum wells and InP barriers. Also shown to scale are the parameters needed to convert the data to band offset energies. (Ref. [6].)

diodes were formed by photolithography with either 100 or 325 μm diameters. The measurements were made in a helium-flow Dewar which could be varied from 20 K to 500 K. The capacitance and AC conductance measurements were made with a digital LCR meter at frequencies between 100 Hz and 1 MHz.

The band-bending diagram of a p^+–n superlattice sample is shown in fig. 3. Part of the superlattice (0 to 0.17 μm) is within the depletion region, with the remainder (0.17 to 0.8 μm) in neutral material. Admittance spectroscopy normally applies to defects located within the depletion region. However, here we make use of the fact that such spectra can also be

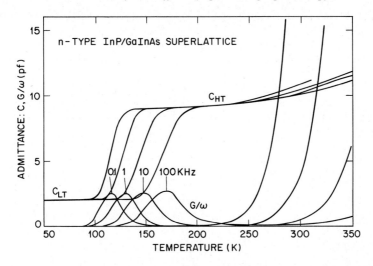

Fig. 4. Admittance spectroscopy data: capacitance C and conductance G divided by angular frequency versus temperature for four measurement frequencies. The temperatures of the conductance peaks used in fig. 6 are noted by arrows. (Ref. [6].)

due to RC time constant effects related to the resistance of the undepleted portion of the superlattice. Thus while the measurement technique is identical to standard admittance spectroscopy, the analysis is somewhat different, as described below. This variation of the method has been used previously in $GaAs/Al_{0.5}Ga_{0.5}As$ by Casey et al. [19] and in hydrogenated amorphous silicon by Lang et al. [20].

Thermal scans of capacitance $C(T)$ and conductance $G(T)$, measured at zero bias, are basic to the measurement technique. Note in fig. 4 that the $C(T)$ and $G(T)$ curves for each measurement frequency have the same general shape: C changes from a low-temperature value C_{LT} to a high-temperature value C_{HT}, and G has a peak at the step in C. In normal admittance spectroscopy [3] the step in C and the peak in G correspond to the thermal emission rate of trapped carriers being equal to the (angular) measurement frequency. However, in our case these features correspond to the inverse of the RC time constant of the sample being equal to the measurement frequency, where C is the depletion-layer capacitance and R is the series resistance of the undepleted portion of the superlattice. This can be understood by recalling that the RF capacitance of a p–n junction is directly related to the inverse of the first moment $\langle x \rangle$ of the small-signal AC charge distribution dQ/dV induced by the capacitance meter, i.e.

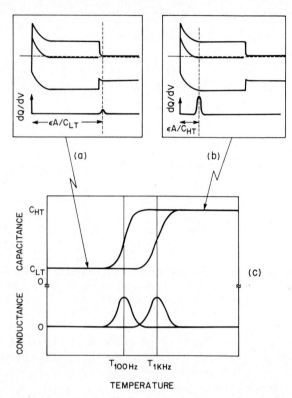

Fig. 5. Band bending diagrams of idealized heterojunction Schottky barrier and corresponding capacitance and conductance curves versus temperature for two different measurement frequencies to illustrate the series resistance effects on AC admittance measurements.

$C = \varepsilon A / \langle x \rangle$, where ε is the permittivity of the material and A is the area of the junction [20]. In fig. 5 we show how this effect appears in an idealized heterojunction sample. We consider a Schottky barrier on a wide-gap semiconductor in which the Fermi level is controlled by deep levels in the upper half of the gap. At the heterojunction the material becomes narrow gap and more strongly n-type. In fig. 5a we show that at low temperatures dQ/dV (and hence C_{LT}) is located at the heterojunction and not at the space-charge layer associated with the Schottky barrier. This is because the resistivity of the wide-gap material with deep levels is too large to allow the AC charge within the space-charge layer to flow through the wide-gap layer on the time scale of the angular measurement frequency ω, i.e. $RC > \omega^{-1}$. At high temperatures, on the other hand (fig. 5b), the thermally activated resistance of the wide-gap material has decreased so that the RC time constant is now short enough to allow dQ/dV to occur at the nominal depletion layer width at the Schottky barrier, i.e. $RC < \omega^{-1}$. Thus $C_{HT} > C_{LT}$ as shown in fig. 5c. Note that the transition between these two capacitance values as a function of temperature depends on the measurement frequency. The capacitance step is accompanied by a conductance peak at the temperature T_{peak} where the RC time constant is equal to the inverse of the angular measurement frequency, i.e. $RC = \omega^{-1}$. By constructing the Arrhenius plot of log ω versus $1000/T_{peak}$, one may obtain the activation energy of the RC time constant, which in this case is dominated by the temperature dependence of the resistance of the wide-gap material. If the example in fig. 5 had not involved the deep level in the wide-gap material, the relevant series resistance would have been that of the depletion region associated with the heterojunction. As we will discuss below, neglect of this series-resistance effect in normal N–n heterojunctions gives rise to anomalous results when the Kroemer C–V profile method to obtain band offsets is used in the C_{LT} regime.

For the superlattice sample shown in fig. 3 we find that C_{LT} corresponds to $\langle x \rangle = W_{LT} = 0.80$ μm, which is the width of the superlattice. Similarly, C_{HT} corresponds to $\langle x \rangle = W_{HT} = 0.17$ μm, which is the zero-bias depletion width for a uniform doping of 6×10^{16} cm^{-3}. The capacitance step corresponds to the temperature above which the perpendicular conductivity of the undepleted portion of the superlattice is large enough so that the AC displacement current can respond on a time scale fast compared to the measurement frequency, as in fig. 5b. For temperatures below the step the superlattice is "frozen out" and the capacitance of the sample is the same as if the superlattice were an insulator, as in fig. 5a. The superlattice can thus be viewed as a multilayer sandwich of metallic $Ga_{0.47}In_{0.53}As$ layers interleaved with leaky InP insulator layers. In effect, we measure the thermally activated dielectric loss peak of a superlattice "dielectric" layer in a capacitor, which is dominated by the resistance of the InP barriers. In similar superlattice samples with narrower barriers we had previously demonstrated the effects of tunneling on the transport properties, e.g. the effective-mass filter effect [5].

Clearly, for the transport of the sample to be dominated by the perpendicular conductivity of the superlattice we must have a low-leakage device structure. For example, most p–n junctions for research are fabricated as mesa structures and at some temperatures and bias voltages surface leakage current may dominate the thermionic emission current over the barriers in the superlattice. This is more likely to be the case for large

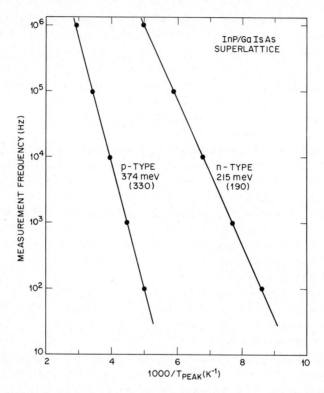

Fig. 6. Arrhenius plot of the inverse temperature of the conductance peaks in fig. 4 versus measurement frequency. The slopes of these data, E_{act}, are shown. The energies in parentheses are equal to $E_{act} - 2kT$, with T the average temperature of the measurements. (Ref. [6].)

offsets or for devices with particularly poor surface passivation. Indeed, we have seen very anomalous admittance spectroscopy results in leaky devices. One way to prepare low-leakage InP mesa structures is discussed in ref. [18]. In this regard InP-based structures are likely to be better than GaAs-based diodes because of the lower surface leakage typically obtained in InP.

The activation energy of the perpendicular conductivity of the superlattice can be simply obtained from an Arrhenius plot of the log of the measurement frequency versus the inverse temperature of the conductance peak (or capacitance step), as shown in fig. 6. The n-type and p-type samples have activation energies E_{act} of 215 ± 5 meV and 374 ± 10 meV, respectively. As shown by Batey et al. [9], thermally activated current transport at a GaAs/AlGaAs heterojunction is well described by simple thermionic emission theory at low electric fields and temperatures above about 100 K. In our case, therefore, we expect the perpendicular conductivity of the superlattice to be given by

$$\sigma = \sigma_0 T^2 \exp[-(\Delta E_c - E_F - \delta)/kT], \quad (1)$$

where σ_0 is a constant of proportionality, ΔE_c (or ΔE_v) is the conduction- (or valence-)band offset, E_F is the Fermi level measured from the bottom of the $Ga_{0.47}In_{0.53}As$ quantum well, and δ is the effective barrier lowering due to tunneling through the space-charge-generated parabolic well on the top of the InP barrier, as shown in fig. 3. Thus the band offset is related to the measured activation energy E_{act} of the superlattice conductivity in fig. 6 by

$$\Delta E_c = E_{act} - 2kT + E_F + \delta, \quad (2)$$

where $-2kT = d(\ln T^2)/d(1/kT)$ which comes from the effect of the prefactor in eq. (1) on the slope of an Arrhenius plot. A similar expression holds for the valence-band offset.

The quantities E_F and δ depend on the dimensions and doping of the superlattice in a straightforward way. In our case the carriers from the $(6 \pm 1) \times 10^{16}$ cm^{-3} uniform doping of the superlattice are redistributed into a degenerate two-dimensional carrier density of 5×10^{11} cm^{-2} in the wells and essentially none in the barriers. The space charge in the InP barriers give rise to a parabolic dip in the top of the barrier (fig. 3). To determine the effective barrier lowering δ we calculated the transmission coefficients for resonant tunneling via the bound states in the parabolic well at the top of the barrier, as shown in fig. 7. We then obtain the average activation energy for thermionic emission by considering the widths of these resonances and the Boltzmann factors for each and obtained $\delta = 2$ meV for electrons and 0.4 meV for holes.

The calculation of δ is performed as follows. The well-known expression for thermionic emission over a barrier ϕ is given by [8]

$$J_0 = AT^2 \exp(-\phi/kT), \quad (3)$$

where the Richardson constant A is given by

$$A = 4\pi q m^* k^2/h^3. \quad (4)$$

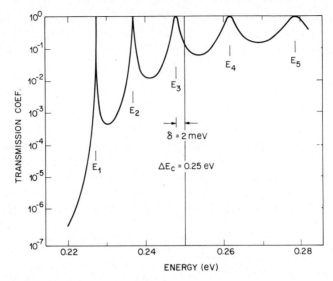

Fig. 7. Calculated transmission coefficient versus energy for resonant tunneling of electrons through and over a parabolic well in the top of an InP barrier formed by a space charge of 6×10^{16} cm^{-3} as shown in fig. 3.

We define $j(E)$ as the thermionic current per unit energy at an energy E over, or through, the barrier. Equation (3) is usually derived by integrating $j(E)$ from ϕ to ∞. Since most of this current flows within kT of the top of the barrier, we may define the effective current density at the barrier top by

$$J_0 = kTj(\Delta E_c - E_F), \qquad (5)$$

where $\phi = \Delta E_c - E_F$ for the case of electron emission as shown in figs. 3 and 7. The effect of resonant tunneling via the energy levels E_i below ΔE_c in fig. 7 is to create additional current paths J_i below the barrier with

$$J_i = \Gamma_i j(E_i - E_F), \qquad (6)$$

where Γ_i is the FWHM of the ith tunneling resonance. Thus the total current J_{tot} is given by

$$J_{\text{tot}} = J_0 + \sum_i J_i, \qquad (7)$$

where the sum is taken over those resonant states below ΔE_c. The effective barrier lowering energy δ is determined by calculating the apparent activation energy E_{act} of J_{tot}, i.e. $\delta = \phi - E_{act}$. By definition, we have

$$E_{act} = d(\ln J_{tot})/d(1/kT). \tag{8}$$

Since the energy dependence of the current density is a simple Boltzmann factor, we have

$$j(E_i - E_F)/j(\Delta E_c - E_F) = \exp(\Delta E_i/kT), \tag{9}$$

where $\Delta E_i = \Delta E_c - E_i$ is the energy of the ith resonance below the top of the barrier. Thus from eqs. (5)–(9) we have

$$E_{act} = \Delta E_c - E_F + 2kT - \frac{\sum_i \Delta E_i \Gamma_i \exp(\Delta E_i/kT)}{kT + \sum_i \Gamma_i \exp(\Delta E_i/kT)}, \tag{10}$$

which combined with eq. (2) gives

$$\delta = \frac{\sum_i \Delta E_i \Gamma_i \exp(\Delta E_i/kT)}{kT + \sum_i \Gamma_i \exp(\Delta E_i/kT)}. \tag{11}$$

As mentioned above, the values of δ obtained in our case are essentially negligible. This may seem surprising in view of the strong resonances apparent below the top of the barrier in fig. 7. However, it is consistent with the very small barrier lowering calculated for Schottky barriers [8] with electric fields similar to that shown in fig. 3. Wider or more heavily doped barriers would not have negligible barrier lowering effects and would thus have an effect on band offset measurements. A key feature of the superlattice admittance spectroscopy method is that the barriers may be made narrow enough to keep δ negligible. This is not the case in the very wide barriers used, for example, by Batey et al. [9].

The position of the Fermi level in the 300 Å $Ga_{0.47}In_{0.53}As$ quantum wells was obtained by first calculating the bound states taking into account the distortion of the well from the excess charge due to carriers originating from the doping in the barriers. Since this calculation considered the charge to be uniformly distributed and was not self-consistent, we estimate an error of approximately ±5 meV in the energy levels. The two-dimensional density of states for each level is constant in energy and equal to $m^*/\pi\hbar^2$, where m^* is the effective mass of the carrier. With a degenerate two-dimensional carrier density of 5×10^{11} cm^{-2} it is then straightforward to show that $E_F = 58 \pm 7$ meV for electrons and $E_F = 16 \pm 5$ meV for holes. Using these results for E_F, δ, and E_{act}, in eq. (2), the conduction-band offset is $\Delta E_c = 250 \pm 10$ meV and the valence-band offset is $\Delta E_v = 346 \pm 10$ meV. The total band-gap discontinuity is the sum of these two values, or $\Delta E_g = 596 \pm 15$ meV. This agrees well with the optically determined band-gap difference between InP and $Ga_{0.47}In_{0.53}As$ [21] which is 613 meV at 4.2 K and 600 ± 10 meV at 300 K. The fractional conduction-band offset is thus given by $\Delta E_c = (0.42 \pm 0.02)\Delta E_g$. This agrees very well with the result of Forrest et al. [11], who report a value of $\Delta E_c = (0.39 \pm 0.01)\Delta E_g$ over the entire InP/GaInAsP lattice-matched heterojunction system.

We can also use this understanding of series-resistance limitations in AC capacitance and conductance measurements to explain the anomalous temperature dependence of ΔE_c reported by both Forrest and Kim [10] and by Ogura et al. [12]. Both of these papers show $C(T)$ curves with a step at about 150 K very similar to our data. In fact, Forrest and Kim [10] performed admittance spectroscopy analysis on their $C(T)$ and $G(T)$ data and obtained an activation energy in n-type material of $E_{act} - 2kT = 0.22 \pm 0.02$ eV from an Arrhenius plot and 0.19 ± 0.01 eV from the theoretical fit to the shape of the conductance peak. This is to be compared with our value of $E_{act} - 2kT = 190$ meV in fig. 6. Without knowing about the series-resistance problems, however, they interpreted this effect as a deep-level interface state which they suggest is due to phosphorus vacancies at the $InP/Ga_{0.47}In_{0.53}As$ interface. They proposed that this state is an acceptor which fills with electrons below 170 K and hence explains the abrupt decrease in ΔE_c to nearly zero which they and Ogura et al. [12] had found below this temperature. We can rule out the trap model in our case because we have used DLTS to study MBE-grown single $InP/Ga_{0.47}In_{0.53}As$ interfaces exactly like those in our superlattices and have seen no evidence for interface traps in this energy range. Furthermore, we can explain the results of these previous workers in a very natural way as a consequence of thermionic emission over a simple band offset without defects. The main point is that the resistance of the depleted InP at a single n-type $InP/Ga_{0.47}In_{0.53}As$ heterojunction with a 0.25 eV offset is sufficient to cause a series-resistance problem with a 1 MHz capacitance measurement at temperatures below about 170 K. This is very clear in our superlattice samples. Under such conditions the Kroemer method [7] of obtaining band offsets from C–V profiles is invalid because the first moment of the AC charge distribution is RC limited by the strongly temperature-dependent series resistance of the heterojunction and is not a true measure of the carrier concentration in the vicinity of the heterojunction. Thus the apparent value of ΔE_c approaches zero for temperatures below the point where the heterojunction freezes out at 1 MHz. Nevertheless, our band offset results agree extremely well with those of Forrest and co-worker [10,11] at higher temperatures where the heterojunction series resistance is not a problem and the Kroemer method is valid.

In summary, we have observed a photo-induced DLTS "giant trap" signal due to a single $InP/Ga_{0.47}In_{0.53}As$ quantum well which we can explain as arising from the emission of both holes and electrons trapped in their respective quantum wells formed by the $Ga_{0.47}In_{0.53}As$ layer in the depletion layer of the p–n junction. The measured activation energy depends strongly on electric field and is roughly equal at low fields to the average of the expected hole and electron quantum well energies calculated from the band offsets determined by admittance spectroscopy. We have also shown that admittance spectroscopy, when properly interpreted taking series-resistance effects into account, can be a powerful new zero-field AC transport method to determine band offsets at semiconductor heterojunctions, especially superlattices. We have applied this to the case of $InP/Ga_{0.47}In_{0.53}As$ superlattices and obtain $\delta E_c = 250 \pm 10$ meV and $\Delta E = 346 \pm 10$ meV. The fractional conduction-band offset is thus $\Delta E_c = (0.42 \pm 0.02)\Delta E_g$, in good agreement with the results of Forrest and co-workers [10,11]. Finally, we show that in cases where series-resistance effects are not properly analyzed, anomalous band offset results can be obtained by the Kroemer C–V method [7] at low temperatures.

Acknowledgement

We wish to acknowledge many helpful discussions and experimental collaborations in the course of this work with colleagues at AT&T Bell Laboratories. In particular we are indebted to A.M. Sergent, F. Capasso, H. Temkin, M.B. Panish, W.T. Tsang, J. Allam, R.A. Hamm, A.L. Hutchinson, A. Savage and S. Sumski. We especially wish to thank A.C. Gossard and G.A. Baraff for the computer program used to calculate tunneling transmission coefficients in multi-quantum well structures.

References

[1] D.V. Lang, in: Thermally Stimulated Relaxation in Solids, ed. R. Braunlich, Topics in Applied Physics, Vol. 37 (Springer, Berlin, 1979) p. 93.
[2] D.V. Lang, J. Appl. Phys. 45 (1974) 3023.
[3] D.L. Losee, J. Appl. Phys. 46 (1975) 2204.
[4] P.A. Martin, K. Meehan, P. Gavrilovic, K. Hess, N. Holonyak Jr and J.J. Coleman, J. Appl. Phys. 54 (1983) 4689.
[5] D.V. Lang, A.M. Sergent, M.B. Panish and H. Temkin, Appl. Phys. Lett. 49 (1986) 812.
[6] D.V. Lang, M.B. Panish, F. Capasso, J. Allam, R.A. Hamm, A.M. Sergent and W.T. Tsang, Appl. Phys. Lett. 50 (1987) 736.
[7] H. Kroemer, Wu-Yi Chein, J.S. Harris, Jr and D.D. Edwal, Appl. Phys. Lett. 36 (1980) 295.
[8] S.M. Sze, Physics of Semiconductor Devices (Wiley, New York, 1981).
[9] J. Batey, S.L. Wright and D.J. DiMaria, J. Appl. Phys. 57 (1985) 484.
[10] S.R. Forrest and O.K. Kim, J. Appl. Phys. 53 (1982) 5738.
[11] S.R. Forrest, P.H. Schmidt, R.B. Wilson and M.L. Kaplan, Appl. Phys. Lett. 45 (1984) 1199.
[12] M. Ogura, M. Mizuta, K. Onaka and H. Kukimoto, Jpn. J. Appl. Phys. 22 (1983) 1502.
[13] K. Steiner, R. Schmitt, R. Zuleeg, L.M.F. Kaufmann, K. Heime, E. Kuphal and J. Wolter, Surf. Sci. 174 (1986) 331.
[14] P.E. Brunemeier, D.G. Deppe and N. Holonyak Jr, Appl. Phys. Lett. 46 (1985) 755.
[15] H. Temkin, M.B. Panish, P.M. Petroff, R.A. Hamm, J.M. Vandenberg and S. Sumski, Appl. Phys. Lett. 47 (1985) 394.
[16] M.B. Panish and S. Sumski, J. Appl. Phys. 55 (1984) 3571.
[17] W.T. Tsang, Appl. Phys. Lett. 45 (1984) 1234.
[18] J. Allam, F. Capasso, M.B. Panish and A.L. Hutchinson, Appl. Phys. Lett. 49 (1986) 707.
[19] H.C. Casey Jr, A.Y. Cho, D.V. Lang, E.H. Nicollian and P.W. Foy, J. Appl. Phys. 50 (1979) 3484.
[20] D.V. Lang, J.D. Cohen and J.P. Harbison, Phys. Rev. B 25 (1982) 5285.
[21] T.P. Pearsall, GaInAsP Alloy Semiconductors (Wiley, New York, 1982) p. 456.

PART II

PHYSICS OF HETEROJUNCTION DEVICES: BAND-GAP ENGINEERING

CHAPTER 10

BAND-GAP ENGINEERING AND INTERFACE ENGINEERING: FROM GRADED-GAP STRUCTURES TO TUNABLE BAND DISCONTINUITIES

Federico CAPASSO

AT&T Bell Laboratories
Murray Hill, NJ 07974, USA

Heterojunction Band Discontinuities: Physics and Device Applications
Edited by F. Capasso and G. Margaritondo
© *Elsevier Science Publishers B.V., 1987*

Contents

1. Band-gap engineering, interface engineering 401
2. Quasi-electric fields in graded-gap materials 402
3. Electron velocity measurements in graded-gap semiconductors 404
4. High-speed graded-base transistors 407
5. Emitter grading in heterojunction bipolar transistors 413
6. Resonant tunneling bipolar transistors and effective mass filtering 417
7. Multilayer sawtooth materials 424
   7.1. Rectifiers 425
   7.2. Electrical polarization effects in sawtooth superlattices 426
   7.3. Staircase structures 428
      7.3.1. Staircase avalanche photodiodes based on impact-ionization assisted by band edge discontinuities 429
      7.3.2. Impact-ionization across the band discontinuity: A new solid-state photo-multiplier 432
      7.3.3. Repeated velocity overshoot 436
8. Superlattice band-gap grading and pseudo-quaternary alloys 438
9. Doping interface dipoles: tunable band-edge discontinuities 442
References 449

1. Band-gap engineering, interface engineering

The advent of molecular beam epitaxy (MBE) [1] has made possible the development of a new class of materials and heterojunctions with unique electronic and optical properties. Most notable among these are: heterojunction and doping superlattices, modulation-doped superlattices, strained layer and variable-gap superlattices. The investigation of the novel physical phenomena made possible by such structures has proceeded in parallel with their exploitation in novel devices. As a result a new approach or philosophy to designing heterojunction semiconductor devices, band-gap engineering, has gradually emerged [2,3].

The starting point of band-gap engineering is the realization of the extremely large number of combinations made possible by the above-mentioned superlattices and heterojunction structures. This allows one to design a large variety of new energy band diagrams. In particular through the use of band-gap grading one can obtain, starting from a basic energy band diagram, practically arbitrary and continuous variation of this diagram. Thus the transport and optical properties of a semiconductor structure can be modified and tailored to a specific device application. One of the most powerful consequences of band-gap engineering is the ability of independently tuning the transport properties of electron and holes, using quasi-electric fields in graded-gap materials and the difference between conduction- and valence-band discontinuities in a given heterojunction.

In this chapter recent advances in band-gap engineering are discussed, particularly in the area of photodetectors, heterojunction transistors and graded-gap structures. Other important developments are covered in the chapters by Chemla and Miller (modulators and bistable quantum well devices), Dutta (quantum well lasers), Luryi (hot-electron and resonant-tunneling devices), and by Hess and Iafrate (heterojunction transport theory). A recent technique capable of modifying band discontinuities at heterojunctions using doping interface dipoles is discussed in the last section of this chapter. This method allows one to modify selectively the band diagram in the vicinity ($\lesssim 100$ Å) of a heterointerface (interface engineering) and therefore has tremendous potential for basic studies and device applications. Alternative techniques to modify band offsets are discussed in the chapter by Margaritondo and Perfetti (ch. 2).

2. Quasi-electric fields in graded-gap materials

Kroemer [4] was the first who considered the problem of transport in a graded-gap semiconductor. As a result of compositional grading electrons and holes experience "quasi-electric" fields of different intensities,

$$F_e = -\frac{dE_c}{dz}, \qquad F_h = +\frac{dE_v}{dz}, \tag{1}$$

where $E_c(z)$ and $E_v(z)$ are the conduction- and valence-band edges. In addition the forces resulting from these fields push electrons and holes in the same direction. This is illustrated in fig. 1a for the case of an intrinsic material. Note that such a graded material can be thought of as a large number of isotype heterojunctions of progressively varying band-gap stacked on top of each other. If the conduction- and valence-band edge discontinuities ΔE_c and ΔE_v of such heterojunctions are known and depend little on the alloy composition then one can also expect that for the structure of fig. 1a the ratio of the quasi-electric fields F_e/F_h is equal to $\Delta E_c/\Delta E_v$.

However, for a p-type graded-gap material the situation is different; the energy band diagram is illustrated in fig. 1b. The valence-band edge is now horizontal so that no effective field acts on the holes while the effective field for the electrons is $F_e = -dE_g/dz$, which can be significantly greater than in the intrinsic case. In other terms all the band-gap grading is transferred

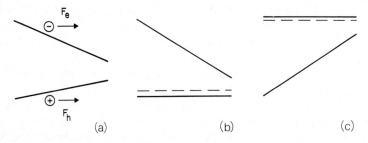

Fig. 1. Energy band diagram of compositionally graded materials: (a) intrinsic; (b) p-type; (c) n-type.

to the conduction band. This can be interpreted physically using the following heuristic argument.

Let us assume to start with the intrinsic material of fig. 1a and to dope it p-type. The acceptor atoms will introduce holes in the valence band which under the action of the valence-band quasi-electric field will be spatially separated from their parent ionized acceptor atoms (negatively charged). This separation produces an electrostatic (space-charge) electric field. Holes accumulate (on the right-hand side of fig. 1b) until this space-charge field equals in magnitude and cancels the hole quasi-electric field F_h thus achieving the thermodynamic equilibrium configuration (flat valence band) of fig. 1b. Note, however, that as a result of this process the equilibrium hole density is spatially nonuniform. The electrostatic field of magnitude $|dE_v/dz|$ produced by the separation of holes and acceptors adds instead to the conduction-band quasi-electric field to give a total effective field acting on an electron

$$F_e = -\left(\frac{dE_c}{dz} + \frac{dE_v}{dz}\right) = -\frac{dE_g}{dz}. \tag{2}$$

Thus in a p-type material the conduction-band field is made up of a non-electrostatic (quasi-electric field) and an electrostatic (space-charge) contribution.

For an n-type material the same type of argument can be applied to electrons, leading to the band diagram of fig. 1c and to an effective field acting on the hole given also by eq. (2).

Consider, for example, the case of an $Al_xGa_{1-x}As$ graded-gap p-type semiconductor. Assuming that in an $Al_xGa_{1-x}As$ heterojunction 62% of the band-gap difference is in the conduction band [5] it follows that 62% of the effective conduction-band field $F = -dE_g/dz$ will be quasi-electric in nature and the rest (38%) electrostatic. The opposite occurs in the case of an n-type $Al_xGa_{1-x}As$ graded material, where 62% of the valence-band effective field is instead electrostatic in nature.

So far we have only considered quasi-electric fields arising from band-gap grading. When the composition of the alloy is changed, however, also the effective mass of the carriers changes, giving rise to additional quasi-electric fields for electrons and holes. These are given by [6]

$$F_e = \frac{d}{dz}\left[\tfrac{3}{2}kT \ \ln(m_e^*/m_0)\right], \tag{3}$$

$$F_\mathrm{h} = \frac{\mathrm{d}}{\mathrm{d}z}\left[\tfrac{3}{2}kT\,\ln(m_\mathrm{h}^*/m_0)\right]. \tag{4}$$

The quasi-electric fields in direct-band-gap graded composition material consisting of $Al_xGa_{1-x}As$ are primarily due to band-gap grading; the quasi-fields due to the effective-mass gradients are in this case negligible [6]. However, effective-mass gradients can make a substantial contribution to the quasi-field for $Al_xGa_{1-x}As$ graded materials in which the composition x is varied through the direct–indirect transition at $x = 0.45$. This is due to the fact that the electron effective mass varies by about one order of magnitude in the direct–indirect transition region. Similar considerations are thought to apply also to other III–V alloys.

3. Electron velocity measurements in graded-gap semiconductors

Quasi-electric fields are particularly important since they can be used to enhance the velocity of minority carriers which would otherwise move by diffusion (a relatively slow process) rather than by drift.

Kroemer [4] was, in fact, the first who proposed the use of a graded-gap p-type layer (fig. 1b) for the base of a bipolar transistor, in order to reduce the minority carrier (electron) transit time in the base.

Recently, Levine and co-workers [7,8], using an all-optical method, measured for the first time the electron velocity in a heavily doped p^+ compositionally graded $Al_xGa_{1-x}As$ layer grown by molecular beam epitaxy.

The energy band diagram of the sample is sketched in fig. 2 along with the principle of the experimental method. The measurement technique is a "pump and probe" scheme. The pump laser beam, transmitted through one of the AlGaAs window layers is absorbed in the first few thousand Å of the graded layers. Optically generated electrons under the action of the quasi-electric field drift towards the right-hand side in fig. 2 and accumulate at the end of the graded layer. This produces a refractive index change at the interface with the second window layer. This refractive index variation produces a reflectivity change that can be probed with the counter propagating laser beam. This reflectivity change is measured as a function of the delay between pump and probe beam using phase sensitive detection techniques. The reflectivity data [7] are shown in fig. 3 for a sample with a 1 µm thick transport layer graded from $Al_{0.1}Ga_{0.9}As$ to GaAs and doped to

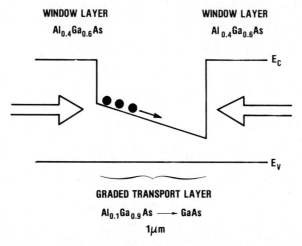

Fig. 2. Band diagram of a sample used for electron velocity measurements and schematic illustration of the pump-and-probe measurement technique.

$p \cong 2 \times 10^{18}$ cm^{-3}. This corresponds to a quasi-field of 1.2 kV/cm. The laser pulse width was 15 ps and the time 0 in fig. 3 represents the center of the pump pulse as determined by two-photon absorption in a GaP crystal

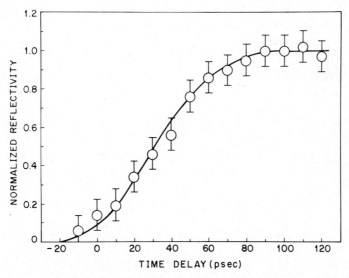

Fig. 3. Normalized experimental results for pump-induced reflectivity change versus time delay obtained in 1 μm thick graded-gap p$^+$ AlGaAs at a quasi-electric field $F = 1.2$ kV/cm.

cemented near the sample. The transit time is approximately given by the shift of the half-height of the reflectivity curve from zero, which is $\tau = 33$ ps. Taking as the drift length the graded layer thickness minus the absorption length of the pump beam ($1/\alpha \cong 2500$ Å) one finds a minority carrier velocity $v \approx L - \alpha^{-1}/\tau \approx 2.3 \times 10^6$ cm/s.

In this relatively thick sample diffusion effects are important and cause a spread in the arrival time of electrons at the end of the sample, given roughly by the 10–90% rise time of the reflectivity curve, i.e. 63 ps.

It is interesting to note that the drift mobility obtained from the measurement is $\mu_d = v_e/F = 1900$ cm^2 V^{-1} s^{-1}, which is comparable with the usual mobility of 2200 cm^2 V^{-1} s^{-1} at the doping level of the graded layer in GaAs.

Electron velocity measurements were also made in a 0.42 μm thick strongly graded ($F_e = 8.8$ kV/cm), highly doped ($p = 4 \times 10^{18}$ cm^{-3}) Al$_x$Ga$_{1-x}$As layer graded from Al$_{0.3}$Ga$_{0.7}$As to GaAs. A transit time of only 1.7 ps was measured, which is more than an order of magnitude shorter than that for $F = 1.2$ kV/cm, as shown in fig. 4, corresponding roughly to a velocity $v_e \cong 2.5 \times 10^7$ cm/s [8]. The velocity can be obtained rigorously and accurately ($\pm 10\%$ error) from the reflectivity data, by solving the drift diffusion equation and taking into account the effects of

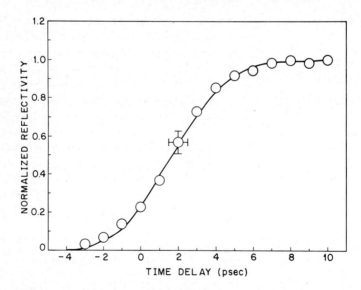

Fig. 4. Normalized experimental reflectivity change versus time delay measured in a 0.4 μm thick graded-gap p$^+$ AlGaAs layer at a quasi-electric field $F = 8.8$ kV/cm.

the pump absorption length (especially important in a thin sample) and the partial penetration of the probe beam in the graded material. Including all these effects, one finds that reflectivity data can be fitted using only one adjustable parameter, the electron drift velocity [8]. This velocity is $v_e = 2.8 \times 10^6$ cm/s for $F = 1.2$ kV/cm and $p = 2 \times 10^{18}$ cm^{-3}; and $v_e = 1.8 \times 10^7$ cm/s for $F = 8.8$ kV/cm and $p = 4 \times 10^{18}$ cm^{-3}.

We see that when we increased the quasi-field from 1.2 to 8.8 kV/cm (a factor of 7.3) the velocity increased from 2.8×10^6 to 1.8×10^7 cm/s (a factor of 6.4). That is, we observe the approximate validity of the relation $v_e = \mu F$. In fact, using $\mu = 1700$ cm^2 V^{-1} s^{-1} (for $p = 4 \times 10^{18}$ cm^{-3}) we calculate $v_e = 1.5 \times 10^7$ cm/s for $F = 8.8$ kV/cm which is in reasonable agreement with the experiment. It is worth noting that this measured velocity of 1.8×10^7 cm/s (in the quasi-field) is significantly larger than that for pure undoped GaAs where $v_e = 1.2 \times 10^7$ cm/s for an ordinary electric field of $F = 8.8$ kV/cm. In fact, our, measured, high velocity is comparable to the peak velocity reached in GaAs for $F = 3.5$ kV/cm before the intervalley transfer occurs from the Γ to the L valley. It is noteworthy that our measured velocity is also comparable to the maximum possible phonon limited velocity in the Γ minimum of GaAs. This is given by $v_{max} = [(E_p/m^*) \tanh(E_p/2kT)]^{1/2} = 2.3 \times 10^7$ cm/s, where $E_p = 35$ meV is the optical phonon energy and the effective mass $m^* = 0.067 m_0$.

This high velocity can be understood without reference to transient effects since the transit time is much larger than the momentum relaxation time of 0.3 ps. The large velocity results from the fact that the electrons spend most of their time in the high-velocity central Γ valley rather than in the low-velocity L valley. This may result from the injected electron density being so much less than the hole density that the strong hole scattering can rapidly cool the electrons without excessively heating the holes. Furthermore, most of the electrons remain in the Γ valley throughout their transit across the graded layer since the total conduction-band edge drop ($\Delta E_g = 0.37$ eV) is comparable to the GaAs Γ–L separation ($\Delta E_{\Gamma L} = 0.33$ eV) and therefore they do not have sufficient excess energy for significant transfer to the L valley.

4. High-speed graded-base transistors

The first device utilizing the high electron velocity of electrons in p-type material was reported by Capasso et al. [9]. The structure is a phototransistor with an AlGaAs graded-gap base with a quasi-field of $\approx 10^4$ kV/cm, as shown in fig. 5b.

The device was grown by molecular beam epitaxy on a Si-doped ($\approx 4 \times 10^{18}$ cm^{-3}) n$^+$-GaAs substrate. A buffer layer of n$^+$-GaAs was subsequently grown followed by a Sn-doped, n-type ($\approx 10^{15}$ cm^{-3}) 1.5 μm thick, GaAs collector layer. The 0.45 μm thick base layer was compositionally graded from GaAs (on the collector layer side) to Al$_{0.20}$Ga$_{0.80}$As ($E_g = 1.8$ eV) and heavily doped with Be ($p^+ \cong 5 \times 10^{18}$ cm^{-3}). The abrupt wide-gap emitter consists of an Al$_{0.45}$Ga$_{0.55}$As ($E_g = 2.0$ eV) 1.5 μm thick window layer n-doped with Sn in the range $\cong 2 \times 10^{15}$–5×10^{15} cm^{-3}. Figure 5b shows the energy band diagram of the phototransistor.

To study the effect of grading in the base on the speed of the device, 4 ps wide laser pulses were used. The wavelength ($\lambda = 6400$ Å) was chosen so that the light is absorbed only in the base layer. The incident power was kept relatively high (100 mW) to minimize the effective emitter charging time. Under these conditions the speed limiting factors are the RC time constant and the base transit time. Figures 6a,b show the pulse response of the device as monitored by a fast sampling scope. In fig. 6b the response was signal-averaged; note the symmetric rise and fall time and the absence of long tails; this is normally very difficult to achieve in picosecond

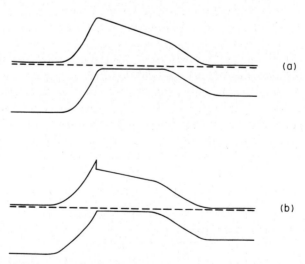

Fig. 5. Band diagram of graded-gap base bipolar transistor: (a) with graded emitter–base interface; (b) with ballistic launching ramp for even higher velocity in the base.

photodetectors. From the observed 10–90% response time of 30 ps a sum of the squares approximation was used to estimate an intrinsic detector response time of $\cong 20$ ps. In absence of a quasi-electric field in the heavily doped p^+ base, a broadening of the response followed by a tail with a square root of time dependence due to slow diffusion processes is expected. The diffusion time t_D is given by $W^2/2D$, where W is the base thickness and D is the diffusion coefficient. For a phototransistor with a GaAs $p^+ = 10^{18}$ cm^{-3} base, $D \cong 16$ cm^2/s. In our structure D is likely to be

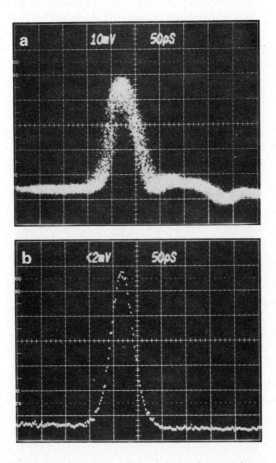

Fig. 6. Pulse response of graded-gap base AlGaAs/GaAs phototransistor to a 4 ps laser pulse displayed on a sampling scope (a); and after signal-averaging the sampling scope signal (b).

smaller because of the AlGaAs graded base, which has a lower mobility than GaAs, and the higher doping. Thus for our structure $t_D \geq 50$ ps. The fact that no such broadening is observed indicates that the quasi-electric field in the base sweeps out the electrons in a time small compared with the diffusion time. From the velocity measurements previously discussed we know that the base transit is $\cong 2$ ps which is indeed much smaller than t_D. Thus the pulse response of this device is consistent with Kroemer's prediction [4] and is the first experimental verification of this effect in an actual device [9].

Finally, the combination of the graded-gap base with the abrupt wide-gap emitter (fig. 5b) suggests a new high-speed transistor [9,10]. In fact, the conduction-band discontinuity can be used to ballistically launch electrons into the base with a high initial velocity; the quasi-field in the base will maintain an average velocity substantially higher than 10^7 cm/s. If no electric field is introduced in the base, the ballistic launching alone, using the abrupt base–emitter heterojunction, would not be sufficient to achieve a very high velocity in the base as a result of collisions with plasmons or coupled plasmon–phonons modes in the heavily doped base, which rapidly relax the initial forward momentum and velocity. It is sufficient for an initial high velocity that the conduction-band discontinuity used for the launching be a few kT (typically 50 meV at 300 K).

Recently, the first bipolar transistor with compositionally graded base has been reported [11,12]. Incorporation of a graded-gap base gives much shorter base transit times due to the induced quasi-electric field for electrons, thus allowing a precious tradeoff against the base resistance. To understand this last point consider a base of width W linearly graded from one alloy with a band gap E_{g1} to another one with band gap E_{g2} ($< E_{g1}$). The quasi-electric field for electrons, $(E_{g1} - E_{g2})/eW$, results in a base transit time (neglecting diffusion effects)

$$\tau_b' = \frac{eW^2}{\mu(E_{g1} - E_{g2})}, \tag{5}$$

where we have made use of the experimental fact that the velocity in the graded base equals μF_e where F_e is the quasi-field [7]. This time must be compared with the diffusion-limited base transit time of a transistor with an ungraded GaAs base of the same thickness and doping level

$$\tau_b = \frac{W^2}{2D}, \tag{6}$$

where D is the ambipolar diffusion coefficient. Comparing eqs. (5) and (6), and using the Einstein relationship ($D = \mu kT/e$), we find that the base transit time is shortened by the factor

$$\frac{\tau_b}{\tau_b'} = \frac{E_{g1} - E_{g2}}{2kT}, \tag{7}$$

using a graded-gap base. Although eq. (7) is rigorous only in the limit $E_{g1} - E_{g2} \gg kT$, it can be employed as a useful "rule of thumb" in cases where $E_{g1} - E_{g2}$ is several times kT. Thus the band-gap difference must be made as large as possible, without exceeding the intervalley energy separation ($\Delta E_{\Gamma L}$) which would result in a strong reduction of the electron velocity. Using $E_{g1} - E_{g2} = 0.2$ eV, the transit time is reduced by a factor of $\cong 4$ at 300 K over a bipolar with an ungraded base of the same thickness. This allows a precious tradeoff against the base resistance (R_b), making possible an increase of the base thickness and a consequent reduction of R_b, while still keeping a reasonable base transit time. Finally, an extra advantage of the quasi-field is an increase of the base transport factor since the short transit time reduces minority carrier recombination in the base.

The devices [11], grown by MBE on an n^+ substrate, had a 1.5 μm GaAs buffer layer followed by a 5000 Å thick collector doped to $n \cong 5 \times 10^{16}$ cm^{-3}. The p-type 2×10^{18} cm^{-3} base was graded from $Al_{0.02}Ga_{0.98}As$ to $Al_{0.2}Ga_{0.8}As$ over 4000 Å. This grading corresponds to a field of $\cong 5.6$ kV/cm. The lightly doped ($n \cong 2 \times 10^{16}$ cm^{-3}) wide-gap emitter consists of an $Al_{0.35}Ga_{0.65}As$ 3000 Å thick layer and a region, adjacent to the base, graded from $Al_{0.02}Ga_{0.98}As$ to $Al_{0.35}Ga_{0.65}As$ over 500 Å. This corresponds to a base/emitter energy gap difference of approximately 0.18 eV. This grading removes a large part of the conduction-band spike allowing most of the band gap difference to fall across the valence band, blocking the unwanted injection of holes from the base. Figure 5a shows the energy band diagram of the structure in the equilibrium (unbiased) configuration. The devices had a current gain of 35 at a base current of 1.6 mA and the collector characteristics were nearly flat with minimum collector–emitter offset voltage.

More recently, high current gain graded-base bipolars, with good high-frequency performance have been reported by Malik et al. [13]. The base layer was linearly graded over 1800 Å from $x = 0$ to $x = 0.1$ resulting in a quasi-electric field of 5.6 kV/cm and was doped with Be to $p = 5 \times 10^{18}$ cm^{-3}. The emitter–base junction was graded over 500 Å from $x = 0.1$ to

$x = 0.25$ to enhance hole confinement in the base. The 0.2 μm thick $Al_{0.25}Ga_{0.75}As$ emitter and the 0.5 μm thick collector were doped n-type at 2×10^{17} cm^{-3} and 2×10^{16} cm^{-3}, respectively. The $Al_xGa_{1-x}As$ layers were grown at a substrate temperature of 700°C. It was found that this high growth temperature resulted in better $Al_xGa_{1-x}As$ quality as determined by photoluminescence. However, it is known that significant Be diffusion occurs during MBE growth at high substrate temperatures and at high doping levels ($p > 10^{18}$ cm^{-3}). SIMS data also indicated a misplacement of the p–n junction into the wide band-gap emitter at 700°C substrate growth temperatures. Therefore, it was determined empirically that the insertion of an undoped setback layer of 200–500 Å between the base and emitter to compensate for the Be diffusion resulted in significantly increased current gains. Zn diffusion was used to contact the base and provides a low base contact resistance.

The common-emitter I–V characteristics of a test transistor with an emitter area of 7.5×10^{-5} cm^2 is shown in fig. 7. It is seen that the current gain increases with higher current levels and that the collector current exhibits flat output characteristics. The maximum differential DC current gain is 1150, obtained at a collector current density of $J_c = 1.1 \times 10^3$ A cm^{-2}. A small offset voltage of about 0.2 V is evident. The maximum V_{ce} for these devices before collector breakdown was about 8 V.

These high gains were obtained with a dopant setback layer in the base of 300 Å and can be compared with previous work which consistently

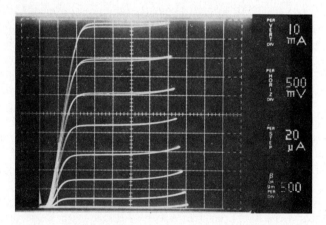

Fig. 7. Common-emitter characteristics of a graded base ($\simeq 2000$ Å) bipolar transistor.

resulted in current gains of < 100 in transistors without the setback layer [11,12]. Several transistor wafers were processed with undoped setback layers in a base of 200–500 Å and all exhibited gain enhancement.

These transistor has a current gain cutoff frequency $f_T \approx 5$ GHz and a maximum oscillation frequency of $f_{max} \approx 2.5$ GHz. Large signal pulse measurements resulted in rise times of $\tau_r \simeq 150$ ps and pulse collector currents of $I_c > 100$ mA which is useful for high-current laser drivers.

Recently AlGaAs/GaAs transistors with graded-gap base having $f_T = 45$ GHz and $f_{max} = 70$ GHz have been reported [14].

5. Emitter grading in heterojunction bipolar transistors

The essential feature of the heterojunction bipolar transistor (HBT) relies upon a wide band-gap emitter wherein part of the energy band-gap difference between the emitter and base is used to suppress hole injection. This allows the base to be more heavily doped than the emitter leading to a low base resistance and emitter–base capacitance both of which are necessary for high-frequency operation, while still maintaining a high emitter injection efficiency [10]. In this section we discuss in detail the grading problem in heterojunction bipolars. The performances of recently developed $Al_{0.48}In_{0.52}As/Ga_{0.47}In_{0.53}As$ bipolars with graded and ungraded emitters are compared [15] and the optimum way to grade the emitter is discussed.

Most of the work on MBE-grown heterojunction bipolar transistors has concentrated on the AlGaAs/GaAs system. Recently, the first vertical N–p–n $Al_{0.48}In_{0.52}As/Ga_{0.47}In_{0.53}As$ heterojunction bipolar transistors grown by MBE with high current gain have been reported by Malik et al. [15].

The (Al, In)As/(Ga, In)As layers were grown by MBE lattice-matched to a Fe-doped semi-insulating InP substrate. Two HBTs structures were grown; the first with an abrupt emitter of $Al_{0.48}In_{0.52}As$ on a $Ga_{0.47}In_{0.53}As$ base, and a second with a graded emitter comprising a quaternary layer of AlGaInAs of a width of 600 Å linearly graded between the two ternary layers. Grading from $Ga_{0.47}In_{0.53}As$ to $Al_{0.48}In_{0.52}As$ was achieved by simultaneously lowering the Ga and raising the Al oven temperatures in such a manner as to keep the total group-III flux constant during the transition.

It should be noted that this is the first use of a graded quaternary alloy in a device structure. The energy band diagram for the abrupt and graded

(a) ABRUPT EMITTER

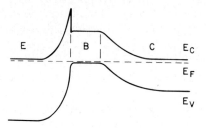

(b) GRADED EMITTER

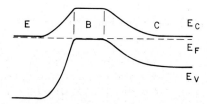

Fig. 8. Band diagrams under equilibrium of heterojunction bipolar with: (a) abrupt emitter, and (b) graded emitter. Note the elimination of the conduction-band spike through the use of a graded emitter and the increase of the emitter–base valence-band barrier.

emitter transistors are shown in fig. 8a and fig. 8b, respectively. It is seen that the effect of the grading is to eliminate the conduction band spike in the emitter junction. This in turn leads to a larger emitter–base valence-band difference under forward bias injection. The following material parameters were used in both types of transistors. The $Al_{0.48}In_{0.52}As$ emitter and $Ga_{0.47}In_{0.53}As$ collector were doped n-type with Sn at levels of 5×10^{17} cm^{-3} and 5×10^{16} cm^{-3}, respectively. The $Ga_{0.47}In_{0.53}As$ base was doped p-type with Be to a level of 5×10^{18} cm^{-3}. Recent experimental determination of the band edge discontinuities in the $Al_{0.48}In_{0.52}As/Ga_{0.47}In_{0.53}As$ heterojunction indicates that $\Delta E_c \cong 0.50$ eV and $\Delta E_v \cong 0.20$ eV [16]. This value of ΔE_v is large enough to allow the use of an abrupt $Al_{0.48}In_{0.52}As/Ga_{0.47}In_{0.53}As$ emitter at 300 K. Nevertheless, a current gain increase by a factor of 2 is achieved through the use of the graded-gap emitter which is attributed to a larger valence-band difference between the emitter and base under forward bias injection. This increase is clearly shown in fig. 9.

It is apparent from figs. 9a,b that there is a relatively large collector–emitter offset voltage. This voltage is equal to the difference

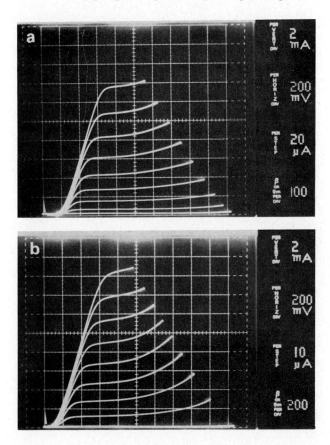

Fig. 9. Common-emitter characteristics of the $Al_{0.48}In_{0.52}As/Ga_{0.53}In_{0.47}As$ heterojunction bipolar transistors with: (a) abrupt emitter, and (b) graded emitter at 300 K.

between the built-in potential for the emitter–base p–n junction and that of the base–collector p–n junction. No such offset is therefore present in homojunction Si bipolars.

We have recently shown that by appropriately grading the emitter near the interface with the base such offset can be reduced and even totally eliminated [17]. The other advantage of grading the emitter is, of course, that the potential spike in the conduction band can be reduced, thus increasing the injection efficiency. The conduction-band potential is the result of the sum of two potentials: the electrostatic potential ϕ_{es} equal to

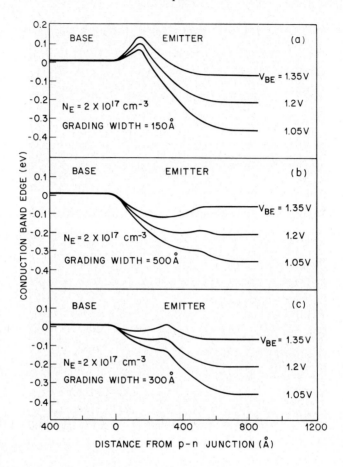

Fig. 10. Conduction-band edge versus distance from the p$^+$-n base–emitter junction for three different linear grading widths at different base–emitter forward bias voltages.

$V_{bi} - V_{be}$ (the built-in potential minus the base–emitter voltage), which varies parabolically with distance and the grading potential ϕ_g. If linear grading is used there is always unwanted structure in the conduction band (spikes or notches, fig. 10). The "notches" can reduce the injection efficiency by promoting carrier recombination. It now become obvious that any structure can be eliminated by grading with the complementary function of the electrostatic potential in the emitter region $(1 - \phi_{es})$ over the depletion layer width at a forward bias equivalent to the base band gap (fig.

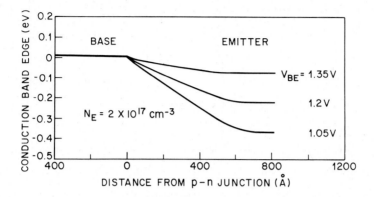

Fig. 11. Conduction-band edge versus distance from the p^+–n junction, using a parabolically graded layer 500 Å wide at different forward bias voltages.

11). Note that in this case if the base–emitter junction is forward biased at 1.42 eV, the two potentials (grading and electrostatic) give rise to a smooth conduction-band edge and one attains the flat-band condition with a built-in voltage for the base emitter equivalent to the band gap in the base ($= 1.42$ eV).

A HBT with such a parabolic grading has been fabricated, using MBE, with a $Ga_{0.7}Al_{0.3}As$ emitter and a GaAs base and collector [17]. A schematic diagram of the transistor structure and the common emitter characteristic are shown in fig. 12a and fig. 12b, respectively. The emitter–base junction was graded from $x = 0$ to $x = 0.3$ on the emitter side over a distance of 600 Å, the parabolic grading function being approximated by linear grading over nine regions. It can be seen from the characteristics shown in fig. 12b that the offset is very small about 0.03 V. Virtually identical characteristics with offsets ≤ 0.03 V were obtained for all devices on the wafer.

6. Resonant tunneling bipolar transistors and effective mass filtering

Serge Luryi in his contribution (ch. 12) discusses in detail the physics of resonant tunneling (RT) as well as several device applications. Here we shall concentrate on the use of RT double barriers in HBTs, particularly in conjunction with band-gap grading.

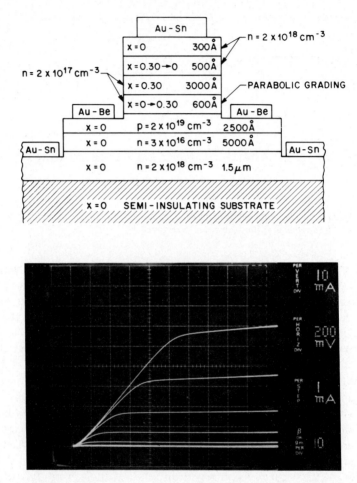

Fig. 12. (a) Schematic diagram of an AlGaAs/GaAs bipolar transistor that has a parabolic grading width of 600 Å at the base–emitter junction; (b) common-emitter characteristics of the transistor shown in (a). Note the negligible offset voltage.

The physical picture of coherent RT has lead to a design strategy intended to optimize the Fabry–Pérot resonator conditions. In particular, Ricco and Azbel [18] pointed out that achievement of a near-unity resonant transmission requires equal transmission coefficients for both barriers at the operating point – a condition not fulfilled for barriers designed to be symmetric in the absence of an applied field. To counter that, a RT

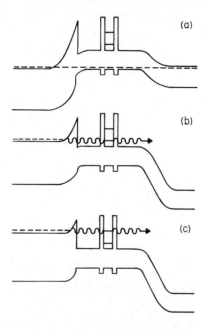

Fig. 13. Band diagram of a resonant-tunneling bipolar transistor (RTBT) with tunneling emitter under different bias conditions: (a) in equilibrium; (b) resonant tunneling through the first level in the well; (c) resonant tunneling through the second level. (Not to scale.)

structure was proposed by Capasso and Kiehl [19] in which a symmetric double barrier was built in the base of a bipolar transistor, and the Fabry–Pérot conditions were maintained through the use of minority-carrier injection (figs. 13 and 14). Thus this novel geometry maintains the crucial, structural symmetry of the double barrier, allowing, in principle, near-unity transmission at all resonance peaks and higher peak-to-valley ratios and currents compared to conventional RT structures. Both tunneling and ballistic injection in the base have been considered (see figs. 13 and 14). Particularly intriguing is the structure of fig. 14b which uses a parabolic well in the base to give rise to equally spaced peaks in the I–V.

Shortly after this initial proposal, Yokoyama et al. [20] reported the low-temperature operation (70 K) of a *unipolar* RT hot-electron transistor (RHET). This structure contains a double barrier in the emitter.

Recently, we have demonstrated the room-temperature operation of the first RT bipolar transistor (RTBT) [21]. The band diagram of the transistor under operating conditions is sketched in fig. 15, along with a schematics of

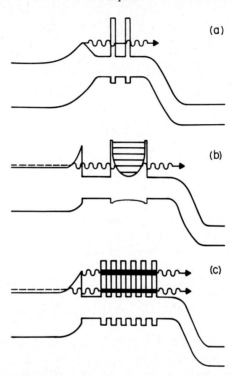

Fig. 14. (a) Band diagram of RTBT with graded emitter (at resonance). Electrons are ballistically launched into the first quasi-eigenstate of the well. (b) RTBT with parabolic quantum well in the base and tunneling emitter. A ballistic emitter can also be used. (c) RTBT with superlattice base. (Not to scale.)

the composition and doping profile of the structure (bottom). The double barrier consists of a 74 Å undoped GaAs QW sandwiched between the two undoped 21.5 Å AlAs barriers and the AlGaAs graded emitter is doped to $\approx 3 \times 10^{17}$ cm^{-3}. The portion of the base (Al$_{0.07}$Ga$_{0.93}$As) adjacent to the emitter was anodically etched off, while the rest of the base was contacted using AuBe. These base processing steps are essential for the operation of the device. The emitter area is $\simeq 2 \times 10^{-5}$ cm^2.

There is an essential difference with respect to the previously discussed RT transistors. These structures rely on *quasi-ballistic or hot-electron* transport through the base. These schemes place stringent constraints on the design and make it difficult to achieve room-temperature operation due to the small electron mean free path ($\simeq 500$ Å at 300 K), since electrons that

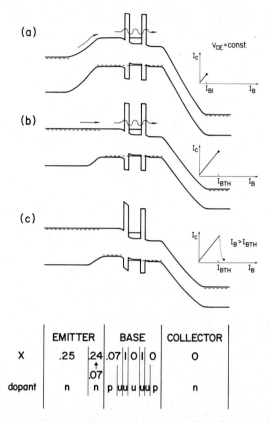

Fig. 15. Energy band diagrams of the RTBT and corresponding schematics of collector current I_C for different base currents I_B at a fixed collector emitter voltage V_{CE} (not to scale). As I_B is increased the device first behaves as a conventional bipolar transistor with current gain (a), until near flat-band conditions in the emitter are achieved (b). For $I_B > I_{BTH}$ a potential difference develops across the AlAs barrier between the contacted and uncontacted regions of the base. This raises the conduction-band edge in the emitter above the first resonance of the well, thus quenching resonant tunneling and the collector current (c). Also shown are the composition and doping profile of the structure; u stands for unintentionally doped.

have suffered a few phonon collisions cannot reach the collector. The key to the present structure (fig. 15) is that electrons are *thermally* injected into and transported through the base, thus making the device operation much less critical. This new approach has allowed us to achieve for the first time RT transistor action at room temperature. Thermal injection is achieved by adjusting the alloy composition of the portion of the base adjacent to the emitter in such a way that the conduction band in this region lines up with the bottom of the ground-state subband of the QW (fig. 15a). For a 74 Å well and 21.5 Å AlAs barriers the first quantized energy level is $E_1 = 65$ meV. Thus the Al mole fraction was chosen to be $x = 0.07$ (corresponding to $E_g = 1.521$ eV) so that $\Delta E_c \cong E_1$. This equality need not be rigorously satisfied for the device to operate in the desired mode, as long as E_1 does not exceed ΔE_c by more than a few kT. The QW is undoped; nevertheless, it is easy to show that there is a high concentration ($\cong 7 \times 10^{11}$ cm^{-2}) two-dimensional hole gas in the well. These holes have transferred from the nearby $Al_{0.07}Ga_{0.93}As$ region, by tunneling through the AlAs barrier, in order to achieve Fermi-level line-up in the base. Consider a common-emitter bias configuration. Initially the collector–emitter voltage V_{CE} and the base current I_B, are chosen in such a way that the base–emitter and the base–collector junctions are, respectively, forward and reversed-biased. If V_{CE} is kept constant and the base current I_B is increased, the base–emitter potential also increases until flat-band condition in the emitter region is reached (fig. 15b, left-hand side). In going from the band configuration of fig. 15a to that of fig. 15b the device behaves like a conventional transistor with the collector current increasing with the base current (figs. 15a and 15b, right-hand side). The slope of this curve is, of course, the current gain β of the device. In this region of operation electrons in the emitter overcome, by thermionic injection, the barrier of the base–emitter junction and undergo RT through the double barrier. If now the base current is further increased above the value I_{BTH} corresponding to the flat-band condition, the additional potential difference drops primarily across the first semi-insulating AlAs barrier (fig. 15c), between the contacted and uncontacted portions of the base, since the highly doped emitter is now fully conducting. This pushes the conduction-band edge in the $Al_{0.07}Ga_{0.93}As$ above the first energy level of the well, thus quenching the RT. The net effect is that the base transport factor and the current gain are greatly reduced. This causes an abrupt drop of the collector current as the base current exceeds a certain threshold value I_{BTH} (fig. 15c, right-hand side). The devices were biased in a common-emitter configuration at 300 K and the I–V characteristics were displayed on a curve tracer. For base currents ≤ 2.5 mA the transistor exhibits normal characteristics, while for $I \geq 2.5$ mA the behavior previously discussed was observed. Figure 16 shows the collector current versus base current at $V_{CE} = 12$ V, as obtained form the common-emitter characteristics. The collector current increases with the base current and there is clear evidence of current gain ($\beta = 7$ for $I_C > 4$ mA). As the base current exceeds 2.5 mA, there is a drop in I_C

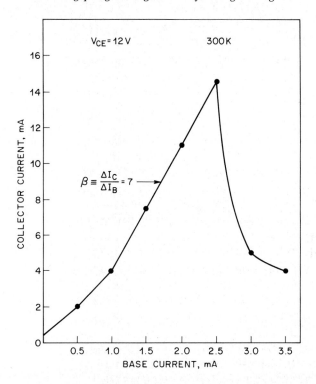

Fig. 16. Collector current versus base current of the RTBT in the common-emitter configuration at room temperature with the collector–emitter voltage held constant. The line connecting the data points is drawn only to guide the eye.

because the current gain is quenched by the suppression of RT. Single frequency oscillations (at 25 MHz, limited by the probe stage) have been observed in these devices when biased in the negative conductance region of the characteristics.

Recently, Palmier et al. [22] have reported a heterojunction bipolar transistor with a superlattice in the base layer, similar to the suggestion of Capasso and Kiehl [19]. The 2500 Å thick superlattice base layer is made of regularly spaced 40 Å wells (doped to $p = 5 \times 10^{16}$ cm^{-3}) and undoped $Al_{0.3}Ga_{0.7}As$ 17 Å or 38 Å thick barriers. Grading of the base–emitter interface was used to match the bottom of the conduction band in the emitter to the electronic quantum states of the superlattice. For the structure with a 17 Å barrier a very high current gain β was observed, which was interpreted in terms of the effective-mass filtering mechanism i.e., electrons

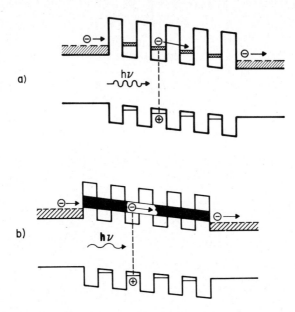

Fig. 17. Band diagram of effective-mass filtering in the case of: (a) phonon-assisted tunneling, and (b) miniband conduction.

traverse the base by miniband conduction while heavy holes remain confined in the wells due to the very small tunneling probability. Effective-mass filtering was first reported by Capasso et al. [23,24] in superlattice photoconductors of $Al_{0.48}In_{0.52}As/Ga_{0.47}In_{0.53}As$ and is illustrated in fig. 17. Photogenerated holes remain relatively localized in the wells (their tunneling probability is very small) while electrons propagate through the superlattice. The photoconductive gain is simply the ratio of the lifetime of the electron–hole pair to the electron transit time. The gain strongly decreased with increased AlInAs barrier layer thickness thus confirming effective-mass filtering as the physical origin of the effect. A similar dependence on the barrier thickness was observed by Palmier et al. [22] in the case of HBTs with superlattice base.

7. Multilayer sawtooth materials

In this section we examine the electronic transport properties of sawtooth structures obtained by periodically varying in an asymmetric fashion the composition of the alloy. The key feature of such structures is the lack of reflection symmetry [25]. This has several important consequences; for example, these devices can be used as rectifying elements or, under suitable conditions, one can optically generate in these structures a macroscopic electrical polarization which gives rise to a cumulative photovoltage across the uniformly doped sawtooth material. In addition, under appropriate bias they give rise to a staircase potential which has several intriguing applications.

7.1. Rectifiers

The basic principle of the device, demonstrated by Allyn and co-workers [26,27], is shown in fig. 18. A sawtooth-shaped potential barrier is created

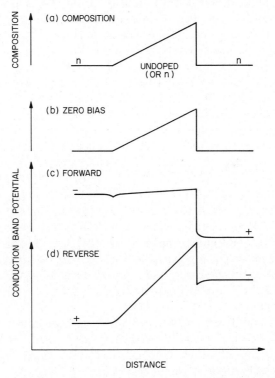

Fig. 18. (a) Compositional structure of a sawtooth-barrier rectifying structure, (b) potential distribution for band-edge conduction electrons at zero bias (undoped barrier case), (c) potential distribution under forward bias, and (d) potential distribution under reverse bias.

by growth of a semiconductor layer of graded chemical composition followed by an abrupt composition discontinuity. The adjoining layers, to which contact is made, are of the same conductivity type. In the present case, the material from which the barrier is constructed is aluminum gallium arsenide ($Al_xGa_{1-x}As$), in which the aluminum content is graded, and the adjoining layers are n-type GaAs.

Near zero bias, conduction in the direction perpendicular to the layer is inhibited by the barrier. When the device is biased in the forward direction [as shown in fig. 18c, labeled forward bias], the voltage drop initially occurs across the graded layer, reducing the slope of the potential barrier, and allowing increased thermionic emission over the reduced barrier. When the applied voltage exceeds the barrier height, the device will conduct completely, as is the case of a Schottky barrier. In the reverse direction (fig. 1), electrons will be attracted to but inhibited from passing over the abrupt potential discontinuity at the sharp edge of the sawtooth. The primary reverse current-carrying mechanism will be tunneling. The barrier can be either doped or undoped, although depletion of carriers from within the barrier (in the case of doped barriers) leads to band bending which will reduce the equilibrium height and width of the barrier. Multiple sawtooth barriers with five periods were are also fabricated [26]. These showed a turn-on voltage equal to five times that of the single barrier, thus demonstrating the additivity of the technique.

7.2. Electrical polarization effects in sawtooth superlattices

The lack of planes of symmetry in such material, compared to conventional superlattices with rectangular wells and barriers can lead to electrical polarization effects. Recently, Capasso et al. [28] have reported for the first time on the generation of a transient macroscopic electrical polarization extending over many periods of the superlattice. This effect is a direct consequence of the above-mentioned lack of reflection symmetry in these structures.

The energy band diagram of a sawtooth p-type superlattice is sketched in fig. 19a, where we have assumed a negligible valence-band offset. The layer thicknesses are typically a few hundred Å. The superlattice is sandwiched between two highly doped p^+ contact regions.

Let us assume that electron–hole pairs are excited by a very short light pulse as shown in fig. 19a. Electrons experience a higher quasi-electric field due to the grading than holes. For this reason and because of their much

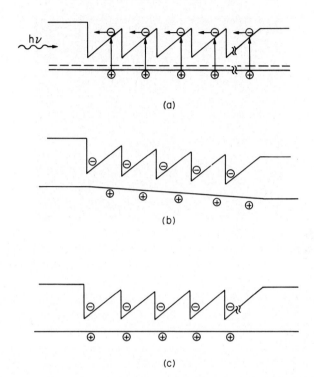

Fig. 19. Formation and decay of the macroscopic electrical polarization in a sawtooth superlattice.

higher mobility, electrons separate from holes and reach the low-gap side in a subpicosecond time ($\simeq 10^{-13}$ s). This sets up an electrical polarization in the sawtooth structure which results in the appearance of a photovoltage across the device terminals (fig. 19b). The macroscopic dipole moment and its associated voltage subsequently decay in time by a combination of: (a) dielectric relaxation, and (b) hole drift under the action of the internal electric field produced by the separation of electrons and holes.

The excess hole density decays by dielectric relaxation to restore a flat valence band (equipotential) condition, as illustrated in fig. 19c. Note that in this final configuration holes have redistributed to neutralize the electrons at the bottom of the wells. Thus also the net negative charge density on the low-gap side of the wells decreases with the same time constant as the positive charge packet (the dielectric relaxation time).

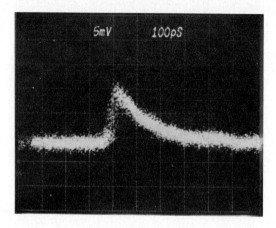

Fig. 20. Pulse response of an unbiased p-type AlGaAs/GaAs sawtooth superlattice to a 6 ps laser pulse.

The other mechanism by which the polarization decays is hole drift caused by the electric field created by the initial spatial separation of electrons and holes.

The above effect was demonstrated in a graded-gap p-type superlattice structure grown by molecular beam epitaxy (MBE). A total of ten graded periods were grown with a period of ≈ 500 Å. The layers were graded from GaAs to $Al_{0.2}Ga_{0.8}As$. A heavily doped GaAs contact layer of ≈ 700 Å was grown on top of the $Al_{0.45}Ga_{0.55}As$ 1 μm thick ($p \approx 5 \times 10^{18}$ cm^{-3}) window layer. The devices were mounted unbiased in a microwave stripline and illuminated with short light pulses (4 ps) of wavelength $\lambda = 6400$ Å. The absorption length is ≈ 3500 Å. The time dependence of the observed photovoltage is shown in fig. 20. In this particular wafer the carrier concentration was 10^{16} cm^{-3}. Note that the rise time is close to the scope limit while the fall time (at the $1/e$ point) is $\cong 200$ ps.

Unlike conventional detectors, the current carried in this photodetector is of displacement rather than conduction nature since it is associated with a time varying polarization. This current by continuity equals the conduction current in the external load.

7.3. Staircase structures

Recently Capasso and co-workers introduced the concept of a staircase potential [29–34]. This innovative structure has several interesting applications. We shall concentrate on multilayer graded-gap avalanche photodiodes and on the repeated velocity overshoot device.

7.3.1. Staircase avalanche photodiodes based on impact-ionization assisted by band edge discontinuities

Figure 21a shows the band diagram of the graded-gap multilayer material (assumed intrinsic) at zero applied field. Each stage is linearly graded in composition from a low (E_{g1}) to a high (E_{g2}) band gap, with an abrupt step back to low band-gap material. The conduction-band discontinuity shown accounts for most of the band-gap difference. The materials are chosen for a conduction-band discontinuity comparable to or greater than the electron ionization energy E_{ie} in the low-gap material following the step. The biased detector is shown in fig. 21b. Consider a photoelectron generated near the p$^+$ contact. The electron does not impact-ionize in the graded region before the conduction-band step because the net electric field is too low. At the step, however, the electron ionizes and the process is repeated at every stage. Note that the steps correspond to the dynodes of a phototube. Holes created by electron impact-ionization at the steps do not impact-ionize, since the valence-band steps are of the wrong sign to assist ionization and the electric field in the valence band is too low to cause hole initiated ionization. Obviously, holes multiply since at every step both and electron and a hole are created. The gain is $M = (2 - \delta)^N$ where δ is the fraction of electrons that do not ionize per stage and N is the number of stages. The noise per unit bandwidth on the output signal, neglecting dark current, is given by $\langle i^2 \rangle = 2eI_{ph}M^2F$ where I_{ph} is the primary photocurrent and F the avalanche excess noise factor. For the staircase APD F is given by [34]

$$F = 1 + \frac{\delta\left[1 - (2-\delta)^{-N}\right]}{2-\delta}. \tag{8}$$

Note that for small δ, $F \cong 1$ and F is practically independent of the number of stages. Thus, the multiplication process is essentially noise free. It is interesting to note that the excess noise of this structure does not follow the McIntyre theory of a conventional APD [35]. In a conventional APD the minimum excess noise factor at high gain (> 10) is 2 if one of the ionization rates is zero. The reason is that in the staircase APD the avalanche noise is lower than in the best conventional APD ($\alpha/\beta = \infty$) and can be understood as follows: in a conventional APD the avalanche is more

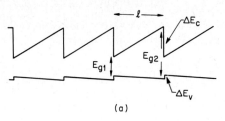

(a)

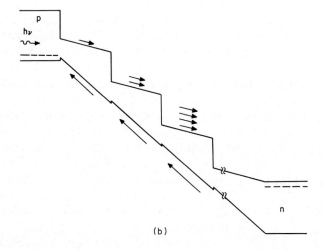

(b)

Fig. 21. Band diagram of staircase solid-state photomultiplier. The arrows in the valence band indicate that holes do not impact-ionize.

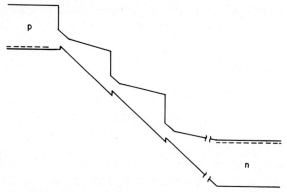

Fig. 22. Band-diagram of staircase avalanche photodiode with ungraded sections after the steps, to maximize the ionization probability.

random because carriers can ionize everywhere in the avalanche region, while in the staircase APD electrons ionize at well-defined positions in space (i.e. the multiplication process is more deterministic). Note that, similarly, in a photomultiplier tube the avalanche is essentially noise free ($F \cong 1$).

Finally, the low voltage operation of this device with respect to conventional APDs should be mentioned. For a five-stage detector and $\Delta E_c \cong E_{g1} \cong 1$ eV, the applied voltage required to achieve a gain $\cong 32$ is slightly greater than 5 V. A possible material system for the implementation of this device in the 1.3–1.6 μm region is HgCdTe. In a practical structure one should always leave an ungraded layer immediately after the step having a thickness of the order of a few ionization mean free paths ($\lambda_i \cong 50$–100 Å) to ensure that most electrons ionize near the step. This modified staircase is shown in fig. 22.

Note that although the staircase APD has not yet been implemented, Capasso et al. [36] have recently demonstrated experimentally an enhancement of the ionization rates ratio α/β ($\cong 8$) in an AlGaAs/GaAs quantum

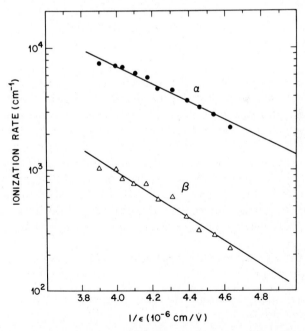

Fig. 23. Ionization rate for electrons, α, and for holes, β, for the AlGaAs/GaAs superlattice. (From ref. [36].)

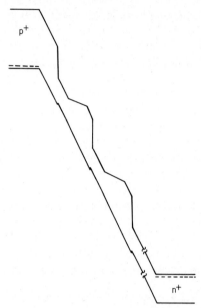

Fig. 24. Band-diagram of a superlattice avalanche photodiode with graded regions to eliminate carrier pile-up in the wells.

well superlattice (fig. 23). The effect has been attributed to the difference between the conduction- and valence-band discontinuities ($\Delta E_c > \Delta E_v$) and was first predicted by Chin et al. [37]. Thus electrons enter the well with a higher kinetic energy than holes and have a higher probability to ionize. Note that the staircase APD is the limiting case of this detector since the whole ionization energy is gained at the band discontinuity. An intermediate case between the superlattice quantum-well APD and the staircase APD is shown in fig. 24. Here we assume that the discontinuity ΔE_c is smaller than the ionization energy after the step. Thus it is necessary to furnish an extra energy to the electron by leaving an ungraded region before the step.

The staircase devices probably represent the best example of the band-gap engineering concept.

7.3.2. Impact-ionization across the band discontinuity: A new solid-state photomultiplier

Recently, we have observed a new avalanche phenomenon in superlattices, namely the impact-ionization across band discontinuities of carriers con-

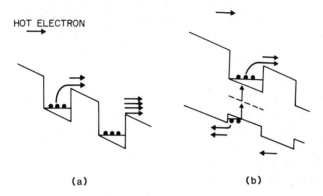

Fig. 25. Impact-ionization across the band-edge discontinuity. (a) Quantum wells are doped n-type. (b) Wells are undoped. Shown is the ionization across band discontinuities of carriers dynamically stored in the wells. These carriers originate from thermal generation processes via midgap centers.

fined in the wells [38]. This phenomenon, independently predicted by Chwang and Hess [39], could lead to a new type of solid-state PMT. This effect is illustrated in fig. 25a. Consider a multiple QW structure with n-type doped wells and undoped barriers. Electrons from the parent donors can remain confined in the wells even in the presence of a relatively strong electric field, provided the barriers are thick enough to minimize tunneling. Consider now a hot electron in a barrier layer. When it enters the well with sufficient energy, it can impact-ionize one of the bound carriers out of the well. In this ionization effect only one type of carrier is created so that the positive feedback of impact ionizing holes is eliminated, leading to the possibility of a quiet avalanche with small excess noise. Of course, in this case, one must constantly supply the electrons in the wells by applying suitable selective contacts to the well regions. From a conceptual point of view this effect has some similarities with the impact-ionization of deep levels in the sense that the QW may be treated as an artificial trap. It is important to point out that doped wells are not required for the observation of the effect. Due to the thermal generation of electrons and holes in the well layers (which gives rise to bulk dark current), relatively large electron and hole densities can be *dynamically* stored in the wells if the band discontinuities are appreciably larger than the average energies of the carriers in the wells and the dark current is relatively large. This situation may occur in the high-field region of certain QW p–i–n photodiodes, such as the ones investigated by us [38], and is illustrated in fig. 25b. These structures contain an $Al_{0.48}In_{0.52}As/Ga_{0.47}In_{0.53}As$ superlattice in the intrinsic region, with barrier and well thicknesses in the 100–500 Å range. A large ratio of the multiplications for holes and electrons was observed (M_h/M_e), implying that holes ionize at a significantly higher rate than the electrons in these structures. A similar effect has been found in p–i–n diodes containing AlSb/GaSb [38] and $InP/Ga_{0.47}In_{0.53}As$ superlattices [40]. A systematic study of the temperature and chopping frequency dependence of the multiplication showed conclusively that the observed effect is not a band-to-band process. Also deep-level ionization could be ruled out since it would require unrealistically large densities of such centers ($\cong 10^{17}$ cm^{-3}). Such densities are also in contrast with DLTS data, which in the case of $InP/Ga_{0.47}In_{0.53}As$ superlattice diodes, give an upper limit of $\approx 10^{14}$ cm^{-3}.

By appropriately grading the interface of the wells, the storage of electrons can be eliminated, while holes are still confined (fig. 26). This

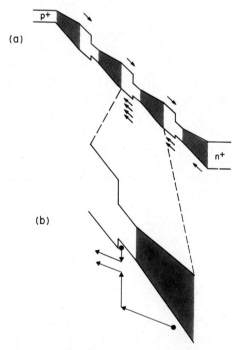

Fig. 26. (a) Band structure of a multiple graded well photomultiplier (the graded regions are shaded) showing multiplication of holes. The small multiplication of electrons is not shown. (b) Mechanism of hole multiplication by impact ionization across the band-edge discontinuity.

should maximize the ionization rates ratio, by minimizing electron-initiated multiplication. This structure was grown by MBE in the AlInAs/GaInAs system [41]. The structures consisted of a three-period superlattice, placed in the intrinsic region of a p–i–n photodiode, with 501 Å $Al_{0.48}In_{0.52}As$ barriers, 292 Å $Ga_{0.47}In_{0.53}As$ wells and 1022 Å AlInGaAs graded regions. These regions were grown linearly graded and lattice-matched to InP by computer-controlled MBE. For hole-initiated multiplication, avalanche gain occurs at a reverse bias of 7 V and reaches ≈ 20 at -12 V, at a temperature ≈ 100 K. For electrons, the multiplication is dominated by ionization across the band gap and is less than 1.4, resulting in a ionization rate ratio β/α $[\approx (M_h - 1)/(M_e - 1)]$ in excess of 50 (fig. 27). This is the highest value measured in a III–V material. Thus we have observed near single-car-

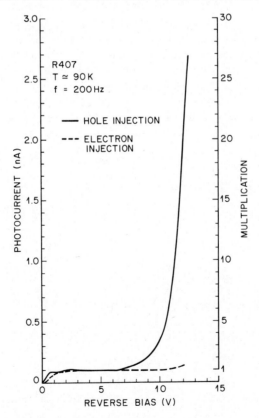

Fig. 27. Reverse bias photocurrent for the device of fig. 26 with 3 periods, under conditions of pure electron injection.

rier-type multiplication of holes by ionization over the discontinuity, with feedback provided by band-to-band ionization of electrons.

7.3.3. Repeated velocity overshoot

Another interesting application of staircase potentials has been proposed, the repeated velocity overshoot device [42]. This structure offers the potential for achieving average drift velocities well in excess of the maximum steady-state velocity over distances greater than 1 μ. Figure 28a shows a general type of staircase potential structure. The corresponding electric field, shown in fig. 28b, consists of a series of high-field regions of value E_1 and width d superimposed upon a background field E_0. To illustrate the electrical behavior and design considerations for a specific case, we consider electrons in the central valley of GaAs. The background field E_0 is chosen so that the steady-state electron energy distribution is not excessively broadened beyond its thermal equilibrium value, but at the same time the average drift velocity is still relatively high. For GaAs, an appropriate value would be around 2.5 kV/cm. At this field, the steady-state drift velocity is 1.8×10^7 cm/s and fewer than 2% of the electrons reside in the satellite valley. The electron distribution immediately downstream from the high field region is shifted to higher energy by an amount $\Delta W = E_1 d$. (Note that

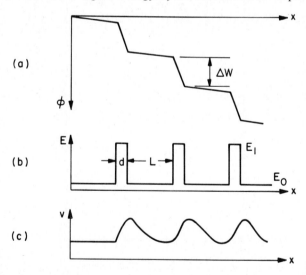

Fig. 28. Principle of repeated velocity overshoot staircase potential and corresponding electric field. The ensemble velocity as a function of position is also illustrated automatically.

while the distribution is shifted uniformly in energy, it is compressed in momentum in the direction of transport.) We choose d so that the transit time across the high-field region is shorter than the mean phonon scattering time, which is about 0.13 ps in GaAs. The energy step ΔW is chosen to maximize the average velocity of the distribution after the step while still keeping most of the distribution below the threshold energy from transfer to the satellite valley. In GaAs, the intervalley separation is about 0.3 eV, so an appropriate value of ΔW would be about 0.2 eV, resulting in an average velocity of approximately 1×10^8 cm/s immediately after the step. The momentum decays rapidly beyond the step due to scattering by polar optical phonons, with the result that the velocity decreases roughly linearly with distance, as shown in fig. 28c. During this time the distribution is broadened considerably in momentum. After the momentum (and velocity) have relaxed, the distribution requires additional time to relax to its original energy. Thus, the spacing L between the high-field regions must be large enough to allow sufficient cooling of the electron distribution before another overshoot can be attempted. This is necessary in order to avoid populating the high-mass satellite valleys. The effect of the resulting repeated velocity overshoot shown in fig. 28c is that average drift velocities greater than the maximum steady-state velocity can be maintained over relatively long distances.

A practical way to achieve this device with graded-gap materials is shown in fig. 29.

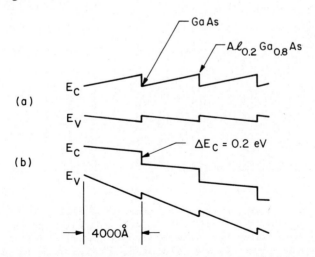

Fig. 29. Band diagram of a graded-gap repeated velocity overshoot device.

8. Superlattice band-gap grading and pseudo-quaternary alloys

The growth of graded-gap structures of very short period, represents a real challenge for the MBE crystal grower. Not only computer-controlled MBE system is required, but also new techniques to achieve such short distance compositional grading are necessary. One such technique is the recently introduced pulsed-beam method [43], by which, e.g., a variable-gap alloy, is grown by alternatively opening (and closing) the aluminum and gallium ovens with the shutters. The result is an AlAs/GaAs superlattice with ultrathin constant period (≈ 20 Å) but varying ratio of AlAs to GaAs layer thicknesses. The local band gap is therefore that of the alloy corresponding to the local average composition determined by the thicknesses of the AlAs and the GaAs. Since the period of the superlattice is smaller than the de Broglie wavelengths of the carriers, the material behaves basically like a *variable-gap ordered alloy*. Such techniques have been used recently to grow parabolic quantum wells [44] (fig. 30).

Another interesting example of superlattice alloys are the pseudo-quarternary materials introduced by Capasso et al. [45]. Such artificial structures are capable of conveniently replacing GaInAsP semiconductors in a variety of applications.

The concept of a pseudo-quaternary GaInAsP semiconductor is easily explained. Consider a multilayer structure of alternated $Ga_{0.47}In_{0.53}As$ and InP. If the layer thicknesses are sufficiently thin (typically a few tens of Å) one is in the superlattice regime. One of the consequences is that this novel

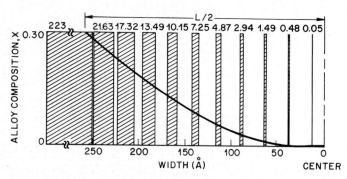

Fig. 30. Compositional structure of parabolic quantum well versus distance from the well center (only half of the well is shown). The parabolic compositional profile (solid line) is obtained by growing a superlattice of alternated $Al_{0.30}Ga_{0.70}As$ and GaAs layers (dashed and white regions respectively) of varying thicknesses. The numbers on the top portion of the figure are the thicknesses of the $Al_{0.30}Ga_{0.70}As$ layers.

material has now its own band gap, intermediate between that of $Ga_{0.47}In_{0.53}As$ and InP. In the limit of layer thicknesses of the order of a few monolayers the energy bandgap can be approximated by

$$E_g = \frac{E_g(Ga_{0.47}In_{0.53}As)\, L(Ga_{0.47}In_{0.53}As) + E_g(InP)\, L(InP)}{L(Ga_{0.47}In_{0.53}As) + L(InP)}, \quad (9)$$

where the L's are the respective layer thicknesses.

These superlattices can be regarded as novel pseudo-quaternary GaInAsP semiconductors. In fact, similarly to $Ga_{1-x}In_xAs_{1-y}P_y$ alloys, they are grown lattice-matched to InP and their band gap can be varied between that of InP and that of $Ga_{0.47}In_{0.53}As$. The latter is done by adjusting the ratio of the $Ga_{0.47}In_{0.53}As$ and InP layer thicknesses. Pseudo-quaternary GaInAsP is particularly suited to replace variable-gap $Ga_{1-x}In_xAs_{1-y}P_y$. Such alloys are very difficult to grow since the mole fraction x (or y) must be continuously varied while maintaining lattice-matching to InP.

Figure 31a shows a schematic of the energy band diagram of undoped (nominally intrinsic) graded-gap pseudo-quaternary GaInAsP. The structure consists of alternated ultrathin layers of InP and $Ga_{0.47}In_{0.53}As$ and was grown by a new vapor phase epitaxial growth technique (levitation epitaxy) [46]. Other techniques such as molecular beam epitaxy or metallorganic chemical vapor deposition may also be suitable to grow such superlattices. From fig. 31a it is clear that the duty factor of the InP and $Ga_{0.47}In_{0.53}As$ layer is gradually varied, while keeping constant the period of the superlattice. As a result the average composition and band gap (dashed lines in fig. 31a) of the material are also spatially graded between the two extreme points (InP and $Ga_{0.47}In_{0.53}As$). In our structure both ten and twenty periods (1 period = 60 Å) were used. The InP layer thickness was linearly decreased with distance from $\cong 50$ Å to $\cong 5$ Å while correspondingly increasing the $Ga_{0.47}In_{0.53}As$ thickness to keep the superlattice period constant ($= 60$ Å).

The graded-gap superlattice was incorporated in a long-wavelength $InP/Ga_{0.47}In_{0.53}As$ avalanche photodiode, as shown in fig. 31b. This device is basically a photodetector with separate absorption ($Ga_{0.47}In_{0.53}As$) and multiplication (InP) layers and a high–low electric field profile (HI-LO SAM APD). This profile (fig. 31c) is achieved by a thin doping spike in the ultralow doped InP layer and considerably improves the device performance compared to conventional SAM APDs [47]. The $Ga_{0.47}In_{0.53}As$ absorption layer is undoped ($n \approx 1 \times 10^{15}$ cm^{-3}) and 2.5 μm thick. The n$^+$ doping spike thickness and carrier concentration were varied in the 500–200

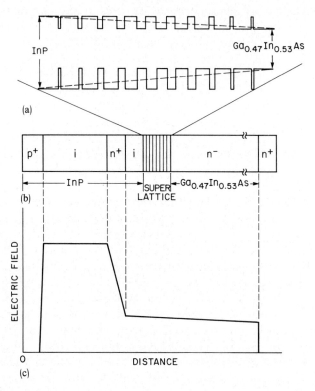

Fig. 31. (a) Band diagram of a pseudo-quaternary graded-gap semiconductor. The dashed lines represent the average band gap seen by the carriers. (b) and (c) are a schematic and the electric field profile of a high–low avalanche photodiode using the pseudo-quaternary layer to achieve high speed.

Å and $5 \times 10^{17} - 10^7$ cm^{-3} ranges, respectively (depending on the wafer), while maintaining the same carrier sheet density ($\cong 2.5 \times 10^{12}$ cm^{-2}). The n^+ spike was separated from the superlattice by an undoped 700–1000 Å thick InP spacer layer. The p^+ region was defined by Zn diffusion in the 3 µm thick low carrier density ($n^- \approx 10^{14}$ cm^{-3}) InP layer. The junction depth was varied from 0.8 to 2.5 µm.

Similar devices, but without the superlattice region, were also grown.

Previous pulse response studies of conventional SAM APDs with abrupt InP/Ga$_{0.47}$In$_{0.53}$As heterojunctions found a long (> 10 ns) tail in the fall time of the detector due to the pile-up of holes at the heterointerface [48]. This is caused by the large valence-band discontinuity ($\cong 0.45$ eV). It has

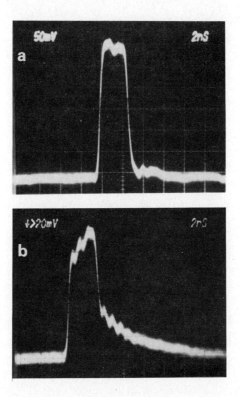

Fig. 32. Pulse response of a high–low avalanche detector with pseudo-quaternary layer (a) and without pseudo-quaternary layer (b) to a 2 ns $\lambda = 1.55$ μm laser pulse. The bias voltage is -65.5 V for both devices. Time scale 2 ns/div.

been proposed that this problem can be eliminated by inserting between the InP and $Ga_{0.47}In_{0.53}As$ region a $Ga_{1-x}In_xAs_{1-y}P_y$ layer of intermediate band gap [49]. This quaternary layer is replaced, in our structure, by the InP/$Ga_{0.47}In_{0.53}As$ variable-gap superlattice. This not only offers the advantage of avoiding the growth of the critical, independently lattice-matched GaInAsP quaternary layer, but also may lead to an optimum "smoothing out" of the valence-band barrier for reproducible high speed operation. This feature is essential for HI–LO SAM APDs since the heterointerface electric field is lower than in conventional SAM devices.

For the pulse response measurement we used a 1.55 μm GaInAsP driven by a pulse pattern generator. Figure 32 shows the response to a 2 ns laser pulse of a HI–LO SAM APD with (a) and without (b) a 1300 Å thick superlattice. Both devices had similar doping profiles and breakdown voltage ($\cong 80$ V) and were biased at -65.5 V. At this voltage the ternary layer was completely depleted in both devices and the measured external quantum efficiency $\cong 70\%$. The results of fig. 32 were reproduced in many devices on several wafers. The long tail in fig. 32b is due to the pile-up effect of holes associated with the abruptness of the heterointerface.

In the devices with the graded-gap superlattice (fig. 32a) there is no long tail. In this case the height of the barrier seen by the holes is no more the valence-band discontinuity ΔE_v but

$$\Delta E = \Delta E_v - e\epsilon_i L, \tag{11}$$

where ϵ_i is value of the electric field at the InP/superlattice interface and L the thickness of the pseudo-quaternary layer.

The devices are biased at voltages such that $\epsilon_i > \Delta E_v/eL$ so that $\Delta E = 0$ and no trapping occurs.

In the devices with no superlattice instead $\Delta E = \Delta E_v$ for every ϵ_i so that long tails in the pulse response are observed at all voltages.

9. Doping interface dipoles: tunable band-edge discontinuities

It is clear from the material presented in the previous sections, that band discontinuities and, in general, barrier heights, play a central role in the design of novel heterojunction devices. For example their knowledge is absolutely essential for devices such as multilayer APDs and HBTs. If a technique were available to artificially and controllably vary band offsets and barrier heights at abrupt heterojunctions, this would give the device physicist tremendous flexibility in device design, as well as many novel opportunities.

Recently Capasso et al. [50,51] demonstrated for the first time that barrier heights and band discontinuities at an abrupt, intrinsic heterojunction can be tuned via the use of a doping interface dipole (DID) grown by MBE.

Compositional grading at the interface is an effective way to control barrier heights, but eliminates the abruptness of the heterojunction. In many cases one would like to preserve it while simultaneously being able to tune the barrier height. This concept is illustrated in fig. 33. The left-hand side of fig. 33 represents the conduction-band diagram of an abrupt heterojunction. The material is assumed to be undoped (ideally intrinsic) so that we can neglect band-bending effects over the short distance (a few hundred Å) shown here.

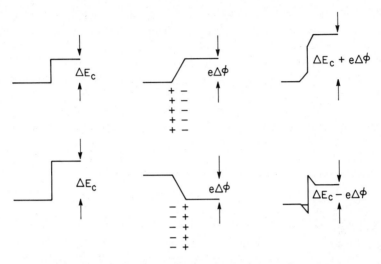

Fig. 33. Tunable band discontinuities formed from doping interface dipoles: (Top). The conduction-band discontinuity is increased. (Bottom) Interchange of the acceptor and donor sheets reduces the band discontinuity. Tunneling through the spike and size quantization in the triangular well plays a key role in the reduction.

We next assume to introduce in situ, during the growth of a second identical heterojunction, one sheet of acceptors and one sheet of donors, of identical doping concentrations, at the same distance $\frac{1}{2}d$ (≤ 100 Å) from the interface (fig. 33, top). The doping density N is in the 1×10^{17}–1×10^{19} cm^{-3} range, while the sheets' thickness t is kept small enough so that both are depleted of carriers ($t \leq 100$ Å). In fact, one monolayer thick doping sheets can be grown by MBE using the impurity growth mode [52]. The DID is therefore a microscopic capacitor. The electric field between the plates is σ/ϵ, where $\sigma = eNt$. There is a potential difference $\Delta\Phi = (\sigma/\epsilon)d$ between the two plates of the capacitor. Thus the DID produces abrupt potential variations across a heterojunction interface by shifting the relative positions of the valence and conduction bands in the two semiconductors outside the dipole region. This is done without changing the electric field outside the DID.

The conduction-band barrier height at the heterojunction is increased by the DID to a value $\Delta E_v + e\Delta\Phi + (\sigma/\epsilon)t$. If $\Delta\Phi$ is dropped over a distance of a few atomic layers and the total potential drop across the charge sheets $[= (\sigma/\epsilon)t]$ is small compared to $\Delta\Phi$, the conduction-band discontinuity has effectively been increased by $e\Delta\Phi$ (fig. 33, top). By interchanging the position of donors and acceptors the DID can be made to reduce the conduction band discontinuity (fig. 33, bottom). On the low-gap side of the heterojunction a triangular quantum well is formed. Since, typically, the electric field in this region is $\geq 10^5$ V/cm and $e\Delta\Phi \approx 0.1$–0.2 eV, the bottom of the first quantum subband E_1 lies near the top of the well. Therefore the thermal activation barrier seen by an electron on the low-gap side of the heterojunction is $\cong \Delta E_c - \frac{1}{2}e\Delta\Phi$. Electrons can also tunnel through the thin (≤ 100 Å) triangular barrier, this further reduces the effective barrier height. In the limit of a DID a few atomic layers thick and of potential $\Delta\Phi$ the triangular barrier is totally transparent and the conduction-band discontinuity is lowered to $\Delta E_c - e\Delta\Phi$.

Note that experimental evidence suggests that "natural" dipoles may occur at polar heterojunction interfaces causing the orientation dependence of band discontinuities, as discussed in the contribution by Grant et al. (ch. 4).

To verify the barrier lowering due to the DID, heterojunction AlGaAs/GaAs p–i–n diodes on p-type (100)GaAs substrates were grown by MBE. Two types of structures were grown: one with and the other without dipole. The one with dipole consists of four GaAs layers: the first with $p^+ > 10^{18}$ cm^{-3} (5000 Å), the second one is undoped (5000 Å); the third one with $p^+ = 5 \times 10^{17}$ cm^{-3} (100 Å), forming the negatively charged sheet of the dipole, and the fourth one is undoped (100 Å); followed by four $Al_{0.26}Ga_{0.74}As$ layers; the first one is undoped (100 Å); the second one with $n^+ = 5 \times 10^{17}$ cm^{-3} (100 Å), forming the positively charged sheet of the dipole; the third one is undoped (5000 Å); and the fourth one with $n^+ \geq 10^{18}$ cm^{-3} (5000 Å). The second type of structure is identical, with the exception that it does not have DID. They were grown consecutively in the MBE chamber without breaking the vacuum to ensure virtually identical growth conditions. Beryllium was used for the p-type dopant and silicon for the n-type. The substrate temperature was held at 590°C during growth. The background doping of the undoped layers is $\leq 10^{14}$ cm^{-3}. The charged sheets were introduced by controlling the aperture of the shutters of the ovens, without interrupting the growth of the GaAs and AlGaAs layers. This minimizes the formation of defects in the interface region.

The solid and dashed lines in fig. 34a are, respectively, the band diagram of the diodes at zero applied bias, with and without dipole (not to scale). In the structure with the DID the electric field inside the dipole layer is strongly increased while it is slightly ($\cong 10\%$) decreased outside the dipole (compared to the structure without dipole) since the potential drop across the depleted intrinsic layer is identical to that of the diodes without dipole.

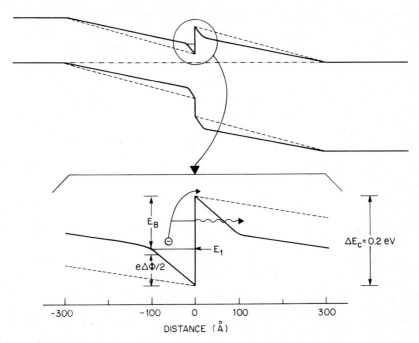

Fig. 34. (a) Solid and dashed lines represent, respectively, the band diagram of the p–i–n diodes with and without interface dipole (not to scale). (b) Band diagram of the conduction band near the heterointerface of the diodes with and without dipole (to scale).

Figure 34b gives the conduction-band diagram (to scale) near the interfaces for the cases with and without dipole. The potential of the dipole is $\Delta\Phi = 0.14$ V; for ΔE_c we have used the value 0.2 eV, following the new band line-ups for AlGaAs/GaAs [5]. The barrier height E_b is about a factor of 2 smaller than in the case without dipole ($\approx \Delta E_c$).

We have measured the photocollection efficiency of the two structures; light chopped at 1 kHz and incident on the AlGaAs side of the diode was used and the short-circuit photocurrent was measured with a lock-in. Absolute efficiency data were obtained by comparing the photoresponse to that of a calibrated Si photodiode. In fig. 35 we have plotted the external quantum efficiency η as a function of wavelength for devices with and without dipole. In the ones without dipole, η is very small ($\leq 2\%$) for $\lambda \gtrsim 7100$ Å; this wavelength corresponds to the band gap of the $Al_{0.26}Ga_{0.74}As$ layer as determined by photoluminescence measurements. At wavelengths longer than this and shorter than ≈ 8500 Å, photons are

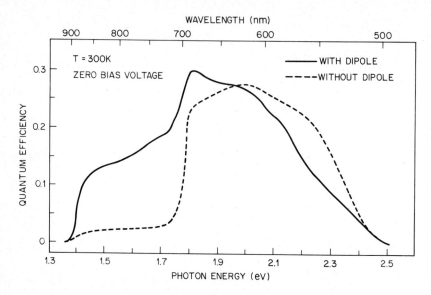

Fig. 35. External quantum efficiency of the heterojunctions with and without dipole at zero bias versus photon energy. Illumination is from the wide-gap side of the heterojunction.

absorbed partly in the GaAs electric field region and partly in the p$^+$ GaAs layer within a diffusion length from the depletion layer. Thus most of the photoinjected electrons reach the heterojunction interface and have to surmount the heterobarrier of height $\Delta E_c = 0.2$ eV to give rise to a photocurrent. Thermionic emission limits therefore the collection efficiency, which is proportional to $\exp[(-\Delta E_c)/kT]$. This explains the low efficiency for $\lambda > 7100$ Å since ΔE_c is significantly greater than kT.

For $\lambda < 7100$ Å the light is increasingly absorbed in the AlGaAs as the photon energy increases and the quantum efficiency becomes much larger than for $\lambda > 7100$ Å since most of the photocarriers do not have to surmount the heterojunction barrier to be collected. For $\lambda < 6250$ Å the quantum efficiency decreases since losses due to recombination of photogenerated holes in the n$^+$ AlGaAs layer and to surface recombination start to dominate [53]. The above behavior of the efficiency is consistent with predictions for abrupt AlGaAs/GaAs heterojunctions without interface charges.

The solid curve in fig. 35 is the photoresponse in the presence of the DID. A striking difference is noted compared to the case without dipole. While the quantum efficiencies for $\lambda \lesssim 7100$ Å are comparable, at longer

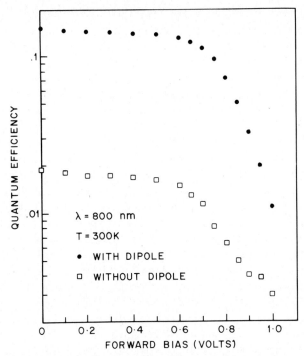

Fig. 36. External quantum efficiency at λ = 8000 Å versus forward bias voltage of the diodes with and without dipoles.

wavelengths it is enhanced by a factor as high as one order of magnitude in the structures with dipoles. This effect was reproduced in four sets of samples.

The physical interpretation is simple. The barrier height E_b has been lowered by ≅ 87 meV (fig. 34), which enhances thermionic emission across the barrier. Tunneling through the thin triangular barrier and hot-electron effects due to the smaller reflection coefficient also contribute to the enhanced collection efficiency.

Figure 36 shows the quantum efficiency versus forward bias at λ ≅ 800 Å for the two structures. It decreases first gradually and then rapidly above a certain cutoff voltage. This behavior, due to band-flattening, is well-known and has been observed previously. The efficiency rapidly increased with reverse voltage in both structures and then saturated. Above 10 V the quantum efficiency in the energy range 1.5–1.7 eV was identical in both structures and ≅ 60%. This is expected, since at fields $> 10^5$ V/cm the electrons acquire so much energy that the barrier height is no more a significant limiting factor to the efficiency.

The smallest size of the DID depends on the diffusion coefficient of the dopants, which depends on the doping density, the substrate temperature and the growth time. For Si and Be in the AlGaAs/GaAs system one should be able to place the doping sheets as close as 10 Å, for substrate temperatures $\leq 600\,°\mathrm{C}$ and growth times of $< \frac{1}{2}$hr without significant interdiffusion.

An important application of DIDs is the enhancement of impact-ionization of one type of carrier in superlattice and staircase avalanche photodiodes. Introducing DIDs at the band steps of these detectors can be used to further enhance the ionization probability of electrons at the steps, since carriers gain ballistically the dipole potential energy in addition to the energy ΔE_c obtained from the band step. In addition, the DID helps by promoting over the valence-band barrier holes created near the step by electron impact-ionization without sacrificing speed. Thus the dipole energy $e\Delta\Phi$ should equal or exceed the valence-band barrier.

In conclusion, DIDs represent a technique to effectively tune barrier heights and band discontinuities and as such have great potential for heterojunction devices. Another technique to tune band offsets using interlayers is discussed in the chapter by Margaritondo and Perfetti (ch. 2).

In conclusion, in this chapter we have discussed in detail the band-gap engineering approach. The key in this approach to designing microstructures and devices is the ability to model the energy-band diagrams of semiconductor structures. With these models, scientists can visualize the behavior of electrons and holes in a device. Variable-gap materials, superlattices, band discontinuities, and doping variations can be used along or in combination to modify the energy bands almost arbitrarily and to tailor these bands for a specific application.

Acknowledgments

It is a pleasure to acknowledge the many colleagues that have collaborated with the author: A.C. Gossard, A.Y. Cho, W.T. Tsang, H.M. Cox, R.J. Malik, S. Sen, B.F. Levine, J.A. Cooper, K.K. Thornber, S. Luryi, G.F. Williams, C.G. Bethea, A.. Hutchinson and R.A. Kiehl. R.C. Miller kindly supplied figure 30.

References

[1] A.Y. Cho and J.R. Arthur, Prog. Solid State Chem. 10 (1975) 157;
A.Y. Cho, Thin Solid Films 100 (1983) 291.
[2] F. Capasso, J. Vac. Sci. & Technol. B 1 (1983) 457;
F. Capasso, Science 235 (1987) 172.
[3] For recent reviews on band-gap engineering see:
F. Capasso, in: Gallium Arsenide Technology, ed. D.K. Ferry (Sams, Indianapolis, IN, 1985) ch. 8;
F. Capasso, in: Picoscecond Electronics and Optoelectronics, eds. G. Mourou, D.M. Bloom and C.H. Lee (Springer, Berlin, 1985) p. 112.
[4] H. Kroemer, RCA Rev. 18 (1957) 332.
[5] R.C. Miller, D.A. Kleinman, A.C. Gossard and O. Munteanu, Phys. Rev. B 29 (1984) 7085.
See also the reviews by G. Duggan (ch. 5) and S.R. Forrest (ch. 8) in this volume.
[6] J.A. Hutchby, J. Appl. Phys. 49 (1978) 4041.
[7] B.F. Levine, W.T. Tsang, C.G. Bethea and F. Capasso, Appl. Phys. Lett. 41 (1982) 470.
[8] B.F. Levine, C.G. Bethea, W.T. Tsang, F. Capasso, K.K. Thornber, R.C. Fulton and D.A. Kleinman, Appl. Phys. Lett. 42 (1983) 769.
[9] F. Capasso, W.T. Tsang, C.G. Bethea, A.L. Hutchinson and B.F. Levine, Appl. Phys. Lett. 42 (1983) 93.
[10] H. Kroemer, J. Vac. Sci. & Technol. B 1 (1983) 126.
[11] J.R. Hayes, F. Capasso, A.C. Gossard, R.J. Malik and W. Wiegmann, Electron. Lett. 19 (1983) 410.
[12] D.L. Miller, P.M. Asbeck, R.J. Anderson and F.H. Eisen, Electron. Lett. 19 (1983) 367.
[13] R.J. Malik, F. Capasso, R.A. Stall, R.A. Kiehl, R. Wunder and C.G. Bethea, Appl. Phys. Lett. 46 (1985) 600.
[14] O. Nakajma, K. Nagato, Y. Yamaguchi, H. Ito and T. Ishibashi, IEDM Technical Digest 1986, p. 266.
[15] R.J. Malik, J.R. Hayes, F. Capasso, K. Alavi and A.Y. Cho, IEEE Electron Devices Lett. EDL-4 (1983) 383.
[16] R. People, K.W. Wecht, K. Alavi and A.Y. Cho, Appl. Phys. Lett. 43 (1983) 118.
[17] J.R. Hayes, F. Capasso, R.J. Malik, A.C. Gossard and W. Wiegmann, Appl. Phys. Lett. 43 (1983) 949.
[18] B. Ricco and M.Ya. Azbel, Phys. Rev. B 29 (1984) 1970.
[19] F. Capasso and R.A. Kiehl, J. Appl. Phys. 58 (1985) 1366.
[20] N. Yokoyama, K. Imamura, S. Muto, S. Hiyamizu and H. Nishi, Jpn. J. Appl. Phys. 24 (1985) L583.
[21] F. Capasso, S. Sen, A.C. Gossard, A.L. Hutchinson and J.E. English, IEEE Electron Device Lett. EDL-7 (1985) 573.
[22] J.F. Palmier, G. Minot, J.L. Lievin, F. Alexander, J.C. Harmand, J. Dangla, C. Dubon-Chevallier and D. Ankri, Appl. Phys. Lett. 49 (1986) 1260.
[23] F. Capasso, K. Mohammed, A.Y. Cho, R. Hull and A.L. Hutchinson, Appl. Phys. Lett. 47 (1985) 420.
[24] F. Capasso, K. Mohammed, A.Y. Cho, R. Hull and A.L. Hutchinson, Phys. Rev. Lett. 55 (1985) 1152.
[25] P.J. Price, IEEE Trans. Electron Devices ED-28 (1981) 911.
[26] C.L. Allyn, A.C. Gossard and W. Wiegmann, Appl. Phys. Lett. 36 (1980) 373.

[27] A.C. Gossard, W. Brown, C.L. Allyn and W. Wiegmann, J. Vac. Sci. & Technol. 20 (1982) 694.
[28] F. Capasso, S. Luryi, W.T. Tsang, C.G. Bethea and B.F. Levine, Phys. Rev. Lett. 51 (1983) 2318.
[29] F. Capasso, G.F. Williams and W.T. Tsang, Tech. Digest IEEE Specialist Conf. on Light Emitting Diodes and Photodetectors, Ottawa-Hull (IEEE, New York, 1986) p. 166.
[30] F. Capasso and W.T. Tsang, Tech. Digest Int. Electron Devices Meeting, Washington, D.C., pp. 334.
[31] G.F. Williams, F. Capasso and W.T. Tsang, IEEE Electron Device Lett. EDL-3 (1982) 284.
[32] F. Capasso, IEEE Trans. Nucl. Sci. NS-30 (1983) 424.
[33] F. Capasso, Surf. Sci. 132 (1983) 527.
[34] F. Capasso, W.T. Tsang and G.F. Williams, IEEE Trans. Electron Devices ED-30 (1983) 381.
[35] R.J. McIntyre, IEEE Trans. Electron Devices ED-13 (1966) 164.
[36] F. Capasso, W.T. Tsang, A.L. Hutchinson and G.F. Williams, Appl. Phys. Lett. 40 (1982) 38–40.
[37] R. Chin, K.N. Holonyak Jr, G.E. Stillman, J.T. Tang and K. Hess, Electron. Lett. 16 (1980) 467.
[38] F. Capasso, J. Allam, A.Y. Cho, K. Mohammed, R.J. Malik, A.L. Hutchinson and D. Sivco, Appl. Phys. Lett. 48 (1986) 1294.
[39] S.L. Chwang and K. Hess, J. Appl. Phys. 48 (1986) 2885.
[40] J. Allam, F. Capasso, M.B. Panish and A.L. Hutchinson, (1986) unpublished.
[41] J. Allam, F. Capasso, K. Alavi and A.Y. Cho, IEEE Electron Device Lett. EDL-81 (1987) 2013.
[42] J.A. Cooper Jr, F. Capasso and K.K. Thornber, IEEE Electron Device Lett. EDL-3 (1982) 402.
[43] M. Kawabe, N. Matsuuza and H. Inuzuka, Jpn. J. Appl. Phys. 21 (1982) L447.
[44] R.C. Miller, D.A. Kleinman and A.C. Gossard, Phys. Rev. B 29 (1984) 7085.
[45] F. Capasso, H.M. Cox, A.L. Hutchinson, N.A. Olsson and S.G. Hummel, Appl. Phys. Lett. 45 (1984) 1193.
[46] H.M. Cox, J. Cryst. Growth 69 (1984) 641.
[47] F. Capasso, A.Y. Cho and P.W. Foy, Electron. Lett. 20 (1984) 635.
[48] S.R. Forrest, O.K. Kim and R.G. Smith, Appl. Phys. Lett. 41 (1982) 95.
[49] J. Campbell, A.G. Dentai, W.S. Holden and B.L. Kasper, Electron. Lett. 19 (1983) 818.
[50] F. Capasso, A.Y. Cho, K. Mohammed and P.W. Foy, Appl. Phys. Lett. 46 (1985) 664.
[51] F. Capasso, K. Mohammed and A.Y. Cho, J. Vac. Sci. & Technol. B 3 (1985) 1245.
[52] See, for example, K. Ploog, A. Fischer and E.F. Schubert, Surf. Sci. 174 (1986) 120.
[53] S.F. Womac and R.H. Rediker, J. Appl. Phys. 43 (1972) 4129.

CHAPTER 11

MODERN ASPECTS OF HETEROJUNCTION TRANSPORT THEORY

K. HESS

*Department of Electrical and Computer Engineering
and the Coordinated Science Laboratory
University of Illinois, Urbana, IL 61801, USA*

and

G.J. IAFRATE

*US Army Electronics Technology and Devices Laboratory
Fort Monmouth, NJ 07703, USA*

*Heterojunction Band Discontinuities: Physics and Device Applications
Edited by F. Capasso and G. Margaritondo
© Elsevier Science Publishers B.V., 1987*

Contents

1. Introduction 453
2. Heterojunction theory pertinent to electronic transport 453
   2.1. Electronic states and the idealized heterojunction 453
   2.2. The self-consistent field at the interface 455
3. Transport parallel to heterointerfaces 457
   3.1. Elements of the theory of electron scattering at interfaces ... 457
   3.2. The low-field mobility for transport parallel to interfaces .. 460
   3.3. High-field transport parallel to interfaces 461
      3.3.1. General aspects 461
      3.3.2. Transient transport in a many-subband system 462
      3.3.3. Classical transport in the presence of confining (transverse) electric fields ... 465
4. Transport perpendicular to the interface 468
   4.1. General analytical considerations 468
   4.2. Hot-electron thermionic emission – real-space transfer (RST) .. 470
   4.3. Aspects of tunneling perpendicular to heterolayers 471
   4.4. Transport over heterolayer wells 473
      4.4.1. Collection of carriers into heterolayer wells 474
      4.4.2. Propagation of carriers over the base of planar-doped barrier transistors . 475
5. Transport in heterolayers devices 478
   5.1. Quantum effects and the high electron mobility transistor (HEMT) 478
   5.2. Discussion of selected device schemes illustrating the physics of parallel and perpendicular nonlinear transport 483
References 486

1. Introduction

This review of electronic transport parallel and perpendicular to heterolayers is an extension of three previous reviews [1–3] and includes some of the more recent developments on ballistic transport, modulation doping and real-space transfer. These subjects are, therefore, only briefly discussed and the assumption is made that the reader is familiar with the basic concepts. One particular topic, the role of built-in and transverse electric fields, is treated in more detail and is discussed from a more basic point of view. These effects are fundamental to nonlinear transport at interfaces and form the theoretical basis for effects such as the ionization enhancement in superlattice avalanche photodiodes. On the other hand, some of the effects of transverse fields on parallel transport can be identified as real-space transfer effects.

In reviewing the literature, it is particularly gratifying to notice that the field of nonlinear transport at interfaces has grown beyond the high expectations which we have expressed in the previous reviews [2,3]. The high electron mobility transistor [4] and quantum-well laser [5] are firmly established and show more than great promise from the applied viewpoint. Real-space transfer transistors [6,7] and superlattice avalanche photodiodes [9] are currently being successfully fabricated and have revealed interesting physical effects. The general use of heterolayers to create boundary conditions on quantum levels is continuing without relaxation to add many unexpected effects which have applications in new forms of devices.

2. Heterojunction theory pertinent to electronic transport

2.1. Electronic states and the idealized heterojunction

In this chapter we idealize the junction by assuming an abrupt change of conduction- and valence-band edge (the conduction/valence-band discontinuity) and also an abrupt change in the dielectric constant. The electrons on each side are treated within the effective mass approximation. At the junction we use the following connection rules [3]:

(i) For the envelope wavefunctions ψ,

$$\psi^L = \psi^R, \quad \text{and} \tag{2.1}$$

$$\frac{\partial \psi^L}{\partial z} \approx \frac{m_L^*}{m_R^*} \frac{\partial \psi^R}{\partial z}. \tag{2.2}$$

We follow throughout this chapter the convention of labelling quantities for the left-hand side of the junction with L and for the right-hand side with R. The direction perpendicular to the junction is the z-direction. m^* is the effective mass and eq. (2.2) follows from the continuity of the current.

(ii) For the electrostatic potential and field: we assume that the band edge discontinuities are predetermined. The external potential just shifts energy bands rigidly. It is often convenient to introduce a vacuum reference energy which is the electron energy (at rest) outside the semiconductor (metal) and to measure the band edge energies from there. We will do this in several instances. Given these facts we can connect the left side ϕ_i^L and right side ϕ_i^R of the interface potential ϕ_i by

$$\phi_i^R - \phi_i^L = \Delta E_c/e, \quad \text{and} \tag{2.3}$$

$$\epsilon_L \partial \phi_i^L / \partial z = \epsilon_R \partial \phi_i^R / \partial z. \tag{2.4}$$

Notice that eq. (2.4) applies only in the absence of an interface charge Q_i which would add a term Q_i/ϵ_i where ϵ_i is the dielectric constant at the interface, i.e.

$$\epsilon_i \approx \tfrac{1}{2}(\epsilon_L + \epsilon_R). \tag{2.5}$$

It is important to notice that the above idealizations are only of limited value. Consider for example the heterojunction of $GaAs-Al_xGa_{1-x}As$. The potential around the various atoms at the interface can schematically be represented by a configuration as shown in fig. 1. Since the aluminum replaces the Ga randomly, a clear line of an interface cannot be drawn. The amount of random fluctuation will, in addition, be dependent on the aluminum mole fraction x. Material parameters and additional details can be found in refs. [10–15].

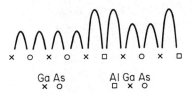

Fig. 1. Schematic representation of atomic potentials at a binary–ternary interface.

2.2. The self-consistent field at the interface

The solution for the potential and field at each side of the interface is by now standard procedure. Because of the band edge discontinuity, electrons transfer to the material with a lower conduction-band edge and accumulate there, leaving a depletion layer behind. This is shown in fig. 2. The solution is usually obtained separately for the accumulation and depletion and then matched by eqs. (2.3) and (2.4). The solution of the Poisson equation for the depletion region is trivial while the accumulation or inversion layers (depending on the type of conduction in the GaAs) present a formidable problem. This problem has been solved by Stern and Howard [10] as well as Ando [11] with great precision including exchange correlation effects. Because of size quantization effects the electron energy levels are grouped

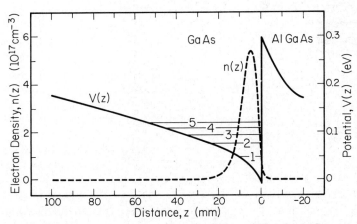

Fig. 2. Subband energies (1–5), interface potential (including exchange correlation effects) and electron density at a GaAs–Al$_{0.3}$Ga$_{0.7}$As interface at room temperature. (After Yokoyama and Hess [13].)

into subbands and the Poisson equation needs to be solved self-consistently with the Schrödinger equation which, according to Stern and Das Sarma [12], can be accomplished as described below.

The wavefunctions parallel to the heterointerface (xy-plane) are assumed to be plane waves. The envelope function normal to the layer interface (z-direction) $F_m(z)$ for the mth subband satisfies the following Schrödinger equation.

$$-\frac{\hbar^2}{2m^*}\frac{d^2 F_m(z)}{dz^2} + V(z)\,F_m(z) = E_m F_m(z). \tag{2.6}$$

The effective potential $V(z)$ is given by

$$V(z) = -e\phi_e(z) + V_h(z) + V_{xc}(z), \tag{2.7}$$

where $\phi_e(z)$ is the electrostatic potential given by the solution of eq. (2.8) below, $V_h(z)$ is the step function describing the interface barrier, and $V_{xc}(z)$ is the local exchange correlation potential. The Poisson equation reads

$$\frac{d^2\phi_e(z)}{dz^2} = \frac{e}{\epsilon_0\epsilon}\left[\sum_{i=1}^{n} N_i F_i^2(z) + N_A(z) - N_D(z)\right], \tag{2.8}$$

where N_i represents the number of electrons in subband i and is given, in equilibrium, by

$$N_i = \frac{m^* k_B T}{\pi \hbar^2} \ln\left[1 + \exp\left(\frac{E_F - E_i}{k_B T}\right)\right], \tag{2.9}$$

$N_A(z)$ and $N_D(z)$ are the position-dependent acceptor and donor concentrations, and E_F is the Fermi energy.

The temperature appearing in eq. (2.9) is the electron temperature which can be different from the temperature of the crystal lattice, e.g. if the electrons are accelerated by strong electric fields. It is well-known that an electron temperature is not always well-defined [1]. Equation (2.9) needs then to be recalculated for a general distribution function. This complicated procedure has never been achieved to the authors' knowledge. We will later use results based on an approximation which calculates the average energy of the electrons (which always can be done) and then relates the temperature to the average energy $\langle E \rangle$ by

$$\langle E \rangle = \frac{1}{8\pi^3} \int \frac{E}{\exp[(E-E_F)/kT]+1} g(E) \, dE, \tag{2.10}$$

where $g(E)$ is the density of states function.

An iterative scheme is usually used to solve the above-described system of eqs. (2.7)–(2.9). The Schrödinger equation is solved by the Numerov method [13]. The initial conditions of the binding energies and the corresponding wavefunctions can be calculated analytically for the triangular potential approximation, where the wavefunctions are expressed by Airy functions. The effective potential calculated from eq. (2.7) is then updated by iteration [13]. The result of this calculation, including five subbands, is shown in fig. 2 for the GaAs–AlGaAs system at room temperature.

3. Transport parallel to heterointerfaces

Electronic transport in semiconductors parallel to interfaces has been treated in the literature, first in connection with the metal-oxide–semiconductor system and later for lattice-matched semiconductor–semiconductor heterolayers. The scattering mechanisms are modified from the three-dimensional scattering in bulk semiconductors by the confinement of the electrons at the interface (their quasi-two-dimensionality) and their propagation in electronic subbands. Scattering agents can be located at some distance from the interface and the scattering mechanisms therefore frequently carry the name "remote". Most important is remote impurity scattering in modulation doped structures. Details can be found in several reviewers' articles [2,3]. Here we describe only some of the most recent and most complete results [13].

3.1. Elements of the theory of electron scattering at interfaces

Interactions of electrons (holes) with phonons at interfaces have been treated in a substantial number of papers [16–19]. The results can be summarized as follows. At interfaces, the confined electrons can interact with interface or bulk-phonons. Bulk-like phonons may still exist at interfaces which are electronically different enough to confine electrons. Even if

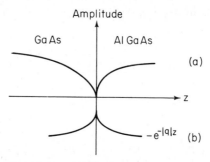

Fig. 3. Schematic representation of phonon modes in two half-spaces decaying exponentially in the neighboring medium and decreasing toward the interface (a) and an interface mode decreasing as $\exp(-|q|z)$ where q is the phonon wavevector parallel to the interface (b).

the two materials are different from a phonon point of view and interface modes are important, the bulk-phonon approximation may still be reasonable for reasons illustrated in fig. 3. The form of the phonon-mode amplitudes at the interface suggests that the sum of all modes may look bulk-like. This in turn may be a reason for the success of the theories which treat the phonons bulk-line in lattice-matched semiconductor heterolayers [16]. A word of caution is in order, however, since the form of the electronic wavefunctions may give preference to certain phonon modes when calculating the matrix-element for the scattering probability (e.g. for scattering between different subbands [20]).

Other changes in the electron–phonon interaction due to the presence of the interface are expected to be small. The deformation-potential constant will change only a few percent [2] and so will the polar and piezoelectric coupling coefficients. Experimental proof of these theoretical predictions is complicated, however, and there are some indications in the literature that the interface deformation potential could differ significantly from the bulk deformation potential. A still unsolved problem is the dynamic screening of the electron–phonon interaction by the quasi-two-dimensional electrons. In view of the high accumulation (inversion) densities of the electrons which are common at interfaces, this problem may be of considerable importance.

Impurity scattering at interfaces has also been treated in detail [16,19,23]. We, therefore, review below the scattering of accumulation (inversion) layer electrons in GaAs by remote impurities located, e.g., in AlGaAs. The envelope functions $F_i(z)$ can numerically be obtained as described in sect. 2 and the matrix element $M_{mm}(Q)$ for impurity scattering of an electron from wavevector (two-dimensional) k to $k + Q$ is then given by

$$|M_{mn}(Q)|^2 = \int M_{mn}^2(z_0) \, N_I(z_0) \, dz_0 \quad \text{with} \tag{3.1}$$

$$M_{mn}(z_0) = \int e\phi(Q, z) \, F_m(z) \, F_n(z) \, dz, \tag{3.2}$$

where $N_I(z_0)$ represents the impurity concentration at $z = z_0$. For scattering within the same subband we have $m = n$. The scattering potential $\phi(Q, z)$ is somewhat difficult to calculate. Several equivalent procedures have been proposed in the literature. For our purpose it is best to solve the following equation [16]

$$\phi(Q, z) = -\int dz_1 \frac{1}{Q} \exp(-Q|z - z_1|) \sum_{i=1}^{\infty} S_i g_i(z_1)$$

$$\times \int dz_2 \, \phi(Q, z_2) \, g_i(z_2) + \frac{e}{2\epsilon_0 \epsilon Q} \exp(-Q|z - z_0|). \tag{3.3}$$

Here $g_i(z) = F_i^2(z)$ and ϵ_0 is the dielectric constant of free space. From this equation, $\phi(Q, z)$ has been iteratively calculated by explicit approximations and also numerically [13]. Some results for the matrix element are shown in fig. 4. Explicit results can be obtained by approximating $g_i(z)$ by

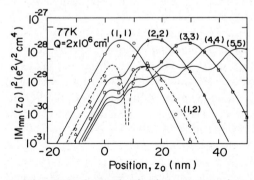

Fig. 4. Absolute square of the matrix element for the electron–remote-impurity interaction at 77 K. The change in wavevector is assumed to be fixed at $Q = 2 \times 10^6$ cm^{-1}. The indices correspond to the subscripts m, n for the subbands. The $\bigcirc$, $\square$, $\diamond$ and $\triangle$ show the results of explicit approximations. (After Yokoyama and Hess [13].)

$g_i(z) \approx \delta(z - z_i)$ where z_i is the position of the maximum of $g_i(z)$. The results for this approximation are also shown in fig. 4. It is interesting to observe the oscillatory structure of the scattering probability which is, of course, caused by the form of the wavefunction. The matrix element is strongly reduced for larger z_0 which explains the effect of modulation doping. The inter-subband matrix elements ($m \neq n$) are generally rather small as can be seen for scattering from the lowest to the next higher subband in fig. 4. The electron–phonon interaction can be treated in a similar manner using the numerically obtained envelope functions. For details, the reader is referred to ref. [13].

Consequences of the scattering theory for electronic transport will be discussed in the next subsections.

3.2. The low-field mobility for transport parallel to interfaces

Experiments performed on GaAs/AlGaAs modulation doped heterostructures at low temperatures have produced the highest values for electron mobilities ever measured in GaAs, in excess of 10^6 cm^2/(V s) [24,25]. Spurred by these results, many theoretical investigations of the mobility in the range 4 K < T < 40 K have been performed [16–22,26]. The scattering agents that have been taken into consideration are the screened impurity ions (both the remote ones in the AlGaAs barrier and the background impurities in the GaAs channel), the acoustic phonons which couple to the electrons through the deformation potential and the piezoelectric field, alloy disorder and surface roughness. The mobility is usually calculated by averaging the momentum relaxation time which is calculated from the various scattering mechanisms.

The main feature that has been observed is the linear decrease of μ with increasing T. Such a dependence, characteristic of the acoustic-phonon-limited mobility, led to the hypothesis that the acoustic phonons are the main factor which determines how μ varies with temperature as it approaches $\mu(0)$, its maximum value at $T \approx 0$ K. Walnkiewicz et al. [22] calculated the mobility limited by acoustic phonons and ionized impurities using the commonly accepted value for the deformation potential in GaAs of 7 eV and found reasonable agreement with the experimental data of Hiyamizu et al. [24]. Later, other researchers [21,26] suggested that screening of the electron–phonon interaction needs to be included and a value of ~ 13.5 eV must be used for the deformation-potential constant. The dependence of the mobility on the electron concentration in the GaAs channel has also been studied [25,26]. Experiments [24] and calculations [22] have shown that, in general, higher sheet concentrations imply higher values for $\mu(0)$, μ being directly proportional to n at low temperatures. If the concentration is increased beyond $\sim 6 \times 10^{11}$ cm^{-2}, the mobility starts to decrease. This indicates that the concentration is high enough to allow scattering into higher subbands. The presence of intersubband scattering puts forth the issue of the validity of the wavefunctions used in the calculations. In most of the calculations to date, a variational function [10] was used. This is a good approximation as long as only the lowest subband is occupied. For multi-subband conduction the use of numerically computed wavefunctions gives quite different results for the electron mobility [27]. The increased interface electron density also makes the neglect of electron–electron interactions questionable.

3.3. High-field transport parallel to interfaces

3.3.1. General aspects

High-field transport at interfaces has attracted attention of researchers for several reasons. With respect to transport parallel to interfaces, one of the best studied systems is the metal-oxide–silicon (MOS) system where it became clear early that high electric fields are involved in the operation of fine-line MOS devices. An additional attraction was provided for the theoretical work by the reduced dimensionality of the electron gas at the interface. If the electron gas is treated as two-dimensional, the density of states is independent of energy and the scattering terms in the Boltzmann equation are greatly simplified [28]. The Boltzmann equation can then be solved analytically for many practical cases.

A new effect, found in connection with the systematic studies of hot electrons, has been the emission of hot electrons from the silicon channel into the silicon dioxide. This effect was noticed by Hess and co-workers [29] and independently by Ning and co-workers [30] who also realized the full implications for device design. A complete theory of this effect has been given by Tang and Hess [31].

Interest in transport in the $Al_xGa_{1-x}As$–GaAs system has been largely spurred by the studies on modulation doping. The high mobilities enhance the possibility of heating the electrons by electric fields. In fact, for the highest mobilities, pronounced nonlinearities occur in the current–voltage characteristic for electric fields as low as 1 V/cm. As the electrons are heated up by electric fields they can gain enough energy to transfer back to the $Al_xGa_{1-x}As$ layers (the perpendicular momentum is supplied by a scattering agent). This transfer of electrons in real space (as opposed to k-space) and the possibility of achieving negative resistance and storage effects has been predicted and experimentally demonstrated by Hess and co-workers [6,7], and has been demonstrated to be useful for device applications [8]. Real-space transfer will be treated in connection with perpendicular transport below, although the electric field which heats the electrons is parallel to the interface. It is this effect which distinguishes most the interface transport characteristics from nonlinear transport in bulk semiconductors. Real-space transfer has pronounced effects on the steady-state current even if the electrons do not transfer into the AlGaAs but only spread farther into the GaAs as they gain random energy from the electric field and subsequent scattering. This effect can be understood easiest from results for transient transport in the quantized many-subband system, which is treated below.

Transient effects are also enhanced by the high mobilities in modulation doped structures. We will consider the case of instantaneously switching on an electric field at the time $t = 0$ and discuss the development of the distribution function to the point of steady state. At very early times the electrons are accelerated ballistically. Then electrons start to interact with scattering centers and begin to redistribute over a considerable range of the energy band structure (in GaAs electrons transfer from the Γ minimum to the X and L valleys). After typically less than 5 ps steady state is reached. The two-dimensional aspects and subband structure add new features to this effect, as is described in the next section.

3.3.2. Transient transport in a many-subband system

In this section we follow the study of Yokoyama and Hess [13,33] of transient transport in single quantum well structures. A uniform field is switched on at $t = 0$ parallel to the heterointerface (y-direction), and the five lowest subbands around the Γ-point of GaAs are included into the calculation. Complete Γ–L–X band structures are used for both GaAs and AlGaAs layers. For details of material parameters we refer the reader to the original references. The steady-state results of this simulation are not too different from the familiar results for bulk GaAs as long as the electron transfer into the AlGaAs is negligible.

For the analysis of transient transport phenomena, ensemble Monte Carlo methods must be used, i.e., many electrons must be simulated simultaneously. The initial distribution of electrons also plays an important role. Yokoyama and Hess [13] have generated the initial subband population and energy (k-vector values) in each subband using the rejection technique of von Neumann. 10 000 electrons were chosen for the simulation and a constant time step discretization scheme was introduced. This scheme allows the tracking of the time evolution of electron transport, and is also advantageous for updating the self-consistent calculation of the electronic states as the distribution function changes with time.

This updating is needed since screening depends on the distribution function (Lindhardt expression). Therefore after a few time steps the Schrödinger equation must be re-solved and the scattering mechanisms re-tabulated since their rates change with the screening. This enormously complicated fully self-consistent procedure can only be accomplished with large computational resources. If the screening is calculated only for the initial equilibrium distribution the calculation (equilibrium method) is much simpler. Both methods are discussed below. Results for the transient behavior of the drift velocity (versus time) are shown in figs. 5 and 6 for lattice temperatures $T_L = 77$ K and $T_L = 300$ K, respectively. Some of the finer details of the curves in these figures are unknown from studies of bulk material. The overshoot for a field of 1 kV/cm and the shoulders (at 3 and 5 kV/cm) arise from the step-like increase of the scattering rate at the threshold of optical phonon emission. In the low-energy range of this near ballistic regime, electrons are heated up by the electric field without undergoing polar optical phonon emission processes. After the electrons reach $\hbar\omega_0$, they lose energy and momentum at a greatly increased rate. The

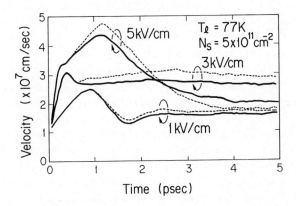

Fig. 5. Velocity overshoot versus time at $T_L = 77$ K for a constant electric field of 5.3 and 1 kV/cm applied at $t = 0$. The dashed curve is calculated by the equilibrium method, the solid curve by the fully self-consistent method. (After Yokoyama and Hess [33].)

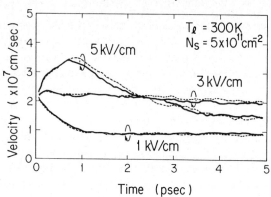

Fig. 6. Velocity overshoot as a function of time for various electric fields. The dashed curves are calculated with the equilibrium method, the full curves are fully self-consistent.

velocity response curves at 300 K are different from those at 77 K, as is evident from figs. 5 and 6. This is because a significant number of electrons already have sufficient energy from the start to emit phonons at 300 K. A significant role in the detailed form of the overshoot is also played by the changing population of the electronic subbands.

Figures 5 and 6 show clearly that the time-dependent screening as described by the fully self-consistent method reduces generally the overshoot phenomena (the maximum values of the transient drift velocity) compared to the results predicted by the equilibrium method (which is less exact). This result has far reaching implications for the ultimate speed of high electron mobility transistors [4]. It suggests that overshoot and ballistic effects are suppressed by the time dependence of the self-consistent field.

There is another remarkable implication hidden in the results of figs. 5 and 6. Since the screening influences mostly the transverse electric field which confines the electrons, these figures (and the details of the calculation) show that the parallel drift velocity depends on the transverse field. One may think that the effect is a quantum-mechanical one as the wavefunctions change their form along with changes in screening. Although this is definitely true, the main effects have classical analogies. Before explaining these analogies, let us consider the case of two subbands with independent electron mobilities μ_1 and μ_2 (both being a function of the electron energy distribution). The power P_s supplied to the electron system is then (neglecting diffusion),

$$P_s = e(n_1\mu_1 + n_2\mu_2)F^2, \qquad (3.4)$$

where n_1 and n_2 are the respective electron concentrations. The power P_d dissipated to the phonons contains, in general, different contributions from the two subbands. In steady state the average energies $\langle E_1 \rangle$ and $\langle E_2 \rangle$ of the electrons in the two subbands can be estimated from the balance equation

$$P_s = P_d, \tag{3.5}$$

if μ_1 and μ_2 are known as functions of these average energies (actually μ_1 and μ_2 are functionals of the complete energy distribution). Equations (3.4) and (3.5) show clearly that the average energies $\langle E_1 \rangle$ and $\langle E_2 \rangle$ become functions of the transverse electric field if μ_1 and μ_2 do. We will show in the next section that this result is basic to nonlinear transport of electrons confined by transverse fields. It is not a consequence of approximations in the above Gedanken experiment.

3.3.3. Classical transport in the presence of confining (transverse) electric fields

We now turn to a classical description of the influence of confining electric fields (such as transverse gate fields). It is clear that such fields by themselves do not cause any heating of the electron gas [34,35] since no net current flows parallel to the field direction. Therefore, it has been customary to assume that changes due to these fields are due to size quantization. In problems of nonlinear transport, however, there are also significant classical consequences for carrier heating. The nonlinearity of high-field transport is essential for these effects to occur and the nonlinearity is strongest at very high electric fields. The influence of the transverse fields can be seen for any part of the energy distribution of electrons. However, the effects manifest themselves most clearly for the high-energy tail. The impact ionization rate samples this high-energy tail since only electrons above the energetic ionization threshold [36] contribute to the ionization. To illustrate these effects more clearly we follow again a Monte Carlo simulation [36].

We describe in the following the influence of transverse fields on the ionization rate α of electrons which has been computed [36] for two different potential configurations which are shown in fig. 7a,b. In both the square well structure and the triangular well, a longitudinal field is also applied. Within the square well structure the electrons are initially launched according to a Maxwellian distribution 500 Å from the origin in the transverse (z-)direction. The electrons drift along the device subject to the longitudinal field and can move within the well until either the step

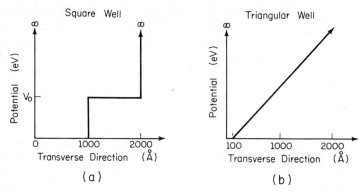

Fig. 7. Diagrams for transversal potential energy: (a) square well potential; (b) triangular well potential.

potential barrier or the infinite potential barrier at the origin is encountered. At the origin the electrons are specularly reflected back into the well. Upon encountering the potential step at $z = 1000$ Å the electrons are either transferred or reflected according to classical theory (depending on their energy).

The calculated ionization rate as a function of inverse field for a 0.4 eV potential step is shown in fig. 8. Of greatest concern here is the relative

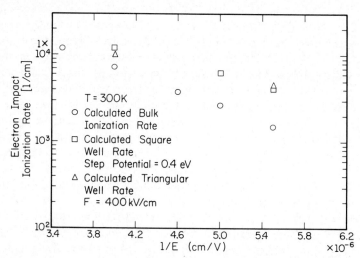

Fig. 8. Ionization rate versus inverse electric field in bulk silicon and for the potential configurations of fig. 7a,b. (After Brennan and Hess [36], © 1986 IEEE.)

difference between the calculated bulk and potential-well ionization rates. As is clearly seen from this figure, the ionization rate in the presence of the step is significantly larger (> 10%) than the calculated bulk rate.

The triangular well of fig. 7b can be viewed as a Riemann-sum approximation of many square wells. Therefore, for a transverse field in the triangular well, corresponding to the ratio of step height to width of the square well, somewhat similar results should be obtained. Calculations of the ionized rate in a triangular potential well for a transverse field of 400 kV/cm are also shown in fig. 8 as a function of inverse longitudinal field. As in the case of the square well structure, the electron ionization rate is significantly larger when the transverse field is present. The step well width, 1000 Å, and the step height, 0.4 eV, of the square well structure correspond to the chosen slope of the triangular well.

The electron-heating effect calculated above can be explained as follows. Under the action of an applied electric field, the electrons move through the bulk semiconductor material until quasi-steady state is established; the average energy gained from the field is balanced by the energy lost to the lattice via phonon scattering. If a potential step is introduced, as in the calculations for the square well potential, the carriers move parallel to the potential interface until scatterings redirect their momentum towards the interface. Provided the electrons have sufficient kinetic energy, gained from the longitudinal field, they can then undergo real-space transfer into the high-potential region [6]. Upon transferring, the electrons lose kinetic energy to the potential offset. All other factors being equal, the longitudinal field will re-heat the electrons in the high-potential region up to an average energy which is equal to the average in the low-potential region but now measured from the minimum potential energy in the high-potential region. In other words, the electrons are heated up to the usual average energy plus the potential step. Of course, the heating to this energy is an upper bound to what is indeed possible since electrons will transfer back into the low-potential energy region and lose their excess energy by phonon scattering. It is clear, however, that the average electron energy measured from the minimum of the conduction-band edge will be larger compared to the case when no step is present and therefore the ionization rate is increased. We would like to emphasize that in this process no net energy has been gained from the potential step itself after the electron returned to the low potential energy region. It is the longitudinal field that re-heats the carriers within the high potential energy region where the carriers will have, in general, a higher mobility (depending on their energy and upon how long they remain within the high-potential range). Consequently, after re-entering the low-potential region, the electrons are hotter than they can become without the step. The increase of the ionization rate in superlattice avalanche photodiodes is a variation of this effect and is described in chapter 10 of this book.

The effect is basic to all nonlinear transport in confining fields and should be included also in less sophisticated transport simulations.

4. Transport perpendicular to the interface

4.1. General analytical considerations

Consider two different neighboring semiconductors as shown in fig. 1 with a given conduction-band discontinuity ΔE_c and assume that the bands are flat, as shown in fig. 9. We are interested in the transport of electrons from the left to the right and vice versa. The calculations can easily be performed by using an approximation introduced by Bethe. Bethe assumed that the distribution function is equal to the Fermi distribution (or Maxwell–Boltzmann distribution) with a quasi-Fermi level E_{QF} which is different in the two materials but constant until approaching the interface. This means that a strong electron–electron interaction must be present which causes the distribution function to look Fermi-like independent of the fact that electrons can propagate at a high rate to the neighboring material. This loss of electrons can be so substantial that it is hard to believe that the electron distribution still keeps the Fermi-shape. Exact calculations of the electron distribution, however, necessitate an ensemble Monte Carlo simulation [32].

The following explicit treatment is instructive and probably correct within a factor of 2 or so in most instances. Classically, all the electrons having a velocity component in the positive z-direction which is large enough to overcome the energy ΔE_c will propagate into the $Al_xGa_{1-x}As$ and all the electrons in the $Al_xGa_{1-x}As$ having a component in the

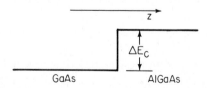

Fig. 9. Idealized heterojunction band edge under flat-band conditions.

negative z-direction will propagate into the GaAs. The current density from the left (GaAs) to the right (Al$_x$Ga$_{1-x}$As) is then

$$j_{LR} = \frac{e}{4\pi^3} \int_{k_x k_y} dk_x \, dk_y \int_{k_z > k_{z_0}} dk_z \, vf(k), \tag{4.1}$$

where v is the velocity of the electrons and k_{z_0} is the minimum k-vector component which is necessary to overcome the band edge discontinuity. This gives the well-known result

$$j_{LR} = A^* T^2 \exp\left[\left(E_F^L - E_c^L - |\Delta E_c|\right)/kT\right], \tag{4.2}$$

where E_c^L is the energy of the band edge in the GaAs and j_{LR} is called the thermionic emission current, $A^* \equiv (e/2\pi^2)(m^*/h^3)k^2$ is the Richardson constant which is, for the free electron ($m^* = m_0$), given by $A = 120$ A/(cm^2 K^2) and T is the temperature of the charge carriers. If the carriers are heated by electric fields to a temperature T_c which is different from the lattice temperature T_L, then T_c has to appear in the exponent of eq. (4.2).

The current from the Al$_x$Ga$_{1-x}$As toward the GaAs, j_{RL}, is given by

$$j_{RL} = A^* T^2 \exp\left[\left(E_F^R - E_c^R\right)/kT\right], \tag{4.3}$$

where E_F^R is the quasi-Fermi level in the Al$_x$Ga$_{1-x}$As. The dynamics of electron transport over a step-like structure (such as shown in fig. 9) can be estimated from the equation of continuity (neglecting generation–recombination),

$$e \frac{\partial n}{\partial t} = \frac{\partial j}{\partial z}. \tag{4.4}$$

Assume for the moment that current flows from the AlGaAs to the GaAs and that this thermionic emission current flows only within a certain range which is of the order of the mean free path L_m of the electrons (beyond this region the average velocity is much reduced because of collisions). Then using $\partial j/\partial z \approx L_m$ we have

$$e \frac{\partial n(t)}{\partial t} \approx -A^* T^2 \exp\left[\left(E_F^R(t) - E_c^R\right)/kT\right]/L_m, \tag{4.5}$$

which gives, using the definition of quasi-Fermi levels,

$$n(t) = n_c \exp\left[-\left(\frac{A^* T^2 C_1}{eL_m}\right)t\right]. \tag{4.6}$$

Here $C_1 = 4\hbar^2(2m^*kT\pi)^{3/2}$ and n_c is the concentration in the $Al_xGa_{1-x}As$ at $t = 0$. The constant $eL_m/A^*T^2C_1$ is typically of the order of picoseconds.

This estimate is, of course, rather crude. Especially the choice of L_m is somewhat arbitrary and needs to be replaced in some instances by a layer width. The time dependence of $n(t)$ implies also that electrons are not replenished from the right by diffusion. In addition, as the electrons leave the AlGaAs and their parent donors, the band edges will not be flat as shown in fig. 9 and need to be calculated self-consistently from the Poisson equation

4.2. Hot-electron thermionic emission – real-space transfer (RST)

The emission of *hot* electrons over barriers (or tunneling through them) is more complicated and more difficult to understand than other effects basic to semiconductor device operation. The reason is that RST can only be visualized by the combination of two concepts concerning the energy distribution of electrons. The first concept is the concept of quasi-Fermi levels as defined in the last section and the second is the concept of a charge carrier temperature T_c (different from the equilibrium lattice temperature T_L). Each concept is usually explained and quantified by solving the Boltzmann equation using the method of moments. The quasi-Fermi levels are calculated from the rate of change of the spherical symmetrical part of the distribution function (generation–recombination rates) and the electron temperature can be obtained from the power balance which also involves the spherical symmetrical part of the energy distribution.

For RST problems both concepts matter and both the carrier temperature and the quasi-Fermi levels are a function of space-coordinate and time. Imagine, for example, electrons residing in a layer of high-mobility GaAs neighboring to two layers of low-mobility AlGaAs. The GaAs equilibrium distribution function f is

$$f \propto \exp(-E/kT_L), \tag{4.7}$$

while in the AlGaAs we have

$$f \propto \exp[-(\Delta E_c + E)/kT_L], \tag{4.8}$$

where E is measured from the conduction-band edge and ΔE_c is the band edge discontinuity. If now the electrons are heated by an external field (e.g.

parallel to the layers) we have to replace T_L in eqs. (4.7) and (4.8) by a space-dependent carrier temperature T_c. From the form of eqs. (4.7) and (4.8) it is clear that for $T_c \to \infty$ the difference between the AlGaAs and the GaAs population density vanishes. In other words, the electrons will spread out to the AlGaAs layers. This also means that even perpendicular to the layers (z-direction) a constant Fermi level cannot exist and E_F has to be replaced by the quasi-Fermi level $E_{QF}(z)$ as the density of electrons becomes a function of $T_c(z)$. This is unusual since commonly the quasi-Fermi levels differ only in the direction of the applied external voltage V_{ext} (by the amount eV_{ext}). In our case, a voltage is applied parallel to the layers, the electrons redistribute themselves perpendicular to the layers and a field (and voltage) perpendicular to the layers develops caused by the redistribution.

The self-consistent inclusion of this field will, in general, require numerical methods. In this discussion we disregard self-consistencies as is justified for extremely low carrier concentrations and transfer over short distances. We than can use the theory developed in sect. 4.1 to calculate the quasi-Fermi levels. This has been described previously and follows directly from the condition

$$j_{LR} = j_{RL}. \tag{4.9}$$

The lattice temperature must, of course, be replaced by the electron temperature T_c at the respective side. The switching speed can also be derived from the formalism presented above, and one obtains for the time constant of electrons propagating out of the well

$$n = n_0 \exp(-t/t_s), \quad \text{with} \tag{4.10}$$

$$t_s = \frac{eN_c L_m}{A^*(T_c)^2} \exp(\Delta E_c/kT_c), \tag{4.11}$$

where N_c is the effective density of states of GaAs. A typical time constant t_s derived in this way can range from picoseconds (for high carrier temperature T_c) to hours for very low temperatures T_c. The short time constants are attractive for device applications.

4.3. Aspects of tunneling perpendicular to heterolayers

Tunneling in heterolayers is treated in chapters 10 and 12. Therefore, we will discuss here only two topics which may have applications in future electron devices.

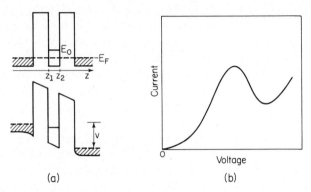

Fig. 10. Double-barrier tunneling structure: (a) energy band diagram without and with applied voltage; (b) schematic current–voltage characteristic.

Tunneling in double-barrier structures has been investigated by Chang and co-workers in 1974 [37]. Since then, the investigations by Sollner and co-workers [38] in the teraherz range have stimulated considerable interest. The current–voltage characteristics through a double heterojunction barrier exhibits pronounced negative differential resistance as shown in fig. 10.

This negative resistance has been originally attributed to resonance tunneling, i.e. to a Fabry–Pérot type resonance which builds up between the two barriers with corresponding variations in the transmission coefficient. Luryi [39] found that the negative differential resistance can be explained without any resonance effect by mere subsequent tunneling. His reasoning involves only conservation of energy and of the parallel (to the interface) component of the electron wavevector. One can easily convince oneself that no electrons can tunnel under these conservation laws as soon as the conduction-band edge at the left-hand side of the two barrier system (fig. 10a) is higher than the energy of the confined state with energy E_0. At this point the current should be much reduced and negative differential resistance will occur.

Other recent investigations of tunneling in heterolayer structures concern the influence of band structure on the tunneling probability. Consider the situation shown in fig. 11. The electrons propagate in GaAs mainly in the Γ minimum. For the AlGaAs barrier the X minimum can be lower than the Γ minimum. Tunneling then becomes a more complicated problem. The parallel wavevector needs to change significantly for the X minimum to be important in the tunneling process. The alloy disorder, phonons and impurities can provide such large wavevector-difference. It has been shown

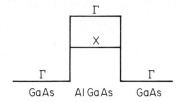

Fig. 11. GaAs–AlGaAs tunneling barrier with lower X and higher Γ minimum in the AlGaAs.

by McGill and co-workers [40] that the barrier height of the X minima is important for the tunneling current which is, however, about a factor of 1000 lower than expected for transport in one and the same conduction band minimum only.

4.4. Transport over heterolayer wells

While transport over barriers and tunneling through them has been treated in the past in considerable detail, work on transport over wells is just starting to receive attention. Figures 12a,b,c show a quantum well under various bias and doping conditions. The simplest case, the intrinsic unbiased situation, is shown in fig. 12a. The electron concentration in the quantum well will deviate from the intrinsic concentration for two reasons. First, the density of states mass in the valence band in the well may be slightly different from the three-dimensional average value and, second, the intrinsic concentration in the AlGaAs is different from the corresponding concentration in "fictitious" GaAs having the same conduction-band edge again because of the differences in effective mass. Therefore, electrons and holes will transfer between the layers causing a deviation from the flat-band condition. If doping is present, the bands will be even more distorted as is shown in fig. 12b for donor-doping in the AlGaAs. The electrons leave their parent donors and accumulate in the GaAs.

Figure 12c shows the consequences of an external bias. Electrons are then removed from and supplied to the well by several mechanisms. Incoming currents and carrier generation will tend to fill the well while phonon-assisted tunneling, thermionic emission and recombination will tend to empty it. Since the electrons are injected at higher energies into the well and quantum effects such as reflections and resonances may be important, the problem is of considerable complexity and has not been solved entirely. Two limiting cases have been treated in some detail and will be described below.

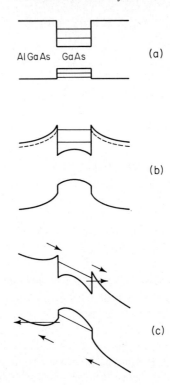

Fig. 12. Schematic diagram of energy bands around a heterolayer well for: (a) intrinsic case; (b) doped heterolayers; and (c) heterolayers plus external bias.

Notice that transport over potential minima occurs in heterolayer devices such as heterolayer avalanche photodiodes and has therefore considerable practical relevance and interest.

4.4.1. Collection of carriers into heterolayer wells

Collection of carriers into a heterolayer well has been treated by Shichijo et al. [41] and by Tang et al. [42] in a semi-classical way. They assumed that the well is wide enough so that the electron will not "see" the other side. They also treated the transition made from the continuum states above the well to the quasi two-dimensional well states as bulk-like. Their calculations are therefore only applicable for rather large wells. Brum and Bastard [43] have calculated the capture in quantum wells accounting for size quantization but using simplified energy distributions of the electrons and assuming

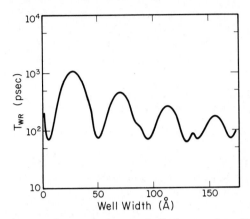

Fig. 13. Relaxation time τ_{wr} (inverse capture rate) of electrons into GaAs quantum wells plotted versus well width (after Brum and Bastard [43]).

that the Golden Rule holds for the transition of electrons into the well. These transitions then depend sensitively on the relative location of the bound states in the well and the continuum states with respect to the phonon energy $\hbar\omega_0$ of the polar optical phonons which have been assumed to cause the transition. A major role is also played by the resonance states and their energetic position with respect to the continuum edge. A consequence of these dependences are oscillations in the rate for the recombination (capture) of the electrons into the well as a function of well width. The results of Brum and Bastard [43] are shown in fig. 13. Notice that the inverse well recombination rate τ_{wr} is of the order of 100 ps, while the classical limit for very large well width is about fifty times smaller. This rather large increase of τ_{wr} is consistent with the observation of a cut-off in the lasing of quantum wells below a certain well width as observed by Shichijo et al. [41]. However, we would like to stress that the experimental situation is more complicated than assumed in the Brum–Bastard calculation [43]. As shown in fig. 1, the well boundaries will not be sharp even in the most ideal situation. Furthermore, the boundaries are in between two different materials and the relevant virtual bound states for real quantum wells may be very different from the idealized situation.

4.4.2. Propagation of carriers over the base of planar-doped barrier transistors

The propagation of electrons over the heavily doped base of a planar-doped barrier transistor [44] structure is shown schematically in fig. 14.

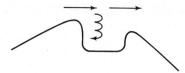

Fig. 14. Schematic diagram of electron propagation in a planar-doped barrier structure; the electron can propagate over the barriers and can also be captured in the base due to phonon emission.

Although electron propagation over a planar-doped barrier is formally very similar to propagation over quantum wells, there are some significant differences. The barriers, although created by thin doping spikes [44], still vary over distances of ≈ 100 Å instead of being abrupt and the base width is ≈ 500 Å in typical experiments. Therefore, resonance quantum effects are probably negligible. Quantum reflection at the barriers may still be of importance but will not be treated here.

Structures as shown in fig. 14 have been produced to investigate the possibility of ballistic transport, i.e. the possibility of electrons traversing the base without losing energy or being deflected into a different direction. A semiclassical Monte Carlo simulation of this situation has been performed by Wang et al. [45] who calculated the energy distribution of electrons at the second barrier for various barrier heights. In these calculations all known scattering mechanisms have been included: scattering by ionized impurities, electron–electron pair scattering and scattering by coupled phonon–plasmons. A typical result for the energy distribution at the second barrier is shown in fig. 15. The large peak at ≈ 0.28 eV in fig. 15 arises from ballistic electrons. The broadening of the peak is due to electron–electron interactions. Notice also the electron–phonon replica peak immediately below the main ballistic peak. At low energy, a broad maximum can be observed which is due to electrons which have been reflected at the barrier. The energy in this range can also be raised by electron–electron interactions with the fast incident electrons. This effect is not included in fig. 15.

The experimental results agree only roughly with the graph in fig. 15. The ballistic peak is broader and phonon replica are not observed. This is probably due to the fact that that the barriers in the experiments are not ideal and exhibit potential fluctuations due to the discreteness of the impurities. The effect of the discreteness of impurities is shown in fig. 16. For a barrier created by a doping spike of 10^{18} cm^{-3} acceptors in n-type material as was used in the experimental work, the fluctuations (shown in

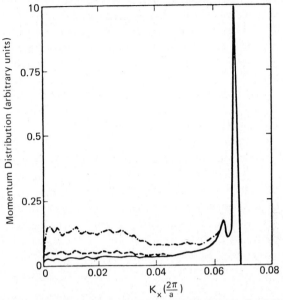

Fig. 15. Forward momentum distribution obtained from Monte Carlo simulations. The solid curve represents the distribution of hot electrons before they reach the collector. The dashed and dash-dotted curves include the contribution from both injected and reflected electrons for two different heights of the "analyzing" barrier. (After Wang et al. [45].)

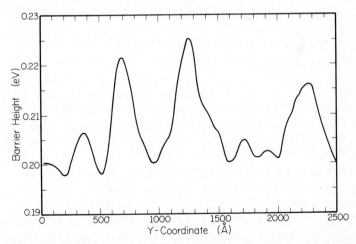

Fig. 16. Sample portions of the barrier height of a planar-doped barrier as a function of position parallel to the barrier for a given position perpendicular to the barrier (close to the average maximum).

fig. 16) correlate roughly to the observed broadening of the ballistic peak [46]. In passing we note that the discreteness of electrons and dopants will in general invalidate the usual continuum approximation when the structures become very small. Also the conventional treatment of plasmons becomes questionable if only about six electrons are present on average across the device.

5. Transport in heterolayer devices

In this section selected heterolayer transport effects are discussed from the perspective of applications in current and future electron devices. The effects of high mobility and overshoot on the characteristic of high-mobility transistors are described in qualitative terms. We also discuss the effects of quantization on the electron distribution and device capacitance and, once more, the effects of built-in fields on device transport. The final section deals with some of the more recent device applications of real-space transfer and ballistic transport.

5.1. Quantum effects and the high electron mobility transistor (HEMT)

The most pronounced effects of size quantization on parallel transport are the effects on the electron mobility which have been described in sect. 3. The electron mobility is influenced by the two-dimensional aspects of the electronic system, the density of state changes, the confinement of phonons, the existence of interface phonons, many-subband effects and the like. Modulation doping has the most pronounced influence on the low-field mobility of all effects related to heterolayer transport. Mimura et al. [4] attempted to make use of the dramatic mobility increase in the device structure (the high electron mobility transistor) which is schematically shown in fig. 17. The transistor represents a mixture of metal–semiconductor (MESFET) and metal–insulator–semiconductor (MISFET) field effect transistors. The metal gate forms a Schottky barrier and the AlGaAs is, in the ideal case, almost free of mobile electrons since the electrons transfer to the GaAs. Therefore, the gate current remains small even if the Schottky barrier is forward biased. If the insulating properties of the AlGaAs were sufficient, one could omit the Schottky barrier gate (and replace it for instance by heavily doped GaAs) and still achieve transistor action. (Notice, however, that the breakdown field in AlGaAs is $\approx 10^5$ V/cm while it is $\approx 10^7$ V/cm in the SiO_2 of MOS-transistors).

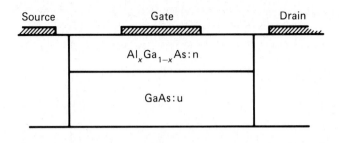

Fig. 17. Sketch of the geometry of the high electron mobility transistor. The GaAs is undoped and the two-dimensional electron gas resides at the $Al_xGa_{1-x}As$ interface.

The effects of the mobility enhancement on device performance is less dramatic than expected for two reasons. First, since the channel is very thin, the source-access resistance (a figure of merit in FET theory) can be rather large even if the mobility is very high and, second, the mobility degrades in rather small electric fields because of hot-electron effects. The record mobilities of 10^6 cm^2/(V s) do not appear during transistor operation where the mobility is much reduced, as can be seen from figs. 5 and 6. A detailed theory has been given by Widiger et al. [47].

As device sizes are reduced to the submicron range, the field degradation of the mobility will increase as the applied voltages cannot be reduced below a certain limit. However, there may exist an advantage of the HEMT over conventional MESFETs even in the limit of small device size. The overshoot effect can be significantly larger in the two-dimensional electron gas than in bulk GaAs. This is because the bulk GaAs will contain a high donor density (for MOSFETs in the submicrometer range) and impurity scattering will randomize the electron distribution earlier than in undoped material and therefore enforce the steady state and reduce the overshoot effect. Two opposing effects, however, reduce this advantage of the HEMT (to an extent which is unclear at this time). The time dependence of the self-consistent field reduces the overshoot significantly even in the absence of impurity scattering. This effect is caused by the spread of the hot electrons towards the bulk. Figure 18 shows the electron density in the lowest subband (dashed curve) and the continuum (denoted by "bulk" in fig. 18). The source contact is on the left-hand side of the figure. It can be seen that at the places of highest electric fields (under the gate and toward the drain contact on the right side of fig. 18) almost no electrons reside in

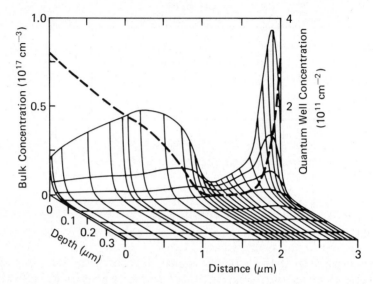

Fig. 18. Carrier density distribution in a HEMT for large drain voltages. The interface is located at a depth = 0, the source contact is at the left-hand side of the figure. (After Widiger et al. [47], © IEEE 1985).

the quantum well. Therefore, possible advantages of the two-dimensionality will be much reduced.

The effects of size quantization on the spacial carrier distribution perpendicular to the interface are not very significant with respect to device operation. The reason is that the total interface charge Q_{tot} can be calculated typically within classical approximations. Since this point has not been clearly presented in the literature we will describe it here in some detail.

Assume for simplicity that the AlGaAs is free of charge and that the potential at the surface of the AlGaAs is fixed. This corresponds to the situation in a metal–insulator semiconductor device where on top of the insulator (AlGaAs) a metal is placed as a "gate" and the gate voltage V_G is fixed. A two-fold integration of the Poisson equation in the charge-free AlGaAs gives then

$$\phi_R = c_1 z + c_2, \tag{5.1}$$

with c_1 and c_2 being integration constants. With the interface at $z = 0$, the boundary condition at the gate gives

$$V_G = c_1 d + c_2, \qquad (5.2)$$

where d is the AlGaAs thickness. The connection rules, eqs (2.3) and (2.4), give

$$\epsilon_L F_i^L = \epsilon_R c_1, \quad \text{and} \qquad (5.3\text{a})$$

$$c_2 = \phi_i^R, \qquad (5.3\text{b})$$

Equation (5.3) together with eq. (5.2) and the Gauss law result in

$$V_G = -\frac{Q_{\text{tot}} d}{\epsilon_0 \epsilon_R} + \phi_i^R, \qquad (5.4)$$

$\epsilon_0 \epsilon_R / d = C_{\text{ins}}$ is the (insulator) capacitance of the AlGaAs layer and ϕ_i^R is the interface potential. Remember that for the classical case one can derive an equation for $\phi_i^L = \phi_i^R$ in terms of $F_i^L = Q_{\text{tot}} / \epsilon_0 \epsilon_L$ and therefore one can calculate Q_{tot} as a function of V_G (the semiconductor potential far away from the interface is assumed to be zero as the second boundary condition). The form of ϕ_i^R as a function of Q_{tot} is shown in fig. 19.

The interface potential rises first linearly with Q_{tot} and saturates at the onset of strong inversion (accumulation) of free charge at the interface. Because of the generality of the screening concept, the form of ϕ_i^R as a function of Q_{tot} is valid from both a quantum and a classical point of view.

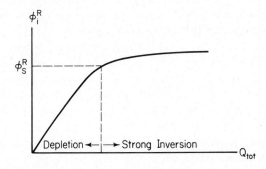

Fig. 19. Interface potential as a function of total charge Q_{tot} at the left-hand side of the interface.

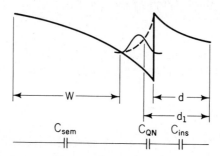

Fig. 20. Insulating regions and corresponding capacitance for the classical and quantum-mechanical calculation at a heterojunction.

For $|\phi_i^R| < |\phi_s^R|$, i.e., for weak inversion, the potential well is not very deep and the potential varies slowly enough to make size quantization effects unimportant. The classical results are therefore valid. Beyond the onset of strong inversion ϕ_i^R saturates for classical and quantum calculations. Therefore, one obtains from eq. (5.4)

$$-Q_{\text{tot}} = C_{\text{ins}}(V_G - \phi_s^R) \tag{5.5}$$

as a general result. This equation is basic to the operation of the important metal–insulator–semiconductor field effect transistor.

The general validity of eq. (5.5) is the precise reason why the inclusion of quantum effects is not necessary in order to obtain a qualitative picture of MOS-transistor operation (with respect to electron concentration). Quantitatively quantum effects influence transistor operation for two reasons (with regard to the interface charge density).

First, it is not the total charge which one needs to know to characterize completely a transistor but rather the spacial distribution of free and fixed charge. This distribution differs somewhat classically and quantum mechanically. The major effect is shown in fig. 20 which illustrates the band structure and carrier density at the interface together with the capacitances in the depletion region of width W, the insulator of width d and the range where the wave function of the electron almost vanishes (because of the band edge discontinuity) of width $d_1 - d_2$. This range creates an additional "quantum" capacitance which does not appear in classical considerations [48].

5.2. Discussion of selected device schemes illustrating the physics of parallel and perpendicular nonlinear transport

In this section, some of the device schemes which are interesting because of their elucidation of the physics of transport at interfaces are described.

Recently a series of interesting devices has been proposed and fabricated by Luryi, Kastalsky and co-workers [49] which are based on real-space transfer [6,7]. Figure 21 shows a typical device geometry and corresponding band structure as used by Kastalsky and co-workers [8].

In this device, the charge residing close to the GaAs(top)–AlGaAs interface is heated by parallel electric fields and after gaining energy and perpendicular momentum (scattering) transfers over the AlGaAs barrier into a highly doped conducting region which is on top of an insulating

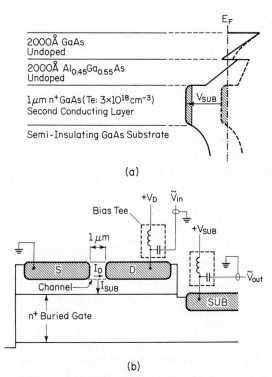

Fig. 21. Real-space transfer device: (a) device structure and energy-band diagram; (b) device contact lay-out and circuit arrangement for the charge injection mode of operation.

substrate. The device operation is best compared with a vacuum tube, the GaAs corresponding to the filament and the conducting layer corresponding to the anode. The difference is, of course, that here only the electrons are heated and thermionically emitted (within picoseconds) while the lattice stays at thermal equilibrium. The intrinsic switching times are extremely short [6] since the AlGaAs layer can be made thin enough to make the transit times of the order of 10^{-13}–10^{-12} s. The devices have indeed shown astonishing properties such as a transconductance of 1000 mS/mm, a negative differential resistance with a peak to valley ratio of over 100 and cut-off frequencies of ≈ 29 GHz with a predicted possible improvement of up to a factor of 10 for the cut-off frequency.

Interesting results are offered by these devices also from the physics point of view. The currents are linked to the transverse momentum distribution in contrast to the ballistic devices [44] which in turn will depend sensitively on the electron–electron and electron–phonon interaction. A comparison of the experiments with detailed Monte Carlo simulations should give valuable information on nonlinear high-field transport.

Even more interesting, from the physics point of view, is the structure of the dual high electron mobility transistor as proposed by Iafrate [50], and shown in fig. 22. In such a structure, the hot-electron thermionic emission currents j_{RL} and j_{LR} (sect. 4) can be controlled independently thus giving more complete information on the momentum distribution of electrons as a function of electric field. In addition, the ultimate solid-state switching speed may be achieved by considering real-space transfer between two interface channels in a single quantum well structure [51].

Perpendicular transport has been discussed in some detail in sect. 4 including "generic" device structures [44]. The planar-doped barrier tran-

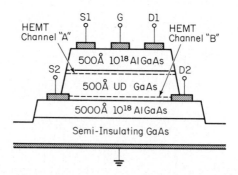

Fig. 22. Schematic diagram of a dual-channel structure (after Iafrate [50]).

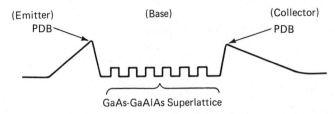

Fig. 23. A proposed millimeter-wave "Bloch" oscillator/transistor structure.

sistor structures are very useful to determine the forward momentum distribution of carriers including the ballistic peak. A new twist in these experiments is the importance of phonons and coupled phonons–plasmons due to the high doping of the base.

Tunneling structures including Fabry–Pérot type of geometries have received increasing attention and promise to uncover interesting physical effects connected to time-dependent tunneling and transient effects. There have also been structures proposed which combine the launching principle of the planar-doped barrier experiment and tunneling structures. These structures underline once more the richness of effects that can be investigated and created by varying the boundary conditions on a quantum level.

A millimeter-wave device which combines the ideas of Bloch oscillator and planar-doped barriers (PDB) (fig. 23) has been designed [52] and is currently being fabricated for millimeter-wave detector applications in the 10–100 GHz range. The device is designed to detect millimeter-wave radiation by phase-modulation methods. A GaAs–GaAlAs superlattice with 1 nm spacing is grown between two PDBs to control the electron injection/collection process. A constant electric field and an alternating millimeter-wave field are then superimposed across the superlattice, resulting in a phase modulation of the electron velocity. When the constant electric field is adjusted so that the Bloch frequency is a multiple of the alternating millimeter-wave field frequency, a DC current is induced across the transistor, the magnitude of which reflects the power of the millimeter-wave radiation. This is the essence of the Bloch millimeter-wave detector. The superlattice in fig. 23 is one-dimensional. Devices have also been proposed that utilize lateral superlattices [53] which consists of a doubly periodic array of metal dots embedded in the MOS-FET oxide. These dots induce a two-dimensional superlattice potential in the surface potential of the inversion layer.

Acknowledgement

We would like to thank M. Artaki, K. Brennan and I. Kizilyalli for valuable discussions, and K. Yokoyama for his contributions to this work. The work was supported by the US Army Research Office, the US Office of Naval Research and the Electronics Technology and Devices Laboratory. Fort Monmouth. The technical assistance by Mrs. E. Kesler, R. MacFarlane and R.T. Gladin is greatly appreciated.

References

[1] K. Hess, Phenomenological physics of hot carriers in semiconductors, in: Physics of Nonlinear Transport in Semiconductors, eds. D.K. Ferry, J.R. Barker and C. Jacoboni (Plenum Press, New York, 1980) p. 1.
[2] K. Hess, Aspects of high-field transport in semiconductor heterolayers and semiconductor devices, in: Advances in Electronics and Electron Physics, Vol. 59, ed. P.W. Hawkes (Academic Press, New York, 1982) p. 239.
[3] K. Hess and G.J. Iafrate, Hot electrons in semiconductor heterostructures and superlattices. in: Topics in Applied Physics, Vol. 58, ed. L. Reggiani (Springer, Berlin, 1985) p. 201.
[4] T. Mimura, S. Hiyamizu, T. Fujii and K. Nanbu, Japan J. Appl. Phys. 19 (1980) 225; A two-dimensional model has been given by D. Widiger, K. Hess and J.J. Coleman, IEEE Electron Device Lett. EDL-5 (1984) 266.
[5] N. Holonyak Jr and K. Hess, Quantum-well heterostructure lasers, in: Synthetic Modulated Structures, eds. L.L. Chang and B.C. Giessen (Academic Press, New York, 1985) p. 257.
[6] K. Hess, J. Phys. (France) C 7 (1981) C7–3; see also p. 32 of ref. [1].
[7] M. Keever, K. Hess and M. Ludowise, IEEE Electron Device Lett. EDL-3 (1982) 297.
[8] A. Kastalsky, J.H. Abeles, R. Bhat, W.K. Chan and M.A. Koza, Appl. Phys. Lett. 48 (1986) 71.
[9] See chapter 10 of this book.
[10] F. Stern and W.E. Howard, Phys. Rev. 163 (1967) 816.
[11] T. Ando, J. Phys. Soc. Jpn. 51 (1982) 3893.
[12] F. Stern and S. Das Sarma, Phys. Rev. B 30 (1984) 840.
[13] We follow here the treatment of K. Yokoyama and K. Hess, Phys. Rev. B 23 (1986) 8.
[14] H.C. Casey Jr and M.B. Panish, Heterostructure Lasers, Part B: Materials and Operating Characteristics (Academic Press, New York, 1978).
[15] See p. 202 of ref. [3].
[16] K. Hess, Appl. Phys. Lett. 35 (1979) 485.
[17] P.J. Price, Ann. Phys. (USA) 133 (1981) 217.
[18] P.J. Price, Surf. Sci. 113 (1982) 199.
[19] S. Mori and T. Ando, Phys. Rev. B 19 (1979) 6433.
[20] A. Pinczuk, private communication.
[21] P.J. Price, Phys. Rev. 32 (1985) 2643.
[22] W. Walukiewicz, H.E. Ruda, J. Lagowski and H.C. Gratos, Phys. Rev. 32 (1985) 2645.

[23] R. Dingle, H.L. Stormer, A.C. Gossard and W. Wiegmann, Appl. Phys. Lett. 33 (1978) 665.
[24] S. Hiyamizu, J. Saito, K. Nambu and T. Ishikawa, Jpn. Appl. Phys. 22 (1983) L609.
[25] E.E. Mendez, P.J. Price and M. Heiblum, Appl. Phys. Lett. 45 (1984) 294.
[26] B.J.F. Lin, D.C. Tsui, M.A. Paalanen and A.C. Gossard, Appl. Phys. Lett. 45 (1984) 695.
[27] N.T. Thang, G. Fishman and B. Vinter, Jpn. Appl. Phys. 59 (1986) 499.
[28] K. Hess and C.T. Sah, Phys. Rev. B 10 (1974) 3375.
[29] K. Hess, A. Neugroschel, C.C. Shiue and C.T. Sah, J. Appl. Phys. 46 (1975) 1721.
[30] T. Ning, Solid-State Electron. 21 (1978) 273.
[31] J.Y. Tang and K. Hess, J. Appl. Phys. 54 (1983) 5145; See also
R.J. Trew, R. Sultan, J.R. Hauser and M.A. Littlejohn, in: The Physics of Submicron Structures, eds. H.L. Grubin, K. Hess, G.J. Iafrate and D.K. Ferry (Plenum Press, New York, 1984) p. 177.
[32] G.J. Iafrate and K. Hess, High-speed transport in ultrasmall dimensions, in: VLSI Electronics: Microstructure Science. Vol. 9, ed. N.G. Einspruch (Academic Press, New York, 1985).
[33] K. Yokoyama and K. Hess, J. Appl. Phys. 59 (1986) 3798.
[34] S.A. Schwarz and K.K. Thornber, IEEE Electron Device Lett. EDL-4 (1983) 11.
[35] W.T. Jones, K. Hess and G.J. Iafrate, Solid-State Electron. 25 (1982) 1017.
[36] K. Brennan and K. Hess, IEEE Electron Devices Lett. EDL-7 (1986) 86.
[37] L.I. Chang, L. Esaki and R. Tsu, Appl. Phys. Lett. 24 (1974) 593.
[38] T.C.L.G. Sollner, W.D. Goodhue, P.E. Tannenwald, C.D. Parker and D.D. Peck, Appl. Phys. Lett. 45 (1984) 1319.
[39] S. Luryi, Proc. IEDM (1985) p. 666.
[40] A.R. Bonnefoi, D.H. Chow, T.C. McGill, R.D. Burnham and F.a. Ponce, J. Vac. Sci. & Technol. B 4 (1986) 988.
[41] H. Shichijo, R.M. Kolbas, N. Holonyak, R.D. Dupuis and P.D. Dapkus, J. Appl. Phys. 53 (1982) 6043.
[42] J.Y. Tang, K. Hess, N. Holonyak Jr, J.J. Coleman and P.D. Dapkus, J. Appl. Phys. 53 (1982) 6043.
[43] J.A. Brum and G. Bastard, Phys. Rev. B 33 (1986) 1920.
[44] J.R. Hayes, A.F.J. Levi and W. Wiegmann, Phys. Rev. Lett. 54 (1985) 1570, and references therein.
[45] T. Wang, K. Hess and G.J. Iafrate, J. Appl. Phys. 59 (1986) 2125.
[46] See the report by A.C. Robinson, Science 231 (1986) 22, and references therein.
[47] D. Widiger, I.C. Kizilyalli, K. Hess and J.J. Coleman, IEEE Trans. Electron Devices ED-32 (1985) 1092.
[48] H. Shichijo, unpublished results.
[49] S. Luryi and A. Kastalsky, Physica B&C 134 (1985) 453.
[50] G.J. Iafrate, to be published.
[51] H. Sakaki, Jpn. J. Appl. Phys. 21 (1982) L381.
[52] G.J. Iafrate, in: Gallium Arsenide Technology, ed. D.K. Ferry (Howard W. Sams & Co., Indianapolis, IN, 1986) p. 443.
[53] R.K. Reich, R.O. Grondin, D.K. Ferry and G.J. Iafrate, IEEE Electron Devices Lett. EDL-3 (1982) 381.

CHAPTER 12

HOT-ELECTRON INJECTION AND RESONANT-TUNNELING HETEROJUNCTION DEVICES

Serge LURYI

AT&T Bell Laboratories
600 Mountain Avenue
Murray Hill, NJ 07974, USA

Heterojunction Band Discontinuities: Physics and Device Applications
Edited by F. Capasso and G. Margaritondo
© *Elsevier Science Publishers B.V., 1987*

Contents

1. Introduction 491
2. Hot-electron injection devices 493
   2.1. Introduction 493
   2.2. Ballistic injection devices 496
      2.2.1. Doped-base transistors 500
      2.2.2. Induced base transistor 504
      2.2.3. Speed of ballistic transistors 508
      2.2.4. Hot-electron spectrometers 510
   2.3. Real-space transfer devices 513
      2.3.1. Three-terminal RST devices 515
         2.3.1.1 Various bias configurations and I–V characteristics 517
      2.3.2. Theory of charge injection in three-terminal RST devices 525
         2.3.2.1. Energy balance equation 527
         2.3.2.2. Discussion of the theory 532
      2.3.3. The ultimate speed of CHINT/NERFET devices 535
      2.3.4. Concluding remarks 537
3. Resonant-tunneling heterojunction devices 539
   3.1. Introduction 539
   3.2. Double-barrier quantum-well structures 543
      3.2.1. Mechanism of operation of quantum well diodes 544
      3.2.2. Frequency limit of DBQW oscillators 548
   3.3. Tunneling in superlattices 551
   3.4. Three-terminal resonant-tunneling devices 553
      3.4.1. Stark-effect transistor 554
      3.4.2. Quantum wire transistor 556
4. Conclusion 559
References 560

1. Introduction

Let us begin by quoting from the well-known out-of-print book by Sze [1]: *"... a useful solid-state device is one which can be used in electronic applications or can be used to study the fundamental physical parameters."*

We shall be mainly concerned with the first category of devices. The topics selected in this chapter are somewhat unusual – in that none of the exotic devices we shall discuss have already found a commercial application, and yet their very raison d'être is to be eventually included in the first category.

Since the invention of the transistor, many advances in semiconductor electronics and physics have been associated with a steady improvement in methods for preparing material structures with precisely controlled composition and dimensions. Over the last decade, the molecular beam epitaxy (MBE) has established itself as the most versatile epitaxy technique for growing single-crystal layers of semiconductors, giving the ultimate spatial resolution (on the scale of a nm) of the composition and the impurity incorporation in these layers. It has been clearly demonstrated that MBE is gainfully applicable to a number of conventional microwave, logic, and opto-electronic devices [2]. In addition, the extraordinary dimensional control of semiconductor layers afforded by MBE has enabled the fabrication of unconventional device structures, such as modulation-doped layers, superlattices, quantum wells, etc., whose new and remarkable properties result from the confinement of electronic states to narrow layers. It is interesting to note that as this field has entered its later stages of maturity, it faces fierce competition from another technique, the metal-organic chemical vapor deposition (MOCVD), which is potentially as powerful and versatile as MBE but may have significant economic advantages [3].

The power of MBE and MOCVD of III–V (and perhaps also II–VI) compound semiconductors for producing new device structures derives in large part from the existence of *lattice-matched* material systems of variable bandgap and refractive index. These systems, notably the GaAs/AlGaAs heterostructures, allow the construction by MBE of virtually arbitrary potential profiles for electrons and holes, including abrupt band discontinuities used to confine carriers in a two-dimensional state. The study of such device structures has led to discoveries of new and often unanticipated phenomena, the most dramatic of which is undoubtedly the fractional quantum Hall effect, discovered by Tsui et al. [4]. At present, the most important new electronic device based on carrier confinement at band discontinuities is the so-called MODFET or modulation-doped field-effect transistor (for recent references see refs. [5,6]). Waveguiding of light by the variable refractive index in compound semiconductor systems as well as direct interband transitions, available in these materials, are extensively used for heterostructure lasers, and two-dimensional confinement of carriers in quantum-well structures grown by MBE or MOCVD further improves the laser characteristics (see, e.g., ref. [7]). A number of other novel electronic and opto-electronic devices available through the "band-gap engineering" of III–V compound semiconductors have been reviewed by Capasso [8].

The present chapter is organised as follows. Section 2 will describe several device concepts which are based on charge injection of hot electrons across potential barriers due to the band discontinuities at a heterointerface. Depending on the employed mode of hot-electron transport, these device structures can be classified as either *ballistic-injection* or *electron-temperature* devices. The former group of devices (reviewed in sect. 2.2) have high-energy electrons traversing narrow base layers with minimum scattering – so that they can then clear a (heterojunction) collector barrier. Most of the proposed devices in this group are all-semiconductor analogs of the well-known metal-base transistor (see, e.g., ref. [1]). In the second group of devices the initial electron transport is parallel to the heterostructure layers – with an electric field heating the electronic system to temperatures T_e well above the ambient temperature. This process, resulting in an efficient charge injection across the interface into the adjacent heterolayers, is usually called the real-space transfer or RST. The RST devices originally proposed by Hess et al. [9] were diodes and the idea was to utilize different electron mobilities in the adjacent layers to generate a negative differential resistance (NDR). More recently, a three-terminal device structure based on the RST effect was proposed by Kastalsky and Luryi [10] and studied experimentally (see the reviews by Luryi and Kastalsky [11] and references therein). As will be discussed in sect. 2.3, this structure exhibits a rather complicated set of useful characteristics – which enable a wide variety of applications.

Section 3 deals with the so-called resonant-tunneling (RT) devices, first studied experimentally by Tsu and Esaki [12]. It begins with a discussion of the NDR effect observed in double-barrier quantum-well diodes. The underlying physics of the NDR will be explained and estimates of the maximum operating frequency will also be presented (sect. 3.2). In this discussion we shall emphasize the difference between a truly resonant tunneling, in which an electron wavefunction is coherent across the entire double-barrier structure, and a sequential resonant tunneling in which transmission is a two-step process. Resonant tunneling in superlattices will be discussed in sect. 3.3, again emphasizing the difference between coherent tunneling through many superlattice periods (miniband conduction) and a sequential tunneling process. The possibility of obtaining an amplification of electromagnetic radiation in superlattices at frequencies tunable by an applied electric field will be briefly described, following the early theoretical work by Kazarinov and Suris [13,14] and recent experiments of Capasso et al. [15,16]. Possible ways of supplying a third controlling terminal to the RT structure – providing a transistor action – will be discussed, including the recent proposal by Luryi and Capasso [17] of a transistor based on resonant tunneling of two-dimensional electrons across a one-dimensional quantum well (the "quantum wire"), as well as the Stark-effect transistor, recently proposed by Bonnefoi et al. [18]. In addition to an NDR in the source-drain circuit, these transistors are expected to possess a negative transconductance.

2. Hot-electron injection devices

2.1. Introduction

As the dimensions of semiconductor devices shrink and the internal fields rise, a large fraction of carriers in the active regions of the device during its operation are in states of high kinetic energy. At a given point in space and time the velocity distribution of carriers may be narrowly peaked, in which case one speaks about "ballistic" electron packets. At other times and locations, the non-equilibrium electron ensemble can have a broad velocity distribution – usually taken to be Maxwellian and parameterized by an effective electron temperature $T_e > T$, where T is the lattice temperature. Hot-electron phenomena have become important for the understanding of all modern semiconductor devices [19]. Moreover, a number of devices have been proposed whose very principle is based on such effects. This group of devices will be reviewed in the present work.

Commercial utilization of hot-electron phenomena began with the Gunn effect, based on the Hilsum–Ridley–Watkins mechanism for a negative differential resistance (see, e.g., ref. [19], chap. 11). The Gunn diode is a bulk device in which the NDR arises due to the transfer of hot electrons from the high-mobility central valley in a direct-gap III–V compound semiconductor to its higher-lying low-mobility satellite valleys. This is undoubtedly the best-known hot-electron device, for which a mature technology has been developed.

Another successful application of a hot-carrier effect has been made in a nonvolatile memory device invented by Frohman-Bentchkowsky [20] and is called FAMOS. It represents a p-channel MOSFET structure with a floating gate electrode. In the process of "writing" the memory, carriers, heated by the drain field, avalanche near the drain junction with hot electrons from the avalanche plasma injected into the floating gate. As the gate is charged up, its potential is lowered and the p-channel conductance increases. The FAMOS bears a conceptual similarity to some of the real-space-transfer (RST) devices discussed below.

We shall be concerned only with the hot-electron *injection* devices, i.e. such devices in which hot carriers are physically transferred between adjacent semiconductor layers. The family tree of the hot-electron injection devices is shown in fig. 1. The family is large and its members often go under different names. In the attempt to represent only distinct ideas, we may well have overlooked some important relatives!

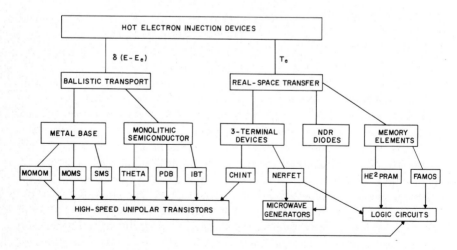

Fig. 1. The family of hot-electron injection devices and their potential applications.

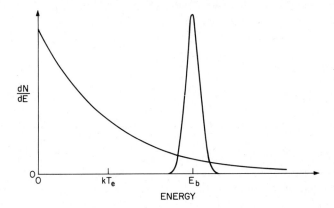

Fig. 2. The typical hot-electron distribution functions. In ballistic devices dN/dE is sharply peaked at $E_b \approx \frac{1}{2}mv_b^2$, with v_b mainly directed perpendicular to the base layer. In the real-space transfer devices, one has, approximately, $dN/dE \propto \exp(-E/kT_e)$.

Two distinct classes of hot-electron injection devices can be identified – depending on which of the two hot-electron regimes is essentially employed (the ballistic or the T_e regime, see fig. 2). In the *ballistic* devices, electrons are injected into a narrow base layer at a high initial energy in the direction normal to the plane of the layer. Performance of these devices is limited by various energy-loss mechanisms in the base and by the finite probability of a reflection at the base-collector barrier. In the *electron-temperature* (RST) devices the heating electric field is applied parallel to the semiconductor layers with hot electrons then spilling over to the adjacent layers over an energy barrier. This process is quite similar to the usual thermionic emission – but at an elevated effective temperature T_e – and the carrier flux over a barrier of height Φ can be assumed proportional to $\exp(-\Phi/kT_e)$. Even though a small fraction of electrons – those in the high-energy tail of the hot-carrier distribution function – can participate in this flux, their number is replenished at a fast rate determined by the energy relaxation time, so that the injection can be very efficient.

Although the first hot-electron injection devices were proposed a quarter century ago, their full potential has become realizable only with the advent of such hetero-epitaxial techniques as the molecular beam epitaxy (MBE) and organometallic chemical vapor deposition (OMCVD). These techniques can now provide abrupt heterointerfaces and the modulation of doping on the scale of a nm – which is essential for the implementation of both classes of hot-electron devices discussed below. In the next sections we shall review the most important hot-electron injection device structures.

2.2. Ballistic injection devices

As discussed above, our main conceptual classification is made according to the kind of a hot-electron ensemble employed in the device operation. Although all injection devices involve a real-space transfer (RST) of hot electrons, we shall, following the established terminology, reserve this term for devices operating in the electron-temperature regime. The first proposal of a hot-electron injection device was made by Mead [21,22]. His device, the MOMOM (metal-oxide–metal-oxide–metal) transistor, belongs to the category of ballistic-transport transistors. Accordingly, we begin by reviewing this group of devices.

Devices of this group represent a unipolar analog of the bipolar junction transistor *. Among themselves they differ by the materials employed and by the physical mechanism of hot-electron injection into the base. The original MOMOM proposal [21] was based on electron tunneling from a metal emitter through a thin oxide barrier into a high-energy state in a metal base, fig. 3a. Another insulating barrier separated the base from a metal collector electrode. Subsequent versions of this device [23] had the second MOM replaced by a metal–semiconductor junction, resulting in a transistor structure called the MOMS (fig. 3b). Attempts have also been made to employ a vacuum collector barrier (MOMVM).

Theoretical estimates of the frequency performance of tunnel-emitter transistors have led several authors to conclude [1] that these devices are inherently inferior to the bipolar transistor. Those conclusions were disputed by Heiblum [24] who, using another set of parameters for evaluation, suggested that certain tunnel-emitter configurations may have an edge. Experimentally, this question is open, although the general consensus is probably reflected in the fact that the tunnel-emitter metal-base transistor concepts have not gained much development in recent years.

Metal-base transistors (MBT), which employ thermionic rather then tunneling injection of hot carriers into the base, were first proposed by Atalla and Kahng [25] and Geppert [26] in the form of a semiconductor–metal–semiconductor (SMS) structure. The basic SMS tran-

---

* For a review of the early developments in the history of the ballistic hot-electron transistors the reader is referred to Sze [1] and Heiblum [24].

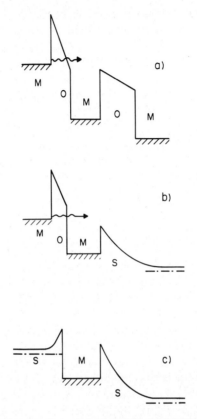

Fig. 3. Metal base transistors: (a) MOMOM; (b) MOMS; (c) SMS.

sistor is illustrated in fig. 3c. Experimental studies of the SMS device are actively pursued to this day: recent advances [27,28] have been associated with the development of epitaxial techniques for the growth of monolithic single-crystal silicon–metal-silicide–silicon structures. This continued interest is explained not only by the scientific usefulness of the SMS structure (it is an excellent tool for studying fundamental properties of hot-electron transport through thin films), but also by lingering hopes to produce a transistor which is faster than the bipolar or FET devices.

The potential merits of the SMS transistor had been appraised long ago by Sze and Gummel [29]. They predicted that despite its possibly superior frequency performance this device would hardly ever replace the bipolar junction transistor. The problem which has plagued the SMS (and all other metal-base) transistors is their poor transfer ratio α (the common-base current gain). Even assuming an ideal monocrystalline SMS structure and extrapolating the base thickness to zero, the typical calculated values of α are unacceptably low – mainly due to the quantum-mechanical (QM) reflection of electrons at the base-collector interface. In our view, these conclusions of Sze and Gummel [29] remain valid today. The origin of the QM reflection problem may be related to the large Fermi energy of electrons in a metal base. Indeed, consider an (over)simplified model of a metal–semiconductor barrier (fig. 4a), in which parabolic energy-momentum relationships are assumed in both materials. The well-known solution of this QM problem gives for the above-barrier reflection coefficient R the following expression

$$R = \left(\frac{k-q}{k+q}\right)^2 = \left(\frac{1-\zeta}{1+\zeta}\right)^2, \quad \text{where} \quad \zeta = \left(1 - \frac{\Phi}{E}\right)^{1/2}, \tag{1}$$

$E = \hbar^2 k^2/2m$ is the hot-electron energy in the base, and $E - \Phi = \hbar^2 q^2/2m$ with Φ being the barrier height. Note that it is not the clearance $E - \Phi$ but the ratio E/Φ which enters the expression for R – and hence one must correctly choose the zero-energy level, including a large Fermi energy E_F. Typically, Φ/E is close to unity and the reflection is large. For a ballistic electron in Al incident on the interface with GaAs at 0.4 eV above the Schottky barrier ($\Phi \approx 12$ eV), the probability of reflection predicted by eq. (1) is $\sim 50\%$. It should be remarked, however, that in the above estimate we assumed an *abrupt* barrier – which is a somewhat pathological case [note, for example, the curious fact that $\hbar$ does not enter in expression (1) for a QM reflection coefficient – in contrast to eq. (2) below]. If we had assumed a "smooth wall" barrier, the expression for R would be qualitatively different. Consider an exactly soluble example of a barrier of height Φ graded over a distance a (fig. 4b), i.e. $\Psi(x) = \Phi[1 + \exp(-x/a)]^{-1}$. In this case the reflection coefficient is given by

$$R = \frac{\sinh^2[\pi a(k-q)]}{\sinh^2[\pi a(k+q)]}. \tag{2}$$

Equation (2) reduces to eq. (1) as $a \to 0$, but for $ka > 1$ (which only means that a is greater than the lattice constant in the metal) it gives

$$R = \left(\frac{1 - \tanh(\pi aq)}{1 + \tanh(\pi aq)}\right)^2.$$

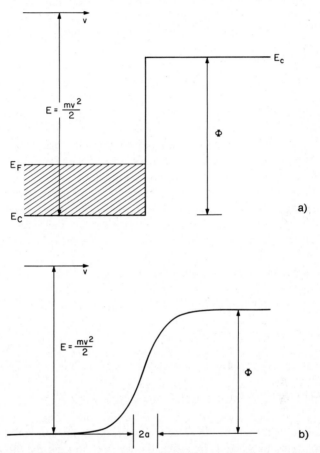

Fig. 4. Free-electron models of the quantum-mechanical above-barrier reflection: (a) abrupt barrier; (b) "smooth wall" barrier.

This implies an almost complete elimination of the QM reflection, if $E - \Phi \gtrsim \hbar^2/ma^2$.

Of course, estimates based on the simplest free-electron models of reflection, eqs. (1) and (2), are certainly invalid for metals with a complicated band structure and indirect-gap semiconductors. One should expect, however, that the band-structure effects will only further inhibit the QM transmission [30]. We are not aware of any metal–semiconductor pair, where the exact solution would predict a *lower* reflection of hot electrons than that predicted by eq. (1). Nevertheless, there have been recent reports [27,28] of a transistor action in monocrystalline $Si/CoSi_2/Si$ structures with α as high as 0.6. One cannot rule out some "accidental" resonance which aids the QM transmission of hot electrons in these devices. Such an interpretation, however, appears in our opinion to be unlikely. A more probable explanation is related to the existence of pinholes in the base metal film, i.e. continuous silicon "pipes" between the emitter and the collector. A careful analysis by Tung et al. [31] of the correlation between the pinhole sizes and the device characterictics revealed no evidence for any hot-electron component of the current through the base. On the other hand, the pinhole conduction in some cases gives α as high as 0.95. The device, therefore, works like a permeable base transistor [32,33], which in our classification is not a hot-electron device. This "natural" version of the PBT had, in fact, been proposed, and experimentally studied by Lindmayer [34] who called his device the metal-gate transistor (see also ref. [35]). Recently, Derkits et al. [36] fabricated by MBE a similar device in GaAs/W/GaAs and obtained a transistor action with $\alpha \approx 0.5$.

If the area of each pinhole is small, then it is often difficult to tell whether the PBT or the MBT mechanism has contributed to the observed I–V characteristics. Pfister et al. [37] suggested that the distinction can be made from a consideration of the collector "transconductance" $(\partial I_E/\partial V_{CB})|_{V_{EB}}$, which, according to their model calculations, should be substantially higher for the case of the pinhole conduction. Using this type of analysis, supported by transmission electron microscopy studies, Delage et al. [38] reported a transfer ratio as high as $\alpha \approx 0.3$ in a supposedly pinhole-free $Si/CoSi_2/Si$ MBT. We remark, however, that the thermionic emission through a permeable base has, in our opinion, a greater device potential than the hot-electron transport through a metal base, and thin silicide films may offer an attractive way of fabricating the PBT – if one learns how to control the statistics of pinhole sizes, making it sharply peaked at a desired area scale.

2.2.1. Doped-base transistors

The problem of QM reflections is largely avoided in the all-semiconductor ballistic hot-electron transistors. Even when the base is degenerately doped, the Fermi energy is typically less than 0.1 eV. It is possible, therefore, to arrange an injection energy so that, say, $\Phi/E \leq \frac{1}{2}$. In this case, even the worst-case eq. (1) gives $R \leq 0.03$. A number of such devices have been manufactured recently, using ion-implanted "camel" barriers [39,40], MBE grown planar-doped barriers [41–43] GaAs/AlGaAs heterostructure barriers grown by MBE [44–48] or OMCVD [49], and MBE-grown InGaAs/InAlAs barriers [50]. Both tunnel-emitter and thermionic-emitter versions of the ballistic hot-electron transistor have been implemented. Figure 5 shows the schematic energy-band diagrams of these devices.

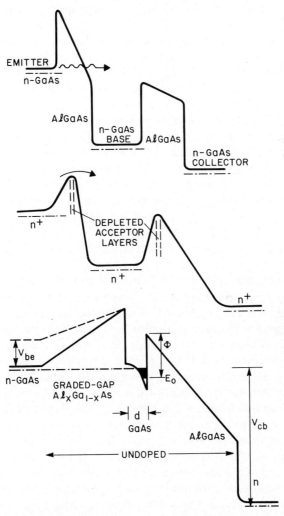

Fig. 5. Ballistic hot-electron transistors with a monolithic all-semiconductor structure: (a) tunnel-emitter (THETA) transistor, after ref. [44]; (b) planar-doped barrier (PDB) transistor, after ref. [41]; (c) induced-base transistor (IBT), after ref. [60].

Using the GaAs/GaAlAs heterostructure technology by MBE, Yokoyama et al. [44] manufactured a tunneling device THETA (tunneling hot-electron transfer amplifier, the name coined by Heiblum [24]) with a 500 Å thick $Al_{0.3}Ga_{0.7}As$ emitter barrier and a 1000 Å thick n-GaAs base. Evidently, in such a structure a significant tunneling current occurs via the Fowler–Nordheim mechanism, fig. 5a. A respectable (for hot-electron transistors) transfer ratio $\alpha = 0.28$ was observed in this device. Subsequently, Yokoyama et al. [45] reported a still higher gain $\alpha = 0.56$ obtained in this structure at lower temperatures (below 77 K). Recently, Heiblum et al. [51] reported ballistic THETA devices with even more impressive DC performance. These devices, implemented with GaAs/AlGaAs heterostructure barriers grown by MBE, had n-doped GaAs base layers of thickness varied from 300 Å to 800 Å. In devices with narrowest base widths the observed transfer ratio was $\alpha = 0.9$ – which implies a differential current gain $\beta \approx 9$, the highest seen in any hot-electron ballistic transistor with a doped base. Interestingly, the highest gain was seen at injection energies just below those corresponding to the energy separation to the nearest satellite (L) valley. At higher injection energies, the gain was reduced (from 9 to about 3) – presumably due to the energy losses in electron scattering into the L valleys. This hypothesis was confirmed by further experiments of Heiblum et al. [52] in which the variation of α was measured at varying hydrostatic pressures. Since it is known that the Γ–L separation decreases with pressure (at the rate of ~ 6.3 meV/kbar, see ref. [53]), application of pressures up to 10 kbar significantly lowers the maximum α. It is natural to expect a higher transport efficiency in ballistic devices implemented using heterostructures with larger separation to satellite valleys. This conclusion is supported by Monte Carlo simulations [54] of hot-electron transport through the base of an InGaAs/InP heterojunction ballistic transistor.

Also impressive are the recent results achieved with thermionic injection. A device of this type is shown in fig. 5b. Woodcock et al. [43] reported an $\alpha = 0.94$ in a planar-doped GaAs transistor with a base thickness of 360 Å. It should be noted, however, that these values were inferred from two-terminal measurements with a floating base. Such measurements are difficult to interpret correctly – not only because of the possible series resistances or extraneous leakage paths between the emitter and the collector, discussed by the authors – but also because of the possible charge trapping in the floating base layer, which may lead to a spurious internal gain in the presence of a hot-electron current. An example of anomalous effects in two-terminal characteristics of a transistor with base floating has been recently discussed by Derkits et al. [55].

One should understand the trade-off involved in the design of all hot-electron transistors with a doped base: cooling of hot-electrons by phonon emission and other inelastic processes (minimized by thin base layers) against the increasing base resistance for thinner layers. It is easy to estimate the RC delay associated with charging the working base–emitter capacitance and the parasitic base–collector capacitance through the lateral base resistance,

$$RC = \tau_B = \frac{\varepsilon L^2}{l\mu\sigma}, \qquad (3)$$

where l is the thickness of the emitter or the collector barriers, $l \sim 10^{-5}$ cm, L the characteristic lateral base dimension (shortest distance to the base contact from the geometric center of the base), $L \sim 10^{-4}$ cm, μ the mobility in the base, σ the mobile charge density per unit base area, and ε the dielectric permittivity. For a hot-electron transistor to be competitive, one must have $\tau_B \approx 1$ psec, which means that the sheet resistance in the base must be $(\mu\sigma)^{-1} \lesssim 1$ k$\Omega/\square$. On the other hand, one cannot make the base too thick, say thicker than 1000 Å, since that would lead to a degradation in α due to various energy-loss mechanisms: for example, hot electrons in GaAs lose energy at the rate of about 0.16 eV/ps due to the emission of optic phonons [56]. The limitation given by eq. (3) is rather severe. The minimum value of L is governed by the lithographic resolution. One cannot really make the barrier thicknesses l much larger than 1000 Å, since this would introduce emitter and collector delays of more than 1 ps. It may appear that increasing the base doping can resolve all the difficulties; however, if it is increased much beyond 10^{18} cm^{-3}, then one can expect a strong degradation in the transfer ratio due to plasmon scattering [57]. Imanaga et al. [58] performed a Monte Carlo simulation of a GaAs/AlGaAs ballistic hot-electron transistor at 77 K, including the plasmon scattering in addition to conventional scattering mechanisms. They concluded that even with a 100 Å thick heavily doped base one cannot expect an α better than 0.9 in a GaAs/AlGaAs device. It may appear that the results of Heiblum et al. [51] are pushing the theoretical maximum – but those results ($\alpha = 0.9$) were obtained at low injection energies, i.e. when the transfer to satellite valleys is reduced. In this regime, which can be more easily realized in materials with higher satellite valleys, such as InGaAs, Imanaga et al. [58] predicted the possibility of obtaining a higher transfer ratio even with the inevitable plasmon scattering.

Similar conclusions were reached by Levi et al. [59], who considered theoretically the mean free path of ballistic electrons in a doped GaAs base – limited at high doping levels by the inelastic interaction with coupled plasmon/phonon modes and the elastic ionized-impurity scattering. For $N_D \gtrsim 10^{18}$ cm^{-3} they obtained a maximum mean free path of order 300 Å. On the basis of these estimates, Levi et al. [59] declared the GaAs/AlGaAs system unsuitable for the fabrication of a viable doped-base ballistic transistor. They have also briefly discussed alternative semiconductor systems, suggesting two routes for possible improvement. One can either look for material with a higher satellite–valley separation (to take advantage of the decrease in the elastic scattering rate with increasing injected-electron energy) or for a material with lower effective mass (the latter corresponds to a lower density of states and therefore also a lower scattering rate).

2.2.2. Induced base transistor

An attempt to circumvent eq. (3) was made in a recent proposal by Luryi [60] of an induced-base transistor (IBT). In this device, shown in fig. 5c, the base conductivity is provided by a two-dimensional electron gas induced by the collector field at an undoped heterointerface. The density of the induced charge is limited by a dielectric breakdown in the collector barrier. For a GaAs/AlGaAs system this means $\sigma/e \lesssim 2 \times 10^{12}$ cm^{-2}. The IBT operation requires little or no lateral electric field in the base, so the device can take a direct advantage of the high electron mobility in a two-dimensional metal at an undoped heterojunction interface (see the review by Ando et al. [61]). The low-field electron mobility parallel to the layers is greatly enhanced (especially at lower temperatures) because of the suppressed Coulomb scattering of electrons by ionized impurities, due to

  (i) increased spatial separation from the scatterers, and
  (ii) a higher than thermal electron Fermi velocity in a degenerate two-dimensional electron gas, which reduces the scattering cross-section in accordance with the Rutherford formula.

At room temperature μ is limited by phonon scattering, $\mu \lesssim 8000$ cm^2/(V s) [6], giving $(\mu\sigma)^{-1} \approx 400$ $\Omega/\square$ at the highest sheet concentrations in the base *. The base sheet resistance is still much lower at 77 K.

---

* The maximum breakdown-limited charge concentration in a two-dimensional electron gas confined at an AlGaAs barrier is not a very well established quantity, as it seems to depend on the way of material preparation. The above quoted value of $\sigma/e \approx 2 \times 10^{12}$ cm^{-2} was based on our experimental results with the charge-injection transistor fabricated by MBE [63]. Recently, Kastalsky et al. [64] were able to apply voltages up to ~ 11 V across a 2000 Å OMCVD-grown AlGaAs barrier – without a breakdown at room temperature – which implies, in principle, the possibility of obtaining an induced charge as high as $\sigma/e \approx 4 \times 10^{12}$ cm^{-2}.

The base conductivity in the IBT is virtually independent of its thickness down to $d \lesssim 100$ Å. At such short distances the loss of hot electrons in the base due to scattering is small. Indeed, injected hot electrons, traveling across the base with a ballistic velocity of order 10^8 cm/s, lose their energy mainly through the emission of polar optic phonons. For $d = 100$ Å the attendant decrease in α is estimated to be about 1%. Energy losses to the collective and single-electron excitations of the two-dimensional electron gas are negligible. The IBT can be regarded as a metal-base transistor – with the notable difference that the base "metal" is two-dimensional. This permits a dramatic improvement in the transfer ratio α – mainly owing to a low quantum-mechanical reflection coefficient R at the collector barrier interface. Model calculations on the basis of eq. (1) of the above-barrier reflection give $R < 0.02$, so that the total α can be as high as 97% [60]. Recently, Chang et al. [62] have manufactured a first induced-base transistor by MBE. Their preliminary results showed a differential $\alpha_{max} \approx 0.96$ (in a relatively narrow range of applied biases).

Let us discuss some alternative possibilities for an implementation of the IBT. An injecting emitter barrier of triangular shape can be organized using either the graded-gap technique, as shown in fig. 5c [65], or using planar-doped barriers [66,67]. The undoped graded-gap version has the advantage in that it avoids the adverse effects of doping fluctuations – both on the barrier injection and the base conductivity (the latter being particularly important for shrinking the *lateral* device dimensions, as pointed out originally by Honeisen and Mead [68]). The second alternative is especially attractive when one wishes to employ heterostructure materials in which lattice-matched grading of the gap is difficult to achieve (such as, e.g., the InGaAs/InAlAs system, which is more desirable than GaAs/AlGaAs from the point of view of intervalley scattering in the base). Possible IBT structures of this type [69] are schematically shown in fig. 6. The structure must contain two built-in charge sheets – planar-doped acceptors and donors. The dopant concentration and the geometry must be designed so as to have both sheets depleted of mobile carriers providing a desired barrier height. The two-dimensional electron gas induced at the heterointerface by the collector field, as shown in fig. 6a (where a GaAs/AlGaAs system is assumed for illustration), should be separated from the donor sheet by an undoped setback layer of at least 50 Å – to ensure the benefit of enhanced mobility in the base.

A similar – though complementary in the dopant polarity – structure can in principle be implemented using Si/Ge heterojunction technology (fig. 6b), provided one is able to achieve a high-quality interface (cf. ref.

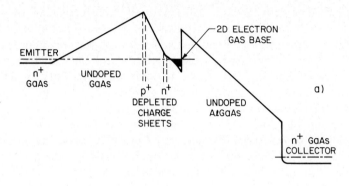
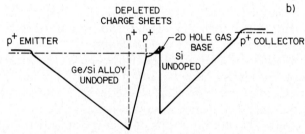

Fig. 6. Induced-base transistor with a planar-doped emitter barrier: (a) Conduction-band diagram for a planar-doped GaAs/AlGaAs IBT. (b) Valence-band diagram of a Ge/Si IBT.

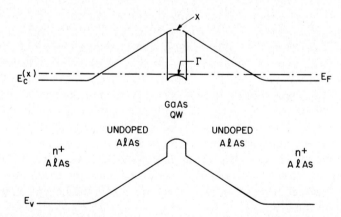

Fig. 7. Hypothetical band-diagram for an AlAs/GaAs heterostructure in which the discontinuities in the Γ valleys of the conduction band and the valence band are assumed split in the proportion 83:17. This rule would imply a continuous X-band minimum.

[70]). In this case one should use injection of hot holes because, unlike the conduction-band minima, valence-band maxima are located at the same $k = 0$ point in both semiconductors. The valence-band offset in a Si/Ge heterojunction is sufficiently large to confine holes in germanium: there is some experimental evidence [71] that depending on the strain conditions the bands can even staggered in these structures – with Ge conduction band being above that of Si.

Some heterostructure combinations may offer a fascinating possibility of employing *different conduction bands* for transporting injected carriers across the base and the lateral conduction in the base [69]. The band offsets in AlGaAs/GaAs heterojunctions at high aluminum concentrations ($x \to 1$) are not as well established at present as those for $x \lesssim 0.45$. This gives us the liberty to illustrate the idea with an AlAs/GaAs structure by assuming that the band discontinuities $\Delta E_c^{(\Gamma)}$ and ΔE_v are split in the proportion 83:17.

Consider a device whose band diagram is illustrated in fig. 7. Conduction-band bottom in the AlAs emitter is in the X valleys. If the above band offset rule is obeyed, then there is no discontinuity in the X band at the GaAs interface and, as far as the X electrons are concerned, the structure is not different from a homogeneous n–i–n diode. In the absence of a base voltage the emitter to collector current is space-charge limited. Electrons sail through the base without noticing the quantum well (apart from a finite probability of making an intervally transfer by spontaneous phonon emission). On the other hand, electrons in the GaAs quantum well (which may or may not be there at equilibrium and, when induced by the collector field, come from the base contact) are in the Γ valley. The collector current can be exponentially quenched by applying a base–emitter reverse bias. The device thus operates much like a bipolar junction transistor (of infinitesimal base thickness but good lateral conductivity) – with carriers in the subsidiary branch of the conduction band replacing the holes of an n–p–n structure. It would be very neat indeed if a heterostructure could be found to implement this idea *. Unfortunately, the $Al_xGa_{1-x}As/GaAs$ combinations would be "ruled-out", if the 60:40 proportion (or close to it) is found to persist for $x \to 1$, since in that case the X-band is strongly discontinuous at all values of x.

---

* Similar ideas have been suggested and independently communicated to the author by G.E. Derkits, R.F. Kazarinov, and S.A. Lyon.

2.3.3. Speed of ballistic transistors

Before leaving the subject of ballistic transistors, let us briefly discuss their potential frequency performance. It is sometimes stated that hot-electron transistors are capable of sub-picosecond operation because such is the time of flight of ballistic electrons across the base. That is a much too often repeated fallacy; the time of flight through the base has little to do with the intrinsic device speed. Like the bipolar, the FET, and most other transistors, hot-electron transistors have a regime in which their output current I rises exponentially with the input (base–emitter) voltage. In this regime, the maximum speed of operation is proportional to I. However, like every exponent in nature, this dependence eventually saturates and goes over into a linear law. One gains no further advantage in speed by increasing I, since the charge stored in all input capacitances will rise proportionally. Ultimately, the speed of a transistor is determined by the current level at which one has a crossover between the exponential and the linear regimes [72]. In transistors with a thermionic emitter this crossover occurs because of the accumulation of the mobile charge diffusing up the emitter barrier and drifting down the collector barrier. A rigorous g_m/C analysis leads to the characteristic delays $\tau_E = l_E/v_T$ and $\tau_C = l_C/v_S$, where l_E and l_C are the thicknesses of the emitter and the collector barriers, respectively, $v_T = (kT/2\pi m)^{1/2}$ is the thermal velocity of carriers, and $v_S \sim 10^7$ cm/s their saturated drift velocity. Of course, neither of the l's can be shrunk much below, say, 1000 Å – because of the complementary limitation given by eq. (3). We conclude that an ideally optimized ballistic transistor will be roughly a three picosecond device.

Speed limitations of thermionic-injection ballistic transistors appear to be quite similar to those of another important heterojunction transistor, the heterojunction bipolar transistor or HBT, shown in fig. 8. This device, first proposed by Kroemer [73] and experimentally studied by a number of groups (for recent references, see refs. [74–76]) is in some sense a unipolar device, as the minority-carrier injection into the emitter is practically suppressed by the band-gap discontinuity between the emitter and base layers. This allows one to use heavily doped base layers without degrading the gain. At the same time, the transit time of injected carriers across the base is reduced by incorporating a quasi-electric field in the base – by grading its bandgap and thus replacing a minority-carrier diffusion transport across the base by a relatively faster drift *. The HBT suffers from

---

* Interestingly, the reciprocity theorem [78,79] shows that the delay time for a signal propagating by diffusion up-hill in a graded-gap base is the same as that for a signal propagating by a drift down-hill.

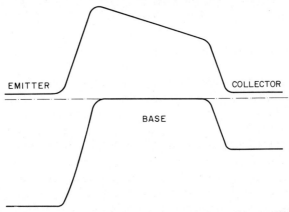

Fig. 8. Schematic illustration of a heterojunction bipolar transistor with a graded-gap base.

manufacturing difficulties similar to those of hot-electron transistors and offers similar potential speed advantages. Compared to induced-base transistors, the HBT does not benefit from an enhanced mobility in the base, but it compensates by offering the possibility of a larger sheet carrier concentration (the Gummel number). Which of the two heterojunction devices will eventually have a greater application potential will probably be decided by technological and circuit considerations, not by fundamental speed limitations.

Speed limitations of tunnel-injection ballistic transistors (THETA) may be somewhat different from those of thermionic-injection devices, since their emitter delay τ_E is no longer given by l_E/v_T and therefore τ_E does not increase with decreasing temperature. The difference, rooted in the physics of the injection process, manifests itself in the fact that the emitter current in THETA devices is only a weak function of the base–emitter bias (it varies only to the extent of the variation in the average tunneling-barrier phase area). Operation of these transistors consists in *switching* of the (approximately constant) emitter current between the base and the collector circuits. Consequently, the maximum transconductance achievable in THETA depends on the degree of collimation of the injected electrons in a narrow energy range, preservation of this narrowness in transit through the base, and on the selectivity of the collector-barrier analyzer. On the other hand, the fact that a tunneling barrier must be thin (≤ 100 Å) puts a more stringent requirement on the base resistance, since the THETA transistors are still subject to the limitation given by eq. (3). In order to be able to operate such transistors at picosecond speeds one must achieve a base sheet resistance $(\mu\sigma)^{-1} \lesssim 100 \, \Omega/\square$.

2.2.4. Hot-electron spectrometers

Ballistic transistor structures allow fundamental studies of the dynamics of non-equilibrium carriers in semiconductors. The idea of hot-electron spectroscopy [77] consists in the following: one measures the dependence of the collector current I_C on the collector–base bias V_{CB} at a fixed emitter–base bias V_{BE} and plots $(\partial I_C/\partial V_{CB})_{V_{BE}}$ versus V_{CB}. If the collector barrier height depends linearly on the bias, $\delta\Phi_C \propto \delta V_{CB}$, and to the extent that the collector bias does not affect the hot-electron energy distribution in the base, the emitter injection efficiency, and the above-barrier QM reflection, the resultant curve is proportional to the number of carriers arriving at the collector barrier with a normal component of the kinetic energy (i.e., the portion of the energy corresponding to the motion normal to the barrier) equal to the barrier height Φ_C.

The spectrometer of Hayes et al. [80,81] illustrated in fig. 9a, was based on a planar-doped-barrier transistor structure. In this structure an approximately linear $\Phi_C(V_{CB})$ dependence can be expected in a range of positive collector biases. The authors were able to observe ballistic transport of hot electrons through the GaAs base and obtain information about the dynamics of the electron energy loss. (For a detailed analysis of the results obtained and interpretation of these results in terms of various electron scattering mechanisms, the reader is referred to the papers by Levi et al. [82,83] and a recent review by Hayes and Levi [84].) The planar-doped barrier spectrometer appears to have the drawback that the width of the electron distribution injected into the base and the breadth of the analyzer filter function are both adversely affected by fluctuations in the barrier doping. These effects have been clearly shown in the work of Wang et al. [85] based on a Monte Carlo analysis of the planar-doped spectrometer structure. Recently, Long et al. [86] reported a ballistic spectrometer in which the planar-doped emitter was replaced by a graded-gap triangular-barrier structure. This structure was used [87] to study the effects of the width and the doping level in the base, as well as of a magnetic field applied in the direction of the current. Results were presented together with Monte Carlo simulations of the hot-electron spectra.

Another type of a hot-electron spectrometer – based on a tunnel emitter and an abrupt heterojunction collector barrier – was used in the works of Yokoyama et al. [45] and Heiblum et al. [46,47,51,52]. In these devices,

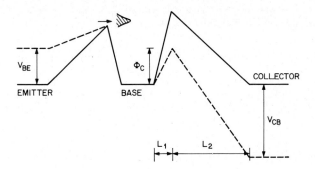

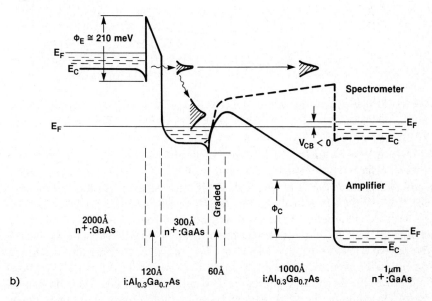

Fig. 9. Hot-electron spectrometers: (a) PDB (after ref. [81]). To a reasonable approximation $\delta\Phi_C = e\delta V_{CB} L_1/(L_1 + L_2)$; (b) THETA (after ref. [46]). For $V_{CB} < 0$ one has $\delta\Phi_C \approx e\delta V_{CB}$.

illustrated in fig. 9b, a linear $\Phi_C(V_{CB})$ dependence occurs at *negative* collector biases – with the analyzer plane located at the collector edge of the base–collector heterojunction barrier. Using such a heterojunction spectrometer, Heiblum et al. [47] were able to make an unambiguous

Energy Distributions

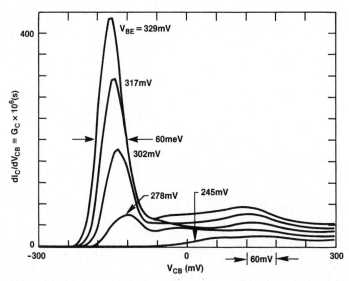

Fig. 9 (continued). (c) Example of a measured energy spectrum [47]. According to these authors, the main peak is due to electrons arriving at the analyzer plane without a single scattering event.

observation of the ballistic electron transport – which is particularly remarkable since the electrons had to traverse without scattering not only the GaAs base but also the entire length of the AlGaAs collector barrier (~ 1000 Å). Examples of the measured hot-electron energy distribution are shown in fig. 9c. The observed main peak increases with the collector current and occurs at the values of V_{CB} which are consistent with the theoretically expected values, assuming the barrier height $\Phi_C(V_{CB})$ calculated from the structure parameters. The authors concluded that the majority of electrons in the peak have arrived to the analyzer without a single scattering event – since otherwise the peak would be displaced by at least 36 mV (the optical phonon energy) to lower voltages, which with the data of fig. 9c could be accounted for only by postulating an unrealistically low barrier height. The sharper distributions obtained in heterojunction spectrometers are at least in part due to the absence of the dopant-fluctuation broadening mentioned above in connection with the planar-doped-barrier devices.

2.3. Real-space transfer devices

The term "real-space transfer" (RST) was coined by Hess et al. [9] to describe a new mechanism for NDR which they proposed and subsequently discovered in layered heterostructures. The original RST structure is shown in fig. 10. In equilibrium the mobile electrons reside in undoped GaAs quantum wells and are spatially separated from their parent donors in AlGaAs layers. Guided by an analogy with the momentum-space intervalley transfer, Hess et al. [9] suggested that carriers, heated by an electric field applied parallel to the layers, will move to the adjacent layers by thermionic emission, causing an enhancement of the mobile charge concentration in one set of layers and depletion in the other. Since the layers had different mobilities, the RST process was predicted to result in an NDR in the two-terminal circuit. This effect was discovered experimentally [88] and used for microwave generation [89]. Several RST diode configurations have been reviewed by Hess [90,91]. If the device is used as an oscillator, electrons must cycle back and forth between the high and low mobility layers. The maximum oscillation frequency is, in our view, limited by the

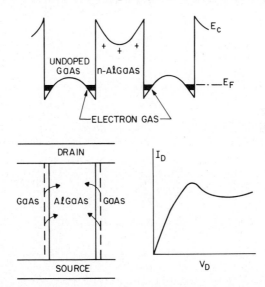

Fig. 10. The real-space transfer diode (after ref. [9]). Electrons, heated by an applied electric field, transfer into the wid-gap layers, where their mobility is substantially lower, giving rise to an overal negative differential mobility. At low temperatures the effect may be hysteretic, since even when the electric field is removed, the transferred electrons remain trapped in the wide-gap layers.

delay due to "cold" electrons returning from the potential "pockets" in the wide-gap layers, cf. fig. 10. This process occurs mainly by thermionic emission over the potential barrier created by the space-charge of ionized donors (the rate of other restoring processes, such as, e.g., electrons in the wide-gap layers drifting to the drain contact while the high-mobility layers being refilled from the source contact, appears to be slower at room temperature, because of the large values of the effective capacitance and resistance involved in such a recharging process). For a modulation-doped AlGaAs/GaAs heterostructure at room temperature one can estimate the return time to be at least 10^{-11} s and still longer at lower temperatures. On the other hand, the time constants involved in the initial transfer of hot electrons are considerably shorter [90].

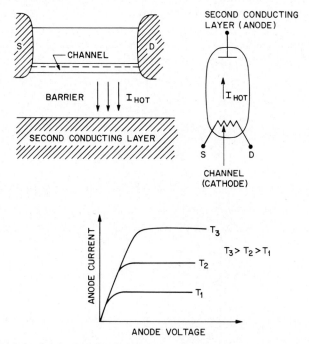

Fig. 11. Illustration of the principle of three-terminal real-space transfer devices (after ref. [63]). The channel serves as a cathode whose effective electron temperature T_e is controlled by the source-to-drain field. The second conducting layer, separated by a potential barrier, serves as an anode and is biased positively. To the extent that the barrier lowering by the anode field can be neglected, the anode current as a function of the anode voltage exhibits quasi-saturation at a value determined by T_e.

The important idea of real-space transfer was taken up in a recent proposal by Kastalsky and Luryi [10] of a *three-terminal* hot-electron device structure. In this structure the RST effect gives rise to charge injection between two conducting layers isolated by a potential barrier and contacted separately. The idea can be best illustrated by a glow-cathode analogy, shown in fig. 11. In a vacuum diode the anode current as a function of the anode voltage saturates at a value determined by the cathode work function and the temperature. One can think of a hypothetical amplifier in which an input circuit controls the cathode temperature and thus the output current, but that would be a slow device. In a three-terminal RST structure the input circuit controls the T_e which, unlike the temperature of a material, can be rapidly varied in one of the conducting layers ("the channel"), resulting in an efficient charge injection into the other layer. Based on this principle, several new device concepts were suggested, most of which by now have been demonstrated experimentally. In the following sections we shall review these devices: the charge-injection transistor or CHINT [63], the negative resistance field-effect transistor or NERFET [92,93], the hot-electron erasable programmable random access memory element or HE^{2}PRAM [94], and some of their possible circuit applications. Our discussion emphasizing the physics of the device operation will be based on the earlier reviews by Luryi and Kastalsky [95,96] as well as on the more recent theoretical work by Grinberg et al. [97].

2.3.1. Three-terminal RST devices

The basic structures used for three-terminal RST devices are illustrated in fig. 12. In the original structure the second conducting layer was implemented as a conducting GaAs substrate separated by a graded-gap AlGaAs barrier from the channel of a modulation-doped FET with source (S) and drain (D) contacts, fig. 12a. Details of the MBE growth and processing are described in ref. [63]. This device had an auxilliary fourth electrode (gate) which concentrated the lateral electric field under a 1 μm wide notch. In the more recent work [98], both the gate electrode and the modulation-doping were eliminated, see fig. 12b, and the channel was induced at the undoped heterointerface by a back-gate action of the second conducting layer. Also in the new structure the rectangular potential barrier provides a better insulation between the two conducting layers. Still more recently, a substantial progress was achieved by Kastalsky et al. [64,99] with a structure similar to that in fig. 12b but grown by OMCVD instead of MBE. (Even though in these structures the second conducting layer is implemented as a heavily

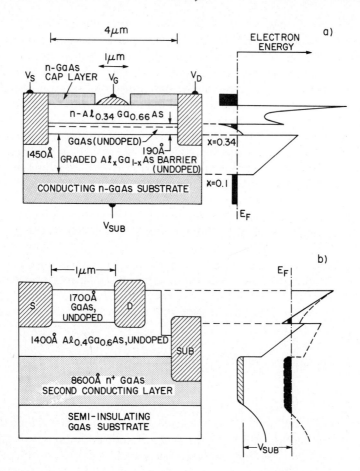

Fig. 12. Cross-section and the energy band diagram of three-terminal RST devices: (a) Type-1 structure, first implemented by MBE [63], had a MODFET-like channel separated from the second conducting layer (the SUB electrode) by a graded $Al_xGa_{1-x}As$ barrier. Two-dimensional electron gas is present in the channel even at $V_{SUB} = 0$ as well as with a floating SUB. Furthermore, the quasi-electric field in the barrier aids the drift of injected electrons toward the second conducting layer. However, the barrier breakdown occurs in this structure at relatively low values of V_{SUB}, which limits the operating range. (b) Type-2 structure has no electrons in the "channel" – until it is induced by a positive $V_{SUB} > V_T$. The barrier is ungraded, which helps a better insulation of the second conducting layer from the channel. This structure was first implemented by MBE [98] and then by OMCVD [64,99].

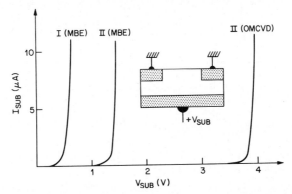

Fig. 13. Barrier insulation between the channel and the second conducting layer at room temperature (after refs. [63,64,96,99]). Characteristics shown correspond to the positive voltage applied to the SUB electrode, which is the situation in the operating regime of both CHINT and NERFET. For the opposite bias polarity, the characteristics were approximately symmetric in the case of a rectangular-barrier type-2 structure and strongly asymmetric for the triangular-barrier type-1 structures, as can be expected from the band diagrams in fig. 12.

doped n-GaAs layer on a semi-insulating substrate, we shall keep the designation SUB for this electrode.)

A critical step in manufacturing the three-terminal RST devices is to provide ohmic contacts to the two-dimensional electron gas in the channel, while preserving the insulation from the SUB layer. Figure 13 shows the characteristics of a diode formed between the collector electrode and S, D terminals tied together and grounded. A better insulation at 300 K in the type-2 device is evident, especially for OMCVD grown structures. For the graded-gap structure the diode characteristics were strongly asymmetric (not shown in fig. 13, cf. refs. [63,95]), as can be expected for thermionic emission over a triangular barrier. The observed current at $V_{SUB} > 0$ is probably due to a combination of barrier lowering and (thermally assisted) tunneling, especially at lower temperatures.

2.3.1.1. Various bias configurations and I–V characteristics

The charge-injection transistor (CHINT) is a solid-state analog of the hypothetical vacuum diode with controlled cathode temperature, discussed above in connection with fig. 11. Application of a voltage V_{SD} produces a lateral electric field which heats the channel electrons and leads to an exponential enhancement of charge injection over the barrier. Figure 14 displays the collector characteristics of the CHINT as a function of the heating voltage V_{SD} with the collector voltage V_{SUB} as a parameter, ob-

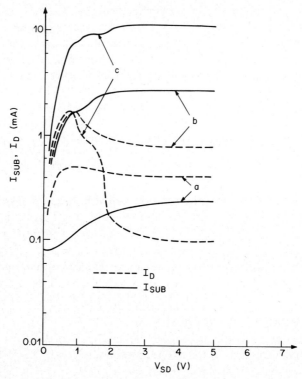

Fig. 14. Typical experimental current–voltage characteristics, I_{SUB}–V_{SD} and I_D–V_{SD}, at room temperature and different collector voltages V_{SUB}–V_T in OMCVD-grown CHINT/NERFET devices (after refs. [64,99]): (a) $V_{SUB} - V_T = 1$ V; (b) $V_{SUB} - V_T = 2$ V; (c) $V_{SUB} - V_T = 2.7$ V. The threshold voltage $V_T \approx 5$ V. With increasing heating voltage V_{SD}, one clearly sees a rapid rise and subsequent saturation of the collector current I_{SUB}, accompanied by an increase and subsequent sharp drop in the drain current I_D.

tained by Kastalsky et al. [99] in OMCVD-grown structures. In the operating regime, this device had a transconductance $g_m \equiv (\partial I_D / \partial V_{SD})|_{V_{SUB}}$ of more than 1000 mS/mm at room temperature.

As evident from fig. 14 (dashed lines), the hot-electron injection in CHINT is accompanied by a strong NDR in the channel circuit. This permits the implementation of a related device called the NERFET (negative-resistance FET), whose characteristics were predicted by Kastalsky and Luryi [10] and experimentally studied by Kastalsky et al. [63,64,92,98]. Typical NERFET characteristics are shown in fig. 15. We note that the NDR is strongly affected by V_{SUB}. It is clear that higher V_{SUB} enhances the

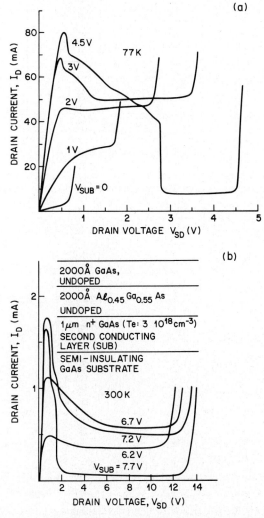

Fig. 15. The NERFET characteristics: (a) Type-1 structure at 77 K (after ref. [92]). Gate dimensions: 1 μm×250 μm; (b) Type-2 structure at 300 K [64]. Channel dimensions: 2 μm×100 μm; Note the existence of a region of a strongly pronounced *negative transconductance* $g_m \equiv (\partial I_D / \partial V_{SD})|_{V_{SUB}}$ in the quasi-saturation regions of the characteristics for both types of structures. This important effect is not yet fully understood, see the discussion in sect. 2.3.2.2.

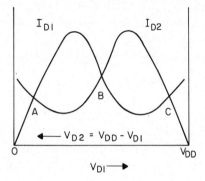

Fig. 16. Graphical construct for determining the operating points of a circuit formed by two identical NDR elements in series. Points A and C are stable, point B is unstable.

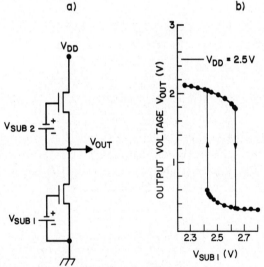

Fig. 17. Simplest NERFET logic circuit [98]. (a) Schematic diagram. Conventional FET circuit symbols are used with the understanding that the SUB electrode plays the role of a gate. (b) Logic transitions at room temperature. The output voltage was measured at fixed V_{SUB2} = 2.5 V as a function of V_{SUB1} slowly varied in the direction shown by the arrows. An interesting observation was that the switch points were sharply defined and repetitive to within less than 1 mV (i.e., to within $\ll kT/e$) in a given device pair. For the displayed example of a fixed V_{SUB2} = 2.5 V, the transition from the low to the high state always occurred at V_{SUB1} = 2.424 V – independently of the value of V_{DD}, provided the latter is ≤ 3.5 V, at which value the new feature of a third stable state appeared, characterized by V_{OUT} being in a range near $\frac{1}{2}V_{DD}$.

electron concentration in the channel (back-gate action), but it also affects the magnitude of the hot-electron flux corresponding to a given T_e (by lowering the collector barrier). The highest peak-to-valley ratio in current obtained in these devices was 160 at room temperature, as observed in OMCVD devices [64].

The main advantage of NERFET over two-terminal microwave generators lies in the possibility of controlling the oscillations by a third electrode. This advantage can also be used in logic applications, as discussed below. When two negative-resistance devices (like Esaki diodes or Gunn diodes) are connected in series and the total applied voltage V_{DD} exceeds roughly twice the critical voltage for the onset of NDR in the single device, then an instability occurs in which one of the devices takes most of the applied voltage, that is to say, contains a high-field domain, while the other is in the low-field mode. This is illustrated by the usual load-line graphical construct, shown in fig. 16. As is well-known, the operating points A and C are stable, while B is unstable. Which of the devices contains the domain is determined by an accidental fluctuation or, if the system is prepared in one of the stable states, by the history. Various schemes have been proposed to utilize this bistability. Due to the existence of a controlling electrode, NERFET offers new possibilities for logic.

Room-temperature operation of a simplest NERFET logic circuit [98] is illustrated in fig. 17. Two type-2 NERFETs with nearly identical characteristics were connected in series, as shown in fig. 17a. One of the controlling voltages was fixed, $V_{SUB2} = 2.5$ V, and the output voltage V_{OUT} was measured as a function of V_{SUB1}, see fig. 17b. As the controlling voltage V_{SUB1} is varied, the system smoothly approaches the switch points (sharply defined and repetitive within 1 mV), at which V_{OUT} jumps between the low and the high values. Such a behaviour with a strong hysteresis is reminiscent of a phase transition.

Two types of logic operation can be thought of in this configuration. Firstly, we can DC pre-bias our input voltage to a value in the middle of the hysteretic loop, say at $V_{SUB1} = 2.5$ V. Applying controlling signals $\Delta V_{SUB1}(t)$ in the form of short low-amplitude ($|\Delta V| \gtrsim 0.15$ V) pulses of varying polarity, we have a *bistable element*: the system will "remember" the sign of the last pulse, i.e. V_{OUT} = high for $\Delta V < 0$ and V_{OUT} = low for $\Delta V > 0$. A second type of logic operation – *inverter* action with amplification – can be obtained by DC pre-biasing V_{SUB1} to high enough voltages ($V_{SUB} \gtrsim 2.7$ V) to ensure a stable low state. The system will then switch to its high state only during a pulse of negative polarity $|\Delta V_{SUB1}| \gtrsim 0.3$ V. Both operations have been demonstrated using pulse-mode experiments.

Similar effects were found [100] in a modified circuit consisting of a single NERFET loaded on a passive resistance in such a way that the load-line diagram allowed for two stable points.

The width of the hysteresis decreases with increasing V_{DD}. At $V_{DD} \gtrsim 3.0$ V, the output swing of the inverter is capable of switching a second inverter without any level shifting, so that direct coupling of inverters, such as in a ring oscillator, is possible. At $V_{DD} > 3.5$ V a new feature was observed [98]: the appearance of a third stable state characterized by V_{OUT} being in a range near $\frac{1}{2}V_{DD}$. Within that range the dependence V_{OUT} versus V_{SUB1} was strictly linear with a voltage gain of nearly four and nonhysteretic. To our knowledge, a tri-stable operation with a stable midpoint has never been observed before with two series-connected voltage-controlled NDR elements. It may become useful for ternary logic.

Another RST logic device can be based on a memory effect observed by Luryi et al. [94] using the same structure (fig. 12a) – but with an unbiased second conducting layer. In this case, the hot-electron injection leads to a charge accumulation in the floating layer and a drop in its electrostatic potential Ψ_{SUB}, which persists for a long time after the heating voltage V_{SD} is set to zero, fig. 18a. The negative Ψ_{SUB} depletes the channel much like it happens in non-volatile memory devices (cf. ref. [19], p. 500). This leads to an NDR in the channel circuit. Such an effect had been predicted by Price [101]. In his words, the disconnected conducting layer "will act as a giant trap causing negative differential mobility to occur". We should note, however, that this is a hysteretic NDR, not capable of generating oscillations – in contrast to the operation of NERFET. It indicates a charge accumulation due to the hot-electron injection into the floating substrate (which remains charged even after the heating voltage is removed). The thermoelectric force developed between the two conducting layers has a characteristic decay time determined by the ambient temperature and the barrier height. Transferred electrons remain mobile and can be rapidly discharged by grounding the second conducting layer (in contrast to the situation with the non-volatile memory devices). Based on this effect, Luryi and Kastalsky [95] proposed a memory device, which allows a fast operation of all functions: *write*, *read*, and *erase*. It received the name HE^{2}PRAM (hot-electron erasable programmable random access memory). The structure (fig. 18b) must be grown on an insulating substrate, followed by a thin conducting GaAs layer. The key new element is the guard-gate MESFET-like structure G and the second, "deep", drain D_2, contacting both the channel and the second conducting layer. Electrically, D_2 is connected to the source S. When the guard-gate voltage is negative, the

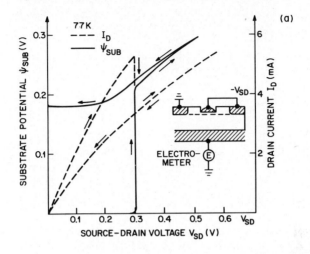

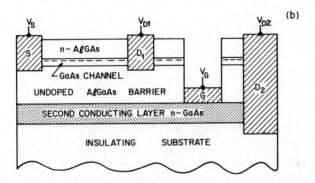

Fig. 18. Memory effect in a CHINT/NERFET structure with floating SUB electrode: (a) Substrate potential Ψ_{SUB} and the channel current I_D as functions of the heating voltage V_{SD}. Arrows indicate the direction of slow (10 mV/s) voltage ramping (after ref. [94]). (b) Schematic cross-section of the proposed HE²PRAM logic element (after ref. [95]). Thickness ($\sim 10^{-5}$ cm) and the doping level in the second conducting layer must be chosen so that this layer can be depleted by the gate field.

substrate conducting layer is isolated from D_2. Applying voltage to D_1 (*write*), we charge up this layer by the hot electron transfer and thus deplete the main channel. Information is *read* by probing the channel resistance. Applying a positive voltage to G, this information can be *erased* (with a characteristic MESFET time). The maximum amount of transferred charge is limited by the "cold" thermionic emission in the triangular-barrier diode forward-biased by Ψ_{SUB}.

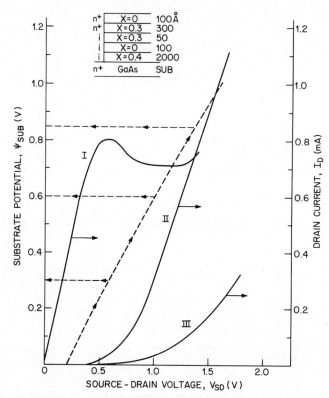

Fig. 19. Memory effects in an OMCVD-grown modulation-doped structure with a "rectangular" barrier separating the channel from the second conducting layer (courtesy of A. Kastalsky, unpublished results). Structure cross-section is shown in the inset; source–drain distance 2 μm, channel width 100 μm. Measurements were carried out in the dark at $T = 77$ K. Solid lines show the channel current I_D. Curve I is obtained prior to collapse (pre-illuminated sample) by a slow (0.5 V/min) ramping of V_{SD}. Curves II and III are measured after the collapse, with curve II corresponding to SUB grounded and III to SUB floating. Transition between these curves occurs without a noticeable delay. The dashed line represents the substrate potential Ψ_{SUB} as the heating voltage V_{SD} is ramped up. If at any point V_{SD} is reduced or set to 0, the potential Ψ_{SUB} stays fixed for a long time at the highest attained value.

One can substantially enhance the memory effect by using thin trapezoidal or rectangular barriers instead of the triangular barrier used in the first demonstration. Clearly, the type-2 structure illustrated in fig. 12b cannot be used – since it has no channel in the absence of a positive SUB bias. Figure 19 shows the results obtained in an OMCVD-grown structure *with* modulation doping. It does give a residual Ψ_{SUB} in excess of 1 V –

which corresponds to a retained charge of order 3×10^{-11} electrons per cm^2. However, the operation of a modulation-doped Al$_x$Ga$_{1-x}$As structure at $V_{SD} \geq 0.5$ V in the dark is associated with the unwelcome *collapse* phenomena (see, e.g., ref. [102]).

2.3.2. Theory of charge injection in three-terminal RST devices

Recently, Grinberg et al. [97] developed an analytical theory of charge injection in the CHINT/NERFET devices. The main simplifying approximation made in that work was the assumption of a uniform lateral electric field in the channel, $F_{SD} = V_{SD}/L$. (This approximation is not entirely satisfactory and results in the inability to give a complete quantitative account of the device behavior, especially in the range of current saturation.) Another approximation made, was that electron scattering into the subsidiary minima (L and X) of the conduction band was neglected. (That approximation may be justified by the fact that for $x \approx 0.4$, the satellite valleys are nearly continuous at the Al$_x$Ga$_{1-x}$As/GaAs interface, so that the transferred electrons face no barrier for a subsequent real-space injection.) Finally, the theory did not include the dynamical screening effect discussed in detail by Luryi and Kastalsky [95], which arises due to the space-charge of injected carriers drifting downhill in the collector barrier.

Hot electrons are injected into the collector either by a thermionic emission over the barrier or by a "thermally assisted" tunneling under the top of the barrier. Grinberg et al. [97] have shown that the tunneling component cannot be neglected in the calculation of device characteristics. Although, for $kT_e \lesssim E_2 - E_1$ (where $E_2 - E_1$ is the subband separation), most of the channel electrons are two-dimensional (i.e. located within the lowest subband E_1), those electrons which participate in the charge injection over the barrier (of height $\Phi \approx 0.3$ eV equal to the conduction-band discontinuity between Al$_x$Ga$_{1-x}$As and GaAs at $x \approx 0.4$) are located in the high-energy tail of the hot-electron distribution function and, therefore, must be treated as three-dimensional. Their flux over the barrier Φ is hence given by a Richardson-like equation

$$J = \gamma A^* T_e^2 \, e^{(E_F - \Phi)/kT_e}, \tag{4}$$

where J is the current density of hot-electron injection, and A^* is the effective Richardson constant containing the electron effective mass in the channel (because the effective electron mass in the channel is lower than that in the AlGaAs barrier). It is assumed in eq. (4) that the shape of the barrier is such that the reverse flux of electrons from the collector into the channel can be neglected (because the collector temperature is low) even when $V_{SD} > V_{SUB}$. Inclusion of tunneling gives rise to an additional factor γ in the current equation given by eq. (4). Although the shape of the potential barrier is trapezoidal, one can regard it as approximately triangular near the top of the barrier – where most of the tunneling occurs. The usual quasiclassical theory for tunneling under a barrier gives in this case

$$\gamma = 1 + \int_0^{\Phi/kT_e} \exp\left(\xi - (\xi/E_{00})^{3/2}\right) d\xi,$$

$$\text{where} \quad E_{00} = \frac{1}{kT_e}\left(\frac{3\hbar e F_\perp}{4(2m)^{1/2}}\right)^{2/3}, \tag{5}$$

and $F_\perp$ is the electric field near the top of the barrier, assumed given by the Gauss law $F_\perp = 4\pi en/\varepsilon$, where $n = n(x)$ is the electron sheet concentration in the channel.

The problem of electron-gas heating and determination of the distribution function of hot electrons is quite involved in general. In the two-dimensional case it is further complicated by the subband structure which requires the consideration of not only intra- but also inter-subband transitions, as discussed in the extensive literature on the subject, see the review by Ando et al. [61]. In a phenomenological treatment of the device characteristics, it is unreasonable to include all the details of the electron heating problem – which would be appropriate in a study of the heating effect itself. Instead, Grinberg et al. [97] used the simplest and most common approximation in which the non-equilibrium distribution is assumed to differ from the equilibrium case only by the temperature T_e, and the latter is evaluated from the energy balance equation. As is well known, the establishment of such a quasi-equilibrium distribution can be ensured by a sufficiently strong electron–electron interaction, which in turn is realized at high enough electron concentrations. Although the electron concentration decreases substantially along the channel due to the injection effect, nevertheless the main portion of the collector current flows from the channel region where the concentration is still high enough ($\geq 10^{11}$ cm^{-2}). One can expect, therefore, that the electron temperature approximation is satisfactory for the purpose of calculating the injection current (see, however, the discussion in sect. 2.3.2.2).

2.3.2.1. Energy balance equation

Within the assumed approximation of a uniform electric field F along the channel, the energy balance equation is of the form:

$$w_1 - w_2 - w_3 + (\mathbf{I} \cdot \mathbf{F})/n = 0, \tag{6}$$

where $I = I(x)$ is the channel current per unit width of the channel. Besides the usual terms corresponding to the energy gain in the external field $(\mathbf{I} \cdot \mathbf{F})$ and the energy loss to the lattice, assumed to be mainly due to the emission of optical phonons by hot electrons, and described by the term $w_3 \equiv n^{-1} \partial \langle E_{ph} \rangle / \partial t$ (where $\langle E_{ph} \rangle$ is the average phonon energy per unit area of the two-dimensional electron gas), this equation contains two additional terms, associated with the electron emission over the barrier. These terms, proportional to the injection current density J, are entirely novel and specific to the three-terminal RST devices.

On the one hand, electrons leaving the channel carry away their energy, at the rate given by (per channel electron),

$$w_2 = \frac{2}{(2\pi)^3 n} \int E f(E_k) \, v_\perp D(E_\perp) \, \mathrm{d}^3 k, \tag{7}$$

where $v_\perp = \hbar k_\perp / m$ is the component of electron velocity normal to the interface, $f(E)$ the electron Fermi distribution function corresponding to the temperature T_e, and $D(E_\perp)$ is a factor describing the barrier transparency, which was taken equal to unity when the transverse kinetic energy component exceeds the barrier height, $E_\perp > \Phi$, i.e., the probability of a quantum-mechanical above-barrier reflection was neglected [as in eq. (5) above].

On the other hand, the nonconservation of channel current results in a peculiar energy transport along the channel, described by the term w_1. Indeed, under stationary conditions the net number of electrons arriving at a point in the channel from the adjacent regions must equal their number leaving the channel at that point. The hot-electron flux per unit area of the channel is given by J/e, with J related to I by the continuity equation $J(x) = -\partial I/\partial x$. In the absence of electron transfer across the barrier ($J = 0$), the divergence of the electron energy-density flux must equal the gradient of the average energy density. Neglecting processes of thermal conductivity, the total variation of the energy of a given elementary volume would be owing only to the difference in the average energy of electrons entering at point x and leaving at point $x+\delta x$. If we neglect that difference (the Thomson heat), then the total variation of energy along δx vanishes. In the presence of a real-space transfer, however, the incoming flux of electrons entering an elementary volume exceeds their outgoing flux after the distance δx by the amount $(J/e)\,\delta x$ and the taken-away energy equals $\langle E \rangle (J/e)\,\delta x$ per unit length of the channel. The rate of energy change per channel electron at a given point is, therefore, of the form

$$w_1 = (J/ne)\langle E \rangle, \tag{8}$$

where $\langle E \rangle$ denotes the mean electron energy of a channel electron. Both J and $\langle E \rangle$ are position dependent, in general, but the value of $\langle E \rangle$ and, therefore, of w_1 can be expressed in terms of the local quasi-Fermi level E_F or, equivalently, in terms of the local concentration $n(x)$ and the temperature $T_e(x)$ (same applies to the other two terms in eq. (6), i.e. w_2 and w_3).

Figure 20 shows the comparison of the calculated loss rates w_1, w_2 and w_3, as functions of the electron temperature at two different channel

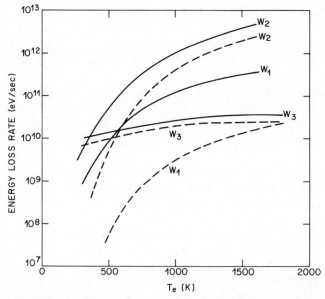

Fig. 20. The calculated energy loss rates per one channel electron for the different processes contributing to the energy balance equation, eq. (6) [97]. Solid lines correspond to the channel concentration $n = 10^{12}$ cm^{-2} and dashed lines to $n = 10^{11}$ cm^{-2}.

concentrations. At high electron temperatures, the energy carried away by hot electrons injected across the barrier (the term w_2) becomes dominant in the balance equation *. It gives rise to a feedback mechanism which stabilizes the T_e and thus limits the injection current. Dependences on the concentration $n(x)$ arise for several reasons, discussed by Grinberg et al. [97]. These include:

(a) variation of the electron quasi-Fermi level;

(b) barrier lowering and the enhancement of its tunneling transparency by a higher $F_\perp \propto n$; and

(c) variation in the efficiency of the electron–phonon interaction with the degree of confinement of the two-dimensional electron gas (the more compressed is the electron wavefunction, the shorter is the wavelength of the phonons effectively involved in the interaction).

Inasmuch as all terms in eq. (6) depend only on the local concentration $n(x)$ and the electron temperature $T_e(x)$, the latter can be expressed as an implicit function of n. Therefore the density of the collector current J can also be regarded as a function of the concentration. To determine the dependence $T_e(n)$, Grinberg et al. [97] solved eq. (6) for T_e at a fixed n. Thus determined electron temperature was then used to calculate $J(n)$ and the latter was substituted in the current continuity equation – which thus becomes a differential equation for the concentration $n(x)$.

The sheet carrier concentration $n(x)$ in the NERFET channel is therefore essentially governed by the process of hot electrons leaving the

---

* It should be noted that in the absence of a real-space transfer (say, if the barrier had "infinite" height, so that both the w_1 and the w_2 terms would vanish) the optical-phonon term w_3 alone can compensate the input power $I \cdot F$ only for fields $F \lesssim 4$ kV/cm. At higher fields our approximation in this situation would lead to a run-away breakdown process. Clearly, the reason for such a behavior is associated with our neglect of the transfer into heavy-mass valleys, which would reduce the rate at which the $I \cdot F$ term increases with the field. We believe, however, that in a GaAs/Al$_x$Ga$_{1-x}$As CHINT/NERFET structure, for $x \approx 0.4$, the momentum-space transfer processes may be neglected (in the balance equation) for the following reason: In this structure the barrier height Φ is approximately equal to the intervalley Γ–L energy separation. The transferred L electrons see virtually no barrier for real-space transfer and their collection efficiency is near unity (the situation with those electrons is quite analogous to that with minority carriers in a bipolar transistor). The energy carried away by an L electron is of the same order as that for Γ electrons. Therefore, this process (transfer into a satellite valley with subsequent diffusion across the barrier) is just another real-space transfer channel, and its contribution to the balance equation is of the same form as w_2. Taking account of satellite valleys would, therefore, lead to a slight increase of the w_2 rate compared to that calculated above.

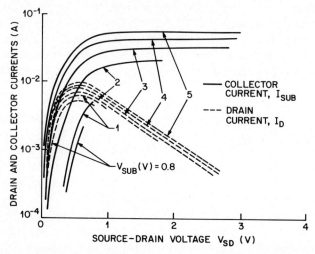

Fig. 21. Calculated current–voltage characteristics of a model CHINT/NERFET device [97]. Parameters assumed are as follows: Lattice temperature $T = 300$ K; barrier thickness $d = 2 \times 10^{-5}$ cm; channel length $L = 2 \times 10^{-4}$ cm; channel width $W = 10^{-2}$ cm; static permittivity of GaAs $\varepsilon_0 = 12.85$; high-frequency permittivity of GaAs $\varepsilon_\infty = 10.9$; barrier height $\Phi = 0.3$ eV; optical phonon energy $\hbar\omega_0 = 35.2$ meV; electron effective mass $m = 0.067\, m_0$; saturation velocity $v_{\text{sat}} = 2 \times 10^7$ cm/s.

channel. Capacitive coupling enters only as a boundary condition at the source end of the channel:

$$en(0) = \frac{\varepsilon}{4\pi d}(V_{\text{SUB}} - V_{\text{T}}), \tag{9}$$

where d is the thickness of the collector barrier and V_{T} is a threshold voltage for the appearance of a two-dimensional electron gas in the channel induced by V_{SUB}. In the actual calculations, Grinberg et al. [97] measured V_{SUB} relative to the threshold level – setting $V_{\text{T}} = 0$.

Figure 21 shows the calculated current–voltage characteristics of a CHINT/NERFET device with assumed parameters listed in the caption. The channel current per unit width was assumed in the form $I(x) = en(x)v_{\text{d}}$, with the drift velocity of channel electrons given by

$$v_{\text{d}} = \mu \frac{V_{\text{SD}}}{L} \quad \text{if} \quad \mu \frac{V_{\text{SD}}}{L} \leq v_{\text{sat}},$$

$$= v_{\text{sat}} \quad \text{if} \quad \mu \frac{V_{\text{SD}}}{L} \geq v_{\text{sat}}, \tag{10}$$

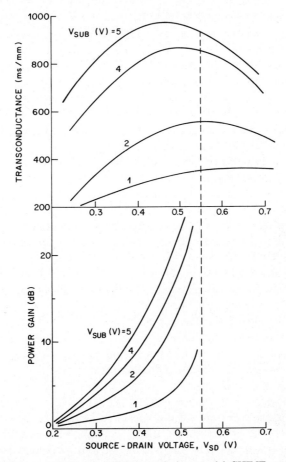

Fig. 22. Calculated transconductance and power gain in a model CHINT structure with the parameters assumed as in fig. 21 for different bias configurations. The dashed line indicates the onset of the NDR in the source–drain circuit of the device, which is approximately independent of V_{SUB}.

where μ is the mobility and v_{sat} the saturation velocity of electrons in the channel. Figure 22 shows the transconductance $g_m \equiv (\partial I_{SUB}/\partial V_{SD})|_{V_{SUB}}$ and the power gain in the CHINT, calculated for different bias configurations. As expected, g_m increases with the collector voltage V_{SUB}. As a function of V_{SD}, the transconductance peaks near the onset of the NDR in the drain circuit. At higher voltages g_m decreases – due to the saturation of the collector current. Peak values of the transconductance and the qualitative functional dependences $g_m(V_{SD}, V_{SUB})$ are in agreement with the experimental data of Kastalsky et al. [99]. It should be mentioned, however, that the experimental curves $g_m(V_{SD})$ exhibit a sharper rise than those seen in the calculated dependences. The calculated power gain in the model CHINT structure is shown in the bottom figure of fig. 22. Note that as a function of V_{SD} the gain becomes infinite when the device is biased into the NDR region. Physically, in this region the system goes over from the small-signal amplification regime into the regime of spontaneous oscillations. The calculated behavior of the power gain in the vicinity of the NDR threshold (the latter is shown in fig. 22 by a dashed line) fits very well the experimental observations.

The calculated characteristics also describe a strong NDR in the drain circuit of the device, i.e., the operation of a NERFET. The theory predicts a current drop by a factor of 30 in the bias range considered, which compares well with the typical experimental data at room temperature (fig. 14). Note that in the NDR region of the NERFET the CHINT collector current saturates, and therefore the operation range of CHINT occurs prior to the onset of the NDR, as discussed above. Figure 21 also shows that even near the peak of I_D the collector current dominates ($I_{SUB} \gg I_D$) – which implies a high current gain. These theoretical results are in good agreement with the experimental observations. Grinberg et al. [97] have shown that inclusion of the tunneling component of the injection current significantly improves agreement between the theory and experiment.

2.3.2.2. Discussion of the theory

Some of the major features of the operation of three-terminal RST devices are adequately described by the theory. Calculations correctly predict the existence of a strong NDR in the drain circuit accompanied by a rapid increase and subsequent saturation of the collector current. Moreover, the theory describes the experimental fact that at high enough heating voltages V_{SD} the collector current substantially exceeds the drain current (inclusion of the tunneling component of the injection current turned out to be crucial in establishing this fact, which is very important for obtaining a high current gain in the operation of CHINT).

However, two important experimental features were not captured by the model. Firstly, the theory does not describe the observed saturation of the drain current from below – after the NDR region. Secondly, although the theory correctly shows the increase in the peak drain current with higher collector voltage V_{SUB} (which is simply a back-gate field effect), it does not describe the salient (and quite puzzling) experimental feature of the *nega-*

tive transconductance in the saturation region. This effect (drop in the saturated I_D with increasing V_{SUB}) is clearly evident in figs. 14 and 15.

Grinberg et al. [97] suggested that the major deficiency of their theory consisted in the assumption of a uniform field in the channel and the neglect of the diffusion component of the channel current. Another process not included in the theory is the dynamical screening effect by the injected charge drifting down-hill in the collector barrier (see ref. [95]). The negative space charge dynamically stored (i.e. stored while in transit) in the AlGaAs barrier layer screens the backgate field and thus depletes the channel. The associated space-charge potential can be regarded as a threshold shift in a field-effect transistor in which V_{SUB} plays the role of a gate bias. This effect becomes important when the collector current exceeds the drain current, i.e., when the areal density of the injected mobile charge becomes comparable to that of channel electrons. Inclusion of this effect may account for the experimentally observed saturation of the drain current and the negative transconductance.

It is more likely, however, that in order to correctly describe these phenomena, one would have to go beyond the assumed electron-temperature approximation – especially in the high-energy tail of the hot-electron distribution, which is responsible for the injection current. In general, an electron temperature T_e different from the lattice temperature T is established when the electron–electron (e–e) scattering time is substantially shorter than the relaxation time associated with the lattice. The low-energy part of the distribution function is well described by a T_e already at low electron concentrations, as soon as the e–e scattering is faster than relaxation on acoustical phonons. The high-energy portion of the distribution, however, can be strongly distorted (depressed) compared to a Maxwellian curve – because of the emission of optical phonons. Recently, Esipov and Levinson [103] considered the electron temperature in a two-dimensional electron gas taking into account the e–e scattering and the optical phonon emission. They showed that although the total distortion of the distribution function is "integrally" weak, as it affects only the tails of the distribution (above the optical phonon threshold, where the number of electrons is small), nevertheless in those very tails the distortion can be quite strong. Similar effects can be even more important in the operation of a CHINT where one is interested only in the tails of the distribution above the collector barrier height $\Phi \sim 0.3$ eV, which are constantly depleted by the injection current.

The effect of channel depletion itself may have a significant influence on the formation of the high-energy tail of the hot-electron distribution. If, as

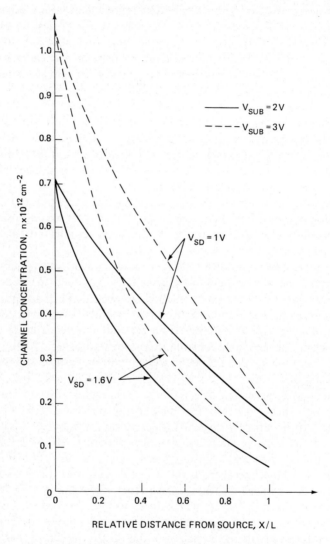

Fig. 23. Calculated variation of the channel concentration $n(x)$ for different drain (V_{SD}) and collector (V_{SUB}) voltages [97]. Note that at a given heating voltage V_{SD} the concentration decay along the channel is faster for higher V_{SUB}. Inclusion of the dynamical screening effect can be expected to strongly amplify the channel depletion. As the channel concentration drops on the way to the drain, a new region may arise, where the assumed T_e is *not established* in the far wings of the electron energy distribution function – above the barrier height – and consequently the real-space transfer process will cease in that region.

we believe, processes of electron–electron interaction are dominant in establishing the T_e, especially in the tail of the distribution, then the channel depletion may have a fundamental self-limiting effect on hot-electron injection. Indeed, consider the variation of electron density in the channel, as calculated by Grinberg et al. [97], shown in fig. 23. We see that the density $n(x)$ rapidly decreases in the direction from the source to the drain *. Strong depletion of the channel should lead to a situation, in which the hot-electron distribution function will deviate from the Maxwellian form – with its high-energy tail further suppressed compared to value predicted by the electron-temperature approximation. This means that one can expect a self limitation of the RST process – with the channel concentration never dropping below a critical level – determined, for a given collector barrier height, by the concentration dependence of the electron–electron interaction. This effect should lead to the experimentally observed saturation from below of the NERFET I_D versus V_{SD} characteristic. Moreover, it may also describe the negative transconductance – since the lower the barrier the smaller is the expected critical concentration for equilibrating the electron distribution at energies above the barrier height. In my opinion, investigation of these effects is of considerable theoretical and practical importance.

2.3.3. The ultimate speed of CHINT/NERFET devices

Fundamental limitations on the intrinsic speed of three-terminal RST devices arise due to the time-of-flight delays characteristic of a space-charge-limited current and because of a finite time required for the establishment of an electron temperature.

Consider the latter limitation first. Energy relaxation of hot carriers in bulk semiconductors has been a subject of considerable number of studies (see the reviews by Jacoboni and Reggiani [104] and Ferry [105], and references therein). The dominant mechanisms for the Maxwellization of the hot-electron energy distribution function are the polar optic phonon scattering and the electron–electron interaction. The phonon mechanism is expected to be not too different in the CHINT/NERFET structure compared to the bulk. Monte Carlo studies by Maloney and Frey [106] and

---

* In a simpler model, Luryi and Kastalsky [95] had shown that the concentration drop along the channel is approximately exponential with a characteristic length which can be as short as 1000 Å at high T_e. (In both calculations the channel depletion is due to the "current stealing" effect of charge injection alone – neglecting the dynamical screening which would further magnify the effect.)

Tang and Hess [107] indicate that the energy loss rate due to polar optic phonon emission by electrons in GaAs is nearly constant for electron energies above 0.1 eV and is of the order 2×10^{11} eV/s. This translates into about 1 ps equilibration time for $T_e \sim 1500$ K. The influence of e–e scattering (which, as discussed above may be of primary importance for the establishment of the quasi-equilibrium in high-energy tails of the electron distribution) is much more difficult to take into consideration. As far as we know, there is no satisfactory treatment of this process in a two-dimensional electron gas [except for the above-mentioned work by Esipov and Levinson [103], which is rather complicated and not immediately adaptable to the charge-injection problem]. The results of Inoue and Frey [108] are hardly applicable to this problem, since they are based on a Debye-like screening, expressed by a length $(\varepsilon k T_e/Ne^2)^{1/2}$, where N is the electron volume concentration. Such a screening presupposes the existence of a positive neutralizing background – a condition not fulfilled in regions of a space-charge-limited current flow, where we would expect a much stronger N dependence of the e–e scattering rate. It is probably a safe bet to assume that at the operating voltages of CHINT and NERFET the hot-electron ensemble equilibrates in less than 1 ps.

The second fundamental limitation of the speed arises due to the space-charge capacitance associated with the mobile charge drifting in the high-field regions of the device. It reduces to the time of flight of electrons over these regions – the high-field portion of the channel and the down-hill slope of the potential barrier. At high T_e both regions are of order 10^{-5} cm and the corresponding delay is about 1 ps.

It should be emphasized that these time-of-flight limitations are different from the time-of-flight-under-the-gate limitation characteristic of an FET. The latter results from charging the channel by the gate field through the output resistance of a previous identical device which necessarily gives $\tau = L/v$ with L being the gate length. In the CHINT the controlling electrode is the drain and L is the total length of the space-charge-limited current regions – which can be substantially shorter than the source-to-drain distance. The same limitation governs the frequency cut-off in NERFET. The dynamical screening mechanism of the NDR is extremely fast – intrinsically limited by the time of flight of injected electrons toward the second conducting layer.

Although in principle the CHINT and NERFET are picosecond devices, their real speed limit at present arises from the RC delay due to large contact pads and the series channel resistance. The main parasitic capacitance – between the D and SUB electrodes – can be reduced by shrinking the contact pads and/or by ion implantation of oxygen underneath the pads. (The latter should also help reducing the leakage in CHINT.) At the same time one should try to minimize the channel resistance (at the peak prior to onset of the NDR), which includes the series contact resistance. Experimentally, Kastalsky et al. [99] have demonstrated the microwave operation of CHINT with a current gain at frequencies up to 32 GHz at room temperature. Microwave generation by NERFET in the gigahertz range had been observed earlier [93]. The OMCVD NERFET devices with improved RC parameters showed microwave generation up to 7.7 GHz [99]. Ideally, the intrinsic RC delays in CHINT/NERFET devices are of the order of several picoseconds. Limits on the switching speed in a two-NERFET logic circuit are uncertain at present and require further study, both experimental and theoretical.

2.3.4. Concluding remarks

We have reviewed the physical principles of several novel devices which employ hot-electron transfer between two conducting layers separated by a potential barrier. Their operation is based on controlling charge injection over the barrier by modulating the electron temperature in one of the layers. The principle was illustrated by a comparison to a hypothetical vacuum diode whose cathode temperature is controlled by an input electrode.

The CHINT (charge-injection transistor), is a general-purpose three-terminal high-speed device. Electrically, its operation is analogous to the bipolar transistor and the ballistic hot-electron transistors discussed in sect. 2.1, if one identifies the terminals as $S \equiv$ emitter, $D \equiv$ base, and SUB $\equiv$ collector. An interesting feature of CHINT is the fact that its differential common-base current gain $\alpha \equiv (\partial I_{\text{SUB}}/\partial I_{\text{S}})$ at $V_{\text{SUB}} = $ const. can substantially exceed unity (due to the NDR in the S–D circuit). By the physical principle involved, the operation of CHINT is different from all previous three-terminal devices – which were based either on the potential effect, i.e., the modulation of a potential barrier by an applied voltage (vacuum triode, bipolar transistor, various analog transistors), or on the field effect, which is the screening of an applied field by a variation of charge in a resistive channel. In CHINT the control of output current is effected by a modulation of the electron temperature resulting in charge injection over a barrier of fixed height. The SUB electrode serves as an anode and the channel as a hot-electron cathode, whose effective temperature is controlled by the source-to-drain field. The existence of power gain in this device has been demonstrated experimentally – both in DC operation and at high frequencies (up to ~ 10 GHz). The value of the mutual conductance g_m obtained in CHINT (over 1000 mS/mm) compares favourably to the best field-effect transistors.

Hot-electron injection in CHINT is accompanied by a strong negative differential resistance in the channel circuit. This gives rise to a related device, called the NERFET (negative resistance FET), which is essentially a two-terminal NDR element, controllable by the voltage on the third (SUB) electrode. The experimentally observed room-temperature NDR in NERFET had the highest peak-to-valley ratio (over 10^2) of any negative-resistance device. The NERFET can be used in several ways. First, it can be used as a controllable amplifier and generator of oscillating signals. Microwave generation in NERFET has been observed, with an efficient DC to AC conversion and a high output power. Second, it can be applied in a variety of logic configurations, the simplest of which have been reviewed in sect. 2.2.

All the three-terminal RST devices discussed above have been experimentally demonstrated with the help of AlGaAs/GaAs heterostructures, fabricated either by MBE or OMCVD techniques. Quantitative analysis of the operation of these devices is rather involved and the theory outlined in sect. 2.2.2 is only a first step toward a complete understanding. Further work is clearly required in this area – especially to include the interplay between the real-space and the momentum-space transfers. Since the latter may considerably slow down the device operation, it had been proposed [63] to employ other heterostructure materials, such as InGaAs/InAlAs – in which the satellite valleys have a higher energy separation. A further bonus in using these materials is the lower effective electron mass, which enhances the heating effects. Preliminary results have been reported by Luryi and Kastalsky [96]. It should be also mentioned that semiconductor heterojunctions are not the only way to implement three-terminal RST devices. An interesting possibility lies in using thin semimetal films forming a Schottky barrier with a semiconductor collector underneath. For example, Bi/Si junctions have a barrier height of 0.63 eV [96]. Because of the reduced electron scattering rates (due to high dielectric permittivity and low carrier concentration), bismuth is known to exhibit strong hot-electron effects. If the electric field is applied laterally to a Bi film on a silicon substrate, one can expect an efficient emission of hot electrons over the Schottky barrier. So far, this effect has not been experimentally verified.

Throughout our discussion of the underlying physics of hot-electron injection in double-layered heterostructures, we emphasized the (many) points where our present understanding is incomplete and where further work, both theoretical and experimental, is required. The most important technological problem in the implementation of all hot-electron heterojunction devices is to provide an abrupt and shallow ohmic contact to the two-dimensional electron gas – with low values of contact resistance and stray capacitance.

3. Resonant-tunneling heterojunction devices

3.1. Introduction

It has been known since the early days of the quantum theory of solids that an electron placed in a lattice field of period a and an additional uniform electric field F will perform a purely oscillatory motion in the direction of the field with a characteristic frequency $\omega = eFa/\hbar$. This follows from the fact that the crystal momentum of such an electron increases linearly with time, while both its energy and velocity are periodic functions of the crystal momentum. The distance an electron travels during one cycle is of the order of I/eF (where I is the width of the allowed band in which the electron is moving) and the turning points can be interpreted as Bragg reflections of the electron by the periodic lattice. In the language of quantum mechanics, the electron energy band in the presence of an electric field splits into a set of discrete levels (the Wannier–Stark ladder) equidistantly separated by energy intervals eFa, each of these levels corresponding to a wave-function centered on a lattice site and extending approximately I/eaF lattice sites along the direction of F. These assertions are elementary consequences of the fact that a translation by one lattice period changes the Hamiltonian of the system by the constant amount eFa.

It is therefore possible, in principle, for a DC electric field to induce an alternating current in the solid. In real solids this effect (sometimes called the Bloch oscillations) is not observed – because collisions usually return an electron to the bottom of the band well before it completes one period (even for the strongest available fields F one has $\omega\tau \ll 1$, where τ is the scattering time). As a possible remedy to this difficulty, Keldysh [109] suggested the "artificial creation of periodic fields" with periods $d \gg a$, which would split the conduction band into several minibands. He also discussed the possibility of using high-frequency ultrasonic waves to establish a "moving" superlattice [110].

The miniband conduction in a superlattice consisting of a large number

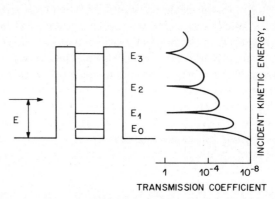

Fig. 24. Schematic illustration of a double-barrier electron resonator. The intensity transmission coefficient plotted against the incident kinetic energy in the direction normal to the resonator layers has a number of sharp peaks. In the absence of scattering, a symmetric resonator is completely transparent for electrons entering at the resonant energies (the Fabry–Pérot effect).

of artificial barriers can be regarded as a generalization of resonant-tunneling transmission through a double-barrier structure. Shortly after Mead [21,22] proposed his MOMOM hot-electron transistor, Davis and Hosack [111] and Iogansen [112] discussed the possible use of a similar structure as an electron interferometer analogous to the optical Fabry–Pérot étalon. In such a device the central M (base) would be replaced by a thin resonator for electron de Broglie waves. The emitter serves as an electron reservoir, from which electrons tunnel through both insulating barriers into the collector. At certain energies the amplitude of the de Broglie waves in the resonator builds up to the extent that these waves leaking in both directions cancel the reflected waves and enhance the transmitted ones. In the absence of scattering, a system of two identical barriers is completely transparent for electrons entering at the resonant energies and the total transmission coefficients plotted against the incident energy has a number of sharp peaks, as shown in fig. 24. Of course, it is not practical to use a metal base as the resonator, since in most metals the de Broglie wavelength $1/k_F$ is too short.

The real excitement in the resonant-tunneling field began after the pioneering works of Esaki and Tsu [113] who proposed the idea of semiconductor superlattices – to be fabricated either by varying the alloy composition (heterojunction superlattices) or the impurity density introduced during epitaxial growth – and discussed their transport properties. They predicted a negative conductivity associated with electron transfer into negative-mass regions of the minizone and the possibility of inducing Bloch oscillations by applying an electric field normal to the superlattice. They have also theoretically considered [12] the transport properties of a *finite* superlattice and showed the possibility of obtaining a negative differential resistance in double-barrier quantum-well structures. These predictions were followed by the first observation of resonant tunneling in double-barrier $GaAs/Al_{0.3}Ga_{0.7}As$ heterojunction diodes [114] and GaAs/AlAs heterojunction superlattices [115].

The field of resonant tunneling in quantum-well structures is in the state of renaissance. The basic physical phenomena anticipated in such structures were qualitatively understood in the earlier period (1970s), but their experimental realization had to wait until the maturity of modern epitaxial techniques. Since the early reports, substantial progress has been achieved in the material quality of heterojunction-barrier structures grown by MBE and OMCVD techniques. The interest in such structures has risen further after the remarkable recent experiments of Sollner and co-workers [116–118] who studied the microwave activity in double-barrier (DB) quantum-well (QW) diodes. These workers have demonstrated a negative differential resistance (NDR) in these diodes directly in the current–voltage characteristics at 77 K (rather than in the derivative of the current as was the case with the first reports) and obtained active oscillations from a DBQW diode mounted in a resonant cavity at frequencies f up to 18 GHz [117]. In their earlier experiments, Sollner et al. [116] used the double-barrier structure as a detector and mixer of far-infrared radiation – operating at $f = 2.5$ THz. The material quality of DBQW diode structures has steadily improved to the point that a pronounced NDR can now be observed at room temperature [119–121]. It had been demonstrated by Shewchuk et al. [119] and confirmed by Tsuchiya and Sakaki [122] that introduction of lightly doped layers immediately outside the barriers improves the characteristics of DBQW diodes, especially at room temperature. Recently, peak-to-valley ratios in the NDR current of nearly 3:1 were obtained at room temperature [123–125]. Typical current–voltage characteristics of a GaAs/AlAs DBQW diode are shown in fig. 25. Most of the recent advances have been achieved with MBE-grown AlGaAs/GaAs systems, but there were also reports of resonant tunneling of electrons in OMCVD-grown DBQW heterojunctions [126,127], strained-layer GaAsP/GaAs systems [128], and resonant tunneling of holes in AlAs/GaAs DBQW heterostructures [129]. There has also been progress in the implementation of superlattice tunneling structures [15,16,130–132], briefly reviewed in sect. 2.3. In addition to

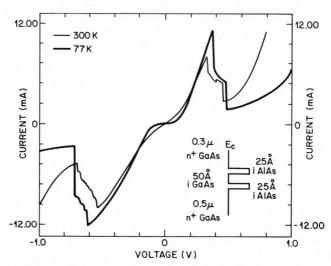

Fig. 25. The I–V characteristics of a symmetric DBQW diode which contains an undoped 50 Å thick GaAs QW clad by two undoped 25 Å thick AlAs barrier layers and two n-doped GaAs layers [123]. Diode area is $\approx 2.8 \times 10^{-7}$ cm^2, corresponding to a peak current density of 30 kA/cm^2 at 300 K.

the diodes, there has been recent interest in incorporating resonant-tunneling structures in three-terminal devices. These devices will be briefly discussed in sect. 3.4. A review of resonant tunneling and other perpendicular quantum transport phenomena in double barriers and superlattices, as well as some of their device applications, was recently given by Capasso et al. [16]. Active research going on in many laboratories can be expected to culminate in the implementation of new and exciting devices to be used in the future high-speed electronics.

In a discussion of the operation of heterojunction tunneling structures it is important to distinguish between truly resonant- and sequential-tunneling processes. In the instance of a superlattice, the former processes correspond to the miniband conduction in which the electronic wave functions remain coherent over many superlattice periods. As discussed by Kazarinov and Suris [13,14], the miniband conduction can be realized only at low applied fields, when the energy acquired by an electron from the field over one period of the superlattice is much less than the miniband width. In the opposite limit, applicable to most experiments in which the NDR effect is observed, electrons are localized within one well – sequentially tunneling into the adjacent wells. The conductance peak is then observed [15] at such values of the applied field when the excited quantum-well states are brought into the resonance with the ground state in the neighboring wells. Similarly, in a double-barrier quantum-well device the transmission through the entire structure can be either truly resonant (in which case one is justified in using the above-mentioned analogy with an optical Fabry–Pérot étalon) or a *sequential* two-step process. Since, as will be discussed in sect. 3.2, the sequential process alone gives rise to the observed structure in the conductance of a DBQW diode, one should be cautious in interpreting the experimental results in terms of coherent resonant tunneling.

3.2. Double-barrier quantum-well structures

The NDR in DBQW diodes is a consequence of the dimensional confinement of states in a QW, and the conservation of energy and lateral momentum in tunneling. In addition to that, following the early work of Davis and Hosack [111] and Iogansen [112], the operation of these structures has often been discussed in connection with a resonant-tunneling effect analogous to that in a Fabry–Pérot resonator. That effect occurs when the energies of incident electrons in the emitter match those of unoccupied states in the QW and the wave function of resonant electrons is coherent across the entire double-barrier structure. Under such conditions, the wave amplitude builds up in the QW and resonant transmission occurs. This physical picture has led to a design strategy intended to optimize the Fabry–Pérot resonator conditions. In particular, Ricco and Azbel [133] pointed out that achievement of a near-unity resonant transmission requires equal transmission coefficients for both barriers at the operating point – a condition not fulfilled for barriers designed to be symmetric in the absence of an applied field. To counter that, Capasso and Kiehl [134] proposed a resonant-tunneling structure in which a symmetric DBQW was built in the base of a bipolar transistor, and the Fabry–Pérot conditions were maintained through the use of minority-carrier injection.

High-frequency operation of DBQW diodes was also considered by Ricco and Azbel [133] on the assumption that the underlying mechanism of NDR requires the Fabry–Pérot resonant enhancement of the tunneling probability. It was found that the dominant delay results from the resonator charging time, which is of the order of the lifetime of the resonant state. For a QW bounded by 50 Å thick AlGaAs barriers, simple estimates [135] gave a frequency limit in the low gigahertz range. At higher frequencies, the amplitude of an electron wavefunction in the QW cannot re-adjust itself in response to an external field variation to provide resonant enhancement of the transmission coefficient. These estimates, contrasted with the experimental results of Sollner et al. [116] in which a DBQW structure was used as a detector and mixer of far-infared radiation at 2.5 THz, have led to the suggestion [135] that the Fabry–Pérot resonant transmission plays only a minor role (if any) in the operation of DBQW diodes. As will be discussed in the next section, the NDR can arise solely due to electron tunneling into a system of states of reduced dimensionality. In this picture, electrons subsequently leave the QW by tunneling through the second (collector) barrier, so that their transport through the entire DBQW structure is described by *sequential* rather than resonant tunneling *.

3.2.1. Mechanism of operation of quantum well diodes

Let us review the mechanism of NDR in double-barrier QW structures – without invoking a resonant Fabry–Pérot effect. This mechanism is illustrated in fig. 26. Consider the Fermi sea of electrons in a degenerately doped emitter (the bottom figure). Assuming that the AlGaAs barrier is free of impurities and inhomogeneities, the lateral electron momentum (k_x, k_y) is conserved in tunneling. This means that for $E_C < E_0 < E_F$ (where E_C is the bottom of the conduction band in the emitter and E_0 is the bottom of the subband in the QW) tunneling is possible only for electrons whose momenta lie in a disk corresponding to $k_z = k_0$ (shaded disk in the figure), where $\hbar^2 k_0^2 / 2m = E_0 - E_C$. Only those electrons have isoenergetic states in the QW with the same k_x and k_y. This is a general feature of tunneling into a two-dimensional system of states. As the emitter–base potential rises, so does the number of electrons which can tunnel: the shaded disk moves downward to the equatorial plane of the Fermi sphere. For $k_0 = 0$ the number of tunneling electrons per unit area equals $mE_F/\pi\hbar^2$. When E_C rises above E_0, then at $T = 0$ there are no electrons in the emitter which can tunnel into the QW while conserving their lateral momentum. Therefore, one can expect an abrupt drop in the tunneling current. The effect is conceptually similar to that in the Esaki tunnel diode. Extension of this picture to the case of several subbands in the QW is straightforward.

---

* Within the sequential-tunneling model, the terahertz results of Sollner et al. [116] can be explained – since rectification of an external signal by a DBQW diode requires the re-adjustment of only the phase of electronic wavefunctions and not their amplitude, so that the operation of the detector is not limited by a Fabry–Pérot charging time. This argument is due to Derkits [136].

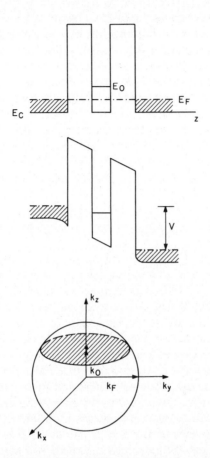

Fig. 26. Illustration of the operation of a double-barrier resonant-tunneling diode [135]. Top figure shows the electron energy diagram in equilibrium. Figure in the middle displays the band diagram for an applied bias V, when the energy of certain electrons in the emitter matches unoccupied levels of the lowest subband E_0 in the QW. The bottom figure illustrates the Fermi surface for a degenerately doped emitter. Assuming conservation of the lateral momentum during tunneling, only those emitter electrons whose momenta lie on a disk $k_z = k_0$ (shaded disk) are resonant. The energy separation between E_0 and the bottom of the conduction band in the emitter is given by $\hbar^2 k_0^2 / 2m$. In an ideal DBD at zero temperature the resonant tunneling occurs in a voltage range during which the shaded disk moves down from the pole to the equatorial plane of the emitter Fermi sphere. At higher V (when $k_0^2 < 0$) resonant electrons no longer exist, which results in a sharp drop in the current.

Similar arguments, of course, apply to systems of lower dimensionality, e.g., to tunneling through a "quantum wire", see sect. 3.4.2.

It is clear from the above that a similar effect should also be observable in various *single-barrier* structures in which tunneling occurs into a two-dimensional system of states. Indeed, according to the described model, in DBQW structures the removal of electrons from the QW occurs via sequential tunneling through the second barrier but other means of electron removal can also be contemplated, for example, *recombination*. Rezek et al. [137] studied electron tunneling through a single barrier into a QW located in a p-type quaternary material. In these experiments the diode current resulted from the subsequent recombination of tunneling electrons with holes in the direct-gap QW. The observed structure in the dependences of the current and the intensity of the recombination radiation on the applied bias can be explained in terms of the above picture based on momentum conservation.

The NDR effect of a similar nature can also be observed in a unipolar single-barrier structure, as was proposed by Luryi [138] and demonstrated by Morkoç et al. [123]. Let the emitter be separated by a thin tunneling barrier from a QW which is confined on the other side by a thin but impenetrable (for tunneling) barrier, as shown in fig. 27. The drain contact to the QW, located outside the emitter area, is electrically connected to a conducting layer underneath. Application of a negative bias to the emitter results in the tunneling of electrons into the QW and their subsequent drift laterally toward the drain contact. There will be no steady-state accumulation of electrons in the QW under the emitter if the drift resistance is made sufficiently small. Since the drain contact is shorted to the conducting layer underneath the collector barrier and its lateral distance from the edge of the emitter much exceeds the combined thicknesses of the two barriers and the QW, application of a drain–emitter voltage results in nearly vertical electric field lines under the emitter, which allows one to control by the applied voltage the potential difference between the emitter and the QW. Of course, this control is much less effective (by the lever rule) than it would be if the second barrier were as thin as the tunnel barrier. Experimentally, Morkoç et al. [123] were able to see a pronounced NDR already at room temperature and at 77 K the observed PTV ratio in current was more than $2:1$. As expected, the NDR was seen only for a negative polarity of the emitter bias – with a peak current occuring at a voltage which is higher than that observed in a control symmetric DBQW structure by a factor given by the ratio of the barrier thicknesses (the lever rule).

The sequential-tunneling mechanism of NDR should be experimentally

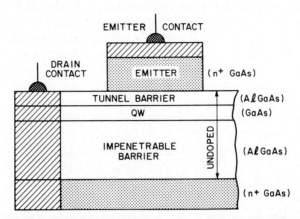

Fig. 27. Illustration of a single-barrier QW structure which exhibits an NDR effect similar to that observed in DBQW diodes [138]. The drain contact is assumed to be concentric with a cylindrical emitter electrode. The "impenetrable" collector barrier separating the QW from the conducting layer underneath (the latter is shorted to the drain) must be thin enough (≤ 1000 Å), so that the emitter-to-QW potential could be effectively controlled by the emitter bias. In the experiments of Morkoç et al. [123] the tunnel barrier and the QW thicknesses and composition were identical to those in the control double-barrier structure (fig. 25), whereas the collector barrier represented a 500 Å thick $Al_xGa_{1-x}As$ layer with $x = 0.3$.

distinguishable from the Fabry–Pérot model. However, it is not clear whether or not such a distinction can be made on the basis of DC current–voltage characteristics. Recently, Weil and Vinter [139] argued against this possibility. They have pointed out that in all practical DBQW diodes, for a given lateral momentum, the energy distribution of electrons in the emitter is much wider than the Fabry–Pérot resonance peak. In this case they showed that both the sequential and the resonant-tunneling pictures lead to similar predictions for the peak current – which is limited by the transmission coefficient of the *single* least transparent barrier. Most of the experiments on the DBQW structure published to date had con-. centrated on the demonstration of NDR and the microwave activity. As described above, all of these data – including dependences of the static current–voltage characterictics on the thickness of the barriers and the QW width [124,125], shift of these characteristics by a fixed charge stored in the barriers in the persistent photoconductivity effect (studied in the DBQW structure by Sollner et al. [118]), as well as the position and shape of the current peaks associated with tunneling into the excited states of the QW – can be adequately explained without invoking resonant transmission, so that caution is required in the interpretation of such data. It remains an open question, what fraction of the diode current in a particular experiment is due to the resonant as opposed to sequential tunneling. It appears, however, that claims of an observation of the "true" resonant tunneling should be accompanied by an experimental proof. An interesting, though still inconclusive, possibility in this regard was recently discussed by Goldman et al. [140] who studied an intrinsic bistability of DBQW diodes resulting from storage of the mobile charge in QW.

The theoretical critera, which can distinguish between the sequential and coherent processes in a given structure, have been recently considered by Capasso et al. [16]. Following the work of Stone and Lee [141], who discussed the resonant tunneling through an impurity center, they suggested that the critical quantity is the ratio τ_0/τ, where τ_0 is the lifetime of the resonant state, as limited by the tunneling process in and out of the QW, and τ is the total phase relaxation time inside the DBQW structure. The latter includes contributions of all inelastic processes, as well as those "elastic" processes which redistribute the electron energy between lateral directions x and y. The Fabry–Pérot resonant enhancement of the tunneling probability is observable when $\tau_0/\tau \lesssim 1$. In the opposite limit, $\tau_0/\tau \gg 1$, electrons will tunnel incoherently through one of the intermediate states in the QW, and only the sequential process is observable.

3.2.2. Frequency limit of DBQW oscillators

A large body of theoretical work has been devoted to the difficult question of the time development in tunneling. As emphasized by Thornber et al. [142], the main difficulty is often in the proper posing of the problem, resulting in physically different delay times associated with the tunneling process. For recent discussions of this problem the reader is referred to refs. [143,144], and references therein.

In the instance of the resonant transmission through double barriers the problem was extensively discussed by Ricco and Azbel [133]. They pointed out that the ideal conductance with a near-unity transmission coefficient occurs only in the steady state – when the wave function inside the QW has already attained its appropriate amplitude. The establishment of such a stationary situation (say, in response to a suddenly imposed external field) must be preceded by a transient process during which the amplitude of the resonant mode is built up inside the well. The transient time τ_0 is of the order of the resonant-state lifetime and hence can be expected to increase exponentially with the barrier phase area. It is this transient time which gives the fundamental speed limit for the active oscillations of most practical DBQW diodes.

A simple estimate for τ_0 can be derived [135] by regarding the transient process as a modulation of charge in the "quantum capacitor" formed between the QW and the controlling electrodes. During the operation of a DBQW diode, this capacitor is being charged or discharged by the tunneling current. Indeed, consider a small-signal oscillation of the diode. As the operating point moves up and down the NDR region of its I–V characteristic, the amount of charge stored in the resonant state inside the QW varies. This charge is re-adjusted all the time by a transient difference between the emitter and the collector currents. Even for a single electronic mode it is permissible to speak of a "capacitor" and its RC time constant since any change in the wave-function amplitude $|\Psi|^2$ is accompanied by a displacement current. It is essential to realize that this capacitor is being charged not by a resonant tunneling current but by the ordinary tunneling through a single barrier. By Gauss' law, in the parallel-plate geometry the variation of charge in the QW is linearly related to the potential variations across each barrier, so that the barrier capacitance per unit area is given by $C = \varepsilon/4\pi d$, where ε is the dielectric permittivity of the barrier material and d the barrier thickness.

The tunnel resistance of a single barrier of average height Φ and thickness d can be estimated from a WKB expression for the current density, giving $R = (2h\lambda d/e^2)\exp(4\pi d/\lambda)$, where $\lambda \equiv h/\sqrt{2m\Phi}$ is the de Broglie wavelength of an electron tunneling under the barrier. Whence we find

$$\tau_0 \equiv RC = \varepsilon \alpha^{-1}(\lambda/c)\, e^{4\pi d/\lambda}, \tag{11}$$

where $\alpha^{-1} \equiv \hbar c/e^2 \approx 137$ and c is the speed of light. The quantity $\Gamma \equiv \hbar/\tau \propto T_{max} = \exp(-4\pi d/\lambda)$, where T_{max} is the highest transmission coefficient of the two barriers, corresponds to a homogeneous broadening of the energy levels in the QW due to electron tunneling in and out; we can interpret τ_0 as a lifetime of the resonant state limited by the tunneling processes. The cut-off frequency is determined from eq. (11) as follows: $f_{max} = 1/2\pi\tau_0$. For simplicity, in the above estimates it was assumed that during the part of the cycle when $|\Psi|^2$ in the well is waxing, the emitter current charges the base–emitter capacitance – ignoring the concomitant electron leakage through the collector barrier as well as the displacement current in that barrier – and vice versa for that part of the cycle when $|\Psi|^2$ in QW is on the wane. This should produce a slightly enhanced estimate for f_{max}.

Let us make numerical estimates assuming the GaAs/AlGaAs DBQW system studied by the Lincoln Lab group [117]. We take $m = 0.096 m_0$ appropriate for $Al_{0.35}Ga_{0.65}As$ barrier material, and find $\lambda \approx 90$ Å for $\Phi \approx 0.2$ eV. (The tunneling barrier height can be estimated as the difference between the conduction-band discontinuity at the $Al_{0.35}Ga_{0.65}As/GaAs$ heterojunction and the zero-point energy in the 50 Å thick QW.) For 50 Å thick barriers, eq. (11) gives $\tau_0 \approx 40$ ps and $f_{max} \approx 4$ GHz. This estimate is not inconsistent with the microwave results of Sollner et al. [117], who observed microwave oscillations with DC to AC conversion efficiency of 2.4% at 9 GHz and less efficient generation at frequencies up to 18 GHz in a tuned coaxial circuit. Recently, the Lincoln group [146] reported millimeter-band oscillations at room temperature and frequencies up to 43 GHz from a DBQW diode with 30 Å thick $Al_{0.3}Ga_{0.7}As$ barriers and a 45 Å thick QW. In that structure, the effective tunneling barrier height is lower ($\Phi \approx 0.15$ eV), resulting in $\lambda \approx 110$ Å. Equation (11) then gives $\tau_0 \approx 2$ ps and $f_{max} \approx 86$ GHz. With sufficiently narrow barriers, DBQW oscillators are capable of operating in a subpicosecond regime *.

It is instructive to compare eq. (11) with the formula $f_{max} = 1/2\pi C \sqrt{R_s R_n}$ (where R_s is a series resistance and R_n is the diode NDR) sometimes used to estimate the maximum oscillation frequency. This expression, similar to those used in the analysis of the high-frequency performance of Esaki diodes, relates only to the capability of a given NDR element to drive an impedance-matched circuit load, and it has nothing to do with the intrinsic frequency limitations of the element itself. Moreover, the information about the intrinsic frequency limit is not contained in the experimentally measured R_n of the diode, since the latter must surely include an inhomogeneous broadening of the resonant peak. It would be helpful to have a reliable equivalent circuit model of the DBQW diode. The simplest of such models have been recently discussed by Coleman et al. [148] and Jogai et al. [149]. It appears likely, however, that the correct equivalent circuit should

---

* Predictions based on eq. (11) seem to fail when applied to DBQW structures with AlAs barriers [147]. It should be noted that all WKB estimates give similar predictions. The linewidth $\Gamma = \Gamma_0 \exp(-4\pi d/\lambda) \equiv \Gamma_0 T_{max}$ is not different from $\Gamma = E_0 T_{max}$ used by Coon and Liu [145] and Weil and Vinter [139]. The quantity E_0, which in those estimates must be understood as the zero-point-motional energy in the QW (so that E_0/h is a quasi-classical "attempt frequency") is practically close to the Γ_0 following from eq. (11). Minor differences in the pre-exponential factor are irrelevant when T_{max} is evaluated quasi-classically. Comparison of the experiment with WKB predictions have led Sollner et al. [147] to question the applicability of the quasi-classical method to estimating the escape time of an electron from a metastable state in a narrow QW.

distinguish between the Fabry–Pérot and the sequential-tunneling modes of the diode operation – or contain both mechanisms somehow in parallel. Such a description, which can only emerge from a full quantum-mechanical theory of the resonant-tunneling diode, should contain in an essential way various mechanisms of the phase relaxation of the wavefunction of tunneling electrons inside the QW. In the limit of sequential-tunneling only, a possible equivalent circuit may contain a variable resistor R (which can be positive and negative) in parallel with the emitter–barrier capacitance, describing tunneling into the QW, in series with the collector–barrier and the load impedances.

3.3. Tunneling in superlattices

We have established that all that is required for the NDR to occur in a DBQW tunneling structure is the reduced dimensionality of electronic states in the tunneling range. One can relax this requirement – by replacing the QW by a superlattice with narrow minibands (a multiple QW structure, for which the tight-binding approximation is a good description). Clearly, we can expect the NDR effect in tunneling from a degenerately doped emitter into a superlattice. Recently, Davies et al. [130] observed a similar effects in tunneling between the minibands of two coupled superlattices. They have also reported NDR-driven oscillations in that structure at room temperature [132].

In the above discussion of the double-barrier QW structure, we have emphasized the essential difference between the coherent transmission and the sequential tunneling through the two barriers. In the instance of a semiconductor superlattice, that difference had been clearly explained by Kazarinov and Suris [13]. In an ideal superlattice consisting of a large number of equally spaced identical quantum wells, one can expect a resonant (miniband) transmission, analogous to the Fabry–Pérot effect, and possibly an NDR due to the Bragg reflections, if the applied field is such that the potential difference, acquired by an electron over many periods of the superlattice, is less than the width of the lowest miniband I. These effects, particularly the Bragg reflections, are extremely difficult to observe because of scattering and Zener tunneling between electron minibands, as well as domenisation of the electric field in the superlattice [115]. The necessary condition for the realization of the miniband-conduction effects is

$$\hbar\tau < eaF \ll I,$$

where τ is the electron mean-free-path time.

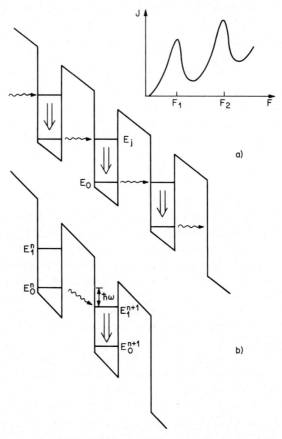

Fig. 28. Illustration of the sequential-tunneling effects in a superlattice (after ref. [13]). (a) Conductivity peaks, associated with the resonance between the excited and the ground states of adjacent wells. Such peaks were observed by Capasso et al. [15] in photocurrent measurements. (b) Photon-assisted sequential tunneling process, predicted to be capable of producing light amplification at frequencies tunable by the applied electric field.

In the opposite limit of a strong electric field, $eaF > I$, an electron belonging to the band I is localized within one well. In this limit an enhanced electron current will flow at sharply defined values of the external field, $eaF_j = E_j - E_0$ ($j = 1, 2, \ldots$), when the ground state in the nth well is degenerate with the first or second excited state in the $(n + 1)$st well, as illustrated in fig. 28a. Under such conditions, the current is due to electron tunneling between the adjacent wells with a subsequent de-excitation in the

($n + 1$)st well. In other words, electron propagation through the entire superlattice involves again a sequential rather than a resonant tunneling. Experimental difficulties in studying this phenomenon are usually associated with the non-uniformity of the electric field across the superlattice and the instabilities generated by negative differential conductivity. To ensure a strictly controlled and spatially uniform electric field, Capasso et al. [15] placed the superlattice in the i-region of a reverse-biased p^+–i–n^+ junction. This structure allowed for the first time to observe the sequential tunneling resonance predicted by Kazarinov and Suris [13]. Two NDR peaks observed in the photocurrent characteristics (at nA currents) had their positions and lineshapes in reasonable agreement with the theoretical predictions.

In the regime when the electric field F exceeds the peak value F_j (i.e. in the NDR portion of the current-field characteristic), Kazarinov and Suris [13] had proposed the possibility of the amplification of electromagnetic waves, brought about by the sequential electron tunneling into adjacent wells with a simultaneous emission of photons, see fig. 28b. The maximum gain was predicted to occur at the frequency $\hbar\omega = ea(F - F_j)$. This exciting effect – a laser action tunable by an applied electric field – has not yet been observed experimentally in superlattices. The fact that this device is supposed to operate in the NDR regime presents a fundamental difficulty, since at high currents the uniform field condition cannot be maintained because of the NDR-driven space-charge instability, resulting in domenisation of the field. It is not clear to me, whether this difficulty can be circumvented even in principle.

3.4. Three-terminal resonant-tunneling devices

Many workers have appreciated the various attractive possibilities which arise from the integration of a resonant-tunneling (RT) structure in a three-terminal transistor device. Bonnefoi et al. [150] discussed the possible integration of a DBQW structure with various analog and field-effect transistors and theoretically analyzed some of the expected characteristics. Yokoyama et al. [151] proposed to employ a DBQW structure for injecting hot electrons in a ballistic transistor. The existence of a peak in the output current with respect to the base–emitter voltage enabled them to implement functional logic gates, such as a frequency multiplier and an exclusive-NOR gate with one transistor. Capasso and Kiehl [134] considered the operation of a bipolar transistor with a resonant-tunneling structure (double-barrier or superlattice) built into its base. That device can be viewed as a switch, in which the input base–emitter voltage controls the transistor α, so that when the RT structure is off resonance most of the emitter current flows into the base.

Jogai and Wang [152] discussed a DBQW device in which an independent contact is made to a heavily doped QW base. Useful operation of such a transistor requires that the base sheet resistance should be less than 100 $\Omega/\square$ (see the discussion in sect. 2.2.), which may be nearly impossible to achieve in practice – without introducing an intolerably large loss of α due to scattering in the base (note that such a device must rely on the "true" resonant transmission, since in a sequential two-step process most of the current will flow into the base contact). It may be worthwhile to consider a variation of this device, in which the base is undoped with the base conductivity provided by a two-dimensional electron gas induced in the QW by the collector field (similar to the IBT device, discussed in sect. 2.2.2).

Bonnefoi et al. [18] proposed a tunnel transistor in which tunneling from a doped emitter into a QW collector is controlled by a third "base" electrode, separated from the well by a non-tunneling barrier. The control is effected by the electric field which shifts the QW energy levels via the Stark effect. Another way of controlling the tunneling current by a third electrode was proposed by Luryi and Capasso [17], who described a transistor in which both the emitter and the collector are two-dimensional and resonant tunneling occurs through a one-dimensional "quantum wire" whose potential relative to the emitter is controlled by a fringing electric field emanating from a planar gate. These devices will be discussed in more detail in the next two sections.

3.4.1. Stark-effect transistor

The concept of this transistor [18] is based on an elegant idea to invert the sequence of the functional transistor layers by placing them in the order *emitter → collector → base*. The basic Stark-effect transistor (SET) structure is illustrated in fig. 29. It consists of an n-doped emitter layer separated from a lightly doped QW "collector" by a thin tunneling barrier. From the other side the QW is bounded by a relatively thick non-tunneling barrier and a heavily doped "base" layer, which serves as the controlling electrode. Independent contacts are provided to all the three layers. As discussed in ref. [18], doping in the emitter should be sufficiently high to provide a large tunneling current and at the same time low enough to optimize the resonant tunneling at room temperature.

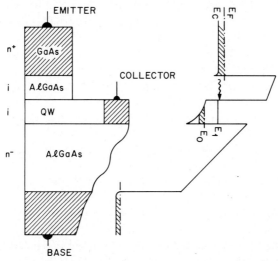

Fig. 29. Illustration of a Stark-effect transistor structure and its energy band diagram. Even though the base field does not penetrate beyond the QW collector layer (so long as the latter is not depleted), it affects the emitter tunneling current by shifting the QW subbands with respect to the Fermi level. As discussed below, this structure can be also expected to possess regions of negative transconductance. The diagram presented here is different in several respects from that given in ref. [18].

Transistor action is achieved by the modulation of the position of the two-dimensional subbands in the QW with respect to the emitter Fermi level by the electric field emanating from the base electrode. So long as the QW is not depleted, the base field does not penetrate beyond the collector layer, and the potential difference between the classical conduction-band edges in the collector and the emitter is not modulated. Nevertheless, the subband position in the collector is shifted – being sensitive to the shape of the QW (the Stark effect) – and depending on the base field, the subbands move relative to the emitter Fermi level. Consequently, the base voltage controls the emitter tunneling current.

Bonnefoi et al. [18] have also proposed a variation of the SET in which a DBQW structure is inserted between the emitter and the collector instead of the single tunneling barrier. That was predicted to result in an NDR in the collector circuit – again controlled by the base field. Potentially, such a characteristic can be expected to give rise to a wealth of useful applications, for example, similar to those described in sect. 2.3.1 in connection with the NERFET switching circuits. As pointed out in ref. [18], an important advantage of this transistor is a negligible base current, and hence a large current transfer ratio. Indeed, the base acts like the gate of an FET – although the transistor operation results not from modulating the amount of charge in the QW but the potential energies of its quantum states.

I would like to mention one interesting feature not discussed in ref. [18], and that is the possibility of obtaining a negative transconductance in the SET device. Indeed, from the discussion in sect. 3.2.1 it is clear that a base-induced monotonic motion of the two-dimensional subbands in the QW collector with respect to the emitter Fermi level can result in a peaked collector current as a function of the base voltage. As has been argued above, to observe such an effect one does not need a DBQW emitter structure (see the discussion in connection with fig. 27). In fact, the single-tunneling-barrier structure illustrated in fig. 27 can be converted into a transistor by separating the contacts to the QW and the conducting layer underneath and using the latter as the controlling electrode. As demonstrated by Morkoç et al. [123] such a structure will exhibit a negative transconductance effect.

3.4.2. Quantum wire transistor

The preceding discussion has dealt with the bulk-carrier tunneling into a two-dimensional density of electronic states. Recently, Luryi and Capasso [17] proposed a novel device structure in which the QW is linear rather than planar and the tunneling is of two-dimensional electrons into a one-dimensional density of states.

New physical phenomena arising from the dimensional confinement of electrons to 1 and 0 degrees of freedom had been anticipated by a number of authors. Sakaki [153] discussed the possibility of obtaining an enhanced mobility along "quantum wires" – because of the suppression of the ionized-impurity scattering. He proposed a V-groove etch of a planar heterojunction QW as means of achieving the one-dimensional confinement. Petroff et al. [154] reported experimental attempts of implementing a one-dimensional confinement by etching techniques. Chang et al. [155] proposed a novel quantum wire structure which can be obtained by an epitaxial overgrowth of a vertical $\langle 110 \rangle$ edge of a pre-grown $\langle 100 \rangle$ heterostructure. Recently, Cibert et al. [156] experimentally demonstrated the carrier confinement in 1 ("wire") and 0 ("box") dimensions by a novel technique involving the electron-beam lithography and laterally confined interdiffusion of aluminum in a GaAs/AlGaAs heterostructure grown by MBE. Also, Reed et al. [157] reported dimensional quantisation effects observed in GaAs/AlGaAs "quantum dots".

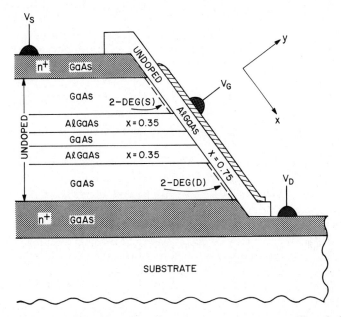

Fig. 30. Schematic cross-section of the proposed surface resonant tunneling device, the quantum-wire transistor [17], assuming a "V-groove" implementation. Thickness of the two undoped GaAs layers outside the double-barrier region should be sufficiently large (≥ 1000 Å) to prevent the creation of a parallel conduction path by the conventional (bulk) resonant tunneling.

We shall describe below the principle of a quantum-wire transistor, assuming a V-groove implementation in GaAs/AlGaAs technology. The structure, shown in fig. 30, consists of an epitaxially grown undoped planar QW and a double AlGaAs barrier sandwiched between two undoped GaAs layers and heavily doped GaAs contact layers. The working surface defined by a V-groove etching is subsequently overgrown epitaxially with a thin AlGaAs layer and gated. Application of a positive gate voltage V_G induces two-dimensional electron gases at the two interfaces with the edges of undoped GaAs layers outside the QW. These gases will act as the source (S) and drain (D) electrodes. The bottom of the two-dimensional subband E_0 is split up from the classical conduction-band edge E_c by the dimensional confinement in the y direction. At the same time, there is a range of V_G in which electrons are not yet induced in the "quantum wire" region (which is the edge of the QW layer) – because of the additional dimensional quantization in the x direction. The operating regime of the quantum-wire transistor is in this range.

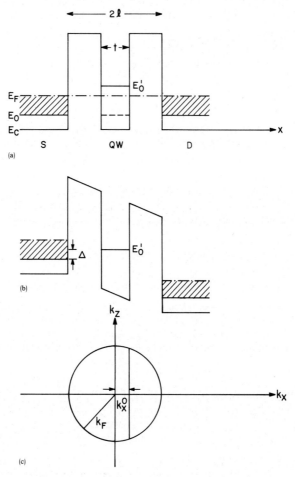

Fig. 31. Illustration of the device operation (after ref. [17]). (a) The band diagram along the channel in the absence of a drain bias. (b) Band diagram for an applied bias V_D, when the energy of certain electrons in the source (S) matches unoccupied levels of the lowest one-dimensional subband E_0' in the quantum wire. (c) Fermi disk corresponding to a two-dimensional degenerate electron gas in the S electrode. Vertical chord (the "resonant segment") at $k_x = k_x^0 \equiv (2m\Delta/\hbar^2)^{1/2}$ indicates the momenta of electrons which can tunnel into the quantum wire while conserving their momentum k_z along the wire. The maximum tunneling current occurs when the resonant segment aligns with the vertical diameter of the Fermi disk. In this case, the number of resonant electrons equals $[2m^*(E_F - E_0)]^{1/2}/\pi\hbar$ per unit length in the z-direction. Further lowering of the level E_0' with respect to E_0 – by increasing either V_D or V_G – results in a sharp drop of the current.

Device characteristics can be understood along the lines described above in connection with fig. 26. In the present case the dimensionality of both the emitter and the base is reduced by 1, so that the emitter Fermi sea becomes a disk and the shaded disk of fig. 26 is replaced by a resonant segment, as shown in fig. 31. Application of a positive drain voltage V_D brings about the resonant tunneling condition and one expects an NDR in the dependence $I(V_D)$. What is more interesting, is that this condition is also controlled by V_G. The control is effected by fringing electric fields: in the operating regime an increasing $V_G > 0$ *lowers* the electrostatic potential energy in the base with respect to the emitter – nearly as effectively as does the increasing V_D. (This has been confirmed by solving the corresponding electrostatic problem exactly with the help of suitable conformal mappings.) At a fixed V_G having established the peak of $I(V_D)$, we can then quench the tunneling current by increasing V_G. This implies the possibility of achieving the *negative transconductance* – an interesting feature in a unipolar device. A negative-transconductance transistor can perform the functions of a complementary device analogous to a p-channel transistor in the silicon CMOS logic. A circuit formed by such a device and a conventional n-channel field-effect transistor can act like a low-power inverter in which a significant current flows only during switching. This feature can find applications in logic circuits.

4. Conclusion

We have reviewed and discussed a number of modern device structures based on the physical processes in semiconductor heterojunctions which in conventional devices are only peripheral to their operation (or even detrimental). Having in mind the assessment of potential electronic applications, we were mainly concerned with three-terminal transistor structures. First, we discussed the wide class of hot-electron devices, which we classified into two groups – the ballistic and the real-space transfer devices – depending on the type of hot-electron ensemble essentially employed in their operation. Next, we have reviewed a class of devices based on the resonant-tunneling phenomena in semiconductor heterojunctions.

So far, none of these rather exotic devices have been used in electronic applications. Most of the work on these devices has concentrated on the demonstrating of the existence of an effect in question, proposals of new structures and effects, and studies of their potential physical limitations. It should be remembered, however, that the main raison d'être of this research is to come up with a device, which will find a commercial electronic application. Are there good reasons to hope that this will happen? We believe that the answer is in the affirmative – provided by the remarkable advancement of the last decade in the techniques of crystal growth (such as MBE and MOCVD) and device processing (submicron lithography, ion implantation, etc.). It is likely, in our view, that the next decade will indeed see a commercial exploitation of both hot-electron injection and resonant-tunneling devices – such as some of those reviewed in this paper and those not yet invented.

Acknowledgements

It is my pleasure to thank F. Capasso, G.E. Derkits, A.A. Grinberg, A. Kastalsky, R.F. Kazarinov, and S.M. Sze for many helpful discussions.

References

[1] S.M. Sze, Physics of Semiconductor Devices, 1st Ed. (Wiley, New York, 1961) Ch. 11.
[2] A.Y. Cho, Thin Solid Films 100 (1983) 291.
[3] R.D. Dupuis, Science 226 (1984) 623.
[4] D.C. Tsui, H.L. Störmer and A.C. Gossard, Phys. Rev. Lett. 48 (1982) 1559.
[5] T. Mimura, S. Hiyamizu, T. Fujii and K. Nanbu, Superlattices and Microstructures 1 (1985) 369.
[6] P.M. Solomon and H. Morkoç, IEEE Trans. Electron Devices ED-31 (1984) 1015.
[7] N. Holonyak Jr and K. Hess, in: Synthetic Modulated Structures, eds. L.L. Chang and B.C. Giessen, (Academic Press, New York, 1985) pp. 257–310.
[8] F. Capasso, in: Gallium Arsenide Technology, ed. D.K. Ferry (Howard Sams, Indianapolis, IN, 1986) ch. 8.
[9] K. Hess, H. Morkoç, H. Shichijo and B.G. Streetman, Appl. Phys. Lett. 35 (1979) 469.
[10] A. Kastalsky and S. Luryi, 1983, IEEE Electron Device Lett. EDL-4 (1983) 334.
[11] S. Luryi and A. Kastalsky, Superlattices and Microstructures 1 (1985) 389;
S. Luryi and A. Kastalsky, Physica B 134 (1985) 453.
[12] R. Tsu and L. Esaki, Appl. Phys. Lett. 22 (1973) 562.
[13] R.F. Kazarinov and R.A. Suris, Sov. Phys.-Semicond. 5 (1971) 707.
[14] R.F. Kazarinov and R.A. Suris, Sov. Phys.-Semicond. 6 (1972) 120.
[15] F. Capasso, K. Mohammed and A.Y. Cho, Appl. Phys. Lett. 48 (1986) 478.
[16] F. Capasso, K. Mohammed and A.Y. Cho, IEEE J. Quant. Electronics QE-22 (1986) 1853.
[17] S. Luryi and F. Capasso, Appl. Phys. Lett. 47 (1985) 1347; erratum: Appl. Phys. Lett. 48 (1986) 1693.
[18] A.R. Bonnefoi, D.H. Chow and T.C. McGill, Appl. Phys. Lett. 47 (1985) 888.
[19] S.M. Sze, Physics of Semiconductor Devices, 2nd Ed. (Wiley, New York, 1981).

[20] D. Frohman-Bentchkowsky, Solid-State Electron. 17 (1974) 517.
[21] C.A. Mead, Proc. IRE 48 (1960) 359.
[22] C.A. Mead, J. Appl. Phys. 32 (1961) 646.
[23] J.P. Spratt, R.F. Schwartz and W.M. Kane, Phys. Rev. Lett. 6 (1961) 341.
[24] M. Heiblum, Solid-State Electron. 24 (1981) 343.
[25] M.M. Atalla and D. Kahng, 1962 IRE-AIEE Solid State DRC (University of New Hampshire, 1962).
[26] D.V. Geppert, Proc. IRE 50 (1962) 1527.
[27] E. Rosencher, S. Delage, Y. Campidelli, and F. Arnaud d'Avitaya, Electron. Lett. 20 (1984) 762.
[28] J.C. Hensel, A.F.J. Levi, R.T. Tung and J.M. Gibson, Appl. Phys. Lett. 47 (1985) 151.
[29] S.M. Sze and H.K. Gummel, Solid-State Electron. 9 (1966) 751.
[30] C.R. Crowell and S.M. Sze, J. Appl. Phys. 7 (1966) 2683.
[31] R.T. Tung, A.F.J. Levi and J.M. Gibson, Appl. Phys. Lett. 48 (1986) 635.
[32] C.O. Bozler and G.D. Alley, 1980 IEEE Trans. Electron Devices ED-27 (1980) 1128; C.O. Bozler and G.D. Alley, Proc. IEEE 70 (1982) 46.
[33] R.A. Murphy, in: Picosecond Electronics, and Optoelectronics, eds. G.A. Mourou, D.M. Bloom and C.-H. Lee (Springer, Berlin, 1985) p. 38.
[34] J. Lindmayer, Proc. IEEE 52 (1964) 1751.
[35] J. Lindmayer and C.Y. Wrigley, Fundamentals of Semiconductor Devices (Van Nostrand, Princeton, 1965) pp. 458–460.
[36] G.E. Derkits Jr, J.P. Harbison, J. Levkoff and D.M. Hwang, Appl. Phys. Lett. 48 (1986) 1220.
[37] J.C. Pfister, E. Rosencher, K. Belhaddad and A. Poncet (1986) to be published.
[38] S. Delage, P.A. Badoz, E. Rosencher and F. Arnaud d'Avitaya (1986) to be published.
[39] J.M. Shannon, IEE J. Solid-State & Electron Devices 3 (1979) 142.
[40] J.M. Shannon and A. Gill, Electron. Lett. 17 (1981) 621.
[41] R.J. Malik, M.A. Hollis, L.F. Eastman, D.W. Woodard, C.E.C. Wood and T.R. AuCoin, in: Proc. 8th Biennial Conf. on Active Microwave Semiconductor Devices and Circuits (Cornell University, 1981) pp. 87–96.
[42] M.A. Hollis, S.C. Palmateer, L.F. Eastman, N.V. Dandekar and P.M. Smith, IEEE Electron Device Lett. EDL-4 (1983) 440.
[43] J.M. Woodcock, J.J. Harris and J.M. Shannon, Physica B 134 (1985) 111.
[44] N. Yokoyama, K. Imamura, T. Ohshima, H. Nishi, S. Muto, K. Kondo and S. Hiyamizu, Jpn. J. Appl. Phys. 23 (1984) L311.
[45] N. Yokoyama, K. Imamura, T. Ohshima, H. Nishi, S. Muto, K. Kondo and S. Hiyamizu, Tech. Digest IEDM-84 (1984) 532.
[46] M. Heiblum, D.I. Thomas, C.M. Knoedler and M.I. Nathan, Appl. Phys. Lett. 47 (1985) 1105.
[47] M. Heiblum, M.I. Nathan, D.I. Thomas and C.M. Knoedler, Phys. Rev. Lett. 55 (1985) 2200.
[48] S. Muto, K. Imamura, N. Yokoyama, S. Hiyamizu and H. Nishi, Electron. Lett. 21 (1985) 555.
[49] I. Hase, H. Kawai, S. Imanaga, K. Kaneko and N. Watanabe, Inst. Phys. Conf. Ser. No. 79: Chap. 11 (Adam Hilger, Bristol, 1985) p. 613.
[50] U.K. Reddy, J. Chen, C.K. Peng and H. Morkoç, Appl. Phys. Lett. 48 (1986) 1799.
[51] M. Heiblum, I.M. Anderson and C.M. Knoedler, Appl. Phys. Lett. 49 (1986) 207.
[52] M. Heiblum, E. Calleja, I.M. Anderson, W.P. Dumke, C.M. Knoedler and L. Osterling, Phys. Rev. Lett. 56 (1986) 2854.

[53] N. Lifshitz, A. Jayaraman, R.A. Logan and H.C. Card, Phys. Rev. B 21 (1980) 670.
[54] H. Ohnishi, N. Yokoyama and H. Nishi, IEEE Electron Device Lett. EDL-6 (1985) 403.
[55] G.E. Derkits Jr, M. Fritze, J.P. Harbison, J. Levkoff and S. Luryi, Appl. Phys. Lett. 51 (1987) to be published.
[56] T.J. Maloney, IEEE Electron Device Lett. EDL-1 (1980) 54.
[57] P. Lugli and D.K. Ferry, IEEE Electron Device Lett. EDL-6 (1985) 25.
[58] S. Imanaga, H. Kawai, K. Kaneko and N. Watanabe, J. Appl. Phys. 59 (1986) 3281.
[59] A.F.J. Levi, J.R. Hayes and R. Bhat, Appl. Phys. Lett. 48 (1986) 1609.
[60] S. Luryi, IEEE Electron Device Lett. EDL-6 (1985) 178.
[61] T. Ando, A.B. Fowler and F. Stern, Rev. Mod. Phys. 54 (1982) 832.
[62] C.Y. Chang, W.C. Liu, M.S. Jame, Y.H. Wang, S. Luryi and S.M. Sze, IEEE Electron Device Lett. EDL-7 (1986) 497.
[63] S. Luryi, A. Kastalsky, A.C. Gossard and R.H. Hendel, IEEE Trans. Electron Devices ED-31 (1984) 832.
[64] A. Kastalsky, R. Bhat, W.K. Chan and M. Koza, Solid-State Electron. 29 (1986) 1073.
[65] C.L. Allyn, A.C. Gossard and W. Wiegmann, Appl. Phys. Lett. 36 (1980) 373;
A.C. Gossard, W. Brown, C.L. Allyn and W. Wiegmann, J. Vac. Sci. & Technol. 20 (1982) 694.
[66] R.J. Malik, K. Board, C.E.C. Wood, L.F. Eastman, T.R. AuCoin and R.L. Ross, Electron. Lett. 16 (1980) 837.
[67] R.F. Kazarinov and S. Luryi, Appl. Phys. Lett. 38 (1981) 810.
[68] B. Honeisen and C.A. Mead, Solid-State Electron. 15 (1972) 891.
[69] S. Luryi, Physica B 134 (1985) 466.
[70] J.C. Bean, Mat. Res. Soc. Symp. Proc. 37 (1985) 245.
[71] G. Abstreiter, H. Brugger, T. Wolf, H. Jorke and H.J. Herzog, Phys. Rev. Lett. 54 (1985) 2441;
Th. Ricker and E. Kasper, Proc. MRS – Europe (1985) p. 193.
[72] R.F. Kazarinov and S. Luryi, Appl. Phys. A 28 (1982) 151.
[73] H. Kroemer, RCA Rev. 18 (1957) 332.
[74] D.L. Miller, P.M. Asbeck, R.J. Anderson and F.H. Eisen, Electron. Lett. 19 (1983) 367.
[75] J.R. Hayes, F. Capasso, A.C. Gossard, R.J. Malik and W. Wiegmann, Electron. Lett. 19 (1983) 410.
[76] R.J. Malik, F. Capasso, R.A. Stall, R.A. Kiehl, R.W. Ryan, R. Wunder and C.G. Bethea, Appl. Phys. Lett. 46 (1985) 600.
[77] P. Hesto, J.-F. Pone and R. Castagne, Appl. Phys. Lett. 40 (1982) 405.
[78] W. Shockley, M. Sparks and G.K. Teal, Phys. Rev. 83 (1951) 151.
[79] Y.H. Kwark and R.M. Swanson, IEEE Trans. Electron Devices ED-33 (1986) 865.
[80] J.R. Hayes, A.F.J. Levi and W. Wiegmann, Electron. Lett. 20 (1984) 851.
[81] J.R. Hayes, A.F.J. Levi and W. Wiegmann, Phys. Rev. Lett. 54 (1985) 1570.
[82] A.F.J. Levi, J.R. Hayes, P.M. Platzman and W. Wiegmann, Phys. Rev. Lett. 55 (1985) 2071.
[83] A.F.J. Levi, J.R. Hayes, P.M. Platzman and W. Wiegmann, Physica B 134 (1985) 480.
[84] J.R. Hayes and A.F.J. Levi, IEEE J. Quant. Electron. QE-22 (1986) 1744.
[85] T. Wang, K. Hess and G.I. Iafrate, 1986, J. Appl. Phys. 59 (1986) 2125.
[86] A.P. Long, P.H. Beton, M.J. Kelly and T.M. Kerr, Electron. Lett. 22 (1986) 130.
[87] A.P. Long, P.H. Beton and M.J. Kelly, Semicond. Sci. & Technol. 1 (1986) 63.
[88] M. Keever, H. Shichijo, K. Hess, S. Banerjee, L. Witkovski, H. Morkoç and B.G. Streetman, Appl. Phys. Lett. 38 (1981) 36.

[89] P.D. Coleman, J. Freeman, H. Morkoç, K. Hess, B.G. Streetman and M. Keever, Appl. Phys. Lett. 40 (1982) 493.
[90] K. Hess, Physica B 117 (1983) 723, and references therein.
[91] K. Hess, Festkörperprobleme 25 (1985) 321.
[92] A. Kastalsky, S. Luryi, A.C. Gossard and R. Hendel, IEEE Electron Device Lett. EDL-5 (1984) 57.
[93] A. Kastalsky, R.A. Kiehl, S. Luryi, A.C. Gossard and R.H. Hendel, IEEE Electron Device Lett. EDL-5 (1984) 321.
[94] S. Luryi, A. Kastalsky, A.C. Gossard and R.H. Hendel, Appl. Phys. Lett. 45 (1984) 1294.
[95] S. Luryi and A. Kastalsky, Superlattices and Microstructures 1 (1985) 389.
[96] S. Luryi and A. Kastalsky, Physica B 134 (1985) 453.
[97] A.A. Grinberg, A. Kastalsky and S. Luryi, IEEE Trans. Electron Devices ED-34 (1987) 409.
[98] A. Kastalsky, S. Luryi, A.C. Gossard and W.K. Chan, IEEE Electron Device Lett. EDL-6 (1985) 347.
[99] A. Kastalsky, J.H. Abeles, R. Bhat, W.K. Chan and M. Koza, Appl. Phys. Lett. 48 (1986) 71.
[100] A. Kastalsky, S. Luryi, A.C. Gossard and W.K. Chan, unpublished (1985).
[101] P.J. Price, IEEE Trans. Electron Devices ED-28 (1981) 911.
[102] A. Kastalsky and R.A. Kiehl, IEEE Trans. Electron Devices ED-33 (1986) 414.
[103] S.E. Esipov and Y.B. Levinson, Zh. Eksp. & Teor. Fiz. 90 (1986) 330 [Sov. Phys.-JETP 63 (1986) 191].
[104] C. Jacoboni and L. Reggiani, Adv. Phys. 28 (1979) 493.
[105] D.K. Ferry, in: Handbook on Semiconductors, Vol. 1, ed. T.S. Moss (North-Holland, Amsterdam, 1982) p. 563.
[106] T.J. Maloney and J. Frey, J. Appl. Phys. 48 (1977) 781.
[107] J.Y.-F. Tang and K. Hess, IEEE Trans. Electron Devices ED-29 (1982) 1906.
[108] M. Inoue and J. Frey, J. Appl. Phys. 51 (1980) 4234.
[109] L.V. Keldysh, Zh. Eksp. & Teor. Fiz. 43 (1962) 661 [Sov. Phys.-JETP 16 (1963) 471].
[110] L.V. Keldysh, Fiz. Tverd. Tela 4 (1962) 2265 [Sov. Phys.-Solid State 4 (1962) 1658].
[111] R.H. Davis and H.H. Hosack, J. Appl. Phys. 34 (1963) 864.
[112] L.V. Iogansen, Zh. Eksp. & Teor. Fiz. 45 (1964) 207 [Sov. Phys.-JETP 18 (1964) 146]; L.V. Iogansen, Usp. Fiz. Nauk 86 (1965) 175 [Sov. Phys.-Usp. 8 (1965) 413].
[113] L. Esaki and R. Tsu, IBM J. Res. Dev. 14 (1970) 61.
[114] L.L. Chang, L. Esaki and R. Tsu, Appl. Phys. Lett. 24 (1974) 593.
[115] L. Esaki and L.L. Chang, Phys. Rev. Lett. 33 (1974) 495.
[116] T.C.L.G. Sollner, W.D. Goodhue, P.E. Tannenwald, C.D. Parker and D.D. Peck, Appl. Phys. Lett. 43 (1983) 588.
[117] T.C.L.G. Sollner, P.E. Tannenwald, D.D. Peck and W.D. Goodhue, Appl. Phys. Lett. 45 (1984) 1319.
[118] T.C.L.G. Sollner, H.Q. Le, C.A. Correa and W.D. Goodhue, Appl. Phys. Lett. 47 (1985) 36.
[119] T. Shewchuk, P.C. Chapin, P.D. Coleman, W. Kopp, R. Fisher and H. Morkoç, Appl. Phys. Lett. 46 (1985) 508.
[120] T. Shewchuk, J.M. Gering, P.C. Chapin, P.D. Coleman, W. Kopp, C.K. Peng and H. Morkoç, Appl. Phys. Lett. 47 (1985) 986.
[121] M. Tsuchiya, H. Sakaki and J. Yoshino, Jpn. J. Appl. Phys. 24 (1985) L853.
[122] M. Tsuchiya and H. Sakaki, IEDM-85 Tech. Digest IEDM-85 (1985) 662.

[123] H. Morkoç, J. Chen, U.K. Reddy, T. Henderson and S. Luryi, Appl. Phys. Lett. 49 (1986) 70.
[124] M. Tsuchiya and H. Sakaki, Jpn. J. Appl. Phys. 25 (1986) L185.
[125] M. Tsuchiya and H. Sakaki, Appl. Phys. Lett. 49 (1986) 88.
[126] A.R. Bonnefoi, R.T. Collins, T.C. McGill, R.D. Burnham, and F.A. Ponce, Appl. Phys. Lett. 46 (1985) 285.
[127] S. Ray, P. Ruden, V. Sokolov, R. Kolbas, T. Boonstra and J. Williams, Appl. Phys. Lett. 48 (1986) 1666.
[128] P. Gavrilovic, J.M. Brown, R.W. Kaliski, N. Holonyak Jr and K. Hess, Solid State Commun. 52 (1984) 237.
[129] E.E. Mendez, W.I. Wang, B. Ricco and L. Esaki, Appl. Phys. Lett. 47 (1985) 415.
[130] R.A. Davies, M.J. Kelly and T.M. Kerr, Phys. Rev. Lett. 55 (1985) 1114.
[131] R.A. Davies, M.J. Kelly, T.M. Kerr, C.J.D. Hetherington and C.J. Humphreys, Nature 317 (1985) 418.
[132] R.A. Davies, M.J. Kelly and T.M. Kerr, Electron. Lett. 22 (1986) 131.
[133] B. Ricco and M.Ya. Azbel, Phys. Rev. B 29 (1984) 1970.
[134] F. Capasso and R.A. Kiehl, J. Appl. Phys. 58 (1985) 1366.
[135] S. Luryi, Appl. Phys. Lett. 47 (1985) 490.
[136] G.E. Derkits Jr (1985) unpublished results.
[137] E.A. Rezek, N. Holonyak Jr, B.A. Vojak and H. Shichijo, Appl. Phys. Lett. 31 (1977) 703.
[138] S. Luryi, Tech. Digest IEDM-85 (1985) 666.
[139] T. Weil and B. Vinter, Appl. Phys. Lett. 50 (1987) 1281.
[140] V.J. Goldman, D.C. Tsui and J.E. Cunningham, Phys. Rev. Lett. 58 (1987) 1256.
[141] A.D. Stone and P.A. Lee, Phys. Rev. Lett. 54 (1985) 1196.
[142] K.K. Thornber, T.C. McGill and C.A. Mead, J. Appl. Phys. 38 (1967) 2384.
[143] M. Büttiker, Phys. Rev. B 27 (1983) 6178;
M. Büttiker and R. Landauer, Phys. Rev. Lett. 49 (1982) 1739.
[144] K.W.H. Stevens, J. Phys. C 16 (1983) 3649.
[145] D.D. Coon and H.C. Liu, Appl. Phys. Lett. 49 (1986) 94.
[146] E.R. Brown, T.C.L.G. Sollner, W.D. Goodhue and C.D. Parker, Appl. Phys. Lett. 50 (1987) 83.
[147] T.C.L.G. Sollner, E.R. Brown, W.D. Goodhue and H.Q. Le, Appl. Phys. Lett. 50 (1987) 332.
[148] P.D. Coleman, S. Goedeke, T.J. Shewchuk, P.C. Chapin, J.M. Gering and H. Morkoç, Appl. Phys. Lett. 48 (1986) 422.
[149] B. Jogai, K.L. Wang and K.W. Brown, Appl. Phys. Lett. 48 (1986) 1003.
[150] A.R. Bonnefoi, T.C. McGill and R.D. Burnham, IEEE Electron Device Lett. EDL-6 (1985) 636.
[151] N. Yokoyama, K. Imamura, S. Muto, S. Hiyamizu and H. Nishi, Jpn. J. Appl. Phys. 24 (1985) L853.
[152] B. Jogai and K.L. Wang, Appl. Phys. Lett. 46 (1985) 167.
[153] H. Sakaki, Jpn. J. Appl. Phys. 18 (1980) L735.
[154] P.M. Petroff, A.C. Gossard, R.A. Logan and W. Wiegmann, Appl. Phys. Lett. 41 (1982) 635.
[155] Y.-C. Chang, L.L. Chang and L.L. Esaki, Appl. Phys. Lett. 47 (1985) 1324.
[156] J. Cibert, P.M. Petroff, G.J. Dolan, S.J. Pearton, A.C. Gossard and J.H. English, Appl. Phys. Lett. 49 (1986) 1275.
[157] M.A. Reed, R.T. Bate, K. Bradshaw, W.M. Duncan, W.R. Frensley, J.W. Lee and H.D. Shih, J. Vac. Sci. & Tech. B 4 (1986) 358.

CHAPTER 13

PHYSICS OF QUANTUM WELL LASERS

N.K. DUTTA

AT&T Bell Laboratories
600 Mountain Avenue
Murray Hill, NJ 07974, USA

Heterojunction Band Discontinuities: Physics and Device Applications
Edited by F. Capasso and G. Margaritondo
© *Elsevier Science Publishers B.V., 1987*

Contents

1. Introduction 567
2. Energy levels 568
3. Gain and radiative recombination 570
   3.1. Density of states 571
   3.2. Spontaneous emission rate 573
   3.3. Optical gain calculation 577
   3.4. Threshold current calculation 578
   3.5. TE and TM mode gain 581
4. Nonradiative recombination 582
   4.1. Auger effect 582
5. Single quantum well and multi-quantum well lasers 586
6. Experimental results 589
   6.1. AlGaAs quantum well lasers 589
   6.2. InGaAsP quantum well lasers 590
References 592

1. Introduction

A double heterostructure laser consists of an active layer sandwiched between two higher gap cladding layers. The active layer thickness is typically in the range of 0.1 to 0.3 μm. In the last few years, double heterostructure lasers with an active layer thickness ~ 100 Å have been fabricated. In these structures the carrier (electron or hole) motion normal to the active layer is restricted. As a result, the kinetic energy of the carriers moving in that direction are quantized into discrete energy levels similar to the well-known quantum mechanical problem of the one-dimensional potential well, and hence these lasers are called quantum well lasers.

When the thickness of the active region (or any low gap semiconductor layer confined between higher gap semiconductors) becomes comparable to the de Broglie wavelength ($\lambda \sim h/p$), quantum mechanical effects are expected to occur. These effects are observed in the absorption and emission (including laser action) characteristics and transport characteristics including phenomena such as tunneling. The optical characteristics of semiconductor quantum well double heterostructures were initially studied by Dingle et al. [1]. Since then extensive work on GaAlAs quantum well lasers have been done by Holonyak et al. [2,3], Tsang [4,5] and Hersee et al. [6]. Quantum well lasers fabricated using the InGaAsP material system have been extensively studied by Dutta and co-workers [7–10].

The physical principles of quantum well double-heterostructure (DH) lasers are described in this chapter. One advantage of quantum well DH lasers over regular DH lasers is that the emission wavelength of the former can be varied simply by varying the width of the quantum wells which form the active region. This phenomenon is discussed is sect. 2. The restriction of the carrier motion normal to the well leads to a modification of the density of states in a quantum well which changes the radiative and nonradiative recombination rates of electrons and holes in a quantum well heterostructure compared to that of a regular DH. These changes may result in several desirable characteristics such as lower threshold current density, higher efficiency, lower temperature dependence of threshold current of suitably designed quantum well lasers compared to regular DH lasers. The calculation of optical gain, radiative recombination rate and nonradiative Auger recombination rates are discussed in sections 3 and 4. The various types of quantum well structures such as the single quantum well, multi-quantum well, modified multi-quantum well and the graded index single quantum well and their principal performance characteristics such as threshold current and efficiency are discussed in sect. 5. Finally, the current status of the experimental results on QW lasers are briefly reviewed in sect. 6.

2. Energy levels

A carrier (electron or hole) in a double heterostructure is confined in a three-dimensional potential well. The energy levels of such carriers are obtained by separating the Hamiltonian into three parts, corresponding to the kinetic energies in the x-, y- and z-directions each of which form a continuum of states. When the thickness of the heterostructure (L_z) is comparable to the de Broglie wavelength, the kinetic energy corresponding to the particle motion along the z-direction is quantized. The energy levels can be obtained by separating the Hamiltonian into energies corresponding to x-, y-, z-directions. For the x, y direction, the energy levels form a continuum of states given by

$$E = \frac{\hbar^2}{2m}\left(k_x^2 + k_y^2\right), \tag{2.1}$$

where m is the effective mass of the carrier and k_x and k_y are the wave vectors along the x and y directions, respectively. Thus the electrons or holes in a quantum well may be viewed as to form a two-dimensional Fermi gas.

The energy levels in the z-direction are obtained by solving the Schrödinger equation for a one-dimensional potential well. It is given by

$$\begin{aligned}-\frac{\hbar^2}{2m}\frac{d\psi^2}{dz^2} &= E\psi &&\text{in the well} &&(0 \le z \le L_z), \\ -\frac{\hbar^2}{2m}\frac{d^2\psi}{dz^2} + V\psi &= E\psi &&\text{outside the well} &&(z \ge L_z;\ z \le 0),\end{aligned} \tag{2.2}$$

where ψ is the Schrödinger wavefunction and V is the depth of the potential well. For the limiting case of an infinite well, the energy levels and the wavefunctions are

$$E_n = \frac{\hbar^2}{2m}\left(\frac{n\pi}{L_z}\right)^2 \quad \text{and} \quad \psi_n = A\sin\frac{n\pi z}{L_z} \quad (n = 1, 2, 3), \tag{2.3}$$

where A is a normalization constant. For very large L_z, eq. (2.3) yields a continuum of states and the system no longer exhibits quantum effects.

For a finite well, the energy levels and wavefunction can be obtained from eq. (2.2) using the boundary conditions that ψ and $d\psi/dz$ are continuous at the interfaces $z = 0$ and $z = L_z$.

The potential well (V) for electron and holes in a double heterostructure depends on the materials involved. Dingle [11] has found the following relation between conduction-band (ΔE_c) and valence-band (ΔE_v) discontinuities for GaAs–AlGaAs double heterostructures

$$\Delta E_c/\Delta E = 0.85 \pm 0.03, \quad \Delta E_v/\Delta E = 0.15 \pm 0.03, \tag{2.4}$$

where ΔE is the band gap difference between the confining layers and active region. A knowledge of ΔE_c and ΔE_v is necessary in order to accurately calculate the energy levels. For an InGaAsP–InP double heterostructure, the values obtained by Forrest et al. [12] are

$$\Delta E_c/\Delta E = 0.39 \pm 0.01, \quad \Delta E_v/\Delta E = 0.61 \pm 0.01. \tag{2.5}$$

The energy eigenvalues for a particle confined in the quantum well are

$$E(n, k_z, k_y) = E_n + \frac{\hbar^2}{2m_n^*}\left(k_x^2 + k_y^2\right), \tag{2.6}$$

where E_n is the nth confined-particle energy level for carrier motion normal to the well and m_n^* is the effective mass. Figure 1 shows schematically the energy levels E_n of the electrons and holes confined in a quantum well. The confined-particle energy levels (E_n) are denoted by E_{1c}, E_{2c}, E_{3c} for electrons, E_{1hh}, E_{2hh} for heavy holes and E_{1lh}, E_{2lh} for light holes. These quantities can be calculated by solving the eigenvalue equations, eqs. (2.2), for a given potential barrier (ΔE_c or ΔE_v) as described earlier.

Electron–hole recombination in a quantum well follows the selection rule $\Delta n = 0$, i.e., the electrons in states E_{1c} (E_{2c}, E_{3c}, etc.) can combine with the heavy holes E_{1hh} (E_{2hh}, E_{3hh}, etc.) and with light holes E_{1lh} (E_{2lh}, E_{3lh}, etc.). Note, however, that since $E_{1lh} > E_{1hh}$ the light-hole transi-

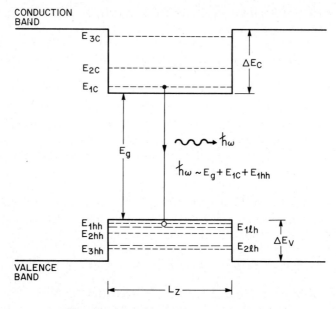

Fig. 1. Energy levels in a quantum well structure.

tions are at a higher energy than the heavy-hole transitions. Since the separation between the lowest conduction-band level and the highest valence-band level is given by

$$E_q = E_g + E_{1c} + E_{1hh} \cong E_g + \frac{h^2}{8L_z^2}\left(\frac{1}{m_c} + \frac{1}{m_{hh}}\right), \qquad (2.7)$$

we see that in a quantum well structure the energy of the emitted photons can be varied simply by varying the well width L_z. Figure 2 shows the experimental results of Temkin et al. [13] for InGaAs quantum well lasers with different well thicknesses bounded by InP cladding layers. For low well thicknesses the laser emission shifts to higher energies.

3. Gain and radiative recombination

The basis of light emission in semiconductors is the recombination of an electron in the conduction band with a hole from the valence band and the excess energy is emitted as a photon (light quantum). The process is called

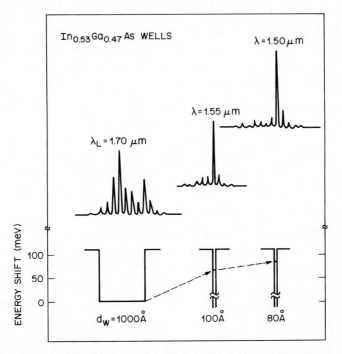

Fig. 2. Lasing wavelength for different well thicknesses of a InGaAs–InP multi-quantum well structure. (Courtesy of H. Temkin).

radiative recombination. Sufficient number of electrons and holes must be excited in the semiconductor for stimulated emission or net optical gain. The condition for net gain at a photon energy E is given by [14]

$$E_{tc} + E_{fv} = E - E_g, \qquad (3.1)$$

where E_{tc} and E_{fv} are the quasi-Fermi levels of electrons and holes, respectively, measured from the respective band edges (positive into the band) and E_g is the band gap of the semiconductor. The quasi-Fermi levels can be calculated as follows from a knowledge of the density of states of electrons and holes in the quantum well heterostructure.

3.1. Density of states

In a quantum well structure, the kinetic energy of the confined carriers (electrons or holes) for velocities normal to the well (z-direction) is quantized into discrete energy levels. This modifies the density of states from the well-known three-dimensional case. Using the principle of box quantization of kinetic energies along the x- and y-direction, the number of electron states per unit area in the x–y direction for the ith subband within an energy interval dE is given by

$$D_i(E)\,dE = 2\frac{d^2k}{(2\pi)^2}, \qquad (3.2)$$

where the factor 2 arises from two spin states and $k = (k_x, k_y)$ is the momentum vector. Using the parabolic band approximation, i.e. $E = \hbar^2 k^2/2m_{ci}$, eq. (3.2) may be written as

$$D_i = \frac{m_{ci}}{\pi \hbar^2}, \qquad (3.3)$$

where m_{ci} is the effective mass of the electrons in the ith subband of the conduction band. Thus the density of states per unit volume is given by

$$g_{ci} = \frac{D_i}{L_z} = \frac{m_{ci}}{\pi \hbar^2 L_z}. \qquad (3.4)$$

A similar equation holds for the holes in the valence band. For the regular three-dimensional case, the density of states is given by

$$\rho(E) = 2(2\pi m_c k_B T/h^2)^{3/2} E^{1/2}. \qquad (3.5)$$

where k_B is the Boltzmann constant and T is the temperature. A comparison of eqs. (3.4) and (3.5) shows that the density of states in a quantum well is independent of the carrier energy and temperature. The modification of the density of states in a quantum well is sketched in fig. 3. This modification can significantly alter the recombination rates in a quantum well double heterostructure compared to that for a regular double heterostructure as discussed later. The simple model for the density of states described above neglects the effect of collisions between carriers which would broaden the discrete energy levels in the z-direction.

If n is the number of electrons in the conduction band, the Fermi level can be obtained from the following equation:

$$n = \sum_i \int g_{ci} f(E_i)\,dE_i, \text{ where } f(E) = \frac{1}{1+\exp[(E_i - E_{fc})/k_B T]}, \qquad (3.6)$$

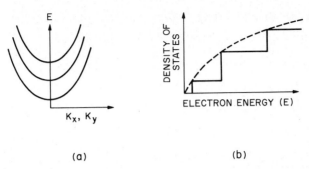

Fig. 3. (a) Energy versus wavevector for each subband; (b) Schematic representation of the density of states in a quantum well.

$f(E)$ is the Fermi function, E_{fc} is the quasi-Fermi energy in the conduction band and E_i is the confined particle energy level of the ith subband. The summation is over all subbands in the conduction band. Equation (3.6) may be rewritten as

$$n = -k_B T \sum_i g_{ci} \ln[\exp(-E_i/k_B T) + \exp(-E_{\text{fc}}/k_B T)]. \qquad (3.7)$$

Under the Boltzmann approximation and assuming that only one subband is occupied eq. (3.7) becomes

$$n = N_c \exp(-E_{\text{fc}}/k_B T) \quad \text{where} \quad N_c = g_c k_B T, \qquad (3.8)$$

and the Fermi factor

$$f_c(E) \cong \exp[(E_{\text{fc}} - E)/k_B T] \cong \frac{n}{N_c} \exp(-E/k_B T). \qquad (3.9)$$

A similar equation holds for the holes in the valence band.

3.2. Spontaneous emission rate

A complete calculation of the spontaneous radiative recombination rate would include summation over all possible transitions between various subbands in the conduction and valence band. Before doing this calculation, we first consider a particular transition, i.e. the transition from the lowest level in the conduction band (1c) to the lowest state in the heavy-hole band (1hh). This transition has the highest gain (due to Fermi factors) under external excitation (e.g., current injection) and hence is responsible for the lasing action in quantum well injection lasers.

The spontaneous emission rate $r_{sp}(E)$ and the absorption coefficient $\alpha(E)$ for electronic transitions between two discrete levels are given by [15]

$$r_{sp}(E) = \frac{4\pi q^2 \bar{\mu} E}{m_0^2 \epsilon_0 c^3 h^2} |M_{if}|^2 \delta(E_i - E_f - E), \tag{3.10}$$

$$\alpha(E) = \frac{q^2 h}{2\epsilon_0 m_0^2 c \bar{\mu} E} |M_{if}|^2 \delta(E_i - E_f - E), \tag{3.11}$$

where q is the electron charge, $\bar{\mu}$ is the refractive index, m_0 is the mass of the free electron, c is the velocity of light, ϵ_0 is the permittivity of free space and h is the Planck constant. M_{if} is the matrix element of the transition from the initial state $|i\rangle$ to the final state $|f\rangle$. Since the photons carry negligible momentum, two cases are possible:

(i) a matrix element obeying k selection rule in the x- and y-direction [16], and

(ii) a constant matrix element [17] (independent of the k vector), which can arise from a band-to-impurity transition. We first consider the case of a matrix element obeying the k selection rule.

Consider an area A of the material in the x–y direction, then the volume V is given by $V = AL_z$. The matrix element $|M_{if}|^2$ is given by [16]

$$|M_{if}|^2 = |M_b|^2 \frac{(2\pi)^2}{A} \delta(k_c - k_v) \delta_{nn'}. \tag{3.12}$$

The last δ-function arises from the selection rule for the confined states in the z-direction. The quantity $|M_b|$ is an average matrix element for the Bloch states of the bands (in the x–y direction), as in the corresponding three-dimensional case. Using the Kane band model, $|M_b|^2$ in bulk semiconductors is given by [18]

$$|M_b|^2 = \frac{m_0^2 E_g (E_g + \Delta)}{12 m_c (E_g + \tfrac{2}{3}\Delta)} = \zeta m_0 E_g, \tag{3.13}$$

where $\zeta = 1.3$ for GaAs. The quantity ζ lies between 1 and 2 for most semiconductors. We assume that $|M_b|^2$ for a quantum well is given by

$$|M_b|^2 = \zeta m_0 E_q,$$

where E_q is the separation between levels 1c and 1hh. The uncertainty in the value of ζ represents the accuracy of our results. Summing over all the states in the band, we get

$$r_{sp}(E) = \frac{4\pi\bar{\mu}q^2 E}{m_0^2\epsilon_0 h^2 c^3}|M_b|^2 \frac{(2\pi)^2}{A}\left(\frac{2A}{(2\pi)^2}\right)^2$$

$$\times \int f_c(E_c) f_v(E_v)\, d^2k_c\, d^2k_v\, \delta(k_c - k_v)\, \delta(E_i - E_f - E), \qquad (3.14)$$

where f_c, f_v denote the Fermi factors for the electron at the energy E_c and hole at the energy E_v. The factor 2 arises from two spin states. The Fermi factors are

$$f_c(E_c) = \frac{1}{1 + \exp[(E_c - E_{fc})/k_B T]},$$

$$f_v(E_v) = \frac{1}{1 + \exp[(E_v - E_{fv})/k_B T]} \qquad (3.15)$$

where E_{fc} and E_{fv} are the quasi-Fermi levels for the electrons and holes, respectively.

We use the same convention for the electron and hole energy levels as before, i.e., the electron energy is measured from the conduction-band edge and is positive into the band and hole energies are measured positive downward into the valence-band from the valence-band edge. The integrals in eq. (3.14) can be evaluated with the following result

$$r_{sp}(E) = \frac{32\pi^2 \mu q^2 E |M_b|^2 m_r}{m_0^2 \epsilon_0 h^4 c^3} A f_c(E_c) f_v(E_v), \qquad (3.16)$$

with

$$E_c = \frac{m_r}{m_c}(E - E_q), \quad E_v = \frac{m_r}{m_h}(E - E_q), \quad m_r = m_c m_v/(m_c + m_v).$$

Since $A = V/L_z$, the total spontaneous emission rate per unit volume is given by

$$R_{sp}(E) = \frac{32\pi^2 \mu q^2 m_r |M_b|^2}{m_0^2 \epsilon_0 h^4 c^3 L_z} I, \qquad (3.17)$$

where

$$I = \int_{E_q}^{\infty} E f_c(E_c) f_v(E_v) \, dE.$$

Following a similar procedure, the expression for the absorption coefficient $\alpha(E)$ is given by [after using eq. (3.11) and summing over the available states]

$$\alpha(E) = \frac{q^2 h}{2\epsilon_0 m_0^2 c \mu E} |M_b|^2 \frac{(2\pi)^2 2}{A} \left(\frac{A}{(2\pi)^2}\right)^2 \left(\frac{1}{V}\right)$$

$$\times \int (1 - f_c - f_v) \, d^2 k_c \, d^2 k_v \, \delta(k_c - k_v) \, \delta(E_i - E_f - E)$$

$$= \frac{q^2 m_r |M_b|^2}{\epsilon_0 m_0^2 c \hbar \mu E L_z} [1 - f_c(E_c) - f_v(E_v)]. \tag{3.18}$$

The factor 2 arises from two spin conserving transitions. The radiative component of the current or the nominal current density J_n at threshold is given by

$$J_R = J_n = q R_{sp}(E). \tag{3.19}$$

Constant matrix element

In this case, we assume that $|M|^2$ in eq. (3.12) is given by

$$|M|^2 = |M_c|^2 \delta_{nn'}, \tag{3.20}$$

where $|M_c|^2$ is independent of the energy and the wavevector of the particles involved. One can now follow the same procedure outlined before and obtain the following expressions for the absorption coefficient and the spontaneous emission rate:

$$\alpha(E) = \frac{4\pi q^2 m_c m_v |M_c|^2}{\epsilon_0 c m_0^2 \hbar^3 \mu E L_z} I_1(E), \tag{3.21}$$

$$R_{sp}(E) = \frac{4\mu q^2 m_c m_v}{\pi \epsilon_0 c^3 \hbar^6 m_0^2 L_z} |M_c|^2 \int_{E_q}^{\infty} E I_2(E) \, dE, \tag{3.22}$$

where

$$I_1(E) = \int_0^{E-E_q} [1 - f_c(E_c) - f_v(E_v)] \, dE_c,$$

$$I_2(E) = \int_0^{E-E_q} f_c(E_c) f_v(E_v) \, dE_c. \tag{3.23}$$

3.3. Optical gain calculation

Using eqs. (3.17)–(3.19), the gain versus nominal current density and gain versus injected current density can be obtained. We first consider the GaAs quantum well and show that the results obtained using the k-selection-rule model agrees well with the experimental results for the threshold current density. For longer wavelength semiconductor lasers, e.g., InGaAsP, additional nonradiative mechanisms (e.g., Auger effect) may be important in determining the threshold current.

Figures 4a,b show the calculated gain versus injected carrier density and gain versus nominal current density at various temperatures for a 200 Å thick undoped GaAs quantum well. The parameters used in the calculation are $m_c = 0.071 m_0$, $m_v = 0.45 m_0$, $m_{1h} = 0.081 m_0$, $\bar{\mu} = 3.5$, and $\zeta = 1.33$. For a 200 Å thick well, the calculated $E_q - E_g \cong 0.011$ eV. The calculation assumes that the band gap E_g varies with temperature at a rate $dE_g/dT =$

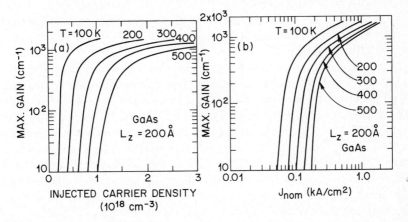

Fig. 4. (a) Calculated maximum gain versus injected carrier density for a GaAs quantum well structure. (b) Calculated gain versus nominal current density at various temperature for a GaAs quantum well structure. (After ref. [16].)

−0.35 meV/K. The nominal current density is calculated by summing the light-hole and heavy-hole transition rates.

3.4. Threshold current calculation

At threshold, the optical gain equals the total optical loss in the cavity. The condition for threshold is

$$\Gamma g_{th} = \alpha_a \Gamma + \alpha_c (1 - \Gamma) + \frac{1}{L} \ln\left(\frac{1}{R}\right), \tag{3.24}$$

where Γ is the fraction of the lasing mode confined in the active region (generally known as the confinement factor), g_{th} is the optical gain at threshold, α_a, α_c are the optical losses such as free carrier absorption and scattering in the active region and cladding layer, respectively, L is the cavity length and R is the reflectivity of the two cleaved facets so that the last term represents the distributed mirror loss.

The evaluation of Γ is, in general, quite tedious. For the fundamental mode, however, a remarkably simple expression,

$$\Gamma \cong D^2/(2 + D^2) \quad \text{with} \quad D = k_0 \left(\mu_a^2 - \mu_c^2\right)^{1/2} d, \tag{3.25}$$

is found to be accurate to within 1.5% [19]. In the above expression, d is the active layer thickness, $k_0 = 2\pi/\lambda$ where λ is the wavelength in free space and μ_a, μ_c are the refractive indices of the active and the cladding layer, respectively. For a single quantum well laser ($d < 300$ Å), eq. (3.25) reduces to a simpler expression $\Gamma \cong \frac{1}{2}D^2$.

We now calculate the threshold current density and compare the calculated value with the measured results. The calculated confinement factor (Γ) for a 250 Å thick GaAs–Al$_{0.52}$Ga$_{0.48}$As quantum well is 0.04. Using an internal loss $\alpha = 10$ cm^{-1}, $R = 0.3$ and $L = 380$ μm, we get $g_{th} \cong 10^3$ cm^{-1} from eq. (3.24). Thus the threshold gain for single quantum well lasers is an order of magnitude larger than that for regular double heterostructure lasers. This difference is principally due to the smaller confinement factor for a single quantum well. For the calculated g_{th} of 10^3 cm^{-1}, the estimated threshold current density using fig. 4b is 550 A/cm^2. This agrees well with the reported value of 810 A/cm^2 for a 200 Å wide GaAs quantum well laser [20]. Equation (3.24) shows that the small value of the confinement factor makes the threshold gain and hence the threshold current density of quantum well lasers very sensitive to internal losses, which is influenced by

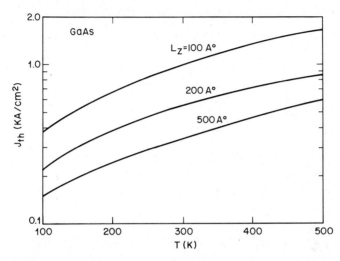

Fig. 5. Calculated temperature dependence of threshold current of a GaAs single quantum well laser. (After ref. [16].)

defects, nonradiative regions and unpumped active regions in an injection laser.

It is generally observed that the threshold current of GaAs quantum well lasers does not increase as rapidly with increasing temperature as that for regular double heterostructure lasers. The threshold current as a function of temperature is $T_0 \sim 250$–400 K for $T > 270$ K for quantum well lasers and $T_0 \sim 200$ K for regular DH lasers. Figure 5 shows the calculated threshold current as a function of temperature of GaAs–$Al_{0.52}Ga_{0.48}As$ quantum wells of different active region thicknesses. The calculated T_0 value from fig. 5 is ~ 300 K for 300 K $< T < 400$ K. The reasons for the high T_0 value is a weaker dependence of quasi-Fermi energies on temperature in this quasi-two-dimensional system than for bulk semiconductors. Since the conditions for gain (at energy E) is $E_{fc} + E_{fv} > E - E_g$, fewer additional carriers at threshold are needed at a higher temperature for a quantum well laser than for a regular DH laser.

Sugimura [21] has calculated the radiative recombination rate in quantum well structures taking into account transitions between all subbands (in the Γ-valley) and the effect of the L-valley. Figure 6 shows the dependence of the gain coefficient on the well thickness (L_z) for different injected carrier densities. The undulations in the value of the gain correspond to recombinations between electrons in different subbands of the conduction

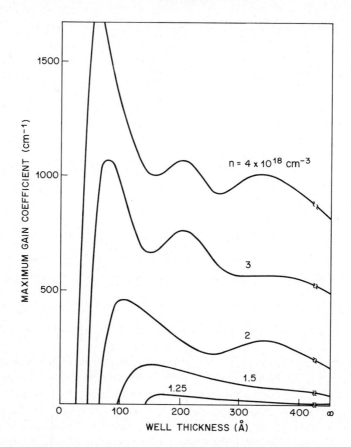

Fig. 6. The maximum gain coefficient for a undoped InGaAsP–InP ($\lambda \sim 1.1$ μm) single quantum well structure as a function of the well thickness L_z. The parameter n is the injected carrier density per unit volume injected into the active region. (After ref. [21].)

band and the heavy or light holes in the valence band. The steep decrease in the gain coefficient for small well thicknesses arises from the distribution of carriers in the L-valley which reduces the number of carriers in the Γ-valley that contribute to optical gain.

Holonyak and co-workers [2,3,22] have proposed phonon-assisted emission in the quantum well structure in order to explain their emission spectrum. Sugimura [23] has calculated the phonon-assisted gain coefficient using LO phonon interactions only. He has found that the phonon-assisted gain coefficient is smaller than the gain coefficient for direct transitions.

3.5. TE and TM mode gain

In the waveguide of a semiconductor laser, two types of modes can propagate. These are the transverse electric (TE) and the transverse magnetic (TM) modes. The TE mode has its electric field parallel to the p–n junction in a semiconductor laser whereas the TM mode has its field normal to the junction. In a QW structure, the optical gain for the TE mode occurs through both light- and heavy-hole transitions whereas for TM mode the optical gain is predominantly through the light-hole transitions [24]. Since the light-hole transitions occur at higher energies than heavy-hole transitions (fig. 1) in a QW structure, the peak optical gain of the TM mode is shifted towards higher energies from that of the TE mode in a QW laser

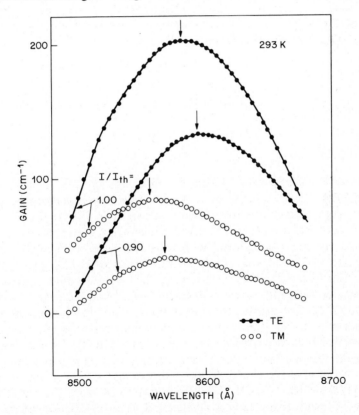

Fig. 7. Measured TE and TM mode gain spectrum for a GaAs quantum well laser. (After ref. [25].)

[25]. Measurements of TE and TM mode gains at two injection currents in a GaAs QW laser are shown in fig. 7. Note that the maximum gain of the TM mode occurs at a higher energy compared to that of the TE mode.

4. Nonradiative recombination

An electron–hole pair can recombine nonradiatively meaning that the recombination can occur through any process that does not emit a photon. In many semiconductors, for example pure germanium or silicon, the nonradiative recombination dominates radiative recombination.

The effect of nonradiative recombination on the performance of injection lasers is to increase the threshold current. If τ_{nr} is the carrier lifetime associated with the nonradiative process, the increase in threshold current density is approximately given by

$$J_{nr} = qn_{th}d/\tau_{nr}, \qquad (4.1)$$

where n_{th} is the carrier density at threshold, d is the active layer thickness and q is the electron charge.

4.1. Auger effect

Since the pioneering work by Beattie and Landsberg [26], it is generally accepted that Auger recombination can be a major nonradiative mechanism in narrow-gap semiconductors. Recent attention to the Auger effect has been in connection with the observed higher temperature dependence of threshold current of long-wavelength InGaAsP lasers compared to short-wavelength AlGaAs lasers. It is generally believed that the Auger effect plays a significant role in determining the observed high temperature sensitivity of threshold current of InGaAsP lasers emitting near 1.3 μm and 1.55 μm [27,28]. In quantum well structures, the modification of the density of states may reduce the Auger recombination rate to allow long-wavelength QW lasers to have a low temperature sensitivity [29–32]. We now outline the formulation for Auger rate calculation in QW structures.

The various band-to-band Auger processes are shown in fig. 8. These processes are labeled CCCH, CHHS, and CHHL in which C stands for the conduction band, H for the heavy-hole band, L for the light-hole band and S for the spin split-off band. The energy versus wavevector (k_x, k_y) diagram is shown for one subband (one state in the z-direction) in each of

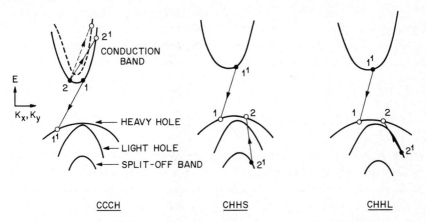

Fig. 8. Band-to-band Auger recombination are shown schematically.

the bands. The dashed curve (for CCCH) shows a second subband in the conduction band. The dashed line shows a process in which the electron makes a transition to a different subband. The CCCH process involves three electrons and a heavy hole and is dominant in n-type material. The CHHS and the CHHL processes are dominant in p-type material. Under high-injection conditions as is present in lasers, all of the above mechanisms must be considered. In this section, we derive expressions for the transition probability of these band-to-band Auger processes.

CCCH Process

The expression for the CCCH Auger rate in a quantum well structure is the following [30]

$$R = \frac{2\pi}{\hbar} \left(\frac{1}{(2\pi)^2} \right)^4 \frac{1}{L_z} \int |M_{\text{if}}|^2 P(1, 1'; 2, 2') \, \delta(E_i - E_f)$$

$$\times \, d^2k_1 \, d^2k_2 \, d^2k_{1'} \, d^2k_{2'}, \tag{4.2}$$

where R denotes the total Auger rate per unit volume. The quantity $|M_{\text{if}}|^2$ denotes the square of the matrix element and $P(1, 1'; 2, 2')$ is the difference between the probability of the Auger process shown in fig. 8 and the inverse process of impact ionization. The factor $1/L_z$ arises from converting the Auger rate per unit area (in two dimensions) to per unit volume.

The quantity M_{if} is the matrix element for the Coulomb interaction between two electrons which results in the Auger transition of fig. 8. It is given by

$$M_{if} = \iint \psi_1'^*(r_1)\, \psi_2'^*(r_2)\, \frac{q^2}{\epsilon |r_1 - r_2|}\, \psi_1(r_1)\, \psi_2(r_2)\, d^3r_1\, d^3r_2, \quad (4.3)$$

where $\psi_m(r)$ are the quantum well wavefunctions.

The Auger recombination rates for the CCCH, CHHS and CHHL processes have been previously calculated. In the nondegenerate approximation, the Auger rate R for the CCCH process varies as [30]

$$R \sim n^2 p \exp(-\Delta E_c / k_B T), \quad \text{with} \quad \Delta E_c = \frac{m_{ct} E_q}{m_v + 2m_c - m_{ct}}, \quad (4.4)$$

where n, p are the electron and the hole concentrations respectively, m_{ct} is the effective mass of the electron at the high energy E_t (electron 2' in fig. 8) and m_c, m_v are the band edge effective masses of the electron and hole, respectively. E_q is the energy difference between the lowest conduction-band subband and the highest heavy-hole subband. For most cases, $E_q \cong E_g$ (band gap). The strong temperature dependence and the band gap dependence of the Auger recombination rate is evident from eq. (4.4). The Auger rate increases rapidly with increasing temperature and with decreasing band gap. For the CHHS and CHHL processes, the Auger rate in the nondegenerate approximation varies as

$$R \sim p^2 n \exp(-\Delta E / k_B T), \quad (4.5)$$

with

$$\Delta E = \frac{m_s(E_q - \Delta_1)}{m_c + 2m_v - m_s} \quad \text{for CHHS, and}$$

$$\Delta E = \frac{m_\ell E_q}{m_c + 2m_v - m_\ell} \quad \text{for CHHL,}$$

where m_s, m_ℓ are the effective masses of the split-off-band hole and light hole, respectively, and Δ_1 is the energy difference between the top of the heavy-hole band and that of the split-off band.

For undoped semiconductors, such as the active region of a semiconductor laser, the Auger rate at an injected carrier density n is given by

$$R = Cn^3, \tag{4.6}$$

where C is known as the Auger coefficient. The Auger lifetime τ_A is given by

$$\tau_A = \frac{n}{R} = \frac{1}{Cn^2}. \tag{4.7}$$

The increase in threshold current due to Auger recombination can now be obtained from eqs. (4.1) and (4.7).

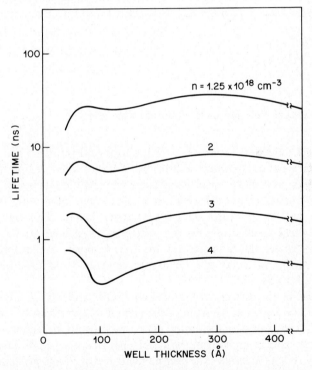

Fig. 9. The calculated Auger lifetime for the CHHS process in an undoped InGaAsP–InP quantum well as a function of the well thickness. The parameter n is the injected carrier density. (After ref. [31].)

The calculated Auger lifetime due to CCCH process depends strongly on the value of ΔE_c in eq. (4.4). For bulk semiconductors, m_{ct}, i.e. the conduction-band mass at the threshold energy E_t, is calculated using the Kane model which gives $m_{ct} \cong 1.7 m_c$. For a quantum well structure, the ratio m_{ct}/m_c is not known. For parabolic bands $m_{ct}/m_c = 1$ and the equation for ΔE_c [eq. (4.4)] reduces to

$$\Delta E_c = \mu E_q / (1 + \mu), \tag{4.8}$$

where $\mu = m_c/m_v$. Dutta [30] has shown that the calculated Auger rate for a given band gap can decrease by more than one order of magnitude if the ratio m_{ct}/m_c increases from 1 to 2. The calculated Auger lifetime for InGaAsP–InP quantum well structure is shown in fig. 9. The calculation assumes the same matrix element as that for bulk semiconductors. Because of the uncertainties of both the matrix elements and effective masses of the bands away from the band edge (i.e. the values of m_{ct}, m_s, m_ℓ), only order of magnitude estimates of the Auger rates in a quantum well structure can be obtained at present.

5. Single quantum well and multi-quantum well lasers

Quantum well injection lasers with both single and multiple active layers have been fabricated. Quantum well lasers with one active region are called single quantum well (SQW) lasers and those with multiple active regions are called multi-quantum well (MQW) lasers. The layers separating the active layers in a MQW structure are called barrier layers. The energy band diagram of these laser structures are schematically shown in fig. 10. The MQW lasers where the band gap of the barrier layers are different from that of the cladding layers are sometimes referred to as modified multi-quantum well lasers.

One of the main differences between the SQW and MQW lasers is that the confinement factor (Γ) of the optical mode is significantly smaller for the former than that for the latter. This can result in higher threshold carrier density and in higher threshold current density of SQW lasers when compared with MQW lasers. The confinement factor of a SQW heterostructure can be significantly increased using a graded-index cladding layer (see fig. 10). This allows the use of the intrinsic advantage of the quantum well structure (high gain at low carrier density) without the penalty of a small

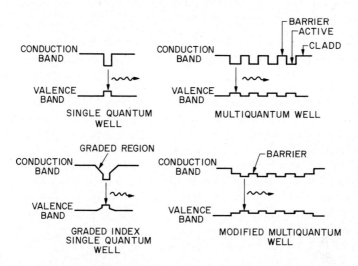

Fig. 10. Single quantum well and multi-quantum well laser structures are shown schematically.

mode-confinement factor. Threshold current densities as low as 200 A/cm² have been reported for a GaAs graded-index (GRIN) SQW laser [33].

We now discuss the calculation of mode-confinement factor in SQW and MQW structures. The confinement factor of the fundamental mode of a double heterostructures is approximately given by eq. (3.25). For small active layer thickness, such as that of a single quantum well the expression for Γ can be further simplified since the normalized waveguide thickness $D \ll 1$ and we obtain [34]

$$\Gamma \cong 2\pi^2 (\mu_a^2 - \mu_c^2) d^2 / \lambda_0^2, \tag{5.1}$$

where μ_a, μ_c are the refractive indices of the active and cladding layer, respectively, d is the active layer thickness and λ_0 is the free space wavelength.

The mode confinement in MQW structures can be analyzed by solving the electromagnetic wave equations for each of the layers with appropriate boundary conditions. This procedure is quite tedious because of a large number of layers involved. Streifer et al. [35] have shown that the following simple formula gives reasonably accurate results

$$\Gamma(\text{MQW}) = \gamma \frac{N_a d_a}{N_a d_a + N_b d_b} \tag{5.2}$$

where

$$\gamma = 2\pi^2(N_a d_a + N_b d_b)^2(\bar{\mu}^2 - \mu_c^2)/\lambda_0^2 \quad \text{and} \quad \bar{\mu} = \frac{N_a d_a \mu_a + N_b d_b \mu_b}{N_a d_a + N_b d_b},$$

N_a, N_b are the number of active and barrier layers in the MQW structure and d_a, d_b (and μ_a, μ_b) are the thicknesses (and refractive indices) of the active and barrier layers, respectively. The formulation of eq. (5.2) may be seen as follows: $\bar{\mu}$ is the average refractive index of a uniform optical mode in the MQW active region (i.e. the active and barrier layers) of total thickness $N_a d_a + N_b d_b$. Hence γ is the confinement factor of the optical mode [same as eq. (5.1)] in both active and barrier layers. $\Gamma(\text{MQW})$ is thus obtained by multiplying γ by the ratio of the total active thickness, $N_a d_a$, to the total thickness of the active and barrier layers, i.e. $N_a d_a + N_b d_b$. For a 1.3 μm InGaAsP–InP multi-quantum well structure with four active layer wells (150 Å thick) and three barrier layers (InP, 150 Å thick), the calculated $\Gamma \cong 0.2$. This is considerably larger than the typical Γ values for SQW lasers with active layer thicknesses in the range 100–150 Å.

Let us now discuss the effect of an Auger process on the threshold current of a single quantum well laser vis-a-vis a multiquantum well laser. The gain at threshold, g_{th}, is given by eq. (3.24). For a SQW laser, the confinement factor is small and hence the threshold gain is very large ($g_{th} \gtrsim 10^3$ cm^{-1}). This makes the carrier density at threshold large ($n_{th} \sim 2.5 \times 10^{18}$ cm^{-3}, see fig. 4a). For a MQW laser, the effective confinement factor is large and hence g_{th} is smaller ($g_{th} \cong 300$ cm^{-1}). The carrier density at threshold is $n_{th} \cong 8 \times 10^{17}$ cm^{-3} at this gain. The Auger rate varies as n^3, thus it follows that compared to the SQW laser, the Auger rate will be significantly smaller (by a factor of 27 in our example) in MQW lasers. Also, for the same reason, any nonradiative mechanism that depends superlinearly on the carrier density will have a smaller effect in MQW lasers than in SQW lasers.

Let us now discuss the effect of increased optical losses with increasing temperature, e.g., intervalence band absorption. For SQW lasers, Γ is small and hence the first term on the right hand side of eq. (3.24) is negligible. Thus the effect of increased intervalence band absorption in determining the temperature dependence of threshold is smaller in SQW lasers than in MQW lasers. Another point to note is that for $g_{th} \gtrsim 10^3$ cm^{-1}, i.e., near the threshold gain region of SQW lasers, the gain varies slowly with increasing current. This will make the threshold current of SQW lasers more sensitive to increased cladding layer losses which may increase with increasing temperature because the free-carrier absorption is larger at higher temperatures.

6. Experimental results

6.1. AlGaAs quantum well lasers

Considerable amount of experimental work has been reported on quantum well lasers fabricated using the AlGaAs material system [2–5,36–42]. The epitaxial layers are grown using molecular beam epitaxy (MBE) or organometallic vapor phase epitaxy (OMVPE or MOCVD) growth tecnique. The performances of AlGaAs lasers grown by MBE and MOCVD techniques have been reviewed by Tsang [5] and Holonyak et al. [2]. These results show that AlGaAs quantum well lasers exhibit lower threshold current, somewhat higher differential quantum efficiency, and a weaker temperature dependence of threshold current when compared with regular double heterostructure lasers.

Some of the results that demonstrate the high performance of AlGaAs quantum well lasers are: fabrication of GRIN SQW lasers with the lowest reported threshold current density of 200 A/cm² [6] (this compares with values in the range 0.6–1 kA/cm² for regular DH lasers), fabrication of modified MQW lasers with threshold current density of 250 A/cm² [4], and

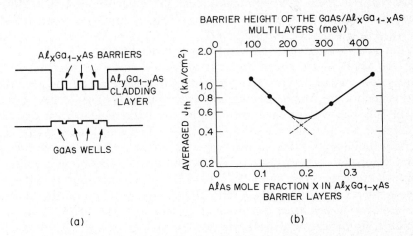

Fig. 11. (a) Schematic of a AlGaAs multi-quantum well (MQW) structure. (b) Threshold current density as a function of barrier energy of a AlGaAs MQW laser. (After ref. [4].)

fabrication of laser arrays with MQW active region which have been operated to highest reported CW power of 2 W [42].

Tsang [4] has reported measurements of threshold current densities of AlGaAs multi-quantum well lasers with different barrier heights. His results are shown in fig. 11. The cladding layers were $Al_{0.35}Ga_{0.65}As$. The results show that there is an optimum value for barrier height ($Al_xGa_{1-x}As$ with $x \sim 0.19$) for lowest threshold current. For larger x, the barrier height is too large for sufficient current injection and for very small x, the effect of the potential well which gives rise to confined two-dimensional like states is reduced [43].

6.2. InGaAsP quantum well lasers

Compared to the literature on QW lasers using AlGaAs material system, there is considerably less amount of work reported using the InGaAsP material system. Rezek and co-workers [44,45] first reported the growth of InGaAsP quantum well structures by liquid phase epitaxy (LPE). Yanase et al. [46] have reported InGaAsP–InP MQW lasers emitting near 1.3 μm and fabricated using the hybride transport VPE growth technique. Threshold current densities were in the range 2–3 kA/cm^2 for well thicknesses in the range 100–300 Å.

Dutta and co-workers [7,8] have reported the fabrication and performance characteristics of InGaAsP double-channel planar-buried heterostructure (DCPBH) laser with multi-quantum well active layer. The MQW structure was grown by LPE. Two types of MQW structures were studied:
  (i) with InP barrier layers [7], and
  (ii) with InGaAsP ($\lambda \sim 1.03$ μm) barrier layers [8].

The MQW active region had four active wells ($\lambda \sim 1.3$ μm InGaAsP) and three barrier layers. The schematic cross sections of the DCPBH structure and the MQW structure are shown in fig. 12. The DCPBH lasers with InP barrier layer had room-temperature threshold currents in the range 40–50 mA and that for InGaAsP ($\lambda \sim 1.03$ μm) barrier layers, the threshold currents were in the range 20–25 mA. The lower threshold current of devices with lower barrier height may be due to more uniform current injection [43]. The CW light–current characteristics at different temperatures of a InGaAsP DCPBH MQW laser are shown in fig. 13. The temperature dependence of threshold current of these MQW lasers are lower than that for regular DH lasers.

Quantum well lasers emitting near 1.55 μm which is the region of lowest optical loss in silica fibers have been reported by several investigators

Physics of quantum well lasers

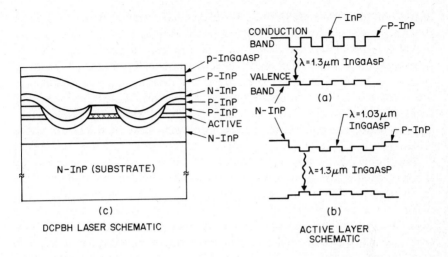

Fig. 12. Schematic cross sections of DCPBH laser structure and the MQW active region are shown. (After ref. [8].)

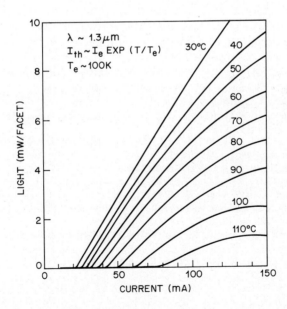

Fig. 13. Light–current characteristics at different temperatures of an InGaAsP MQW laser. (After ref. [8].)

[9,10,47,48]. InGaAs/AlInAs MQW structures emitting in the range 1.5–1.6 μm have been fabricated by MBE growth technique [47]. InGaAsP–InP MQW injection lasers emitting near 1.55 μm and fabricated by the LPE growth technique have been reported by Dutta et al. [9]. Both the threshold current and the temperature dependence of threshold are comparable to those for regular DH lasers. Single frequency distributed feedback-type InGaAsP MQW lasers emitting near 1.55 μm with CW performance characteristics comparable to regular DH lasers have also been reported [10].

For lightwave system applications, the laser light is modulated by modulating the current injected into the laser. The current modulates the injected carrier density and hence the refractive index of the guided wave which leads to a wavelength chirp. Low chirp width is desirable for high-data-rate optical communication systems [49]. Dutta et al. [7] have reported that the measured dynamic linewidth of quantum well lasers is smaller (by a factor of ~ 2) than that for regular double heterostructure lasers. This is consistent with the theoretical results of Burt [50] and Arakawa et al. [51] who showed that the modification of the density of states in a quantum well structure reduces the free-carrier contribution to the refractive index. Thus the advantages of quantum well lasers over regular DH lasers lie in the possibility of lower threshold current, lower temperature dependence of threshold and lower chirp.

References

[1] R. Dingle, W. Wiegmann and C.H. Henry, Phys. Rev. Lett. 33 (1974) 827;
  R. Dingle and C.H. Henry, US Patent 3 982 207, September 21, 1976.
[2] N. Holonyak Jr, R.M. Kolbas, R.D. Dupuis and P.D. Dapkus, IEEE J. Quantum Electron. QE-16 (1980) 170.
[3] N. Holonyak Jr, R.M. Kolbas, W.D. Laidig, B.A. Vojak, K. Hess, R.D. Dupuis and P.D. Dapkus, J. Appl. Phys. 51 (1980) 1328.
[4] W.T. Tsang, Appl. Phys. Lett. 39 (1981) 786.
[5] W.T. Tsang, IEEE J. Quantum Electron. QE-20 (1986) 1119.
[6] S.D. Hersee, B. DeCremoux and J.P. Duchemin, Appl. Phys. Lett. 44 (1984) 476.
[7] N.K. Dutta, S.G. Napholtz, R. Yen, R.L. Brown, T.M. Shen, N.A. Olsson and D.C. Craft, Appl. Phys. Lett. 46 (1985) 19.
[8] N.K. Dutta, S.G. Napholtz, R. Yen, T. Wessel and N.A. Olsson, Appl. Phys. Lett. 46 (1985) 1036.
[9] N.K. Dutta, T. Wessel, N.A. Olsson, R.A. Logan, R. Yen and P.J. Anthony, Electron. Lett. 21 (1985) 571.
[10] N.K. Dutta, T. Wessel, N.A. Olsson, R.A. Logan and R. Yen, Appl. Phys. Lett. 46 (1985) 525.
[11] R. Dingle, Confined carrier quantum states in ultrathin semiconductor heterostructures, Festkörper probleme 15 (1975) 21.
[12] S.R. Forrest, P.H. Schmidt, R.B. Wilson and M.L. Kaplan, Appl. Phys. Lett. 45 (1984) 1199.

[13] H. Temkin, M.B. Panish, P.M. Petroff, R.A. Hamm, J.M. Vandenberg and S. Sumski, Appl. Phys. Lett. 47 (1985) 394.
[14] M.G.A. Bernard and G. Duraffourg, Phys. Status Solidus 1 (1961) 699.
[15] G. Lasher and F. Stern, Phys. Rev. 133 (1964) A553.
[16] N.K. Dutta, J. Appl. Phys. 53 (1982) 7211.
[17] K. Hess, B.A. Vojak, N. Holonyak Jr, R. Chin and P.D. Dapkus, Solid-State Electron. 23 (1980) 585.
[18] H.C. Casey Jr and M.B. Panish, Heterostructure Lasers, Part A (Academic Press, New York, 1978) ch. 3.
[19] D. Botez, IEEE J. Quantum Electron. QE-17 (1981) 178.
[20] W.T. Tsang and J.A. Ditzenberger, Appl. Phys. Lett. 39 (1981) 193.
[21] A. Sugimura, IEEE J. Quantum Electron. QE-20 (1984) 336.
[22] N. Holonyak Jr, R.M. Kolbas, W.D. Laidig, M. Altarelli, R.D. Dupuis and P.D. Dapkus, Appl. Phys. Lett. 34 (1979) 502.
[23] A. Sugimura, Appl. Phys. Lett. 43 (1983) 728.
[24] W.T. Tsang, C. Weisbuch, R.C. Miller and R. Dingle, Appl. Phys. Lett. 35 (1979) 673.
[25] H. Kobayashi, H. Iwamura, T. Saku and K. Otsuka, Electron. Lett. 19 (1983) 156.
[26] A.R. Beattie and P.T. Landsberg, Proc. R. Soc. London 249 (1959) 16.
[27] N.K. Dutta and R.J. Nelson, J. Appl. Phys. 53 (1982) 74.
[28] R.J. Nelson and N.K. Dutta, in: Semiconductors and Semimetals, Vol. 22, Part C, ed. W.T. Tsang (Academic Press, New York, 1985) ch. 1.
[29] L.C. Chiu and A. Yariv, IEEE J. Quantum Electron. QE-18 (1982) 1406.
[30] N.K. Dutta, J. Appl. Phys. 54 (1983) 1236.
[31] A. Sugimura, IEEE J. Quantum Electron. QE-19 (1983) 932.
[32] C. Smith, R.A. Abram and M.G. Burt, J. Phys. C 16 (1983) L171.
[33] W.T. Tsang, Appl. Phys. Lett. 39 (1981) 134.
[34] W.P. Dumke, IEEE J. Quantum Electron. QE-11 (1975) 400.
[35] W. Streifer, D.R. Scifres and R.D. Burnham, Appl. Opt. 18 (1979) 3547.
[36] W.T. Tsang, Appl. Phys. Lett. 38 (1981) 204.
[37] K. Woodbridge, P. Blood, E.D. Fletcher and P.J. Hulyer, Appl. Phys. Lett. 45 (1984) 16.
[38] W.T. Tsang, R.A. Logan and J.A. Ditzenberger, Electron. Lett. 18 (1982) 845.
[39] D. Kasemset, C.S. Hong, N.B. Patel and P.D. Dapkus, Appl. Phys. Lett. 41 (1982) 912.
[40] S.D. Hersee, M.A. Poisson, M. Baldy and J.P. Duchemin, Electron. Lett. 18 (1982) 618.
[41] R.D. Dupuis, R.L. Hartman and F.R. Nash, IEEE Electron Device Lett. EDL-4 (1983) 286.
[42] D.R. Scifres, R.D. Burnham, C. Lindstrom, W. Streifer and T.L. Paoli, Appl. Phys. Lett. 42 (1983) 645.
[43] N.K. Dutta, IEEE J. Quantum Electron. QE-19 (1983) 794.
[44] E.A. Rezek, R. Chin, N. Holonyak Jr, S.W. Kirchofer and R.M. Kolbas, J. Electron. Mater. 9 (1980) 1.
[45] E.A. Rezek, N. Holonyak Jr and B.K. Fuller, J. Appl. Phys. 51 (1980) 2402.
[46] T. Yanase, Y. Kato, I. Mito, M. Yamoykuchi, K. Nishi, K. Kobayashi and R. Lang, Electron. Lett. 14 (1983) 700.
[47] H. Temkin, K. Alavi, W.R. Wagner, T.P. Pearsall and A.Y. Cho, Appl. Phys. Lett. 42 (1983) 845.
[48] W.T. Tsang, Appl. Phys. Lett. 44 (1984) 288.
[49] N.K. Dutta, R.B. Wilson, D.P. Wilt, P. Besomi, R.L. Brown, R.J. Nelson and R.W. Dixon, AT&T Tech. Journal 64 (1985) 1857.
[50] M.G. Burt, Electron. Lett. 20 (1984) 27.
[51] Y. Arakawa, K. Vahala and A. Yariv, Appl. Phys. Lett. 45 (1984) 950.

CHAPTER 14

PHYSICS AND APPLICATIONS OF EXCITONS CONFINED IN SEMICONDUCTOR QUANTUM WELLS

D.S. CHEMLA and D.A.B. MILLER

AT&T Bell Laboratories
Holmdel, NJ 07733, USA

Heterojunction Band Discontinuities: Physics and Device Applications
Edited by F. Capasso and G. Margaritondo
© *Elsevier Science Publishers B.V., 1987*

Contents

1. Introduction 597
2. Excitonic resonances in quantum well structures 599
3. Electroabsortion in quantum well structures 602
4. Nonlinear optical effects 613
5. Conclusions 621
References 622

1. Introduction

Many potential applications of optical signal processing rely on the modification of the dielectric constant of a material by electromagnetic perturbations. Of particular interest are the mechanisms by which an electrostatic E_0 or an optical E_ω field can be used to change the refractive index n or absorption coefficient α of a medium. In the case of semiconductors these effects have received much attention both from the fundamental and the applied point of view. Extensive documentation on electro-optic and nonlinear optical processes in bulk semiconductor compounds has been published, and numerous demonstrations of modulation or switching of light by application of electrostatic or optical fields have been reported. However, the development of very efficient devices is presently limited by the small magnitude of the currently known optical nonlinearities. To assess the appropriate requirements on the nonlinearities more quantitatively let us consider the (complex) phase shift experienced by an optical field travelling through a length l of a nonlinear medium. It is given by

$$\phi + \Delta\phi = \left[\frac{2\pi}{\lambda}n + \tfrac{1}{2}i\alpha\right] + \left[\frac{2\pi}{\lambda}\Delta n + \tfrac{1}{2}i\Delta\alpha\right], \qquad (1)$$

where the nonlinearities produce the terms in the second pair of brackets. If the nonlinear phase shift produced over the effective interaction length $l_a = [1 - \exp(-\alpha l)]/\alpha$, is such that $\text{Re}(\Delta\phi) \approx \tfrac{1}{2}\pi$ or $\text{Im}(\Delta\phi) \approx \tfrac{1}{2}$ then the nonlinear effects are large enough to be utilized in devices. Note that this material criterion is not relaxed by the use of resonant cavities; less phase shift is required in proportion to the finesse, but the loss per pass must be reduced by the same factor because the beam now makes multiple passes through the medium. The overall size of the phase shifts can be increased by using long interaction lengths in highly transparent media, thus working far from the resonances of the material (small nonlinearities). In this case the limitation from diffraction can be avoided in waveguides or fibres. Conversely, it is possible to enhance the nonlinearities significantly by working close to a resonance, but in this case the price to pay is a reduced interaction length. Therefore, the use of resonant nonlinearities is only interesting if the enhancement of the nonlinear response is large enough to obtain sizable real or imaginary phase changes in a length of the order of the reciprocal of the absorption coefficient.

In bulk semiconductors, near-band-gap resonances [1,2] as well as the sharp excitonic or excitonic-complex resonances [3] have been used to demonstrate highly efficient nonlinear optical interactions. However, these processes are seen clearly only at low temperature where the band edge is abrupt or the exciton resonances are still well resolved. In addition, although elegant demonstrations have been achieved, the magnitude of the nonlinearities reported or the total change in phase available in an absorption length are still too small for many applications.

In semiconductors, the electric field dependence of the band-to-band absorption has also been extensively studied. It is known as the Franz–Keldysh effect [4,5]. Here an absorption tail is induced below the gap and oscillations are produced above the gap [6]. When excitonic resonances are present additional effects are seen. The resonances are broadened by field ionization and they shift slightly. This corresponds to a "Stark effect" on the exciton ground state [7–9]. These effects have been utilized in modulation spectroscopy to characterize semiconductors [6] but again their applications have been limited in practice by the small magnitude of the change in absorption coefficient observed for reasonable applied fields [10].

It is apparent from this brief discussion that the development of materials able to meet the strict requirements for devices necessitates the exploitation of novel physical processes that can enhance the optical nonlinearities well beyond those of "natural" compounds. A promising approach is to turn to "artificial" materials prepared by modern techniques of growth [11,12] and fabrication [13], and to use characteristics not encountered in the natural materials to tailor electronic and optical properties.

An example of such artificial materials are semiconductor nanostructures [14]. Let us first see how engineering of some properties can be realized in in these structures. Within the effective mass approximation (EMA) the dynamics of carriers in a semiconductor mimics that of free particles provided the electron mass m_0 is replaced by the effective masses $m_{e,h}$ and the electrostatic interaction is screened by the dielectric constant of the compound. Thus the characteristic lengths, such as the carrier de Broglie wavelengths and the Bohr radii, are increased by several orders of magnitude over those of the free particles (1 Å → 100 Å–1000 Å). Correspondingly, the energies of the electronic levels are also reduced (10 eV →

0.1 eV–0.001 eV). Modern techniques of semiconductor growth enable us now to control the composition of heterostructures on the scale of one monoatomic layer, and the sample preparation techniques can define structures with dimensions in the 100 Å range. It is thus possible to change the electronic energy structure locally, and to construct semiconductor samples with quantum size effects that drastically modify their electronic and optical properties. Once the physics of these modifications is understood, it should be possible to control them and hopefully optimize the sample composition and structure for a particular purpose.

In this chapter we review some of the recent work performed along the directions outlined in the previous paragraph. The simplest example of a system where size effects produce fundamental modifications of the optical and electronic properties are semiconductor quantum well structures (QWS). They consist of ultrathin layers of semiconductors (e.g. 10 Å to 500 Å thick) with different compositions, grown alternately one on the other. As the band gaps depend on the composition, these structures exhibit a strong modulation of the potential seen by the carriers in the direction normal to the layers. If the depth of the potential step between the low-gap compound and the high-gap one is sufficiently high and the separation between the low-gap layers is large enough, then the carriers are substantially confined in the low-gap layers of the QW. The loss of one degree of freedom results in a reduced dimensionality and introduces new selection rules in the optical transitions. The novel nonlinear optical properties and electro-absorption effects seen close to quasi-two-dimensional exciton resonances in III–V QWS have been extensively investigated, and will be discussed here.

The chapter is organized as follows. In sect. 2, we briefly summarize the modification of linear optical effects induced by the confinement of carriers in layers whose thickness is smaller than the bulk exciton Bohr radius. We limit our review to the changes that are relevant to nonlinear optics and electro-optics. In sect. 3, we discuss the electro-absorption effects seen when electrostatic fields are applied parallel or perpendicular to the plane of the layers, and we present some prototype devices that utilize the novel aspects of quantum well (QW) electroabsorption. In sect. 4 the processes observed when cw, picosecond and femtosecond optical excitations are applied to QWS are reviewed. Again new physical processes are seen and we will point out their special properties as well as their potential for applications.

2. Excitonic resonances in quantum well structures

The linear optical properties of III–V QWS have been reviewed recently in refs. [15] and [16]. In QWS the effective mass approximation (EMA) wave functions of the carriers in the direction perpendicular to the layers are sinusoidal in the QW and decay exponentially outside the QW, whereas along the plane the EMA wave functions are still plane waves. Therefore, the bulk energy bands split into a set of subbands which can have quite complex in-plane dispersion. In particular, the degeneracy of the uppermost valence band is lifted even at the center of the Brillouin zone, producing the separate heavy-hole (hh) and light-hole (lh) subbands with highly non-parabolic dispersion. The density of states of the various subbands is two-dimensional with a step-like profile. The optical transitions between these subbands obey selection rules due to parity and angular momentum conservation, and the bulk oscillator strengths are modified by the overlap of the EMA wave functions. Close to the fundamental edge the oscillator strengths of the transitions from the hh and lh subbands to the conduction subbands are in the ratio $\frac{3}{4}$ and $\frac{1}{4}$ for optical field polarized parallel to the layers and 0 and 1 for polarization normal to the layers. Correspondingly, the band to band absorption spectrum consists of a series of steps [15] whose relative heights follow these ratios.

As in three-dimensional materials, the correlation between the photo-excited electron and hole produces a set of excitonic resonances below each inter-subband transition edge and a strong absorption enhancement above. It is important to note that in most III–V semiconductor QWS there is little change in dielectric constant between the QW material and the barrier material. Thus the Coulomb interaction has the usual $1/r$ dependence, i.e. there is negligible "dielectric confinement". Only the motion and the density of states of the carriers are sensitive to the confinement. If the layer thickness L_z is smaller than, or the order of, the bulk exciton Bohr diameter $2a_0$ (e.g., ≈ 300 Å for GaAs), this induces strong modifications of the exciton structure and hence of the absorption resonances. For large band gap discontinuities and for the lowest energy resonances, the extension of the wave function in the direction normal to the layer is mostly governed by the QW potential. Conversely, in the plane the electron–hole (e–h) Coulomb interaction plays the major role. (Strictly speaking the motions in and normal to the plane are correlated but we shall neglect this small effect in our qualitative introduction.) The overall result of the artificial reduction of the average e–h distance is an increase in the exciton binding energy. It can reach a value 4 times that of the bulk exciton for an infinitely narrow QW, and is 2.5–3 times larger in actual QWs with finite thickness. For example in the case of GaAs, the bulk Rydberg constant $R = 4.2$ meV and typical binding energies in QWS are of the order $E_{1S} \approx 10$ meV. Because the QW thickness has unavoidable fluctuations of the order of one atomic layer, even in the best growth conditions, the exciton resonances are inhomogeneously broadened at low temperature, with typical half-width at half-maximum of the order of 1–2 meV in the case of GaAs [17–19].

The excitons associated with the low-energy transitions are well confined in the QW and therefore they mostly interact with the phonons of the QW medium which are not too much modified in the QWS. This has been confirmed by resonant Raman scattering experiments [20,21]. Extended exciton states have also been observed just above the band gap discontinuity [22]. These states correspond to continuum resonances [23] and interact with the phonons of both the QW and the barrier materials [22].

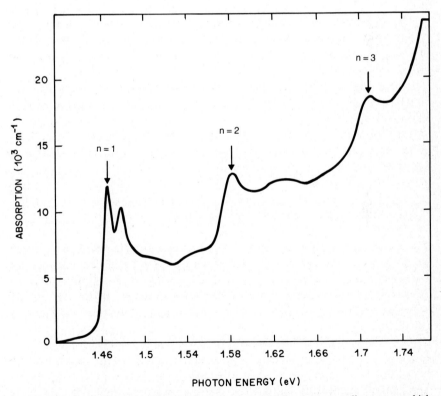

Fig. 1. Room-temperature absorption spectrum of a multiple quantum well structure which consists of 65 periods of 96 Å GaAs and 98 Å $Al_{0.3}Ga_{0.7}As$ layers. The three transitions between the $n = 1, 2$ and 3 valence and conduction subbands produce the three plateaus at the onset of which exciton resonances are clearly resolved.

The net result of an increased binding energy and an interaction with the thermal phonons not too different from that in the bulk, results in one of the most interesting properties of QW for applications in optics and opto-electronics, namely, the observation of well-resolved excitonic resonances at room temperature (see ref. [15] for a discussion, and also refs. [24–26]). This is shown in fig. 1 where the absorption spectrum of an $L_z \approx 100$ Å GaAs/Al$_{0.3}$Ga$_{0.7}$As QWS is presented. The three plateaus of the transitions between the $n = 1, 2, 3$ subbands are clearly seen and preceding each one clear exciton resonance is resolved. The first transition shows a very abrupt edge and the resonances are sharp enough that the hh-exciton and the lh-exciton peaks are separated. In the bulk such clear excitonic features can only be seen at low temperature. Measurements performed in waveguides containing one or two QWs have also shown well-resolved exciton peaks and have also verified the strong dichroism of QWs. For polarization of the optical field normal to the layers the hh-exciton has all but disappeared and all the oscillator strength is seen in the lh-exciton peak [27].

These observations have, of course, very important consequences for applications because any perturbation of the excitons will produce modification of this sharp profile and hence induce large modulation of the optical constants. In addition, because at room temperature kT is 2 to 3 times larger than the binding energy, these excitons have novel and interesting transient nonlinear optical properties.

3. Electroabsorption in quantum well structures

The effects induced by application of electrostatic fields on the absorption of QWS have been extensively studied and are now well documented. An extensive set of references can be found in ref. [28]. As shown in fig. 2, because of the symmetry of the QWS, one can expect contrasted effects when an electrostatic field is applied parallel or perpendicular to the plane of the layers.

Let us consider first the case of a field parallel to the layers because it is qualitatively similar to the case of bulk material [29]. In this geometry, the only force binding the electron and the hole is the Coulomb force. Thus the applied field induces the Stark effects on the exciton ground state. The only difference compared to the case of bulk samples is that the exciton binding energy and the in-plane wave function are slightly modified. In both cases, the dominant effect is a broadening of the exciton resonance with field that

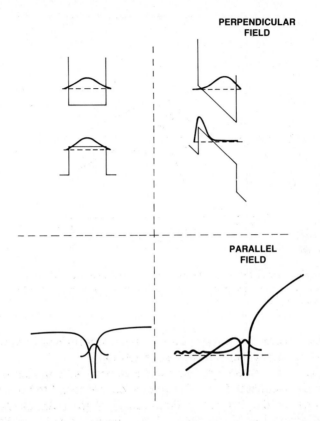

Fig. 2. Illustration of the potential seen by an e–h pair in a quantum well: without an external field applied (left-hand side), and in an external field (right-hand side). For a large field directed along the plane of the layer the Coulomb potential is not strong enough to sustain even quasi-bound states. The exciton field ionizes by tunneling though a rather shallow barrier. For a field directed perpendicular to the layer the large band gap discontinuities hold the e–h pair together more effectively, although the orbit in the plane becomes larger because the Coulomb interaction is reduced somewhat.

results from the shortening of the exciton lifetime due to field ionization. With readily attainable fields (e.g. 10^4 V/cm) this field ionization lifetime can become comparable or shorter than the exciton classical orbit time so that the absorption resonance ceases to be resolvable. The actual Stark shift is limited to a small fraction (e.g. 10%) of the exciton binding energy, and is consequently not normally resolved. This field ionization is a little more

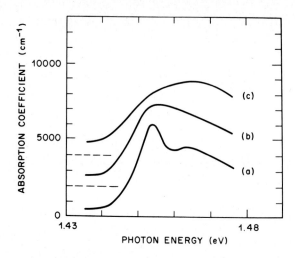

Fig. 3. Absorption spectra for electric fields parallel to the layers. The three spectra have been shift vertically for clarity. They correspond to fields of: (a) $E \approx 0$; (b) $E \approx 1.6 \times 10^4$ V/cm; (c) $E \approx 4.8 \times 10^4$ V/cm.

difficult to achieve in the QW than in three dimensions because of the larger binding energy and geometrical effects [30].

Examples of the absorption spectrum seen with field applied parallel to the layer are presented in fig. 3. The resonances are found to broaden easily and to disappear altogether at moderate fields of the order of $(2-5) \times 10^4$ V/cm. No significant shift is observed even when the sensitive differential spectra (unperturbed absorption minus perturbed absorption) are directly recorded. Accurate comparison to theory is complicated by the convolution of ionization broadening with the natural broadening of the resonance (due to the QW thickness fluctuation and the effect of thermal phonons). However, the evolution of the spectra is in qualitative agreement with theory [29].

One interesting feature of this geometry is that the cross-sectional area $(1 \times 25 \ \mu m^2)$ and thus the capacitance of the sample is very small. It can therefore be used for high-speed modulation and also to test how fast an excitonic absorption change can be induced in a semiconductor. Both of these aspects have been tested recently [31]. In the experiment described in ref. [31], a potential was applied across the electrodes of the sample used to measure the spectra of fig. 3 and the transmission was probed with a 120 fs broad continuum light pulse. A separate strong 50 fs red pulse produced a

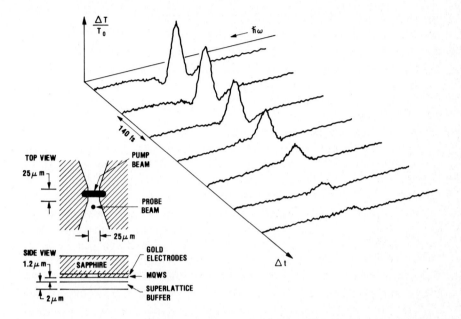

Fig. 4. Time dependence of the electro-absorption for a field parallel to the plane of the layer. The spectra shown are the difference between the spectra with and without voltage initially applied to the structure, so that disappearance of the peak in these spectra corresponds to the excitonic absorption returning to its zero-field state. These absorption differential spectra are shown as a function of the delay between the 140 fs probe pulse and the 50 fs pump pulse that shorts the electrostatic field. The spectrum of the probe pulse is truncated by a bandpass filter, as shown by the dashed lines, so that only the heavy-hole exciton absorption changes are shown.

high density electron–hole plasma in the inter-electrode gap, thus rapidly shorting the field. The differential absorption spectra observed at various delays are shown in fig. 4 and the beam configuration is illustrated in the inset. The $1/e$ recovery time of the absorption was measured to be 330 fs and seems to be still limited by electric field reflections in these non-optimized electrodes.

This result, besides its technological interest for high-speed sampling, approaches several characteristic times in the exciton dynamics. For example, the exciton–LO-phonon collision time, which is dominant in setting the exciton lifetime at room temperature in the absence of fields, is only of the order of 300 fs as will be discussed in sect. 4. In the presence of parallel field, the mean time to field ionization (i.e. how long it takes to the two particles to tunnel away through the Coulomb well one from another) is ≈ 380 fs for an 8 kV/cm field [29]. Perhaps the most fundamental time constant in the exciton dynamics is the classical orbit time of ≈ 150 fs (how long it takes to the e–h pair to "feel" its Coulomb correlation). All these values fall in the same range, and the extension of the work of ref. [31] to shorter time-resolution should enable tests of our understanding of the dynamics of simple quantum systems.

The electro-absorption in the configuration where the field is perpendicular to the layers presents features that are specific to QWS and are not seen in bulk materials at any temperature [29]. From fig. 2 we see that the electron and the hole charge distributions are pulled apart by the field. If the thickness of the QW is much larger than the three-dimensional Bohr radius, the field destroys the excitonic correlation, because the electron and hole are pulled so far apart that their Coulomb attraction is negligible. The exciton resonance then will be destroyed by field ionization just as in the bulk. Conversely, if the QW thickness is less than the three-dimensional Bohr radius, when the particles are pushed against the walls of the well, they are stopped by a potential barrier much larger than the exciton binding energy through which it is relative difficult for them to tunnel. Consequently, the exciton field ionization is inhibited and the exciton resonance is not destroyed. The long tunneling lifetimes are confirmed by tunneling resonances calculations, which show that indeed the width of the resonances and thus the tunneling time can be very long compared to the exciton lifetime from other causes [29]. Therefore, in a narrow QW, even when the electron and the hole are strongly pushed against the opposite sides of the well, they are still close enough to be strongly Coulomb-correlated and thus to exhibit excitonic behavior. Since the exciton is not destroyed by a field, very large Stark shifts can be observed, exceeding the exciton binding energy in magnitude. This mechanism is called the quantum-confined Stark effect (QCSE) [29,34].

These shifts are clearly seen in GaAs QWs [29,33], as shown for example by the data presented in fig. 5, which was obtained using a waveguide structure containing two 94 Å QWs. First of all the strong dichroism is clearly apparent both with and without field. Both excitonic absorptions are active for polarization parallel to the layers whereas in the perpendicular polarization only the lh exciton is present and it then acquires a larger oscillator strength. Even more spectacular are the shifts of the resonance that are induced by the static field. The whole absorption profile shifts toward lower energies as the height of the resonance decreases somewhat and the peaks broaden.

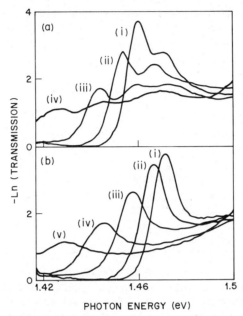

Fig. 5. Absorption spectra of a waveguide containing two 94 Å GaAs quantum wells, as a function of the electric field applied perpendicular to the layers. (a) Optical field polarization parallel to the layers; (i) 1.6×10^4 V/cm, (ii) 10^5 V/cm, (iii) 1.3×10^5 V/cm, (iv) 1.8×10^5 V/cm. (b) Optical field polarization perpendicular to the layers; (i) 1.6×10^4 V/cm, (ii) 10^5 V/cm, (iii) 1.4×10^5 V/cm, (iv) 1.8×10^5 V/cm, (v) 2.2×10^5 V/cm.

The magnitude of the QCSE shifts are surprising. For example, shifts of 40 meV are seen in these spectra at a field of 2.2×10^5 V/cm. This is ten times the bulk exciton binding energy and corresponds to 100 times the exciton classical ionization field. The overall behavior is now well understood. The simplest complete theoretical method is to approximate the exciton wavefunction as a separable wavefunction [29,34]. The shifts of the resonances can then be approximated by the sum of the shifts of the single-particle states (neglecting the Coulomb interaction) and the change of the exciton binding energy due to the compression of the wavefunctions against the walls of the well [29,34]. At low fields, the latter contribution is important whereas at high fields the variations of the single-particle states are dominant. Comparison of the experimental shifts with the model is shown in fig. 6. There are no adjustable parameters in the theory [29]; the two sets of curves correspond to the two splits of the band gap discontinuities that have been proposed for the GaAs/AlGaAs system.

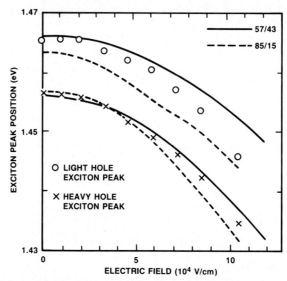

Fig. 6. Energy of the excitonic peaks with applied field perpendicular to the layers. The open circles and crosses are experimental values. The lines result from the model discussed in the text, which involves no adjustable parameters. The solid lines correspond to a 0.57:0.43 split of the band gap discontinuity between the conduction and the valence bands, the dashed lines are for a 0.85:0.15 split.

The changes in height of the resonances and of the continua are well explained by several sum rules that have been demonstrated recently [28]. According to these rules: (i) the total absorption with and without the applied field are equal, (ii) the total absorption below some maximum energy is only governed by the overlap in the direction normal to the layers and is independent of the details of the in-plane wave functions, (iii) the sum of the height of all the transitions between a given conduction subband and one type of valence band (hh or lh) in the presence of the field is equal to the height of the allowed transition in the absence of field.

To illustrate the relation between the QCSE and bulk electroabsorption, we will consider the problem neglecting the electron–hole Coulomb interaction. Although this neglects several important aspects, it is tractable and illuminating. This is the approach normally taken to analyze the Franz–Keldysh effect, the usual electroabsorptive effect in bulk semiconductors. As the QW thickness increases we should expect a continuous transition from a strongly quantized behavior to the three-dimensional Franz–Keldysh behavior. Indeed this conclusion has been analytically

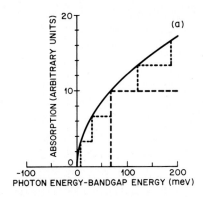 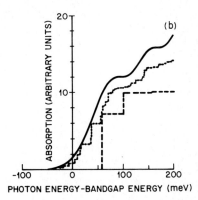

Fig. 7. Illustration of the transition between the three-dimensional Franz–Keldysh effect and quantum well electro-absorption by comparison of the theoretical absorption of GaAs slabs: (a) without field, and (b) with a 10^5 V/cm perpendicular field. Three thicknesses are compared; 100 Å (long dashed lines), 300 Å (short dashed line) and infinitely thick slab (i.e. Franz–Keldysh effect) (solid line).

demonstrated recently [32]. The transition from the QW to the three-dimensional behavior occurs over a narrow range of thickness for a given field and is accompanied by the growth of symmetry-forbidden optical transitions from which the Franz–Keldysh oscillations originate. These trends are illustrated in fig. 7 where numerical results for absorption with and without field are shown for three GaAs slabs of increasing thickness. For 100 Å, the QW behavior is evident. At 300 Å the spectrum already shows precursor signs of the oscillations and tail of the three-dimensional system.

An important aspect of the QCSE for applications results from the sample structure. In order to apply large fields and yet avoid the flow of large currents, the samples are grown as p–i–n diodes with the QWs in the intrinsic region. Then by reverse biasing the diodes the field across the QWs is changed. Because of this geometry the sample also behaves as a photodetector, with an internal quantum efficiency close to unity [29]. As an example, the responsivity of such a diode at various bias voltage is shown in fig. 8; it faithfully reproduces the absorption profile seen under the same conditions. This aspect is important because the photocurrent gives an electrical measure of the absorption that can be utilized in an external circuit to react on the bias voltage, thus providing a mechanism for feedback. We will discuss below several applications of this property [35–37].

The most obvious application of the QCSE is modulation of the trans-

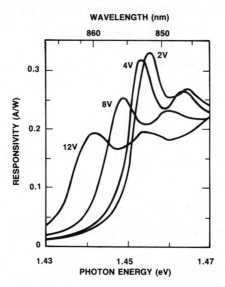

Fig. 8. Responsivity of a p–i–n QW diode as a function of the reverse bias voltage.

mission of a light beam whose wavelength is near or just below the gap. High-speed modulation in the 100 ps range has been demonstrated both in the configuration where the light propagates perpendicular to the plane of the QWs and in waveguide geometries [38–40], with modulation depths of 3 dB and 10 dB, respectively. Recently, the extension to other III–V compounds has been successfully demonstrated [41–43]. The speed is presently limited by circuit considerations and can be further increased. The ultimate response time of the QCSE has not yet been tested but it should also be in the subpicosecond domain. The waveguide geometry has a number of advantages, in particular the possibility of choosing arbitrarily long absorption lengths. It is also a natural geometry for integrated optics and the first steps toward optical integrated devices have been reported [44].

As mentioned before, the possibility of measuring the absorption of p–i–n QWS by measuring the photocurrent leads to new devices. A simple example of such a device is a wavelength-selective voltage-tunable photodetector [45], which operates on the following principle. Light, with photon energy less than that of the hh exciton peak at zero field, incident of the p–i–n QWS produces a photocurrent that is maximized for an applied field just large enough to shift the exciton to the photon energy. If the wavelength is changed then the photodetector response can be maximized again by changing the applied voltage, thus giving a means to follow the frequency of the light source electrically. This operation has indeed been demonstrated [45] and has potential applications in coherent light wave communications where it is often necessary to determine simultaneously the intensity and the wavelength of an optical beam.

The self electro-optic effect devices (SEED) form as novel class of devices that utilize the p–i–n QWS capacity to operate as closely coupled photodetector/modulator. The principle of the operation is quite simple. The p–i–n QWS is connected to an external circuit that imposes a $I(V)$ relation between the voltage applied to the SEED and the current passing through it. On the other hand these two quantities are also related by the response function of the SEED $I(V, P_{in}, \lambda)$, where P_{in} and λ are the power and the wavelength of the light incident on the device, respectively. These two relations have to be satisfied simultaneously, hence the possibility of electronic feedback on the optical behavior. The SEEDs are truly opto-electronic devices in the sense that their optical and electronic responses are intimately related. The feedback can be positive or negative dependent on the operating wavelength. Both have been exploited in optically bistable devices, self-linearized modulators and optical level shifters. These devices have been demonstrated in the propagation mode normal to the QWS as well as in waveguides [33, 35–37].

One interesting SEED configuration for potential applications in optical switching is that of the low switching energy optically bistable gate. It operates as follows. A light beam of variable intensity at a wavelength close to the peak of the exciton at zero field is sent through the SEED, and a bias voltage is applied through a load resistor to red-shift the absorption. At low intensity the transmission of the SEED is relatively high, but as the intensity rises the photocurrent increases inducing a voltage drop in the load resistor and thus reducing the field across the SEED. The absorption therefore increases and consequently so does the photocurrent for the same light intensity, producing a further voltage drop in the load resistor. This positive feedback can operate until the SEED switches to the maximum absorption. Then if the intensity of the light beam is decreased starting from the strong absorption state, the switching to the transmissive state occurs along another path in the P_{out}/P_{in} plane because the absorption is now high, hence the optical hysteresis loops shown in fig. 9. In this figure are plotted the P_{out}/P_{in} characteristics for six wavelengths close to the absorption edge. The positive feedback is effective in causing bistability between 850 nm and 860 nm. Similar operation can be observed with

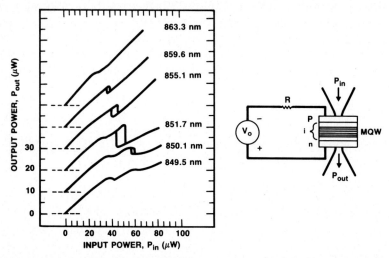

Fig. 9. P_{out}/P_{in} characteristics of a SEED operating as an optical gate, for six wavelengths in the vicinity of the hh exciton peak. The external circuit consists of a 20 V voltage supply and a 10 MΩ load resistor. The positive feedback gives bistability for the four intermediate wavelengths.

different external circuits, for example with a constant current injection [37]. The switching as well as oscillations can be followed electronically on the $I(V)$ of the SEED. The contrast between the two states is large because of the change in absorption coefficient is typically of the order of 0.5×10^4 cm^{-1}. 20:1 contrast ratios have been measured in waveguide SEEDs [28,33].

Some of the attractive features of the SEED for optical switching include the extremely low operating energy per unit area of the device despite the absence of optical cavity, the tolerance of the SEED to fluctuations in the optical and electrical operating parameters, the room-temperature operation, and the natural structure for large planar arrays. These kinds of devices are also compatible with laser diode light sources in wavelengths, powers and materials, and with semiconductor electronics. Switching energies as low as 20 fJ/μm² (15 fJ/μm² electrical and 5 fJ/μm² optical) have been demonstrated. The fundamental limit is expected to be in the few fJ/μm² range. Switching times of 30 ns have been measured on SEEDs 100 μm in diameter. This time is limited by the RC constant of the device and should scale with its area. Therefore, switching times of a few ns could be expected from SEEDs with diameters in the 10 μm range. Recently, operation of planar arrays of SEED including integrated photodiodes as the current load has been shown [46], demonstrating interesting potential application to optical logic in the near future.

4. Nonlinear optical effects

The fact that QWS exhibit well-resolved exciton resonances at room temperature (RT) might be viewed just as a curious feature that allows us to study quasi-two-dimensional exciton physics under convenient conditions. This is however an over-simplified analysis [15] as we will show below. RT quasi-two-dimensional excitons have dynamical properties qualitatively different from those of low temperature quasi-two-dimensional excitons and both exhibit strong differences compared to the usual three-dimensional excitons.

In GaAs QWs, the exciton binding energy (8–10 meV) is 2 to 3 times smaller than kT at room temperature ($kT \approx 25$ meV). The classic analogy between excitons and hydrogen atoms provides a clue to understanding the differences between low- and high-temperature QW exciton physics. At a temperature corresponding to three hydrogenic Rydbergs (about half a million Kelvin) direct atomic collisions very rapidly transform a hydrogen gas into a plasma. Moreover, in a semiconductor the thermal reservoir is the lattice which interacts with the excitons via the thermal phonons. Thus even in the absence of exciton collisions a very dilute exciton gas is ionized by the phonon bath. In fact, a single scattering event with the highly energetic LO-phonons (36 meV for the GaAs LO-phonon) will ionize an exciton (X) and release a e–h pair with substantial excess energy,

$$\text{LO-phonon} + X \rightarrow e + h + (\approx 27 \text{ meV}).$$

Thus the exciton ionization time is simply given by the mean time between exciton–thermal LO-phonon collisions. Thermodynamic arguments show that the excitons have very little probability of reforming from the RT plasma [47]. The mean time for thermal phonon scattering can be inferred from the temperature dependence of the exciton linewidth. It is found to be well described by a constant term equal to the inhomogeneous broadening due to the fluctuations of the layer thickness, plus a term proportional to the density of LO-phonons [47,48]. If this last term is interpreted as a reduction of life time due to phonon scattering it is found that excitons only live 0.4 ps in GaAs QWS at RT. This ionization time sets the time scale for the transients in the nonlinear optical response of QW.

The generation of e–h pairs (bound or unbound) modifies the optical transitions through several mechanisms [49]. First, the exclusion principle blocks transitions originating from empty states or ending in filled states. This phase-space filling (PSF) is a blocking mechanism that is well documented for free electrons and holes for which it gives rise to the Burstein–Moss shift. Although less well known for bound electron–hole pairs, it is also effective for excitons, which are bosons only in the dilute limit. Furthermore, the single-particle states, and their combinations that form the bound states, are changed by screening. The screening for fermions splits into the classical Coulomb long range direct screening (DS) and the short range exchange interaction (EI). In three dimensions, the DS by free e–h pairs is relatively strong, making large contributions to the band gap renormalization and the instability of bound pairs i.e. the so-called Mott transition. DS is more efficient at low temperature than at high temperature, as indicated by the expression of the Debye screening parameter, i.e. the ratio of a plasma frequency to an average kinetic energy. Screening by excitons is much weaker owing to the s–p gap in the bound state excitation spectrum [50,51], and EI is only important in dense plasmas. These general trends are also true for quasi-two-dimensional QWs although the relative magnitudes of the various contributions are modified.

In agreement with the above discussion, the nonlinear optical effects seen in RT QWS with excitations long compared to the ionization time are independent of the wavelength and of the duration of the excitation [47]. They depend only on the density of free e–h pairs generated directly by excitation above the resonances or indirectly via ionization of the resonantly excited bound states. Even absorption well below (50 meV) the hh-exciton peak in the band tail or in defect states is found to produce the same effects, once phonon-assisted transfer promotes e–h pairs into the bands [52]. A typical example of the change of absorption observed under these long excitations is shown in fig. 10. The generation of $N_{eh} \approx 2 \times 10^{12}$ cm^{-2} free e–h pairs in the 100 Å thick GaAs QW by direct excitation 32 meV above the gap with a cw diode laser completely bleaches the two exciton resonances, leaving a step like absorption edge.

This behavior is interpreted as follows. The absolute exciton energy is relatively insensitive to the e–h density because, as in three dimensions, the effects of the plasma on the electron almost exactly cancel those on the companion hole. Conversely, the single-particle state energies are renormalized, leading to a red shift of the gap. Thus the binding energy of the exciton with respect to the renormalized gap decreases. Furthermore, the absorption weakens both because the e–h correlation diminishes and be-

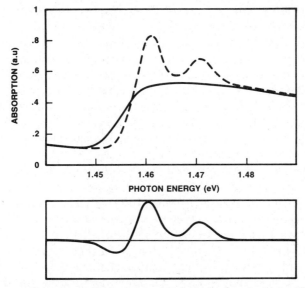

Fig. 10. Absorption spectra of a GaAs QWS with (solid line) and without (dashed line) continuous excitation 32 meV above the gap. The differential spectrum at the bottom of figure shows the difference between the unexcited and the excited spectra.

cause the single-particle states out of which excitons are built up are more and more occupied. This effect is very sensitive to the plasma density. If this density is further increased, the bound states eventually disappear leaving only the two-dimensional renormalized gap. The resulting step-like absorption can be saturated or even converted into gain as the states are filled, but this generally requires considerably higher densities, partly because the renormalization continues with increasing density. Accordingly, measurements of the saturation intensity found that about 50% of the absorption saturates very easily ($I_s \approx 500$ W/cm^2) and the remainder is much harder to saturate ($I_s \approx 5$ kW/cm^2) [48].

The plasma-induced changes in the refractive index have also been determined by four-wave mixing at low densities [53, 47]. The changes in refractive index and in absorption coefficient induced by one e–h pair per cm^3 are large; $n_{eh} \approx 3.7 \times 10^{-19}$ cm^3 and $\sigma_{eh} \approx 7 \times 10^{-14}$ cm^2. Expressed in term of the usual optical nonlinearity they correspond to the huge $\chi^{(3)} \approx 0.06$ esu. However, nonlinear optical susceptibilities have to be used with caution since the relevant parameter is the carrier density N_{eh} and not the optical fields. In particular, the effects of the e–h plasma remain as long as the plasma is present, e.g. about 20 ns. The recombination time can be reduced (e.g. to ≈ 150 ps) without too much modification of the other optical properties by introducing effective recombination centers, for example by ion bombardment [54]. The effective optical nonlinearity will reduce in proportion to the carrier lifetime. Also, this is only a small signal result. As the carrier density is increased, the changes in absorption (and, consequently, refractive index) become progressively more difficult as the "excitonic" part of the absorption saturates. The small signal experimental results have been quantitatively explained by a model based on many-body theory [55].

The nonlinear optical response of QWS is completely different if ultrashort optical pulses are used to excite and probe the sample in times short compared to the exciton ionization time or to the carrier scattering times [56,57]. For selective generation of excitons by resonant excitation, the resulting bound pairs induce certain types of changes in the optical spectrum. Then as they transform into free pairs, these changes should evolve toward those seen using long duration excitations. This is indeed what is observed. As an example differential absorption spectra are shown in fig. 11, measured with a wide band (1.45–1.65 eV) 80 fs continuum at various

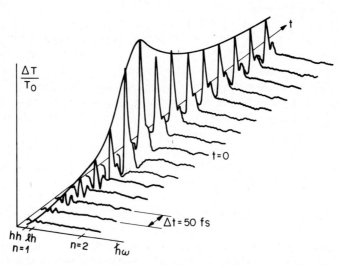

Fig. 11. Absorption differential spectra measured with a broad band (1.45–1.65 eV) 80 fs continuum at various delays after excitation with a 80 fs pump pulse resonant with the hh-exciton. The peak of the exciton shows a very large bleaching that lasts 300 fs, i.e. the ionization time due to collision by thermal phonons.

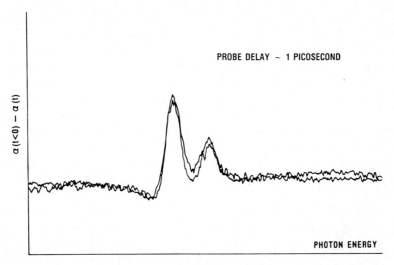

Fig. 12. Comparison of the absorption differential spectra for resonant and non-resonant excitation 1 ps after the pump pulse. The two spectra are indistinguishable, showing that at that time only free e–h pairs are present in the QWS.

delays after the resonant excitation of the hh-exciton by a 80 fs pump pulse [57]. The density of excitons created is approximately $N_X \approx 2 \times 10^{10}$ cm^{-2}, the interval between each spectrum is 50 fs, and each curve corresponds to the integration of 10^6 spectra giving a signal to noise ratio of 10^4. A fast transient lasting about 300 fs is seen close to zero delay where the hh-exciton absorption bleaches very effective, then the bleaching of the resonance reduces and stabilizes at a level about 2.5 times smaller after about 800 fs. From then on the change in the absorption spectra remain constant for tens of ps.

The profile of the differential spectra at long delays (>1 ps) does not depend on whether the excitation is resonant or not. This is shown in fig. 12 where we compare the differential absorption spectra measured with a 120 fs continuum centered at the $n = 1$ exciton 1 ps after excitation by resonant and non-resonant pump generating the same density of e–h pairs [56]. Within our experimental accuracy the two spectra are identical. From these observations one can deduce that the ionization time of the hh-excitons at RT is $t_i \approx 300$ fs. However, the strong bleaching of the exciton peak for resonant pumping is in contradiction with the behavior in three dimensions from which one would expect a weaker effect on the absorption spectrum when bound pairs rather than free pairs are generated [50,51].

This unusual behavior has been explained recently [58]. The theory utilizes two ingredients: (i) in two dimensions, the strength of DS is strongly weakened as compared to three dimensions [59] and its relative importance in bleaching the absorption is reduced relative to the effects of the Pauli principle (EI and PSF); (ii) when excitons are generated with ultrashort optical pulses and are observed before their first interaction with the thermal reservoir, their temperature (as far as it can be defined) is essentially very low, but, upon being ionized by LO-phonon collisions, they release a warm plasma.

In the theory the relative change of the exciton oscillator strength is expanded in the first order in the density of pairs (bound or unbound)

$$\delta f_{1S}/f_{1S} = N_{eh}/N_s, \tag{2}$$

in terms of the saturation density N_s. The saturation density is then calculated from many-body theory for excitons and as a function of temperature for free e–h pairs taking into account EI and PSF. The key element in the model is to realize that an exciton is equivalent to special distributions of electrons and holes equal to the exciton occupation probability distribution in k-space [49], i.e.

$$f_e(k) = f_h(k) = \tfrac{1}{2}|U(k)|^2, \tag{3}$$

where $U(k)$ is the exciton relative motion wave function in k-space. The results are illustrated in fig. 13. The low-temperature plasma that fills up states close to $k=0$ [where $U(k)$ is centered] is more efficient than the excitons in bleaching the exciton resonance absorption. However, as the temperature increases, the thermal e–h distributions expand more and more toward high energy and the effectiveness of the plasma decreases as E_{1S}/kT. For $E_{1S} \approx 9$ meV, the model gives an exciton bleaching due to excitons slightly larger than that due to a RT plasma. It also explains the effects seen at the lh-exciton. Finally, the theoretical value of the nonlinear cross section, $\sigma_X = 2.5 \times 10^{-13}$ cm^2, is found to be in good agreement with the experimental one, $\sigma_X = 1.6 \times 10^{-13}$ cm^2.

The assumption that DS is smaller than EI and PSF in QWS was tested directly by generation of a non-thermal carrier distribution with density $N_{eh} \approx 2 \times 10^{10}$ cm^{-2}, in the $n=1$ continuum, less than one LO-phonon energy above the gap [57], and measuring the evolution of the absorption spectrum including the $n=1$ and the $n=2$ resonances as a function of the pump-probe delay between -100 fs and 200 fs. The experimental spectra

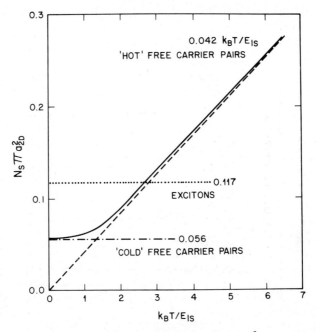

Fig. 13. Theoretical pair density (normalized to the exciton area πa_X^2) for saturation of the exciton resonance: (i) when the pairs are excitons (dotted line), and (ii) when the pairs form a plasma (solid line) versus the temperature (normalized to the exciton binding energy). The dashed line and dotted-dashed line give the limit for "hot" and "cold" free e–h pairs (plasma). The solid line is a sketch of the total effect.

are shown in fig. 14. At small delays the carriers do not occupy states at $k = 0$ in the $n = 1$ subband so that only the weak DS is effective. As they thermalize they occupy states at the bottom of the $n = 1$ subband and the effects of the Pauli principle (EI + PSF) are turned on and strongly bleach the $n = 1$ resonance. Meanwhile the $n = 2$ excitons, which are at higher energy and in any case made of single-particle states nearly orthogonal to those occupied by the carriers, only feel the DS. The effects at the $n = 2$ exciton do not vary in time, showing that DS remains essentially constant as the carriers thermalize. By numerical integration it was shown that the average carrier energy as well as the carrier density remains constant, thus confirming that no interaction with the lattice occurs on this time scale. These experimental results put an upper limit on the magnitude of DS,

$$\text{EI} + \text{PSF} > 6 \times \text{DS}.$$

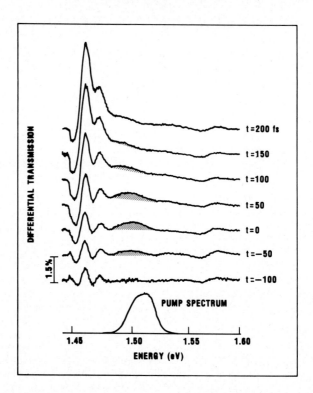

Fig. 14. Differential absorption spectra for a non-thermal carrier distribution generated by a 80 fs pump centered at 1.509 eV and observed at 50 fs interval with a broad 80 fs continuum. The carriers first bleach a hole in the interband absorption and change the absorption at the $n=1$ and $n=2$ exciton resonances through the weak two-dimensional direct screening. Then as they thermalize at the $n=1$ subband gap the consequences of the Pauli principle become effective at the $n=1$ exciton, i.e., phase-space filling and exchange interactions further bleach the resonance.

In addition, these measurements provide important information on the carrier scattering times in these heterostructures. It is found that the carriers leave the states in which they are created in a time less than (but of the order of) 80 fs and that they thermalize by carrier–carrier scattering in about 200 fs.

Investigations of subpicosecond dynamics in narrower wells (e.g. ≈ 50 Å) [60] have revealed an interesting "blue shift" phenomenon of the exciton peaks. This is seen either when excitons are created resonantly, or with non-resonant excitation after the excited carriers have had time to cool. A possible explanation of this [58] is that the DS is so weak in this strongly confined structure that it no longer compensates the blue shift expected from exchange interactions. Thus either cold carriers or excitons would lead to a blue shift of the exciton peaks.

The very large optical nonlinearities seen in QWS at RT have found application to third-order nonlinear optical processes with a laser diode as sole light source [61]. They have also been used as a saturable absorber for modelocking of laser diodes [54], in experiments where stable 1.6 ps pulses at about 1 GHz have been produced. QWS have also been investigated for applications in optical bistability with Fabry–Pérot cavities [60,62,63]. These devices rely on changes in refractive index associated with absorption saturation, and QWS devices have shown bistability at room temperature with a laser diode as the light source [63]. It appears, however, that there is little advantage in using the QWS in this application [60] compared to bulk GaAs. Although the changes in absorption at the exciton peak are larger than in GaAs at room-temperature, hence enabling the modelocking [54], it does not appear possible to work sufficiently close to the exciton resonance in the QWS to take advantage of the better resonant behavior for refractive effects in these structures. The problem is the background absorption for photon energies just below the band gap. A further problem with these devices in both GaAs and QWS structures is that they require relatively strong saturation of the exciton absorption in order to achieve sufficient refractive index change for switching. Depending on the cavity design, this may take them beyond the true excitonic saturation into the much more difficult saturation of the interband absorption.

5. Conclusions

In this review, we have shown that room exciton resonances in QWS present unusual electro-optic and nonlinear optical effects. These effects arise from the strong quantum size effects that appear in nanostructures whose dimensions become of the order of the characteristic quantum lengths of carriers in semiconductors. Some of these effects such as the QCSE are specific to QWS. The dominant mechanisms in the nonlinear optical response of QWS are different from those of bulk semiconductors. Finally, the fact that exciton resonances are seen at room temperature results in ultrafast transients that have no counterpart in bulk semiconductors. These novel properties have been used to demonstrate interesting devices with potential applications to optical signal processing, optical logic and convenient sources of ultrashort optical pulses. The general trends of the underlying physics are more or less understood in the case of the QWS based on the GaAs/AlGaAs system. Development of novel devices in this system is a matter of time and innovation. As for the other semiconductor systems and the other nanostructures the field is still wide open and promises to be very active in the near future.

Acknowledgements

The authors are pleased to have the opportunity of acknowledging the contributions of their many colleagues at AT&T Bell Laboratories and Bell Communications Research to the work with which the authors have been associated in this field: C.A. Burrus, A.Y. Cho, T.C. Damen, M.C. Downer, P.M. Downey, D.J. Eilenberger, J.H. English, R.L. Fork. A.H. Gnauck, A.C. Gossard, H.A. Haus, J.E. Henry, J.P. Heritage, C. Hirlimann, W.H. Knox, A. Pinczuk, J. Shah, S. Schmitt-Rink, C.V. Shank, Y. Silberberg, D. Sivco, P.W. Smith, B. Tell, R.S. Tucker, A. von Lehmen, J.S. Weiner, J.M. Wiesenfeld, W. Wiegmann, T.H. Wood and J.E. Zucker.

References

[1] A. Miller, D.A.B. Miller and S.D. Smith, Adv. Phys. 30 (1981) 697.
[2] D.S. Chemla and A. Maruani, Progr. Quantum Electron. 8 (1982) 1.
[3] D.S. Chemla and J. Jerphagnon, Nonlinear Optical Properties, in: Handbook on Semiconductors, Vol. 2: Optical Properties of Solids, ed. M. Balkanski (North-Holland Amsterdam, 1980) pp. 545–598.
[4] W. Franz, Z. Naturforsch. a 13 (1958) 484.
[5] L.V. Keldysh, Zh. Eksp. & Teor. Fiz. 34 (1958) 1138.
[6] See, for example, M. Cardona, in: Modulation Spectroscopy, Solid State Physics S 11 (Academic Press, New York, 1969) ch VII, and references therein.
[7] J.D. Dow and D. Redfield, Phys. Rev. B 1 (1970) 3358.
[8] D.F. Blossey, Phys. Rev. B 2 (1970) 3976.
[9] Q.H.F. Vrehen, J. Phys. & Chem. Solids 29 (1968) 129.
[10] See, for example, G.E. Stillman and C.M. Wolfe, in: Semiconductors and Semimetals, eds R.K. Willardson and A.C. Beer (Academic Press, New York, 1977) p. 380.
[11] See, for example, Molecular Beam Epitaxy and Heterostructures, eds L.L. Chang and K. Ploog, NATO Advanced Sciences Institute Series (Nijhoff, The Hague, 1985).
[12] See, for example, Metal Organic Vapor Phase Epitaxy, eds. J.B. Mullin, S.J.C. Irvine, R.H. Moss, P.N. Robson and D.R. Wight (North-Holland, Amsterdam, 1984).

[13] See, for example, R.E. Howard, L.D. Jackel, P.M. Mankiewich, W.J. Skocpol, Science 231 (1986) 345; and
R.E. Howard and D.E. Prober, in: VLSI Electronics-Microstructures, Vol. V, ed. N.G. Einspruch (Academic Press, New York, 1982) pp. 145–189; and references therein.
[14] For recent reviews see, for example, F. Capasso and B.F. Levine, J. Lumin. 30 (1985) 144; and
D.S. Chemla, J. Lumin. 30 (1985) 502.
[15] D.S. Chemla and D.A.B. Miller, J. Opt. Soc. Am. B 2 (1985) 1155.
[16] R.C. Miller and D.A. Kleinman, J. Lumin. 30 (1985) 144.
[17] C. Weisbuch, R.C. Miller, R. Dingle and A.C. Gossard, Solid State Commun. 37 (1981) 219.
[18] R.C. Miller, D.A. Kleinman, W.A. Nordland and A.C. Gossard, Phys. Rev. B 22 (1980) 863.
[19] J. Hegarty and M.D. Sturge, J. Opt. Soc. Am. B 2 (1985) 1143.
[20] J.E. Zucker, A. Pinczuk, D.S. Chemla, A.C. Gossard and W. Wiegmann, Phys. Rev. Lett. 51 (1983) 1293.
[21] J.E. Zucker, A. Pinczuk, D.S. Chemla, A.C. Gossard and W. Wiegmann, Phys. Rev. Lett. 53 (1984) 1280.
[22] J.E. Zucker, A. Pinczuk, D.S. Chemla, A.C. Gossard and W. Wiegmann, Phys. Rev. B 29 (1984) 7065.
[23] G. Bastard, Phys. Rev. B 30 (1984) 3547.
[24] D.A.B. Miller, D.S. Chemla, P.W. Smith, A.C. Gossard and W. Wiegmann, Appl. Phys. B 28 (1982) 96.
[25] J.S. Weiner, D.S. Chemla, D.A.B. Miller, T.H. Wood, D. Sivco and A.Y. Cho, Appl. Phys. Lett. 46 (1985) 619.
[26] H. Temkin, M.B. Panish, P.M. Petroff, R.A. Hamm, J.M. Vandenberg and S. Sunski, Appl. Phys. Lett. 47 (1985) 394.
[27] J.S. Weiner, D.S. Chemla, D.A.B. Miller, H.A. Haus, A.C. Gossard, W. Wiegmann and C.A. Burrus, Appl. Phys. Lett. 47 (1985) 664.
[28] D.A.B. Miller, J.S. Weiner and D.S. Chemla, IEEE J. Quantum Electron. QE-22 (1986) 1816.
[29] D.A.B. Miller, D.S. Chemla, T.C. Damen, A.C. Gossard, W. Wiegmann, T.H. Wood and C.A. Burrus, Phys. Rev. B 32 (1985) 1043.
[30] F.L. Lederman and J.D. Dow, Phys. Rev. B 13 (1976) 1633.
[31] W.H. Knox, D.A.B. Miller, T.C. Damen, D.S. Chemla and C.V. Shank, Appl. Phys. Lett. 48 (1986) 864.
[32] D.A.B. Miller, D.S. Chemla and S. Schmitt-Rink, Phys. Rev. B 34 (1986) 6976.
[33] J.S. Weiner, D.A.B. Miller, D.S. Chemla, T.C. Damen, C.A. Burrus, T.H. Wood, A.C. Gossard and W. Wiegmann, Appl. Phys. Lett. 47 (1985) 1148.
[34] D.A.B. Miller, D.S. Chemla, T.C. Damen, A.C. Gossard, W. Wiegmann, T.H. Wood and C.A. Burrus, Phys. Rev. Lett. 53 (1984) 2173.
[35] D.A.B. Miller, D.S. Chemla, T.C. Damen, A.C. Gossard, W. Wiegmann, T.H. Wood and C.A. Burrus, Appl. Phys. Lett. 45 (1984) 13.
[36] D.A.B. Miller, D.S. Chemla, T.C. Damen, A.C. Gossard, W. Wiegmann, T.H. Wood and C.A. Burrus, Opt. Lett. 9 (1984) 567.
[37] D.A.B. Miller, D.S. Chemla, T.C. Damen, A.C. Gossard, W. Wiegmann, T.H. Wood and C.A. Burrus, IEEE, J. Quantum Electron. QE-21 (1985) 1462.
[38] T.H. Wood, C.A. Burrus, D.A.B. Miller, D.S. Chemla, T.C. Damen, A.C. Gossard and W. Wiegmann, Appl. Phys. Lett. 44 (1984) 16.

[39] T.H. Wood, C.A. Burrus, R.S. Tucker, J.S. Weiner, D.A.B. Miller, D.S. Chemla, T.C. Damen, A.C. Gossard and W. Wiegmann, IEEE J. Quantum Electron. QE-21 (1985) 117.
[40] T.H. Wood, C.A. Burrus, R.S. Tucker, J.S. Weiner, D.A.B. Miller, D.S. Chemla, T.C. Damen, A.C. Gossard and W. Wiegmann, Electron. Lett. 21 (1985) 693.
[41] K. Wakita, Y. Kawamura, Y. Yoshikuni and A. Asahi, Electron. Lett. 21 (1985) 338.
[42] K. Wakita, Y. Kawamura, Y. Yoshikuni and A. Asahi, Electron. Lett. 21 (1985) 574.
[43] T. Miyazawa, S. Tarucha, Y. Ohmori and H. Okamoto, Surf. Sci. 174 (1986) 238.
[44] S. Tarucha and H. Okamoto, Appl. Phys. Lett. 48 (1986) 1;
Y. Arakawa, A. Larsson, J. Paslaski and A. Yariv, Appl. Phys. Lett. 48 (1986) 561.
[45] T.H. Wood, C.A. Burrus, A.H. Gnauck, J.M. Wiesenfeld, D.A.B. Miller, D.S. Chemla and T.C. Damen, Appl. Phys. Lett. 47 (1985) 190.
See also A. Larsson, A. Yariv, R. Tell, J. Maserjian and S.T. Eng, Appl. Phys. Lett. 47 (1985) 866.
[46] D.A.B. Miller, J.E. Henry, A.C. Gossard and J.H. English, Appl. Phys. Lett. 49 (1986) 821.
[47] D.S. Chemla, D.A.B. Miller, P.W. Smith, A.C. Gossard and W. Wiegmann, IEEE J. Quantum Electron. QE-20 (1984) 265.
[48] D.A.B. Miller, D.S. Chemla, D.J. Eilenberger, P.W. Smith, A.C. Gossard and W.T. Tsang, Appl. Phys. Lett. 41 (1982) 679.
[49] For a recent review of semiconductor optical nonlinearities close to the band gap see: H. Haug and S. Schmitt-Rink, Prog. Quantum Electron. 9 (1984) 3.
[50] G.W. Fehrenbach, W. Schafer, J. Treusch and R.G. Ulbrich, Phys. Rev. Lett. 49 (1982) 1281.
[51] G.W. Fehrenbach, W. Schafer and R.G. Ulbrich, J. Lumin. 30 (1985) 154.
[52] A. Von Lehmen, J.E. Zucker, J.P. Heritage and D.S. Chemla, Phys. Rev. B 35 (1987) 6479.
[53] D.A.B. Miller, D.S. Chemla, D.J. Eilenberger, P.W. Smith, A.C. Gossard, W. Wiegmann, Appl. Phys. Lett. 42 (1982) 925.
[54] P.W. Smith, Y. Silberberg, D.A.B. Miller, J. Opt. Soc. Am. B 2 (1985) 1228, and references therein.
[55] S. Schmitt-Rink and C. Ell, J. Lumin. 30 (1985) 585.
[56] W.H. Knox, R.L. Fork, M.C. Downer, D.A.B. Miller, D.S. Chemla, C.V. Shank, A.C. Gossard and W. Wiegmann, Phys. Rev. Lett. 54 (1985) 1306.
[57] W.H. Knox, C. Hirlimann, D.A.B. Miller, J. Shah, D.S. Chemla and C.V. Shank, Phys. Rev. Lett. 56 (1986) 1191.
[58] S. Schmitt-Rink, D.S. Chemla and D.A.B. Miller, Phys. Rev. B 32 (1985) 6601.
[59] T. Ando, A.B. Fowler and F. Stern, Rev. Mod. Phys. 54 (1982) 437.
[60] N. Peyghambarian and H.M. Gibbs, J. Opt. Soc. B 2 (1985) 1215, and references therein.
[61] D.A.B. Miller, D.S. Chemla, P.W. Smith, A.C. Gossard and W. Wiegmann, Opt. Lett. 8 (1983) 477.
[62] H.M. Gibbs, S.S. Tarng, J.L. Jewell, D.A. Weinberger, K. Tai, A.C. Gossard, S.L. McCall, A. Passner and W. Wiegmann, Appl. Phys. Lett. 41 (1982) 221.
[63] S.S. Tarng, H.M. Gibbs, J.L. Jewell, N. Peyghambarian, A.C. Gossard, T. Venkatesan and W. Wiegmann, Appl. Phys. Lett. 44 (1984) 360.

AUTHOR INDEX

Abeles, J.H. [8] 462, 483, 486; [99] 515, 516, 517, 518, 532, 537, 563
Abram, R.A. [32] 582, 593
Abstreiter, G. [48] 121, 122, 124, 156, 164; [78] 246, 262; [88] 248, 249, 262; [90] 254, 255, 256, 262; [91] 256, 262; [71] 507, 562
Adams, M.J. [16] 120, 132, 163
Alavi, K. [31] 68, 69, 112; [51] 74, 81, 112; [68] 74, 113; [59] 234, 235, 261; [15] 413, 449; [16] 414, 449; [41] 435, 450; [24] 343, 346, 374; [47] 592, 593
Aleshin, V.G. [26] 178, 206
Alexandre, F. [32] 69, 73, 112; [25] 219, 220, 230, 260; [22] 423, 424, 449
Allam, J. [125] 110, 114; [6] 379, 380, 385, 386, 387, 390, 396; [18] 385, 390, 396; [38] 433, 434, 450; [40] 434, 450; [41] 435, 450
Allan, G. [104] 84, 114
Allen, R.E. [35] 121, 124, 154, 164; [45] 122, 153, 164
Alley, G.D. [32] 500, 561
Allyn, C.L. [31] 68, 69, 112; [40] 364, 375; [26] 425, 426, 449; [27] 425, 450; [65] 505, 562
Altarelli, M. [115] 110, 114; [10] 215, 260; [46] 231, 261; [47] 232, 261; [54] 233, 261; [89] 254, 262; [17] 287, 288, 309; [35] 298, 310; [22] 580, 593
Anastassakis, E. [35] 271, 281
Anderson, I.M. [51] 502, 503, 510, 561; [52] 502, 510, 561
Anderson, J. [32] 121, 164
Anderson, R.J. [80] 73, 113; [12] 410, 413, 449; [74] 508, 562
Anderson, R.L. [6] 11, 17, 21, 56; [2] 61, 76, 77, 80, 97, 111; [9] 119, 132, 163; [3] 169, 201, 205; [4] 322, 355, 356, 357, 364, 373; [42] 364, 375
Andersson, T.G. [104] 155, 166
Ando, T. [85] 247, 262; [86] 247, 262; [1] 265, 272, 280; [11] 454, 455, 486; [19] 457, 458, 460, 486; [61] 504, 526, 562; [59] 618, 624
Andreoni, W. [16] 216, 260
Ankri, D. [22] 423, 424, 449
Anthony, P.J. [9] 567, 592
Antypas, G.A. [30] 344, 374
Appelbaum, J.A. [1] 11, 12, 14, 56; [37] 38, 57; [19] 120, 163; [20] 174, 206
Arakawa, Y. [51] 592, 593; [44] 610, 624
Arch, D.K. [29] 292, 310; [30] 292, 302, 308, 310
Arnaud d'Avitaya, F. [27] 497, 500, 561; [38] 500, 561
Arnold, D. [40] 72, 112; [56] 73, 112; [39] 361, 364, 365, 374
Arthur, J.R. [27] 267, 281; [1] 401, 449
Asahi, A. [41] 610, 624; [42] 610, 624
Asbeck, P.M. [12] 410, 413, 449; [74] 508, 562
Askenasy, S. [31] 224, 225, 233, 260
Asonen, H. [120] 110, 114
Atalla, M.M. [25] 496, 561
AuCoin, T.R. [41] 501, 561; [66] 505, 562
Auleytner, J. [15] 171, 205
Azbel, M.Ya. [18] 418, 449; [133] 543, 548, 564

Bachrach, R.Z. [89] 144, 145, 166
Badoz, P.A. [38] 500, 561
Baer, Y. [16] 171, 205
Bafleur, M. [54] 123, 164
Bajaj, K.K. [33] 225, 231, 232, 233, 260; [38] 274, 277, 282

Baldereschi, A. [16] 216, 260; [52] 233, 261
Baldy, M. [25] 266, 281; [40] 589, 593
Ballingall, J.M. [67] 73, 113
Banerjee, S. [88] 513, 562
Baraff, G.A. [1] 11, 12, 14, 56; [41] 43, 57; [19] 120, 163; [20] 174, 206
Bassani, F. [42] 231, 261
Bastard, G. [32] 69, 73, 112; [43] 75, 112; [5] 213, 259; [8] 214, 253, 260; [25] 219, 220, 230, 260; [28] 222, 260; [43] 231, 261; [64] 237, 243, 261; [3] 285, 287, 288, 295, 296, 309; [8] 285, 287, 293, 294, 309; [15] 287, 289, 309; [23] 291, 310; [43] 474, 475, 487; [23] 601, 623
Bate, R.T. [157] 556, 564
Batey, J. [54] 72, 112; [79] 73, 113; [57] 124, 165; [39] 229, 260; [75] 243, 259, 261; [12] 265, 279, 281; [13] 265, 279, 281; [37] 361, 362, 372, 374; [38] 361, 362, 363, 364, 365, 374; [9] 380, 383, 390, 393, 396
Bauer, R.S. [22] 19, 56; [31] 28, 51, 57; [46] 73, 112; [47] 73, 112; [48] 73, 84, 98, 100, 112; [64] 73, 113; [65] 73, 113; [72] 73, 113; [101] 72, 73, 78, 79, 114; [24] 121, 163; [28] 121, 124, 139, 145, 147, 148, 150, 151, 155, 164; [46] 122, 136, 144, 156, 164; [47] 121, 122, 164; [56] 123, 164; [76] 136, 145, 165; [78] 137, 149, 151, 165; [81] 137, 138, 139, 141, 142, 165; [84] 137, 140, 165; [89] 144, 145, 166; [92] 145, 166; [95] 145, 148, 149, 166; [96] 145, 148, 149, 155, 166; [102] 154, 166; [105] 157, 166; [39] 190, 193, 206
Baukus, J.P. [10] 285, 309
Bean, J.C. [70] 507, 562
Beard, W.T. [53] 72, 112
Beattie, A.R. [26] 582, 593
Bechstedt, F. [13] 61, 76, 77, 88, 89, 97, 111
Belhaddad, K. [37] 500, 561
Belle, G. [46] 231, 261
Ben Daniel, D.J. [7] 214, 260
Bergmark, T. [14] 171, 205
Bernard, M.G.A. [14] 571, 593
Berndtsson, A. [16] 171, 205

Berroir, J.M. [12] 285, 293, 297, 309; [13] 285, 302, 304, 309; [22] 290, 291, 299, 310; [31] 293, 310
Bertoni, C.M. [30] 121, 124, 164; [37] 121, 124, 136, 153, 164; [72] 136, 165
Besomi, P. [49] 592, 593
Bethea, C.G. [7] 404, 410, 449; [8] 404, 406, 407, 449; [9] 407, 410, 449; [13] 411, 449; [28] 426, 450; [76] 508, 562
Beton, P.H. [86] 510, 562; [87] 510, 562
Bhat, R. [8] 462, 483, 486; [59] 504, 562; [64] 504, 515, 516, 517, 518, 519, 521, 562; [99] 515, 516, 517, 518, 532, 537, 563
Biefeld, R.M. [17] 266, 281
Bimberg, D. [53] 233, 261; [54] 233, 261
Blondau, R. [25] 266, 281
Blood, P. [37] 589, 593
Blossey, D.F. [8] 598, 622
Board, K. [66] 505, 562
Boch, R. [24] 291, 310
Bolmont, D.J. [31] 121, 124, 164; [38] 121, 124, 153, 164
Bonnefoi, A.R. [40] 473, 487; [18] 493, 554, 555, 556, 560; [126] 541, 564; [150] 553, 564
Boonstra, T. [127] 541, 564
Botez, D. [19] 578, 593
Bottka, N. [53] 72, 112
Boukerche, M. [14] 287, 290, 309; [25] 291, 310; [37] 302, 310
Bozler, C.O. [32] 500, 561
Bradley, J.A. [94] 258, 262; [23] 266, 267, 281; [28] 268, 269, 272, 281; [29] 268, 269, 272, 276, 281; [31] 268, 273, 281; [36] 272, 282; [37] 273, 276, 279, 282
Bradshaw, K. [157] 556, 564
Braslau, N. [111] 114
Brennan, K. [36] 465, 466, 487
Brillson, L.J. [100] 108, 114; [42] 121, 164; [43] 121, 127, 153, 164; [106] 159, 166; [18] 174, 205; [49] 201, 206
Brown, E.R. [146] 550, 564; [147] 550, 564
Brown, J.M. [128] 541, 564
Brown, K.W. [149] 550, 564
Brown, R.L. [49] 592, 593; [7] 567, 590, 592
Brown, R.N. [32] 296, 310

Brown, W. [31] 68, 69, 112; [40] 364, 375; [27] 425, 450; [65] 505, 562
Brucker, C.F. [42] 121, 164
Brugger, H. [48] 121, 122, 124, 156, 164; [71] 507, 562
Brum, J.A. [22] 290, 291, 299, 310; [43] 474, 475, 487
Brundle, C.R. [67] 126, 165
Brunemeier, P.E. [59] 74, 113; [67] 238, 261; [14] 380, 396
Büttiker, M. [143] 548, 564
Burnham, R.D. [32] 268, 271, 272, 276, 281; [40] 473, 487; [35] 587, 593; [42] 589, 590, 593; [126] 541, 564; [150] 553, 564
Burrus, C.A. [27] 602, 623; [29] 602, 604, 606, 607, 609, 623; [33] 606, 611, 612, 623; [34] 606, 607, 623; [35] 609, 611, 623; [36] 609, 611, 623; [37] 609, 611, 612, 623; [38] 610, 623; [39] 610, 624; [40] 610, 624; [45] 610, 611, 624
Burt, M.G. [32] 582, 593; [50] 592, 593

Caine, E.J. [62] 73, 113
Calandra, C. [30] 121, 124, 164; [37] 121, 124, 136, 153, 164
Caldas, M.J. [25] 20, 26, 54, 57; [8] 61, 76, 77, 85, 98, 111
Calleja, J.M. [74] 73, 113; [52] 502, 510, 561
Campbell, J. [49] 441, 450
Campidelli, Y. [27] 497, 500, 561
Camras, M.D. [32] 225, 260
Capasso, C. [28] 67, 75, 107, 111; [36] 74, 112; [97] 106, 114; [74] 136, 165
Capasso, F. [33] 28, 57; [98] 108, 114; [125] 110, 114; [93] 257, 262; [6] 379, 380, 385, 386, 387, 390, 396; [18] 385, 390, 396; [2] 401, 449; [3] 401, 449; [7] 404, 410, 449; [8] 404, 406, 407, 449; [9] 407, 410, 449; [11] 410, 411, 413, 449; [13] 411, 449; [15] 413, 449; [17] 415, 417, 449; [19] 419, 423, 449; [21] 419, 449; [23] 424, 449; [24] 424, 449; [28] 426, 450; [29] 428, 450; [30] 428, 450; [31] 428, 450; [32] 428, 450; [33] 428, 450; [34] 428, 429, 450; [36] 431, 450; [38] 433, 434, 450; [40] 434, 450; [41] 435, 450; [42] 436, 450; [45] 438, 450; [47] 439, 450; [50] 442, 450; [51] 442, 450; [8] 492, 560; [15] 493, 541, 543, 552, 553, 560; [16] 493, 541, 542, 548, 560; [17] 493, 554, 556, 557, 558, 560; [75] 508, 562; [76] 508, 562; [134] 543, 553, 564; [14] 598, 623
Capozi, M. [99] 108, 109, 114
Car, R. [16] 216, 260
Card, H.C. [5] 117, 118, 120, 163; [53] 502, 562
Cardona, M. [116] 110, 114; [65] 125, 165; [35] 271, 281; [39] 276, 282; [42] 306, 310; [6] 598, 622
Carelli, J. [49] 201, 206
Carlsson, A.E. [18] 14, 56
Caruthers, E. [17] 216, 260
Casey Jr, H.C. [53] 123, 164; [4] 265, 279, 280; [19] 387, 396; [14] 454, 486; [18] 574, 593
Casselman, T.N. [38] 302, 310
Castagne, R. [77] 510, 562
Cerrina, F. [50] 75, 112; [97] 106, 114; [39] 121, 124, 153, 164
Chadi, D.J. [90] 88, 114
Chan, K.S. [48] 232, 233, 261
Chan, W.K. [8] 462, 483, 486; [64] 504, 515, 516, 517, 518, 519, 521, 562; [98] 515, 518, 520, 521, 522, 563; [99] 515, 516, 517, 518, 532, 537, 563; [100] 563
Chandra, A. [41] 364, 375
Chandrasekhar, H.R. [18] 266, 268, 271, 272, 276, 281
Chandrasekhar, M. [18] 266, 268, 271, 272, 276, 281
Chang, C.A. [74] 73, 113; [86] 139, 141, 142, 166; [63] 236, 261; [43] 364, 375
Chang, C.Y. [62] 505, 562
Chang, L.L. [115] 110, 114; [5] 117, 118, 120, 163; [86] 139, 141, 142, 166; [43] 231, 261; [16] 266, 281; [34] 356, 364, 374; [43] 364, 375
Chang, L.I. [37] 472, 487; [114] 541, 563; [115] 541, 551, 563; [155] 556, 564; [11] 598, 622
Chang, Y.C. [12] 214, 215, 216, 260; [15] 216, 260; [15] 287, 289, 309; [21] 290, 291, 294, 296, 310; [155] 556, 564
Chapin, P.C. [119] 541, 563; [120] 541, 563; [148] 550, 564

Charasse, M.N. [62] 235, 236, 261
Chelikowsky, J.R. [25] 177, 206
Chemla, D.S. [60] 74, 113; [56] 233, 234, 261; [2] 598, 622; [3] 598; 622; [14] 598, 623; [15] 599, 600, 602, 613, 623; [20] 601, 623; [21] 601, 623; [22] 601, 623; [24] 602, 623; [25] 602, 623; [27] 602, 623; [28] 602, 608, 612, 623; [29] 602, 604, 606, 607, 609, 623; [31] 604, 606, 623; [32] 609, 623; [33] 606, 611, 612, 623; [34] 606, 607, 623; [35] 609, 611, 623; [36] 609, 611, 623; [37] 609, 611, 612, 623; [38] 610, 623; [39] 610, 624; [40] 610, 624; [45] 610, 611, 624; [47] 613, 614, 615, 624; [48] 613, 615, 624; [52] 614, 624; [53] 615, 624; [56] 616, 617, 624; [57] 616, 617, 618, 624; [58] 618, 621, 624; [61] 621, 624
Chen, J. [50] 501, 561; [123] 541, 542, 546, 547, 556, 564
Chen, P. [31] 121, 124, 164; [38] 121, 124, 153, 164
Chen, Y.S. [57] 234, 261
Chen, Z.H. [12] 61, 76, 77, 88, 89, 111
Cheung, J.T. [102] 75, 114; [32] 180, 206; [11] 285, 309; [43] 286, 307, 308, 310
Chiang, C. [17] 14, 56
Chiang, T.-C. [29] 180, 206
Chiaradia, P. [31] 28, 51, 57; [46] 73, 112; [64] 73, 113; [28] 121, 124, 139, 145, 147, 148, 150, 151, 155, 164; [47] 121, 122, 164; [76] 136, 145, 165; [89] 144, 145, 166; [92] 145, 166; [102] 154, 166; [105] 157, 166
Chien, W.Y. [25] 65, 73, 111; [58] 124, 165; [14] 334, 336, 337, 339, 343, 348, 372, 373; [16] 337, 373; [7] 379, 394, 395, 396
Chin, R. [85] 74, 113; [37] 432, 450; [17] 574, 593; [44] 590, 593
Ching, W.Y. [108] 94, 97, 114
Chiu, L.C. [29] 582, 593
Cho, A.Y. [33] 28, 57; [31] 68, 69, 112; [51] 74, 81, 112; [60] 74, 113; [62] 73, 113; [68] 74, 113; [98] 108, 114; [53] 123, 164; [48] 201, 206; [56] 233, 234, 261; [58] 234, 261; [59] 234, 235, 261; [93] 257, 262; [24] 343, 346, 374; [19] 387, 396; [1] 401, 449; [15] 413, 449; [16] 414, 449; [23] 424, 449; [24] 424, 449; [38] 433, 434, 450; [41] 435, 450; [47] 439, 450; [50] 442, 450; [51] 442, 450; [2] 491, 560; [15] 493, 541, 543, 552, 553, 560; [16] 493, 541, 542, 548, 560; [25] 602, 623; [47] 592, 593
Cho, Y. [105] 157, 166
Chow, D.H. [40] 473, 487; [18] 493, 554, 555, 556, 560
Chow, P.P. [10] 285, 309
Christensen, N.E. [116] 110, 114
Chu, X. [26] 292, 310; [27] 292, 310; [28] 292, 310; [47] 308, 310
Chung, Y.W. [61] 75, 113
Chwang, S.L. [39] 433, 450
Chye, P.W. [45] 51, 57; [40] 121, 122, 127, 128, 147, 153, 160, 164
Cibert, J. [156] 556, 564
Claessen, L.M. [115] 110, 114
Claxton, P.A. [73] 242, 261
Cohen, H.J. [80] 73, 113
Cohen, J.D. [20] 387, 388, 396
Cohen, M.L. [2] 11, 12, 13, 14, 15, 53, 56; [20] 61, 76, 105, 111; [21] 61, 76, 105, 106, 111, 124, 127, 165; [22] 61, 76, 105, 111; [20] 120, 163; [63] 124, 127, 165; [21] 174, 206; [25] 177, 206; [18] 216, 260
Colavita, E. [35] 72, 112; [99] 108, 109, 114
Coleman, J.J. [32] 225, 260; [33] 268, 281; [4] 379, 381, 396; [4] 453, 464, 478, 486; [42] 474, 487; [47] 479, 480, 487
Coleman, P.D. [89] 513, 563; [119] 541, 563; [120] 541, 563; [148] 550, 564
Collins, R.T. [126] 541, 564
Coluzza, C. [112] 108, 114
Comas, J. [53] 72, 112
Coon, D.D. [123] 110, 114; [145] 550, 564
Cooper Jr, J.A. [42] 436, 450
Correa, C.A. [118] 541, 547, 563
Couder, Y. [32] 296, 310
Cox, H.M. [45] 438, 450; [46] 439, 450
Craft, D.C. [7] 567, 590, 592
Crowell, C.R. [30] 499, 561
Cserveny, S.I. [2] 319, 373
Cunningham, J.E. [140] 548, 564

Damen, T.C. [29] 602, 604, 606, 607, 609, 623; [31] 604, 606, 623; [33] 606, 611, 612, 623; [34] 606, 607, 623; [35] 609, 611, 623; [36] 609, 611, 623; [37] 609, 611, 612, 623; [38] 610, 623; [39] 610, 624; [40] 610, 624; [45] 610, 611, 624
Dandekar, N.V. [42] 501, 561
Dangla, J. [22] 423, 424, 449
Daniels, R.R. [38] 75, 112; [50] 75, 112; [34] 121, 124, 134, 164; [36] 121, 124, 164; [19] 174, 205
Dapkus, P.D. [52] 123, 164; [32] 225, 260; [66] 237, 261; [33] 268, 281; [41] 474, 475, 487; [42] 474, 487; [2] 567, 580, 589, 592; [3] 567, 580, 589, 592; [17] 574, 593; [22] 580, 593; [39] 589, 593
Das Sarma, S. [12] 454, 456, 486
Davies, R.A. [130] 541, 551, 564; [131] 541, 564; [132] 541, 551, 564
Davis, G.A. [87] 75, 113
Davis, R.H. [111] 540, 543, 563
Daw, M.S. [44] 122, 153, 164
Dawson, P. [33] 69, 73, 112; [22] 219, 224, 226, 227, 228, 229, 231, 233, 257, 260; [35] 229, 260; [36] 229, 260; [74] 243, 244, 245, 246, 252, 258, 259, 261; [76] 246, 261; [96] 219, 259, 262; [7] 265, 279, 281
DeCremoux, B. [25] 266, 281, 343, 374; [6] 567, 589, 592
Delage, S. [27] 497, 500, 561; [38] 500, 561
Delalande, C. [32] 69, 73, 112; [25] 219, 220, 230, 260; [28] 222, 260; [63] 236, 261; [64] 237, 243, 261; [68] 239, 241, 242, 261
Dench, W.P. [17] 171, 205
Denley, D. [73] 73, 113; [23] 121, 163; [33] 121, 164
Dentai, A.G. [49] 441, 450
Deppe, D.G. [59] 74, 113; [67] 238, 261; [14] 380, 396
Derkits Jr, G.E. [36] 500, 561; [55] 502, 562; [136] 544, 564
Devoldere, P. [94] 145, 166
DeSouza, M. [37] 302, 310
DiMaria, D.J. [54] 72, 112; [79] 73, 113; [57] 124, 165; [12] 265, 279, 281; [37] 361, 362, 372, 374; [9] 380, 383, 390, 393, 396; [12] 265, 279, 281; [37] 361, 362, 372, 374
Dingle, R. [31] 68, 69, 112; [12] 119, 124, 163; [2] 213, 217, 218, 219, 220, 234, 246, 250, 257, 259; [3] 211, 213, 217, 218, 219, 220, 221, 222, 223, 234, 237, 246, 256, 257, 259; [21] 217, 260; [2] 265, 272, 279, 280; [4] 265, 279, 280; [27] 267, 281; [23] 458, 487; [1] 567, 592; [11] 569, 592; [24] 581, 593; [17] 601, 623
Ditzenberger, J.A. [20] 578, 593; [38] 589, 593
Dixon, R.W. [53] 123, 164; [49] 592, 593
Dobson, P.J. [93] 145, 166; [34] 228, 260; [96] 219, 259, 262
Dolan, G.J. [156] 556, 564
Dollinger, G. [34] 271, 281
Donnelly, J.P. [11] 327, 328, 373; [35] 358, 359, 374; [45] 368, 375
Dow, J.D. [11] 61, 76, 77, 88, 89, 95, 97, 101, 111; [35] 121, 124, 154, 164; [45] 122, 153, 164; [7] 598, 622; [30] 604, 623
Downer, M.C. [56] 616, 617, 624
Drickamer, H.G. [33] 268, 281
Drummond, T.J. [15] 265, 279, 281
Dubon-Chevallier, C. [22] 423, 424, 449
Duc, T.M. [119] 110, 114; [44] 286, 308, 310
Duchemin, J.P. [25] 266, 281; [6] 567, 589, 592; [40] 589, 593
Duggan, G. [29] 27, 48, 52, 57; [33] 69, 73, 112; [64] 124, 165; [22] 219, 224, 226, 227, 228, 229, 231, 233, 257, 260; [35] 229, 260; [37] 219, 229, 230, 252, 257, 260; [48] 232, 233, 261; [72] 241, 261; [96] 219, 259, 262; [3] 265, 279, 280; [7] 265, 279, 281; [5] 403, 445, 449
Duke, C.B. [4] 61, 76, 82, 111; [49] 201, 206; [7] 214, 260
Dumke, W.P. [52] 502, 510, 561; [34] 587, 593
Duncan, W.M. [157] 556, 564
Dupuis, R.D. [52] 123, 164; [66] 237, 261; [24] 266, 274, 277, 281; [41] 474, 475, 487; [3] 491, 560; [2] 567, 580, 589, 592; [3] 567, 580, 589, 592; [22] 580, 593; [41] 589, 593

Durán, J.C. [122] 110, 114
Duraffourg, G. [14] 571, 593
Dutta, N.K. [7] 567, 590, 592; [8] 567, 590, 591, 592; [9] 567, 592; [10] 567, 592; [16] 574, 577, 579, 593; [27] 582, 593; [28] 582, 593; [30] 582, 583, 584, 586, 593; [43] 590, 593; [49] 592, 593

Eastman, D.E. [29] 180, 206
Eastman, L.F. [63] 74, 113; [67] 73, 113; [4] 213, 239, 240, 241, 259; [41] 364, 375; [41] 501, 561; [42] 501, 561; [66] 505, 562
Edelman, H.S. [72] 136, 165
Edwall, D.D. [25] 65, 73, 111; [58] 124, 165; [14] 334, 336, 337, 339, 343, 348, 372, 373; [7] 379, 394, 395, 396
Eilenberger, D.J. [48] 613, 615, 624; [53] 615, 624
Eisen, F.H. [12] 410, 413, 449; [74] 508, 562
Eizenberg, M. [30] 68, 72, 81, 111; [111] 114; [60] 124, 165; [92] 256, 262; [11] 265, 279, 281
Ekenberg, U. [89] 254, 262
Ell, C. [55] 616, 624
Elliott, R.J. [48] 232, 233, 261
Enderlein, R. [13] 61, 76, 77, 88, 89, 97, 111
Eng, S.T. [45] 610, 611, 624
English, J.E. [21] 419, 449
English, J.H. [77] 246, 252, 253, 254, 259, 262; [156] 556, 564; [46] 613, 624
Esaki, L. [16] 266, 281; [21] 19, 56; [74] 73, 113; [75] 73, 113; [115] 110, 114; [86] 139, 141, 142, 166; [43] 231, 261; [63] 236, 261; [43] 364, 375; [46] 364, 375; [37] 472, 487; [12] 492, 541, 560; [113] 540, 563; [114] 541, 563; [115] 541, 551, 563; [129] 541, 564; [155] 556, 564
Esipov, S.E. [103] 533, 536, 563
Esso, P. [88] 144, 166
Etienne, B. [68] 239, 241, 242, 261

Fabre, N. [54] 123, 164
Fahlman, A. [14] 171, 205
Fang, F.F. [75] 136, 144, 165; [35] 181, 206

Fano, V. [50] 75, 112
Fasolino, A. [46] 231, 261; [47] 232, 261; [89] 254, 262; [35] 298, 310
Faurie, J.P. [43] 75, 112; [119] 110, 114; [7] 285, 291, 309; [8] 285, 287, 293, 294, 309; [9] 285, 305, 307, 309; [10] 285, 309; [12] 285, 293, 297, 309; [13] 285, 302, 304, 309; [14] 287, 290, 309; [24] 291, 310; [25] 291, 310; [26] 292, 310; [27] 292, 310; [28] 292, 310; [29] 292, 310; [30] 292, 302, 308, 310; [37] 302, 310; [38] 302, 310; [39] 304, 310; [40] 305, 307, 310; [44] 286, 308, 310; [45] 308, 309, 310; [46] 308, 310; [47] 308, 310; [48] 309, 310
Fazzio, A. [25] 20, 26, 54, 57; [8] 61, 76, 77, 85, 98, 111
Fedorus, G.A. [84] 75, 113
Fehrenbach, G.W. [50] 614, 617, 624; [51] 614, 617, 624
Ferry, D.K. [13] 216, 252, 253, 260; [41] 229, 231, 260; [53] 485, 487; [57] 503, 562; [105] 535, 563
Feucht, D.L. [1] 61, 65, 76, 81, 100, 101, 111; [1] 117, 118, 120, 163; [1] 317, 318, 322, 373
Feuerbacher, B. [65] 125, 165
Fischbach, J.U. [54] 233, 261
Fischer, A. [52] 443, 450
Fischer, R. [42] 72, 112; [55] 123, 164; [40] 229, 260; [18] 266, 268, 271, 272, 276, 281; [119] 541, 563
Fishman, G. [84] 247, 262; [27] 461, 487
Fitton, B. [65] 125, 165
Fletcher, E.D. [37] 589, 593
Flores, F. [11] 11, 24, 30, 36, 41, 54, 56; [44] 49, 51, 57; [5] 61, 76, 77, 82, 83, 94, 95, 99, 103, 111; [103] 83, 92, 114; [122] 110, 114
Fork, R.L. [56] 616, 617, 624
Forrest, S.R. [57] 74, 81, 113; [77] 74, 113; [78] 74, 113; [3] 320, 321, 322, 364, 367, 373; [9] 326, 349, 353, 373; [17] 339, 343, 350, 351, 352, 358, 366, 373; [21] 341, 343, 344, 374; [22] 341, 374; [10] 380, 383, 385, 394, 395, 396; [11] 380, 385, 394, 395, 396; [5] 403, 449; [48] 440, 450; [12] 569, 592

Fortunato, C.   [112] 108, 114
Fowler, A.B.   [1] 265, 272, 280; [61] 504, 526, 562; [59] 618, 624
Foxon, C.T.   [100] 150, 166; [49] 233, 261; [96] 219, 259, 262
Foy, P.W.   [33] 28, 57; [98] 108, 114; [19] 387, 396; [47] 439, 450; [50] 442, 450
Frankel, D.   [32] 121, 164
Franz, W.   [4] 598, 622
Freeman, A.J.   [118] 110, 114
Freeman, J.   [89] 513, 563
Freeouf, J.L.   [10] 11, 24, 26, 27, 51, 54, 56; [6] 61, 76, 77, 81, 84, 100, 101, 111
Frensley, W.R.   [8] 11, 23, 24, 26, 27, 56; [14] 61, 76, 77, 85, 90, 97, 111; [14] 120, 132, 133, 140, 163; [2] 169, 201, 202, 205; [28] 343, 374; [157] 556, 564
Frey, J.   [106] 535, 563; [108] 536, 563
Frijlink, P.M.   [64] 237, 243, 261; [70] 239, 261
Fritz, I.J.   [15] 265, 279, 281; [17] 266, 281
Fritze, M.   [55] 502, 562
Frohman-Bentchkowsky, D.   [20] 494, 561
Fry, K.   [28] 268, 269, 272, 281
Fujii, T.   [4] 453, 464, 478, 486; [5] 492, 560
Fuller, B.K.   [45] 590, 593
Fulton, R.C.   [8] 404, 406, 407, 449

Gant, H.   [47] 73, 112; [70] 73, 113; [29] 121, 124, 149, 155, 164; [46] 122, 136, 144, 156, 164; [101] 153, 166
Gates, J.V.   [22] 341, 374
Gavrilovic, P.   [4] 379, 381, 396; [128] 541, 564
Gelius, U.   [27] 178, 206
Gell, M.A.   [19] 216, 260; [94] 258, 262; [21] 266, 279, 281; [22] 266, 268, 279, 281; [37] 273, 276, 279, 282
Geppert, D.V.   [26] 496, 561
Gering, J.M.   [120] 541, 563; [148] 550, 564
Gibbons, G.   [18] 340, 373
Gibbs, H.M.   [60] 620, 621, 624; [62] 621, 624; [63] 621, 624
Gibson, J.M.   [28] 497, 500, 561; [31] 500, 561
Gill, A.   [40] 500, 561
Glembocki, O.J.   [53] 72, 112

Gnauck, A.H.   [45] 610, 611, 624
Goddard, W.A.   [21] 120, 163
Goedeke, S.   [148] 550, 564
Goldman, V.J.   [140] 548, 564
Gonda, S.   [52] 72, 112; [62] 124, 165; [23] 342, 343, 344, 374
Goodhue, W.D.   [38] 472, 487; [116] 541, 544, 563; [117] 541, 550, 563; [118] 541, 547, 563; [146] 550, 564; [147] 550, 564
Gorbik, P.P.   [84] 75, 113
Gossard, A.C.   [61] 124, 165; [48] 201, 206; [21] 217, 260; [23] 219, 220, 222, 229, 233, 252, 257, 260; [24] 220, 231, 233, 260; [28] 222, 260; [29] 219, 223, 224, 227, 228, 229, 231, 233, 252, 254, 257, 259, 260; [77] 246, 252, 253, 254, 259, 262; [79] 246, 250, 251, 252, 262; [81] 246, 252, 262; [5] 265, 279, 280; [6] 265, 279, 281; [27] 342, 362, 374; [40] 364, 375; [5] 403, 445, 449; [11] 410, 411, 413, 449; [17] 415, 417, 449; [21] 419, 449; [26] 425, 426, 449; [27] 425, 450; [44] 438, 450; [4] 492, 560; [63] 504, 514, 515, 516, 517, 518, 538, 562; [65] 505, 562; [75] 508, 562; [92] 515, 518, 519, 563; [93] 515, 537, 563; [94] 515, 522, 523, 563; [98] 515, 518, 520, 521, 522, 563; [100] 563; [154] 556, 564; [156] 556, 564; [23] 458, 487; [26] 460, 487; [17] 601, 623; [18] 601, 623; [20] 601, 623; [21] 601, 623; [22] 601, 623; [24] 602, 623; [27] 602, 623; [29] 602, 604, 606, 607, 609, 623; [33] 606, 611, 612, 623; [34] 606, 607, 623; [35] 609, 611, 623; [36] 609, 611, 623; [37] 609, 611, 612, 623; [38] 610, 623; [39] 610, 624; [40] 610, 624; [46] 613, 624; [47] 613, 614, 615, 624; [48] 613, 615, 624; [53] 615, 624; [56] 616, 617, 624; [61] 621, 624; [62] 621, 624; [63] 621, 624; [31] 68, 69, 112; [34] 69, 73, 112; [44] 72, 112; [45] 72, 112
Gourley, P.L.   [17] 266, 281
Gowers, J.P.   [91] 145, 149, 166
Grant, R.W.   [20] 19, 24, 56; [41] 73, 112; [49] 75, 112; [58] 74, 113; [66] 73, 113; [69] 72, 78, 113; [71] 73, 113; [81] 73, 113; [86] 72, 75, 81, 113; [89] 77, 114;

[102] 75, 114; [25] 121, 124, 138, 139, 163; [26] 121, 124, 136, 144, 163; [68] 126, 128, 131, 165; [73] 136, 137, 139, 141, 142, 144, 165; [77] 137, 165; [79] 137, 165; [82] 138, 165; [85] 141, 142, 165; [103] 155, 166; [10] 170, 173, 181, 182, 205; [11] 170, 173, 181, 183, 205; [12] 170, 177, 205; [13] 170, 177, 178, 179, 180, 205; [22] 177, 206; [30] 180, 190, 193, 203, 206; [31] 180, 203, 206; [32] 180, 206; [33] 180, 200, 203, 206; [36] 181, 206; [37] 185, 186, 187, 188, 189, 206; [40] 190, 193, 194, 195, 196, 199, 206; [42] 191, 193, 203, 206; [43] 191, 192, 193, 203, 206; [46] 194, 195, 206; [47] 194, 197, 198, 206; [43] 286, 307, 308, 310

Gratos, H.C. [22] 460, 486
Greene, R.L. [33] 225, 231, 232, 233, 260; [38] 274, 277, 282
Grinberg, A.A. [97] 515, 525, 526, 528, 529, 530, 532, 533, 534, 535, 563
Grondin, R.O. [53] 485, 487
Groves, S.H. [32] 296, 310
Grunthaner, F.J. [97] 106, 114; [113] 108, 114; [39] 121, 124, 153, 164
Grynberg, M. [32] 296, 310; [33] 296, 310
Gubanov, A.I. [6] 118, 163
Guldner, Y. [43] 75, 112; [8] 285, 287, 293, 294, 309; [12] 285, 293, 297, 309; [13] 285, 302, 304, 309; [15] 287, 289, 309; [23] 291, .310; [31] 293, 310; [33] 296, 310; [36] 301, 310
Gulino, D.A. [33] 268, 281
Gummel, H.K. [29] 497, 498, 561

Halperin, B.I. [38] 38, 57
Hamann, D.R. [1] 11, 12, 14, 56; [17] 14, 56; [37] 38, 57; [40] 43, 57; [19] 120, 163; [20] 174, 206
Hamm, R.A. [125] 110, 114; [6] 379, 380, 385, 386, 387, 390, 396; [15] 380, 396; [13] 570, 593; [26] 602, 623
Hamrin, K. [14] 171, 205
Hansson, G.V. [89] 144, 145, 166
Harbison, J.P. [20] 387, 388, 396; [36] 500, 561; [55] 502, 562
Hark, S.K. [32] 268, 271, 272, 276, 281
Harmand, J.C. [22] 423, 424, 449

Harris, J.J. [43] 501, 502, 561
Harris, J.S. [25] 65, 73, 111; [58] 124, 165; [14] 334, 336, 337, 339, 343, 348, 372, 373; [7] 379, 394, 395, 396
Harrison, W.A. [9] 11, 13, 23, 24, 25, 27, 47, 54, 56; [20] 19, 24, 56; [27] 21, 57; [28] 25, 36, 54, 57; [9] 61, 76, 77, 85, 87, 88, 89, 95, 97, 111; [10] 61, 76, 77, 85, 87, 88, 89, 95, 97, 111; [15] 120, 132, 133, 163; [79] 137, 165; [1] 169, 201, 205; [37] 185, 186, 187, 188, 189, 206; [38] 190, 198, 206; [5] 285, 307, 309
Hartman, R.L. [53] 123, 164; [41] 589, 593
Hase, I. [49] 501, 561
Haug, H. [49] 614, 618, 624
Haus, H.A. [27] 602, 623
Hauser, J.R. [31] 461, 487
Hayashi, I. [51] 123, 164
Hayes, J.R. [11] 410, 411, 413, 449; [15] 413, 449; [17] 415, 417, 449; [44] 475, 476, 484, 487; [59] 504, 562; [75] 508, 562; [80] 510, 562; [81] 510, 511, 562; [82] 510, 562; [83] 510, 562; [84] 510, 562
Hecht, M.H. [113] 108, 114
Hedman, J. [14] 171, 205; [16] 171, 205
Hegarty, J. [19] 601, 623
Heiblum, M. [30] 68, 72, 81, 111; [111] 114; [60] 124, 165; [92] 256, 262; [11] 265, 279, 281; [25] 460, 487; [24] 496, 502, 561; [46] 501, 510, 511, 561; [47] 501, 510, 511, 512, 561; [51] 502, 503, 510, 561; [52] 502, 510, 561
Heim, U. [26] 266, 281
Heimann, P. [29] 180, 206
Heime, K. [13] 380, 396
Heine, V. [92] 92, 114
Heinrich, H. [26] 20, 54, 57; [7] 61, 76, 77, 85, 86, 87, 111
Heinrich, O. [13] 61, 76, 77, 88, 89, 97, 111
Heisinger, P. [26] 266, 281
Hendel, R.H. [63] 504, 514, 515, 516, 517, 518, 538, 562; [92] 515, 518, 519, 563; [93] 515, 537, 563; [94] 515, 522, 523, 563
Henderson, T. [40] 72, 112; [56] 73, 112; [39] 361, 364, 365, 374; [123] 541, 542, 546, 547, 556, 564

Henry, C.H.   [31] 68, 69, 112; [12] 119, 124, 163; [2] 213, 217, 218, 219, 220, 234, 246, 250, 257, 259; [1] 567, 592
Henry, J.E.   [46] 613, 624
Hensel, J.C.   [28] 497, 500, 561
Henzler, M.   [35] 72, 112
Heritage, J.P.   [52] 614, 624
Hersee, S.D.   [25] 266, 281; [6] 567, 589, 592; [40] 589, 593
Herzog, H.J.   [71] 507, 562
Hess, K.   [20] 217, 260; [54] 233, 261; [33] 268, 281; [4] 379, 381, 396; [37] 432, 450; [39] 433, 450; [1] 453, 456, 486; [2] 453, 457, 458, 486; [3] 453, 457, 486; [4] 453, 464, 478, 486; [5] 453, 486; [6] 453, 462, 467, 483, 484, 486; [7] 453, 462, 483, 486; [13] 454, 455, 457, 459, 460, 462, 486; [16] 457, 458, 459, 460, 486; [28] 461, 487; [29] 461, 487; [31] 461, 487; [32] 468, 487; [33] 462, 463, 487; [35] 465, 487; [36] 465, 466, 487; [42] 474, 487; [45] 476, 487; [47] 479, 480, 487; [7] 492, 560; [9] 492, 513, 560; [85] 510, 562; [88] 513, 562; [89] 513, 563; [90] 513, 514, 563; [91] 513, 563; [107] 536, 563; [128] 541, 564; [3] 567, 580, 589, 592; [17] 574, 593
Hesto, P.   [77] 510, 562
Hetherington, C.J.D.   [131] 541, 564
Hetzler, S.   [10] 285, 309
Hickmott, T.W.   [42] 72, 112; [40] 229, 260
Himpsel, F.J.   [29] 180, 206
Hirlimann, C.   [57] 616, 617, 618, 624
Hirtz, J.P.   [68] 239, 241, 242, 261
Hiyamizu, S.   [20] 419, 449; [24] 460, 487; [5] 492, 560; [44] 501, 502, 561; [45] 501, 502, 510, 561; [48] 501, 561; [151] 553, 564
Hjalmarson, H.P.   [11] 61, 76, 77, 88, 89, 95, 97, 101, 111
Hörnfeldt, O.   [15] 171, 205
Hojo, A.   [55] 72, 112; [83] 138, 142, 165; [14] 265, 279, 281
Holden, W.S.   [49] 441, 450
Hollis, M.A.   [41] 501, 561; [42] 501, 561
Holonyak Jr, N.   [59] 74, 113; [85] 74, 113; [20] 217, 260; [32] 225, 260; [66] 237, 261; [67] 238, 261; [33] 268, 281; [4] 379, 381, 396; [14] 380, 396; [37] 432, 450; [7] 492, 560; [128] 541, 564; [137] 546, 564; [4] 453, 464, 478, 486; [5] 453, 486; [41] 474, 475, 487; [42] 474, 487; [2] 567, 580, 589, 592; [3] 567, 580, 589, 592; [17] 574, 593; [22] 580, 593; [44] 590, 593; [45] 590, 593

Honeisen, B.   [68] 505, 562
Hong, C.S.   [39] 589, 593
Horning, R.D.   [30] 292, 302, 308, 310
Hosack, H.H.   [111] 540, 543, 563
Howard, R.E.   [13] 598, 623
Howard, W.E.   [75] 136, 144, 165;[35] 181, 206; [10] 454, 455, 461, 486
Hsieh, S.J.   [87] 75, 113; [88] 74, 81, 113
Hsu, C.   [119] 110, 114; [14] 287, 290, 309; [26] 292, 310; [27] 292, 310; [44] 286, 308, 310
Hull, R.   [23] 424, 449; [24] 424, 449
Hulyer, P.J.   [37] 589, 593
Hummel, S.G.   [45] 438, 450
Humphreys, R.G.   [39] 276, 282; [131] 541, 564
Hunter, A.T.   [10] 285, 309
Hutchby, J.A.   [6] 403, 404, 449
Hutchinson, A.L.   [18] 385, 390, 396; [9] 407, 410, 449; [21] 419, 449; [23] 424, 449; [24] 424, 449; [36] 431, 450; [38] 433, 434, 450; [40] 434, 450; [45] 438, 450
Hwang, D.M.   [36] 500, 561

Iafrate, G.J.   [3] 453, 457, 486; [32] 468, 487; [35] 465, 487; [45] 476, 487; [50] 484, 487; [52] 485, 487; [53] 485, 487; [85] 510, 562
Ihm, J.   [2] 11, 12, 13, 14, 15, 53, 56
Imamura, K.   [20] 419, 449; [44] 501, 502, 561; [45] 501, 502, 510, 561; [48] 501, 561; [151] 553, 564
Imanaga, S.   [49] 501, 561; [58] 503, 562
Inada, M.   [80] 137, 165
Inkson, J.C.   [14] 216, 260
Inoue, M.   [108] 536, 563
Inuzuka, H.   [43] 438, 450
Iogansen, L.V.   [112] 540, 543, 563
Irvine, S.J.C.   [12] 598, 622
Ishibashi, T.   [14] 413, 449

Ishikawa, T. [24] 460, 487
Ishiko, M. [120] 110, 114
Ito, H. [14] 413, 449
Iwamura, H. [25] 581, 582, 593
Iwasa, Y. [45] 231, 261

Jackel, L.D. [13] 598, 623
Jaboni, C. [104] 535, 563
Jain, A.K. [31] 224, 225, 233, 260
Jame, M.S. [62] 505, 562
James, L.W. [30] 344, 374
Jaros, M. [19] 216, 260; [94] 258, 262; [20] 266, 268, 279, 281; [21] 266, 279, 281; [22] 266, 268, 279, 281; [37] 273, 276, 279, 282
Jayaraman, A. [30] 268, 281; [53] 502, 562
Jerhot, J. [5] 117, 118, 120, 163
Jerphagnon, J. [3] 598, 622
Jewell, J.L. [62] 621, 624; [63] 621, 624
Jogai, B. [149] 550, 564; [152] 554, 564
Johansson, G. [14] 171, 205
Johnson, N.M. [113] 108, 114
Johnson, W.C. [15] 334, 373
Jones, C.E. [38] 302, 310
Jones, W.T. [35] 465, 487
Jorke, H. [71] 507, 562
Joyce, B.A. [91] 145, 149, 166; [93] 145, 166; [100] 150, 166; [34] 228, 260

Kahn, A. [41] 190, 206; [49] 201, 206
Kahng, D. [25] 496, 561
Kaliski, R.W. [128] 541, 564
Kallin, C. [38] 38, 57
Kanani, D. [49] 201, 206
Kane, E.O. [106] 94, 98, 114; [107] 93, 94, 114; [9] 214, 260
Kane, W.M. [23] 496, 561
Kaneko, J. [69] 239, 261; [71] 240, 261
Kaneko, K. [49] 501, 561; [58] 503, 562
Kanerva, H.K.J. [5] 322, 330, 331, 364, 373
Kanski, J. [104] 155, 166
Kaplan, M.L. [57] 74, 81, 113; [9] 326, 349, 353, 373; [21] 341, 343, 344, 374; [22] 341, 374; [11] 380, 385, 394, 395, 396; [12] 569, 592
Karlsson, S.-E. [14] 171, 205
Kasemset, D. [39] 589, 593

Kasper, B.L. [49] 441, 450
Kasper, E. [71] 507, 562
Kastalsky, A. [49] 483, 487; [10] 492, 515, 518, 560; [11] 492, 560; [63] 504, 514, 515, 516, 517, 538, 562; [64] 504, 515, 516, 517, 518, 519, 521, 562; [92] 515, 518, 519, 563; [93] 515, 537, 563; [94] 515, 522, 523, 563; [95] 515, 517, 522, 523, 525, 533, 535, 563; [96] 515, 517, 538, 563; [97] 515, 525, 526, 528, 529, 530, 532, 533, 534, 535, 563; [98] 515, 518, 520, 521, 522, 563; [99] 515, 516, 517, 518, 532, 537, 563; [100] 563; [102] 525, 563
Katnani, A.D. [22] 19, 56; [23] 19, 20, 54, 56; [31] 28, 51, 57; [29] 67, 68, 72, 73, 74, 75, 77, 78, 85, 103, 111; [46] 73, 112; [64] 73, 113; [96] 103, 114; [101] 72, 73, 78, 79, 114
Kato, Y. [46] 590, 593; [27] 121, 122, 124, 126, 127, 128, 134, 136, 137, 138, 139, 140, 147, 153, 155, 164; [28] 121, 124, 139, 145, 147, 148, 150, 151, 155, 164; [34] 121, 124, 134, 164; [35] 121, 124, 154, 164; [36] 121, 124, 164; [36] 121, 124, 164; [37] 121, 124, 136, 153, 164; [42] 121, 164; [47] 121, 122, 164; [72] 136, 165; [76] 136, 145, 165; [81] 137, 138, 139, 141, 142, 165; [92] 145, 166; [102] 154, 166; [105] 157, 166; [106] 159, 166; [19] 174, 205; [24] 177, 206; [49] 201, 206
Kaufmann, L.M.F. [13] 380, 396
Kawabe, M. [43] 438, 450
Kawai, H. [69] 239, 261; [71] 240, 261; [49] 501, 561; [58] 503, 562
Kawamura, Y. [41] 610, 624; [42] 610, 624
Kazarinov, R.F. [13] 493, 542, 551, 552, 553, 560; [14] 493, 542, 560; [67] 505, 562; [72] 508, 562
Kazmierski, K. [25] 343, 374
Keever, M. [7] 453, 462, 483, 486; [88] 513, 562; [89] 513, 563
Keldysh, L.V. [109] 539, 563; [110] 539, 563; [5] 598, 622
Kelly, M.K. [35] 72, 112; [37] 75, 112; [86] 510, 562; [87] 510, 562; [130] 541, 551, 564; [131] 541, 564; [132] 541, 551, 564

Keramidas, U.G. [94] 81, 114
Kerr, T.M. [86] 510, 562; [130] 541, 551, 564; [131] 541, 564; [132] 541, 551, 564
Ketterson, A. [40] 72, 112; [56] 73, 112; [39] 361, 364, 365, 374
Keuch, T.F. [94] 258, 262
Kiehl, R.A. [13] 411, 449; [19] 419, 423, 449; [76] 508, 562; [93] 515, 537, 563; [102] 525, 563; [134] 543, 553, 564
Kim, O.K. [77] 74, 113; [78] 74, 113; [57] 234, 261; [3] 320, 321, 322, 364, 367, 373; [17] 339, 343, 350, 351, 352, 358, 366, 373; [10] 380, 383, 385, 394, 395, 396; [48] 440, 450
King, H.E. [28] 268, 269, 272, 281
Kirchoefer, S.W. [85] 74, 113; [33] 268, 281; [44] 590, 593
Kizilyalli, I.C. [47] 479, 480, 487
Klasson, M. [16] 171, 205
Klein, J. [40] 72, 112
Klein, M.V. [55] 123, 164
Kleinman, D.A. [34] 69, 73, 112; [44] 72, 112; [45] 72, 112; [61] 124, 165; [23] 219, 220, 222, 229, 233, 252, 257, 260; [24] 220, 231, 233, 260; [26] 220, 223, 260; [29] 219, 223, 224, 227, 228, 229, 231, 233, 252, 254, 257, 259, 260; [44] 231, 261; [50] 233, 261; [5] 265, 279, 280; [6] 265, 279, 281; [27] 342, 362, 374; [5] 403, 445, 449; [8] 404, 406, 407, 449; [44] 438, 450; [16] 599, 623; [18] 601, 623
Klem, J. [56] 73, 112; [55] 123, 164; [59] 124, 165; [39] 361, 364, 365, 374
Knoedler, C.M. [46] 501, 510, 511, 512, 561; [47] 501, 510, 511, 561; [51] 502, 503, 510, 561; [52] 502, 510, 561
Knox, R.D. [30] 292, 302, 308, 310
Knox, W.H. [31] 604, 606, 623; [56] 616, 617, 624; [57] 616, 617, 618, 624
Kobayashi, H. [25] 581, 582, 593
Kobayashi, K. [46] 590, 593
Kohn, W. [39] 40, 57
Kolbas, R.M. [85] 74, 113; [65] 237, 238, 261; [66] 237, 261; [41] 474, 475, 487; [127] 541, 564; [2] 567, 580, 589, 592; [3] 567, 580, 589, 592; [22] 580, 593; [44] 590, 593

Komashchenko, V.N. [84] 75, 113
Kondo, K. [44] 501, 502, 561; [45] 501, 502, 510, 561
Kopp, W. [59] 124, 165; [119] 541, 563; [120] 541, 563
Korol'kov, V.I. [44] 366, 375
Kote, G. [11] 285, 309
Kowalczyk, S.P. [41] 73, 112; [49] 75, 112; [58] 74, 113; [66] 73, 113; [69] 72, 78, 113; [71] 73, 113; [81] 73, 113; [86] 72, 75, 81, 113; [102] 75, 114; [68] 126, 128, 131, 165; [73] 136, 137, 139, 141, 142, 144, 165; [77] 137, 165; [82] 138, 165; [85] 141, 142, 165; [103] 155, 166; [12] 170, 177, 205; [13] 170, 177, 178, 179, 180, 205; [22] 177, 206; [30] 180, 190, 193, 203, 206; [31] 180, 203, 206; [32] 180, 206; [36] 181, 206; [40] 190, 193, 194, 195, 196, 199, 206; [43] 191, 192, 193, 203, 206; [46] 194, 195, 206; [43] 286, 307, 308, 310
Koza, M.A. [8] 462, 483, 486; [64] 504, 515, 516, 517, 518, 519, 521, 562; [99] 515, 516, 517, 518, 532, 537, 563
Krakowski, M. [25] 266, 281
Kraut, E.A. [20] 19, 24, 56; [41] 73, 112; [49] 75, 112; [58] 74, 113; [66] 73, 113; [69] 72, 78, 113; [71] 73, 113; [81] 73, 113; [86] 72, 75, 81, 113; [102] 75, 114; [26] 121, 124, 136, 144, 163; [68] 126, 128, 131, 165; [73] 136, 137, 139, 141, 142, 144, 165; [77] 137, 165; [79] 137, 165; [82] 138, 165; [85] 141, 142, 165; [99] 149, 166; [103] 155, 166; [10] 170, 173, 181, 182, 205; [11] 170, 173, 181, 183, 205; [12] 170, 177, 205; [13] 170, 177, 178, 179, 180, 205; [22] 177, 206; [30] 180, 190, 193, 203, 206; [31] 180, 203, 206; [32] 180, 206; [33] 180, 200, 203, 206; [36] 181, 206; [37] 185, 186, 187, 188, 189, 206; [38] 190, 198, 206; [40] 190, 193, 194, 195, 196, 199, 206; [43] 191, 192, 193, 203, 206; [46] 194, 195, 206; [47] 194, 197, 198, 206; [43] 286, 307, 308, 310
Kressel, H. [10] 119, 163
Kroemer, H. [8] 11, 23, 24, 26, 27, 56; [15] 11, 12, 19, 26, 27, 52, 56; [16] 13, 14, 47,

48, 56; [14] 61, 76, 77, 85, 90, 97, 111; [24] 65, 98, 106, 111; [25] 65, 73, 111; [62] 73, 113; [2] 117, 118, 163; [8] 118, 119, 163; [13] 120, 163; [14] 120, 132, 133, 140, 163; [58] 124, 165; [80] 137, 149, 165; [2] 169, 201, 202, 205; [6] 169, 170, 189, 193, 205; [7] 169, 205; [8] 169, 193, 205; [95] 259, 262; [14] 334, 336, 337, 339, 343, 348, 372, 373; [16] 337, 373; [28] 343, 374; [33] 352, 374; [7] 379, 394, 395, 396; [4] 402, 404, 410, 449; [10] 410, 413, 449; [73] 508, 562
Kuan, T.S. [21] 19, 56; [75] 73, 113
Kucherenko, Yu.N. [26] 178, 206
Kuech, T.F. [82] 75, 113; [83] 72, 78, 113; [23] 266, 267, 281; [31] 268, 273, 281; [37] 273, 276, 279, 282
Kukimoto, H. [26] 343, 346, 347, 352, 374; [12] 380, 394, 396
Kunc, K. [20] 19, 24, 56
Kunz, C. [66] 126, 165
Kuphal, E. [13] 380, 396
Kurado, M. [3] 117, 118, 163
Kwark, Y.H. [79] 508, 562

Lagowski, J. [22] 460, 486
Laidig, W.D. [32] 225, 260; [3] 567, 580, 589, 592; [22] 580, 593
Lamont, G. [77] 246, 252, 253, 254, 259, 262
Landau, L.D. [1] 210, 240, 259
Landauer, R. [143] 548, 564
Landgren, G. [16] 266, 281
Landsberg, P.T. [26] 582, 593
Lang, D.V. [125] 110, 114; [8] 324, 373; [32] 352, 374; [1] 379, 381, 396; [2] 379, 381, 396; [5] 379, 381, 382, 383, 389, 396; [6] 379, 380, 385, 386, 387, 390, 396; [19] 387, 396; [20] 387, 388, 396
Lang, R. [46] 590, 593
Langer, J.M. [26] 20, 54, 57; [7] 61, 76, 77, 85, 86, 87, 111
Lannoo, M. [104] 84, 114
Lapeyre, G.J. [32] 121, 164; [95] 145, 148, 149, 166
Larson, P.K. [90] 145, 166; [91] 145, 149, 166; [93] 145, 166
Larsson, A. [44] 610, 624; [45] 610, 611, 624

Lasher, G. [15] 574, 593
Lau, S.S. [83] 72, 78, 113
Lawaetz, P. [51] 233, 261; [19] 288, 310
Le Toullec, R. [32] 296, 310
Le, H.Q. [118] 541, 547, 563; [147] 550, 564
Lederman, F.L. [30] 604, 623
Lee, J.W. [157] 556, 564
Lee, P.A. [141] 548, 564
Lee, R.J. [69] 132, 165
Leonhardt, G. [16] 171, 205
Leotin, J. [31] 224, 225, 233, 260
Levi, A.F.J. [44] 475, 476, 484, 487; [28] 497, 500, 561; [31] 500, 561; [59] 504, 562; [80] 510, 562; [81] 510, 511, 562; [82] 510, 562; [83] 510, 562; [84] 510, 562
Levine, B.F. [7] 404, 410, 449; [8] 404, 406, 407, 449; [9] 407, 410, 449; [28] 426, 450; [14] 598, 623
Levinson, Y.B. [103] 533, 536, 563
Levkoff, J. [36] 500, 561; [55] 502, 562
Levy, F. [38] 75, 112
Ley, L. [65] 125, 165
Lievin, J.L. [32] 69, 73, 112; [25] 219, 220, 230, 260; [22] 423, 424, 449
Lifshitz, E.M. [1] 210, 240, 259
Lifshitz, N. [53] 502, 562
Lilja, J. [120] 110, 114
Lin Liu, Y.R. [16] 287, 289, 309
Lin, B.J.F. [26] 460, 487
Lin-Chung, P.J. [17] 216, 260
Linb, N.T. [94] 145, 166
Lindau, I. [45] 51, 57; [121] 110, 114; [40] 121, 122, 127, 128, 147, 153, 160, 164; [67] 126, 165
Lindberg, B. [14] 171, 205
Lindgren, I. [14] 171, 205
Lindmayer, J. [34] 500, 561; [35] 500, 561
Lindstrom, C. [42] 589, 590, 593
Lipari, N.O. [52] 233, 261; [54] 233, 261
List, R.S. [39] 72, 78, 112; [121] 110, 114
Littlejohn, M.A. [31] 461, 487
Liu, H.C. [123] 110, 114; [145] 550, 564
Liu, W.C. [62] 505, 562
Liu, Y.L. [80] 73, 113
Logan, R.A. [27] 267, 281; [53] 502, 562; [154] 556, 564; [9] 567, 592; [10] 567, 592; [38] 589, 593

Long, A.P. [86] 510, 562; [87] 510, 562
Look, D.C. [59] 124, 165
Losee, D.L. [3] 379, 385, 387, 396
Louie, S.G. [2] 11, 12, 13, 14, 15, 53, 56; [20] 61, 76, 105, 111; [18] 216, 260
Louis, E. [44] 49, 51, 57; [103] 83, 92, 114
Louis, S.G. [20] 120, 163
Ludeke, R. [86] 139, 141, 142, 166; [16] 266, 281; [43] 364, 375
Ludowise, M. [7] 453, 462, 483, 486
Lugli, P. [57] 503, 562
Luryi, S. [28] 426, 450; [39] 472, 487; [49] 483, 487; [10] 492, 515, 518, 560; [11] 492, 560; [17] 493, 554, 556, 557, 558, 560; [55] 502, 562; [60] 501, 504, 505, 562; [62] 505, 562; [63] 504, 514, 515, 516, 517, 518, 538, 562; [67] 505, 562; [69] 505, 507, 562; [72] 508, 562; [92] 515, 518, 519, 563; [93] 515, 537, 563; [94] 515, 522, 523, 563; [95] 515, 517, 522, 523, 525, 533, 535, 563; [96] 515, 517, 538, 563; [97] 515, 525, 526, 528, 529, 530, 532, 533, 534, 535, 563; [98] 515, 518, 520, 521, 522, 563; [100] 563; [123] 541, 542, 546, 547, 556, 564; [135] 543, 544, 545, 549, 564; [138] 546, 547, 564
Luttinger, J.M. [34] 298, 310

Maan, J.C. [115] 110, 114; [46] 231, 261
Maenpaa, M. [83] 72, 78, 113
Mahowald, P.H. [39] 72, 78, 112; [121] 110, 114; [105] 157, 166
Mailhiot, C. [4] 61, 76, 82, 111
Malik, R.J. [11] 410, 411, 413, 449; [13] 411, 449; [15] 413, 449; [17] 415, 417, 449; [38] 433, 434, 450; [41] 501, 561; [66] 505, 562; [75] 508, 562; [76] 508, 562
Maloney, T.J. [56] 503, 562; [106] 535, 563
Maluenda, J. [70] 239, 261
Manghi, F. [30] 121, 124, 164; [37] 121, 124, 136, 153, 164
Mankiewich, P.M. [13] 598, 623
Margalit, S. [12] 61, 76, 77, 88, 89, 111
Margaritondo, G. [23] 19, 20, 54, 56; [24] 20, 45, 56; [43] 45, 53, 57; [3] 61, 66, 67, 111; [28] 67, 75, 107, 111; [29] 67, 68, 72, 73, 74, 75, 77, 78, 85, 103, 111; [35] 72, 112; [36] 74, 112; [37] 75, 112; [38] 75, 112; [91] 92, 114; [96] 103, 114; [97] 106, 114; [99] 108, 109, 114; [105] 84, 98, 100, 114; [109] 102, 114; [112] 108, 114; [27] 121, 122, 124, 126, 127, 128, 134, 136, 137, 138, 139, 140, 147, 153, 155, 164; [34] 121, 124, 134, 164; [35] 121, 124, 154, 164; [36] 121, 124, 164; [36] 121, 124, 164; [37] 121, 124, 136, 153, 164; [39] 121, 124, 153, 164; [42] 121, 164; [72] 136, 165; [74] 136, 165; [87] 139, 143, 166; [88] 144, 166; [106] 159, 166; [5] 169, 205; [9] 169, 205; [19] 174, 205; [24] 177, 206; [49] 201, 206
Maria, D.J. [39] 229, 260
Marsh, A.C. [14] 216, 260
Marsh, J.H. [73] 242, 261
Marshak, A.M. [70] 132, 165
Martin, P.A. [4] 379, 381, 396
Martin, R.M. [3] 11, 12, 13, 14, 15, 19, 53, 56; [4] 11, 12, 13, 14, 15, 19, 47, 53, 56; [5] 11, 13, 15, 25, 47, 54, 56; [20] 19, 24, 56; [34] 33, 57
Maruani, A. [2] 598, 622
Marzin, J.Y. [62] 235, 236, 261
Maserjian, J. [45] 610, 611, 624
Mashita, M. [55] 72, 112; [83] 138, 142, 165; [14] 265, 279, 281
Masselink, W.T. [18] 266, 268, 271, 272, 276, 281
Massies, J. [94] 145, 166
Matsuuza, N. [43] 438, 450
McAfee, S.R. [32] 352, 374
McCaldin, J.O. [82] 75, 113; [4] 285, 307, 309; [29] 343, 374
McCall, S.L. [62] 621, 624
McGill, T.C. [32] 28, 51, 57; [49] 121, 122, 139, 156, 164; [19] 266, 281; [1] 285, 290, 304, 309; [2] 285, 309; [4] 285, 307, 309; [10] 285, 309; [20] 290, 310; [29] 343, 374; [40] 473, 487; [18] 493, 554, 555, 556, 560; [126] 541, 564; [142] 548, 564; [150] 553, 564
McGill, T.G. [21] 120, 163
McIntyre, R.J. [35] 429, 450
McMenamin, J.C. [72] 73, 113; [24] 121, 163

Mead, C.A. [4] 285, 307, 309; [29] 343, 374; [21] 496, 540, 561; [22] 496, 540, 561; [68] 505, 562; [142] 548, 564
Mears, A.L. [27] 222, 260
Meehan, K. [4] 379, 381, 396
Mendez, E.E. [21] 19, 56; [26] 66, 72, 111; [74] 73, 113; [75] 73, 113; [43] 231, 261; [82] 247, 262; [8] 265, 279, 281; [16] 266, 281; [25] 460, 487; [129] 541, 564
Menendez, J. [77] 246, 252, 253, 254, 259, 262; [42] 306, 310
Merz, J.L. [62] 73, 113
Meseguer, F. [74] 73, 113
Meuller, R.S. [5] 117, 118, 120, 163
Meyer, R.J. [49] 201, 206
Meynadier, N.H. [32] 69, 73, 112; [25] 219, 220, 230, 260; [64] 237, 243, 261
Mikkelsen, J.C. [95] 145, 148, 149, 166
Milano, R.A. [80] 73, 113
Miller, A. [1] 598, 622
Miller, D.A.B. [60] 74, 113; [56] 233, 234, 261; [1] 598, 622; [15] 599, 600, 602, 613, 623; [24] 602, 623; [25] 602, 623; [27] 602, 623; [28] 602, 608, 612, 623; [29] 602, 604, 606, 607, 609, 623; [31] 604, 606, 623; [32] 609, 623; [33] 606, 611, 612, 623; [34] 606, 607, 623; [35] 609, 611, 623; [36] 609, 611, 623; [37] 609, 611, 612, 623; [38] 610, 623; [39] 610, 624; [40] 610, 624; [45] 610, 611, 624; [46] 613, 624; [47] 613, 614, 615, 624; [48] 613, 615, 624; [53] 615, 624; [54] 616, 621, 624; [56] 616, 617, 624; [57] 616, 617, 618, 624; [58] 618, 621, 624; [61] 621, 624
Miller, D.L. [69] 72, 78, 113; [12] 410, 413, 449; [74] 508, 562
Miller, R.C. [34] 69, 73, 112; [44] 72, 112; [45] 72, 112; [61] 124, 165; [21] 217, 260; [23] 219, 220, 222, 229, 233, 252, 257, 260; [24] 220, 231, 233, 260; [26] 220, 223, 260; [29] 219, 223, 224, 227, 228, 229, 231, 233, 252, 254, 257, 259, 260; [44] 231, 261; [74] 243, 244, 245, 246, 252, 258, 259, 261; [5] 265, 279, 280; [6] 265, 279, 281; [24] 266, 274, 277, 281; [27] 342, 362, 374; [5] 403, 445, 449; [44] 438, 450; [24] 581, 593; [16] 599, 623; [17] 601, 623; [18] 601, 623

Million, A. [43] 75, 112; [7] 285, 291, 309; [8] 285, 287, 293, 294, 309; [24] 291, 310
Mills, K.A. [73] 73, 113; [23] 121, 163; [33] 121, 164
Milnes, A.G. [1] 61, 65, 76, 81, 100, 101, 111; [1] 117, 118, 120, 163; [4] 117, 118, 119, 163; [10] 326, 367, 369, 370, 373; [11] 327, 328, 373; [35] 358, 359, 374; [45] 368, 375
Mimura, T. [4] 453, 464, 478, 486; [5] 492, 560
Minot, G. [22] 423, 424, 449
Misawa, S. [52] 72, 112; [62] 124, 165; [23] 342, 343, 344, 374; [31] 350, 352, 374
Mito, I. [46] 590, 593
Miura, N. [45] 231, 261
Miyazawa, T. [43] 610, 624
Mizuta, M. [26] 343, 346, 347, 352, 374; [12] 380, 394, 396
Mohammed, K. [33] 28, 57; [98] 108, 114; [93] 257, 262; [23] 424, 449; [24] 424, 449; [38] 433, 434, 450; [50] 442, 450; [51] 442, 450; [15] 493, 541, 543, 552, 553, 560; [16] 493, 541, 542, 548, 560
Mönch, W. [47] 73, 112; [70] 73, 113; [29] 121, 124, 149, 155, 164; [46] 122, 136, 144, 156, 164; [101] 153, 166
Montegomery, V. [41] 121, 147, 164
Moon, R.L. [30] 344, 374
Moore, K.J. [36] 229, 260; [37] 219, 229, 230, 252, 257, 260; [96] 219, 259, 262
Mori, S. [19] 457, 458, 460, 486
Morkoç, H. [40] 72, 112; [42] 72, 112; [56] 73, 112; [55] 123, 164; [59] 124, 165; [40] 229, 260; [18] 266, 268, 271, 272, 276, 281; [20] 341, 373; [39] 361, 364, 365, 374; [6] 492, 504, 560; [9] 492, 513, 560; [50] 501, 561; [88] 513, 562; [89] 513, 563; [119] 541, 563; [120] 541, 563; [123] 541, 542, 546, 547, 556, 564; [148] 550, 564
Moss, R.H. [12] 598, 622
Mrstik, B. [97] 149, 166
Mullin, J.B. [12] 598, 622
Muñoz, A. [122] 110, 114
Munoz Yague, A. [54] 123, 164
Munteanu, O. [45] 72, 112; [23] 219, 220, 222, 229, 233, 252, 257, 260; [44] 231, 261; [6] 265, 279, 281; [5] 403, 445, 449

Murphy, R.A. [33] 500, 561
Murschall, R. [47] 73, 112; [46] 122, 136, 144, 156, 164
Muto, S. [20] 419, 449; [44] 501, 502, 561; [45] 501, 502, 510, 561; [48] 501, 561; [151] 553, 564
Mycielski, A. [33] 296, 310

Nagato, K. [14] 413, 449
Nagle, J. [55] 233, 235, 261
Naik, S.S. [13] 329, 373
Nakajima, O. [14] 413, 449
Nakanishi, T. [55] 72, 112; [83] 138, 142, 165; [14] 265, 279, 281
Nambu, K. [4] 453, 464, 478, 486; [24] 460, 487; [5] 492, 560
Nannarone, S. [30] 121, 124, 164
Napholtz, S.G. [7] 567, 590, 592; [8] 567, 590, 591, 592
Nash, F.R. [41] 589, 593
Nathan, M.I. [30] 68, 72, 81, 111; [111] 114; [60] 124, 165; [92] 256, 262; [11] 265, 279, 281; [46] 501, 510, 511, 561; [47] 501, 510, 511, 512, 561
Neave, J.H. [91] 145, 149, 166; [93] 145, 166; [34] 228, 260
Nelson, H. [10] 119, 163; [50] 121, 122, 164
Nelson, P.O. [104] 155, 166
Nelson, R.J. [27] 582, 593; [28] 582, 593; [49] 592, 593
Neugroschel, A. [29] 461, 487
Nicholas, R.J. [49] 233, 261
Nicolet, M.-A. [83] 72, 78, 113
Nicollian, E.H. [19] 387, 396
Nikitin, V.G. [44] 366, 375
Niles, D.W. [35] 72, 112; [99] 108, 109, 114; [105] 84, 98, 100, 114
Nilsson, R. [16] 171, 205
Ning, T. [30] 461, 487
Ninno, D. [94] 258, 262; [37] 273, 276, 279, 282
Nishi, H. [20] 419, 449; [44] 501, 502, 561; [45] 501, 502, 510, 561; [48] 501, 561; [54] 502, 562; [151] 553, 564
Nishi, K. [46] 590, 593
Nordberg, R. [14] 171, 205
Nordland, W.A. [18] 601, 623
Nordling, C. [14] 171, 205; [16] 171, 205

Norris, G.B. [59] 124, 165
Northrop, J.E. [98] 149, 166
Norton, N. [34] 228, 260
Nussbaum, A. [17] 61, 76, 77, 94, 95, 97, 111; [18] 61, 76, 77, 94, 111; [124] 110, 114; [16] 120, 132, 163

Ogura, M. [26] 343, 346, 347, 352, 374; [12] 380, 394, 396
Ohmori, Y. [43] 610, 624
Ohnishi, H. [54] 502, 562
Ohshima, T. [44] 501, 502, 561; [45] 501, 502, 510, 561
Okamoto, H. [45] 231, 261; [43] 610, 624; [44] 610, 624
Okumura, H. [52] 72, 112; [62] 124, 165; [23] 342, 343, 344, 374; [31] 350, 352, 374
Oldham, W.G. [4] 117, 118, 119, 163; [10] 326, 367, 369, 370, 373; [13] 329, 373
Olego, D.J. [9] 285, 305, 307, 309; [40] 305, 307, 310
Olivier, J. [23] 177, 206
Olshanetsky, B.Z. [28] 178, 206
Olsson, N.A. [45] 438, 450; [7] 567, 590, 592; [8] 567, 590, 591, 592; [9] 567, 592; [10] 567, 592
Onaka, K. [26] 343, 346, 347, 352, 374; [12] 380, 394, 396
Ong, N.P. [11] 285, 309
Osbourn, G.C. [17] 266, 281
Osterling, L. [52] 502, 510, 561
Otsuka, K. [25] 581, 582, 593
Ousset, J.C. [31] 224, 225, 233, 260
Owen, S.J. [36] 359, 360, 374

Paalanen, M.A. [26] 460, 487
Palmateer, S.C. [42] 501, 561
Palmier, J.F. [22] 423, 424, 449
Panish, M.B. [125] 110, 114; [51] 123, 164; [5] 379, 381, 382, 383, 389, 396; [6] 379, 380, 385, 386, 387, 390, 396; [15] 380, 396; [16] 385, 396; [18] 385, 390, 396; [40] 434, 450; [14] 454, 486; [26] 602, 623; [13] 570, 593; [18] 574, 593
Panousis, P.T. [15] 334, 373
Pantelides, S. [22] 120, 163
Paoli, T.L. [42] 589, 590, 593
Papenhuizen, J.M.P. [12] 328, 367, 373

Parker, C.D. [38] 472, 487; [116] 541, 544, 563; [146] 550, 564
Paslaski, J. [44] 610, 624
Passner, A. [62] 621, 624
Pastori Parravicini, G. [42] 231, 261
Patel, N.B. [39] 589, 593
Patella, F. [28] 67, 75, 107, 111; [36] 74, 112; [97] 106, 114; [30] 121, 124, 164; [37] 121, 124, 136, 153, 164; [72] 136, 165; [74] 136, 165; [88] 144, 166
Paton, A. [49] 201, 206
Patten, E.A. [87] 75, 113; [88] 74, 81, 113
Pearsall, T.P. [21] 394, 396; [47] 592, 593
Pearton, S.J. [156] 556, 564
Peck, D.D. [38] 472, 487; [116] 541, 544, 563; [117] 541, 550, 563
Peng, C.K. [50] 501, 561; [120] 541, 563
People, R. [31] 68, 69, 112; [51] 74, 81, 112; [68] 74, 113; [59] 234, 235, 261; [24] 343, 346, 374; [16] 414, 449
Perfetti, P. [28] 67, 75, 107, 111; [36] 74, 112; [38] 75, 112; [73] 73, 113; [97] 106, 114; [99] 108, 109, 114; [112] 108, 114; [23] 121, 163; [30] 121, 124, 164; [33] 121, 164; [37] 121, 124, 136, 153, 164; [74] 136, 165; [88] 144, 166
Perkowitz, S. [38] 302, 310
Pessa, M. [120] 110, 114
Petroff, P.M. [48] 201, 206; [24] 266, 274, 277, 281; [15] 380, 396; [154] 556, 564; [156] 556, 564; [13] 570, 593; [26] 602, 623
Peyghambarian, N. [60] 620, 621, 624; [63] 621, 624
Pfister, J.C. [37] 500, 561
Phelps, D.E. [33] 225, 231, 232, 233, 260
Phillips, J.C. [11] 119, 141, 142, 163; [44] 193, 206; [6] 322, 373
Phillips, P. [25] 343, 374
Piaguet, J. [7] 285, 291, 309
Pianetta, P. [39] 72, 78, 112; [105] 157, 166
Pickett, W.E. [2] 11, 12, 13, 14, 15, 53, 56; [20] 61, 76, 105, 111; [21] 61, 76, 105, 106, 111; [20] 120, 163; [21] 174, 206; [18] 216, 260
Pidgeon, C.R. [32] 296, 310
Pinczuk, A. [77] 246, 252, 253, 254, 259, 262; [79] 246, 250, 251, 252, 262; [80] 246, 247, 248, 249, 250, 262; [81] 246, 252, 262; [87] 248, 262; [20] 458, 460, 486; [20] 601, 623; [21] 601, 623; [22] 601, 623
Piqueras, J. [54] 123, 164
Platzman, P.M. [82] 510, 562; [83] 510, 562
Ploog, K. [46] 231, 261; [60] 234, 261; [78] 246, 262; [52] 443, 450; [11] 598, 622
Poirier, R. [23] 177, 206
Poisson, M.A. [40] 589, 593
Pollak, F.H. [16] 266, 281
Pollman, J. [23] 61, 76, 105, 111; [22] 120, 163
Ponce, F.A. [40] 473, 487; [126] 541, 564
Poncet, A. [37] 500, 561
Pone, J.-F. [77] 510, 562
Porod, W. [13] 216, 252, 253, 260
Potz, W. [13] 216, 252, 253, 260; [41] 229, 231, 260
Poulain, P. [25] 343, 374
Prechtel, U. [90] 254, 255, 256, 262; [91] 256, 262
Price, P.J. [25] 425, 449; [17] 457, 460, 486; [18] 457, 460, 486; [21] 460, 486; [25] 460, 487; [101] 522, 563
Priester, C. [104] 84, 114
Prober, D.E. [13] 598, 623

Quaresima, C. [28] 67, 75, 107, 111; [36] 74, 112; [38] 75, 112; [97] 106, 114; [99] 108, 109, 114; [112] 108, 114; [30] 121, 124, 164; [37] 121, 124, 136, 153, 164; [74] 136, 165; [88] 144, 166

Raccah, P.M. [9] 285, 305, 307, 309
Ralph, H.I. [33] 69, 73, 112; [22] 219, 224, 226, 227, 228, 229, 231, 233, 257, 260; [35] 229, 260; [37] 219, 229, 230, 252, 257, 260; [48] 232, 233, 261; [7] 265, 279, 281
Ralph, R.J. [96] 219, 259, 262
Ray, S. [127] 541, 564
Razeghi, M. [55] 233, 235, 261; [68] 239, 241, 242, 261
Reddi, V.G.K. [7] 323, 324, 373
Reddy, U.K. [50] 501, 561; [123] 541, 542, 546, 547, 556, 564
Redfield, D. [7] 598, 622

Rediker, R.H. [53] 446, 450
Reed, M.A. [157] 556, 564
Reggiani, L. [104] 535, 563
Rehr, J.J. [39] 40, 57
Reich, R.K. [53] 485, 487
Reinhart, F.K. [48] 201, 206
Reno, J. [13] 285, 302, 304, 309; [14] 287, 290, 309; [25] 291, 310; [28] 292, 310; [30] 292, 302, 308, 310; [39] 304, 310; [46] 308, 310
Repinsky, S.M. [28] 178, 206
Rezek, E.A. [85] 74, 113; [137] 546, 564; [44] 590, 593; [45] 590, 593
Riben, A.R. [1] 317, 318, 322, 373
Ricco, B. [18] 418, 449; [129] 541, 564; [133] 543, 548, 564
Ricker, Th. [71] 507, 562
Rigaux, C. [33] 296, 310
Roberts, J.S. [73] 242, 261
Robinson, A.C. [46] 478, 487
Robson, P.N. [12] 598, 622
Rogers, D.C. [49] 233, 261
Rosencher, E. [27] 497, 500, 561; [37] 500, 561; [38] 500, 561
Ross, R.L. [66] 505, 562
Rossler, U. [39] 276, 282
Rowe, J.E. [3] 61, 66, 67, 111
Ruan, Y.-C. [108] 94, 97, 114
Ruda, H.E. [22] 460, 486
Ruden, P. [127] 541, 564
Ryan, R.W. [76] 508, 562

Sadana, D.K. [83] 72, 78, 113
Sah, C.T. [7] 323, 324, 373; [28] 461, 487; [29] 461, 487
Sai-Halasz, G.A. [86] 139, 141, 142, 166; G.A. [43] 364, 375
Saito, J. [24] 460, 487
Sakaki, H. [43] 364, 375; [51] 484, 487; [121] 541, 563; [122] 541, 563; [124] 541, 547, 564; [125] 541, 547, 564; [153] 556, 564
Sakaki, J. [86] 139, 141, 142, 166
Saku, T. [25] 581, 582, 593
Sanderson, R.T. [114] 109, 114
Sang Jr, H.W. [31] 28, 51, 57; [46] 73, 112; [64] 73, 113; [65] 73, 113; [28] 121, 124, 139, 145, 147, 148, 150, 151, 155, 164;
[47] 121, 122, 164; [56] 123, 164; [76] 136, 145, 165; [84] 137, 140, 165; [92] 145, 166; [102] 154, 166
Savoia, A. [28] 67, 75, 107, 111; [36] 74, 112; [30] 121, 124, 164; [37] 121, 124, 136, 153, 164; [74] 136, 165
Schafer, W. [50] 614, 617, 624; [51] 614, 617, 624
Schaffer, W.J. [58] 74, 113; [73] 136, 137, 139, 141, 142, 144, 165; [31] 180, 203, 206
Schaffler, F. [48] 121, 122, 124, 156, 164
Schlapp, W. [90] 254, 255, 256, 262; [91] 256, 262
Schluter, M. [17] 14, 56; [41] 43, 57
Schmidt, P.H. [57] 74, 81, 113; [9] 326, 349, 353, 373; [21] 341, 343, 344, 374; [22] 341, 374; [11] 380, 385, 394, 395, 396; [12] 569, 592
Schmitt, R. [13] 380, 396
Schmitt-Rink, S. [32] 609, 623; [49] 614, 618, 624; [55] 616, 624; [58] 618, 621, 624
Schubert, E.F. [52] 443, 450
Schulman, J.N. [12] 214, 215, 216, 260; [15] 216, 260; [19] 266, 281; [1] 285, 290, 304, 309; [2] 285, 309; [15] 287, 289, 309; [21] 290, 291, 294, 296, 310; [38] 302, 310
Schuurmans, M.F.H. [11] 215, 216, 253, 260
Schwartz, R.F. [23] 496, 561
Schwarz, S.A. [34] 465, 487
Scifres, D.R. [35] 587, 593; [42] 589, 590, 593
Seah, M.P. [17] 171, 205
Sebenne, C.A. [38] 121, 124, 153, 164; [31] 121, 124, 164
Segmuller, A. [63] 236, 261
Sen, S. [21] 419, 449
Sergent, A.M. [125] 110, 114; [5] 379, 381, 382, 383, 389, 396; [6] 379, 380, 385, 386, 387, 390, 396
Sermage, B. [62] 235, 236, 261
Sette, F. [28] 67, 75, 107, 111; [36] 74, 112; [74] 136, 165; [88] 144, 166
Sevensson, S.P. [104] 155, 166
Sevoia, E. [88] 144, 166

Shah, J. [79] 246, 250, 251, 252, 262; [57] 616, 617, 618, 624
Sham, L.J. [6] 214, 260; [16] 287, 289, 309
Shanabrook, B.V. [53] 72, 112
Shank, C.V. [31] 604, 606, 623; [56] 616, 617, 624; [57] 616, 617, 618, 624
Shannon, J.M. [39] 500, 561; [40] 500, 561; [43] 501, 502, 561
Shay, J.L. [11] 119, 141, 142, 163; [6] 322, 373
Shen, T.M. [7] 567, 590, 592
Shewchuk, T.J. [119] 541, 563; [120] 541, 563; [148] 550, 564
Shichijo, H. [66] 237, 261; [41] 474, 475, 487; [48] 482, 487; [9] 492, 513, 560; [88] 513, 562; [137] 546, 564
Shih, H.D. [157] 556, 564
Shirley, D.A. [73] 73, 113; [23] 121, 163; [33] 121, 164
Shiue, C.C. [29] 461, 487
Shklyaev, A.A. [28] 178, 206
Shockley, W. [7] 118, 119, 163; [78] 508, 562
Siegbahn, K. [14] 171, 205
Silberberg, Y. [54] 616, 621, 624
Singleton, J. [49] 233, 261
Sivananthan, S. [14] 287, 290, 309; [26] 292, 310; [27] 292, 310; [28] 292, 310; [47] 308, 310
Sivco, D. [60] 74, 113; [56] 233, 234, 261; [38] 433, 434, 450; [25] 602, 623
Skeath, P.R. [40] 121, 122, 127, 128, 147, 153, 160, 164; [45] 51, 57
Skocpol, W.J. [13] 598, 623
Skolnick, M.S. [31] 224, 225, 233, 260
Smith, C. [32] 582, 593
Smith, D.L. [32] 28, 51, 57; [44] 122, 153, 164; [49] 121, 122, 139, 156, 164; [1] 285, 290, 304, 309
Smith, P.M. [42] 501, 561
Smith, P.W. [24] 602, 623; [47] 613, 614, 615, 624; [48] 613, 615, 624; [53] 615, 624; [54] 616, 621, 624; [61] 621, 624
Smith, R.G. [48] 440, 450
Smith, S.D. [1] 598, 622
Snejdar, V. [5] 117, 118, 120, 163
Sokolov, V. [127] 541, 564
Sollner, T.C.L.G. [38] 472, 487; [116] 541, 544, 563; [117] 541, 550, 563; [118] 541, 547, 563; [146] 550, 564; [147] 550, 564; [42] 72, 112
Solomon, P.W. [40] 229, 260; [6] 492, 504, 560
Sou, I.K. [13] 285, 302, 304, 309; [30] 292, 302, 308, 310; [46] 308, 310
Sparks, M. [78] 508, 562
Spicer, W.E. [45] 51, 57; [39] 72, 78, 112; [121] 110, 114; [40] 121, 122, 127, 128, 147, 153, 160, 164; [67] 126, 165
Spratt, J.P. [23] 496, 561
Stall, R.A. [13] 411, 449; [76] 508, 562
Staudenmann, J.L. [29] 292, 310; [30] 292, 302, 308, 310
Steiner, K. [13] 380, 396
Stern, F. [26] 66, 72, 111; [27] 66, 72, 111; [82] 247, 262; [83] 247, 262; [1] 265, 272, 280; [8] 265, 279, 281; [9] 265, 279, 281; [10] 454, 455, 461, 486; [12] 454, 456, 486; [61] 504, 526, 562; [15] 574, 593; [59] 618, 624
Stevens, K.W.H. [144] 548, 564
Stillman, G.E. [37] 432, 450; [10] 598, 622
Stoffel, N.G. [34] 121, 124, 134, 164; [36] 121, 124, 164; [37] 121, 124, 136, 153, 164; [42] 121, 164; [72] 136, 165; [106] 159, 166; [19] 174, 205
Stolz, W. [60] 234, 261
Stone, A.D. [141] 548, 564
Störmer, H.L. [81] 246, 252, 262; [23] 458, 487; [4] 492, 560
Stradling, R.A. [27] 222, 260; [31] 224, 225, 233, 260
Streetman, B.G. [9] 492, 513, 560; [88] 513, 562; [89] 513, 563
Streifer, W. [35] 587, 593; [42] 589, 590, 593
Stucki, F. [95] 145, 148, 149, 166
Sturge, M.D. [19] 601, 623
Su, C.Y. [45] 51, 57; [40] 121, 122, 127, 128, 147, 153, 160, 164
Subbanna, S. [62] 73, 113
Sugimura, A. [61] 235, 261; [21] 579, 580, 593; [23] 580, 593; [31] 582, 585, 593
Sultan, R. [31] 461, 487
Sumski, S. [51] 123, 164; [15] 380, 396; [16] 385, 396; [13] 570, 593

Sun, Y.L. [55] 123, 164
Sunski, S. [26] 602, 623
Suris, R.A. [13] 493, 542, 551, 552, 553, 560; [14] 493, 542, 560
Swanson, R.M. [79] 508, 562
Swarts, C.A. [21] 120, 163
Sze, S.M. [42] 44, 57; [1] 117, 118, 163; [18] 340, 373; [8] 379, 381, 391, 393, 396; [1] 491, 492, 496, 560; [19] 493, 494, 522, 560; [29] 497, 498, 561; [30] 499, 561; [62] 505, 562
Szuber, J. [55] 123, 164

Taguchi, A. [61] 235, 261
Tai, K. [62] 621, 624
Takatani, S. [61] 75, 113
Tang, J.T. [37] 432, 450
Tang, J.Y. [31] 461, 487; [42] 474, 487
Tang, J.Y.-F. [107] 536, 563
Tannenwald, P.E. [38] 472, 487; [116] 541, 544, 563; [117] 541, 550, 563
Tansley, T.L. [36] 359, 360, 374
Tarng, S.S. [62] 621, 624; [63] 621, 624
Tarucha, S. [45] 231, 261; [43] 610, 624; [44] 610, 624
Teal, G.K. [78] 508, 562
Tejayadi, O. [55] 123, 164
Tejedor, C. [11] 11, 24, 30, 36, 41, 54, 56; [44] 49, 51, 57; [5] 61, 76, 77, 82, 83, 94, 95, 99, 103, 111; [74] 73, 113; [103] 83, 92, 114
Tell, R. [45] 610, 611, 624
Temkin, H. [94] 81, 114; [5] 379, 381, 382, 383, 389, 396; [15] 380, 396; [13] 570, 593; [47] 592, 593; [26] 602, 623
Tersoff, J. [11] 11, 24, 30, 36, 41, 54, 56; [12] 11, 13, 20, 24, 33, 36, 44, 47, 53, 54, 56; [13] 11, 42, 43, 49, 56; [14] 11, 13, 20, 49, 51, 56; [19] 16, 56; [27] 21, 57; [28] 25, 36, 54, 57; [30] 27, 43, 44, 57; [35] 33, 50, 57; [36] 36, 42, 49, 57; [16] 61, 76, 77, 82, 83, 85, 92, 93, 94, 95, 97, 98, 99, 100, 101, 103, 110, 111; [93] 92, 100, 114; [95] 102, 114; [109] 102, 114; [110] 103, 114; [18] 120, 132, 163; [4] 169, 201, 205; [6] 285, 307, 309
Thang, N.T. [27] 461, 487
Theis, T.N. [79] 73, 113

Thomas, D.I. [46] 501, 510, 511, 561; [47] 501, 510, 511, 512, 561
Thome, H. [32] 296, 310
Thompson, J. [28] 268, 269, 272, 281
't Hooft, G.W. [33] 69, 73, 112; [11] 215, 216, 253, 260; [22] 219, 224, 226, 227, 228, 229, 231, 233, 257, 260; [7] 265, 279, 281
Thornber, K.K. [8] 404, 406, 407, 449; [42] 436, 450; [34] 465, 487; [142] 548, 564
Tissot, J.L. [24] 291, 310
Tomlinson, R.D. [37] 75, 112
Tove, P.A. [19] 341, 373
Tret'yakov, D.N. [44] 366, 375
Treusch, J. [50] 614, 617, 624
Trew, R.J. [31] 461, 487
Trommer, R. [35] 271, 281
Tsang, W.T. [125] 110, 114; [44] 231, 261; [32] 352, 374; [6] 379, 380, 385, 386, 387, 390, 396; [17] 385, 396; [7] 404, 410, 449; [8] 404, 406, 407, 449; [9] 407, 410, 449; [28] 426, 450; [29] 428, 450; [30] 428, 450; [31] 428, 450; [34] 428, 429, 450; [36] 431, 450; [4] 567, 589, 590, 592; [5] 567, 589, 592; [20] 578, 593; [24] 581, 593; [33] 587, 593; [36] 589, 593; [38] 589, 593; [48] 592, 593; [48] 613, 615, 624
Tsu, R. [37] 472, 487; [12] 492, 541, 560; [113] 540, 563; [114] 541, 563
Tsuchiya, M. [121] 541, 563; [122] 541, 563; [124] 541, 547, 564; [125] 541, 547, 564
Tsui, D.C. [26] 460, 487; [4] 492, 560; [140] 548, 564
Tu, C.W. [74] 243, 244, 245, 246, 252, 258, 259, 261
Tu, D.-W. [41] 190, 206
Tuchendler, J. [32] 296, 310
Tucker, R.S. [39] 610, 624; [40] 610, 624
Tung, R.T. [28] 497, 500, 561; [31] 500, 561
Turowski, M. [37] 75, 112

Ulbrich, R.G. [50] 614, 617, 624; [51] 614, 617, 624
Unlu, H. [17] 61, 76, 77, 94, 95, 97, 111; [124] 110, 114

Vahala, K. [51] 592, 593
Vandenberg, J.M. [15] 380, 396; [13] 570, 593; [26] 602, 623
Van der Veen, J.F. [90] 145, 166; [91] 145, 149, 166; [93] 145, 166
Van de Walle, C.G. [3] 11, 12, 13, 14, 15, 19, 53, 56; [4] 11, 12, 13, 14, 15, 19, 47, 53, 56; [5] 11, 13, 15, 25, 47, 54, 56
Van Opdorp, C. [5] 322, 330, 331, 364, 373
Van Ruyven, L.J. [12] 328, 367, 373
Van Vechten, J.A. [7] 11, 56; [15] 61, 76, 77, 85, 91, 111
Varma, R.R. [41] 121, 147, 164
Varshni, Y.P. [41] 305, 310
Venkatesan, T. [63] 621, 624
Venkateswaran, U. [18] 266, 268, 271, 272, 276, 281
Verhoeven, A.C.J. [12] 328, 367, 373
Veuhoff, E. [23] 266, 267, 281
Vieren, J.P. [43] 75, 112; [8] 285, 287, 293, 294, 309; [12] 285, 293, 297, 309; [13] 285, 302, 304, 309
Vinter, B. [27] 461, 487; [139] 547, 550, 564
Vogl, P. [11] 61, 76, 77, 88, 89, 95, 97, 101, 111
Voisin, P. [63] 236, 261
Vojak, B.A. [32] 225, 260; [137] 546, 564; [3] 567, 580, 589, 592; [17] 574, 593
Von Lehmen, A. [52] 614, 624
Von Ross, O. [19] 61, 76, 77, 94, 111; [17] 120, 132, 163; [71] 132, 165
Voos, M. [32] 69, 73, 112; [43] 75, 112; [25] 219, 220, 230, 260; [28] 222, 260; [63] 236, 261; [64] 237, 243, 261; [68] 239, 241, 242, 261; [8] 285, 287, 293, 294, 309; [12] 285, 293, 297, 309; [15] 287, 289, 309; [23] 291, 310; [31] 293, 310
Vrehen, Q.H.F. [9] 598, 622
Vuoristo, A. [120] 110, 114

Wagner, J. [60] 234, 261
Wagner, S. [11] 119, 141, 142, 163; [6] 322, 373
Wagner, W.R. [47] 592, 593
Wakita, K. [41] 610, 624; [42] 610, 624

Waldrop, J.R. [41] 73, 112; [49] 75, 112; [69] 72, 78, 113; [71] 73, 113; [81] 73, 113; [86] 72, 75, 81, 113; [89] 77, 114; [25] 121, 124, 138, 139, 163; [26] 121, 124, 136, 144, 163; [68] 126, 128, 131, 165; [77] 137, 165; [79] 137, 165; [82] 138, 165; [85] 141, 142, 165; [103] 155, 166; [10] 170, 173, 181, 182, 205; [11] 170, 173, 181, 183, 205; [12] 170, 177, 205; [13] 170, 177, 178, 179, 180, 205; [22] 177, 206; [30] 180, 190, 193, 203, 206; [33] 180, 200, 203, 206; [36] 181, 206; [37] 185, 186, 187, 188, 189, 206; [40] 190, 193, 194, 195, 196, 199, 206; [42] 191, 193, 203, 206; [43] 191, 192, 193, 203, 206; [46] 194, 195, 206; [47] 194, 197, 198, 206
Waldrop, J.W. [66] 73, 113
Waldrup, D.J. [20] 19, 24, 56
Walukiewicz, W. [22] 460, 486
Wang, K.L. [149] 550, 564; [152] 554, 564
Wang, T. [45] 476, 487; [85] 510, 562
Wang, W. [31] 268, 273, 281
Wang, W.I. [21] 19, 56; [26] 66, 72, 111; [27] 66, 72, 111; [75] 73, 113; [82] 247, 262; [83] 247, 262; [8] 265, 279, 281; [9] 265, 279, 281; [10] 265, 279, 281; [129] 541, 564
Wang, Y.H. [62] 505, 562
Watanabe, M.O. [55] 72, 112; [83] 138, 142, 165; [14] 265, 279, 281
Watanabe, N. [69] 239, 261; [71] 240, 261; [49] 501, 561; [58] 503, 562
Waugh, J.L.T. [34] 271, 281
Weaver, J.H. [3] 61, 66, 67, 111
Wecht, K.W. [31] 68, 69, 112; [51] 74, 81, 112; [68] 74, 113; [59] 234, 235, 261; [24] 343, 346, 374; [16] 414, 449
Weil, T. [139] 547, 550, 564
Weiler, M.H. [18] 288, 291, 305, 309
Weimann, G. [90] 254, 255, 256, 262; [91] 256, 262
Weinberger, D.A. [62] 621, 624
Weiner, J.S. [60] 74, 113; [56] 233, 234, 261; [25] 602, 623; [27] 602, 623; [28] 602, 608, 612, 623; [33] 606, 611, 612, 623; [39] 610, 624; [40] 610, 624
Weinstein, B.A. [32] 268, 271, 272, 276, 281

Weisbuch, C. [21] 217, 260; [55] 233, 235, 261; [24] 581, 593; [17] 601, 623
Welch, D.F. [63] 74, 113; [4] 213, 239, 240, 241, 259
Werder, D.J. [77] 246, 252, 253, 254, 259, 262
Wessel, T. [8] 567, 590, 591, 592; [9] 567, 592; [10] 567, 592
White, S.R. [6] 214, 260
Wicks, G.W. [63] 74, 113; [4] 213, 239, 240, 241, 259
Widiger, D. [4] 453, 464, 478, 486; [47] 479, 480, 487
Wiegmann, W. [31] 68, 69, 112; [12] 119, 124, 163; [48] 201, 206; [2] 213, 217, 218, 219, 220, 234, 246, 250, 257, 259; [21] 217, 260; [28] 222, 260; [81] 246, 252, 262; [79] 246, 250, 251, 252, 262; [4] 265, 279, 280; [40] 364, 375; [11] 410, 411, 413, 449; [17] 415, 417, 449; [26] 425, 426, 449; [27] 425, 450; [23] 458, 487; [44] 475, 476, 484, 487; [4] 265, 279, 280; [1] 567, 592; [20] 601, 623; [21] 601, 623; [22] 601, 623; [24] 602, 623; [27] 602, 623; [29] 602, 604, 606, 607, 609, 623; [33] 606, 611, 612, 623; [34] 606, 607, 623; [35] 609, 611, 623; [36] 609, 611, 623; [37] 609, 611, 612, 623; [38] 610, 623; [39] 610, 624; [40] 610, 624; [47] 613, 614, 615, 624; [53] 615, 624; [56] 616, 617, 624; [61] 621, 624; [62] 621, 624; [63] 621, 624
Wiesenfeld, J.M. [45] 610, 611, 624
Wight, D.R. [12] 598, 622
Wijewarnasuriya, P.S. [46] 308, 310
Williams, G.F. [29] 428, 450; [31] 428, 450; [34] 428, 429, 450; [36] 431, 450
Williams, J. [127] 541, 564
Williams, R.H. [41] 121, 147, 164
Willis, R.F. [65] 125, 165
Wilson, B.A. [74] 243, 244, 245, 246, 252, 258, 259, 261
Wilson, R.B. [57] 74, 81, 113; [9] 326, 349, 353, 373; [21] 341, 343, 344, 374; [11] 380, 385, 394, 395, 396; [12] 569, 592; [49] 592, 593
Wilt, D.P. [49] 592, 593
Witkovski, L. [88] 513, 562

Woicik, J. [39] 72, 78, 112; [121] 110, 114
Wolf, T. [71] 507, 562
Wolfe, C.M. [87] 75, 113; [88] 74, 81, 113; [10] 598, 622
Wolford, D.J. [94] 258, 262; [22] 266, 268, 279, 281; [23] 266, 267, 281; [28] 268, 269, 272, 281; [29] 268, 269, 272, 276, 281; [31] 268, 273, 281; [36] 272, 282; [37] 273, 276, 279, 282
Wolfram, T. [18] 266, 268, 271, 272, 276, 281
Wolter, J. [13] 380, 396
Womac, S.F. [53] 446, 450
Wong, K.B. [19] 216, 260; [20] 266, 268, 279, 281; [21] 266, 279, 281; [22] 266, 268, 279, 281
Wood, C.E.C. [67] 73, 113; [41] 501, 561; [66] 505, 562
Wood, T.H. [60] 74, 113; [56] 233, 234, 261; [25] 602, 623; [29] 602, 604, 606, 607, 609, 623; [33] 606, 611, 612, 623; [34] 606, 607, 623; [35] 609, 611, 623; [36] 609, 611, 623; [37] 609, 611, 612, 623; [38] 610, 623; [39] 610, 624; [40] 610, 624; [45] 610, 611, 624
Woodall, J.M. [10] 11, 24, 26, 27, 51, 54, 56; [6] 61, 76, 77, 81, 84, 100, 101, 111
Woodard, D.W. [41] 501, 561
Woodbridge, K. [33] 69, 73, 112; [22] 219, 224, 226, 227, 228, 229, 231, 233, 257, 260; [35] 229, 260; [49] 233, 261; [96] 219, 259, 262; [7] 265, 279, 281
Woodcock, J.M. [43] 501, 502, 561; [37] 589, 593
Worlock, J.M. [80] 246, 247, 248, 249, 250, 262
Wright, S.L. [54] 72, 112; [79] 73, 113; [57] 124, 165; [80] 137, 165; [39] 229, 260; [75] 243, 259, 261; [12] 265, 279, 281; [13] 265, 279, 281; [37] 361, 362, 372, 374; [38] 361, 362, 363, 364, 365, 374; [9] 380, 383, 390, 393, 396
Wrigley, C.Y. [35] 500, 561
Wu, C.M. [76] 73, 113; [38] 229, 260
Wu, G.Y. [20] 290, 310
Wunder, R. [13] 411, 449; [76] 508, 562
Wyder, P. [115] 110, 114

Yamada, S. [61] 235, 261
Yamaguchi, Y. [14] 413, 449
Yamoykuchi, M. [46] 590, 593
Yanase, T. [46] 590, 593
Yang, E.S. [76] 73, 113; [38] 229, 260
Yariv, A. [12] 61, 76, 77, 88, 89, 111; [29] 582, 593; [51] 592, 593; [44] 610, 624; [45] 610, 611, 624
Yeh, J.L. [49] 201, 206
Yen, R. [7] 567, 590, 592; [8] 567, 590, 591, 592; [9] 567, 592; [10] 567, 592
Yndurain, F. [44] 49, 51, 57
Yokoyama, K. [13] 454, 455, 457, 459, 460, 462, 486; [33] 462, 463, 487
Yokoyama, N. [20] 419, 449; [44] 501, 502, 561; [45] 501, 502, 510, 561; [48] 501, 561; [54] 502, 562; [151] 553, 564
Yoshida, J. [55] 72, 112; [3] 117, 118, 163; [83] 138, 142, 165; [14] 265, 279, 281
Yoshida, S. [52] 72, 112; [62] 124, 165; [23] 342, 343, 344, 374; [31] 350, 352, 374
Yoshikuni, Y. [41] 610, 624; [42] 610, 624
Yoshino, J. [121] 541, 563

Zeiderbergs, G. [9] 119, 132, 163; [42] 364, 375
Zhao, Te-Xiu [50] 75, 112; [34] 121, 124, 134, 164; [36] 121, 124, 164; [19] 174, 205
Ziemelis, U.O. [28] 222, 260; [68] 239, 241, 242, 261
Zucker, J.E. [20] 601, 623; [21] 601, 623; [22] 601, 623; [52] 614, 624
Zuleeg, R. [5] 117, 118, 120, 163; [13] 380, 396
Zunger, A. [25] 20, 26, 54, 57; [8] 61, 76, 77, 85, 98, 111; [117] 110, 114
Zur, A. [32] 28, 51, 57; [49] 121, 122, 139, 156, 164
Zurcher, P. [46] 73, 112; [48] 73, 84, 98, 100, 112; [65] 73, 113; [28] 121, 124, 139, 145, 147, 148, 150, 151, 155, 164; [32] 121, 164; [78] 137, 149, 151, 165; [84] 137, 140, 165; [95] 145, 148, 149, 166; [39] 190, 193, 206

SUBJECT INDEX

Ab initio calculations, *see* first-principle models
admittance spectroscopy 385–395
affinity rule, *see* electron affinity rule
AgI 89
Al 108, 157–160
AlAs 44, 81, 86, 89, 90, 93, 100, 104, 214, 420, 422
 –CuBr 139
 –GaAs 13, 15, 52, 56, 72, 78, 79, 86, 87, 96, 97, 101, 129, 131, 137–139, 141–143, 200, 201, 203, 204, 438, 506, 507, 541, 542
 –Ge 72, 79, 96, 97, 101, 129, 131, 138, 139, 141–143, 203
(Al,Ga)As, *see* $Ga_xAl_{1-x}As$
$Al_xGa_{1-x}As$, *see* $Ga_xAl_{1-x}As$
$Al_xIn_{1-x}As$
 –$In_yGa_{1-y}As$ 74, 79, 233–235, 240, 241, 346, 413–415, 424, 434, 435, 592
 –InP 73
AlN 89
AlP 44, 89, 93
AlSb 44, 81, 89, 90, 93, 104
 –GaSb 73, 83, 96, 236, 434
antiphase disorder 198, 200, 204
Au 101
avalanche photodiodes 429–436, 438–442, 465–468

BAs 89
band bending 7, 9, 63, 66, 68, 127, 128, 131, 151, 384, 386, 388
barrier height, *see* Schottky barrier *and* built-in potential
BeO 89
BeS 89

BeSe 89
BeTe 89
BN 89
BP 89
branch point, *see* charge neutrality level
built-in potential 63, 65, 153
bulk traps, *see* traps

C 89
capacitance–voltage characteristics, *see* $C-V$ characteristics
CdS 81, 86, 89, 90, 100, 104
 –Ge 74, 78, 96, 97, 101, 108, 136, 139
 –InP 141, 142, 322
 –Si 75, 78, 96, 97, 101, 108, 139
CdSe 81, 86, 89, 90, 95, 100, 104
 –Ge 75, 78, 96, 97, 101, 139
 –Si 75, 78, 96, 97, 101, 139
CdTe 44, 81, 86, 89, 90, 93, 100, 104, 180, 286–291, 295
 –HgTe 75, 285–310
 –(Hg,Cd)Te 308
 –(Hg,Mn)Te 308
 –Ge 75, 78, 96, 97, 101, 139, 141, 142
 –Si 75, 78, 96, 101, 139, 142
 –Sn 75, 96
charge accumulation 9
charge neutrality level 6, 7, 21, 30–36, 40–51, 83, 92–104, 134, 143, 169
charge-transfer method 66, 72–75
Co 85
common anion rule 27, 87, 307, 343
commutativity 77, 78, 135, 137, 138
compositional grading, *see* graded gap
Cr 85
Cs 108
CuBr 81, 89, 104, 191
 –AlAs 139

647

Subject index

  –GaAs 75, 79, 97, 139, 203
  –Ge 75, 79, 97, 138, 191–193, 203
CuCl 89
CuF 89
CuI 89
CuGaSe$_2$ 81, 104
  –Ge 75
CuInSe$_2$ 81, 104
  –Ge 75, 78
  –Si 75, 78
current–voltage characteristics, *see* I–V characteristics
Cu$_2$S 81
  –ZnS 75
C–V characteristics 65, 69, 72–75, 124, 138, 313–354, 383, 394, 395
  depletion method 332–354
  intercept method 315–332
cyclotron resonance, *see* magneto-optical measurements

Debye length 335, 336, 340
deep impurity levels 20, 21, 54, 85–87
deep level transient spectroscopy (DLTS) 379–385
defects, *see* interface defects
density of states 571–573
deuterium 108
dielectric constant 14, 34, 35
dielectric electronegativity 92
dipole, *see* interface dipole
DLTS, *see* deep level transient spectroscopy

effective mass 68, 124
effective-mass filtering 424
effective work function 84
electroabsorption 602–613
electron affinity 6, 17–29, 33, 35, 47, 50, 51, 61, 76, 80–87, 96, 97, 101, 119–120, 132, 133, 139, 140, 151
electron affinity rule, *see* electron affinity
electron mean free path 125, 126, 171, 174, 184
electronic states 453–455
escape depth, *see* electron mean free path
excitons 599–622

Fe 85, 86, 87
field ionization, *see* tunneling
first-principle models 12, 52, 55
forbidden gaps (table) 81
free-carrier concentration profile 331, 333, 347

(Ga,Al)As, *see* Ga$_x$Al$_{1-x}$As
Ga$_x$Al$_{1-x}$As 62, 81, 86, 87, 403–408, 411, 412, 420–423, 426, 428, 437, 444–446, 458, 460–462, 468–472, 478–481, 483, 484, 504, 512–516, 519, 525, 533, 538, 543, 547, 550, 555, 567, 589, 590
  –GaAs 62, 69, 70, 72, 73, 79, 86, 87, 101, 108, 138, 210, 211, 217–232, 237–258, 265–280, 287, 336, 337, 339, 342–344, 348, 352, 361–365, 380, 391, 404–413, 418, 431, 444–446, 448, 454, 455, 457, 460, 461, 483–485, 496, 501–507, 514, 529, 541–545, 550, 556, 557, 569, 578, 579, 597–622.
GaAs 44, 81, 86, 89, 90, 93, 95, 100, 104, 145, 146, 180, 191, 214, 341, 342, 404–431, 442–448, 454–486, 498–559, 577–590, 600–621
  –AlAs 13, 15, 52, 56, 62, 72, 78, 79, 86, 87, 96, 97, 101, 129, 131, 137–143, 200, 201, 203, 204, 438, 506, 507, 541
  –CuBr 75, 79, 97, 139, 203
  –Ga$_x$Al$_{1-x}$As 62, 69, 70, 72, 73, 79, 86, 101, 108, 138, 210, 211, 217–232, 237–259, 265–280, 287, 336, 337, 339, 342–344, 348, 352, 361–365, 380, 391, 404–413, 418, 431, 444–446, 448, 454, 455, 457, 460, 461, 483–485, 496, 501–507, 514, 529, 541–545, 550, 556, 557, 569, 578, 579, 597–622
  –Ge 13, 19, 24, 70, 73, 74, 78, 79, 83, 84, 96, 97, 101, 108, 122, 123, 127, 129, 135–162, 174, 181–191, 194–200, 203, 318, 322, 357, 364
  –InAs 13, 74, 96, 97, 139, 141–143, 203
  –In$_x$Ga$_{1-x}$As 235, 236
  –In$_x$Ga$_{1-x}$P 74
  –Si 73, 78, 97, 101, 139, 143

Subject index

–ZnSe 13, 75, 96, 97, 101, 139, 141, 142, 144, 154, 203
–ZnSnP$_2$ 75
GaAs$_{1-x}$P$_x$
  –Ge 364
Ga$_{1-x}$In$_x$As, see In$_x$Ga$_{1-x}$As
Ga$_{1-x}$In$_x$As$_{1-y}$P$_y$, see In$_x$Ga$_{1-x}$As$_{1-y}$P$_y$
GaInAsP, see InGaAsP
GaN 89
GaP 44, 81, 86, 89, 90, 93, 95, 100, 104
  –Ge 74, 78, 96, 97, 101, 108, 138, 139, 143, 154
  –Si 74, 78, 83, 96, 97, 101, 138–140, 143, 144, 161, 364
gap, see forbidden gaps
gap states 38, 42, 83, 92, 120, 132
GaSb 44, 81, 89, 90, 93, 95, 100, 104
  –AlSb 73, 83, 96, 101, 236, 434
  –Ge 74, 78, 96, 101, 138, 141–143
  –InAs 13, 83, 139, 141–143
  –Si 74, 78, 96, 139, 143
GaSb$_{1-y}$As$_y$
  –In$_x$Ga$_{1-x}$As 364
GaSe 81, 104
  –Ge 75, 78, 97
  –Si 75, 78, 97
Ge 44, 81, 89, 90, 92, 94, 95, 100, 104, 177–180, 191
  –AlAs 72, 79, 96, 97, 101, 129, 131, 138, 139, 141–143, 203
  –CdS 74, 78, 96, 97, 101, 108, 136, 139
  –CdSe 75, 78, 96, 97, 101, 139
  –CdTe 75, 78, 96, 97, 101, 139, 141, 142
  –CuBr 75, 79, 97, 138, 191–193, 203
  –CuInSe$_2$ 75, 78
  –GaAs 13, 19, 24, 70, 73, 74, 78, 79, 83, 84, 96, 97, 101, 108, 122, 123, 127, 129, 135–162, 174, 181–191, 194–200, 203, 318, 322, 357, 364
  –GaAs$_{1-x}$P$_x$ 364
  –GaP 74, 78, 96, 97, 101, 108, 138, 139, 143, 154
  –GaSb 74, 78, 96, 101, 139, 141–143
  –GaSe 75, 78, 97
  –InAs 74, 78, 95, 96, 139, 143
  –InP 74, 78, 95, 96, 97, 101, 139, 143
  –InSb 74, 78, 95, 96, 101, 139, 141, 142

  –PbTe 75
  –Si 13, 19, 72, 78, 90, 96, 97, 101, 121, 137, 138, 139, 143, 318, 322, 326–331, 358, 359, 364, 368, 369
  –ZnSe 13, 24, 67, 75, 78, 84, 96, 97, 101, 106–109, 121, 128, 130, 136, 137–141, 144, 157, 161, 193, 203
  –ZnTe 75, 78, 97, 101, 139
graded-gap 401–442, 501, 508–510, 515–517
graded-index 586–587

(Hg,Cd)Te 298
  –CdTe 308
(Hg,Mn)Te
  –CdTe 308
HgTe 44, 81, 93, 104, 180, 286–291, 295
  –CdTe 75, 285–310
  –ZnTe 308
hot electrons 432–433, 461–472, 476–478, 483–484, 493–507, 510–537
hydrogen 79, 108
hydrostatic pressure, see pressure

ideality factor 360
impact ionization, see avalanche photodiodes
In$_y$Al$_{1-y}$As
  –In$_x$Ga$_{1-x}$As 74, 79, 233–235, 240, 241, 346
InAs 44, 81, 89, 90, 93, 104, 180
  –GaAs 13, 74, 96, 97, 139, 141–143, 203
  –GaSb 13, 83, 139, 141–143
  –Ge 74, 96, 139, 143
  –Si 74, 96, 97, 139, 143
infrared absorption, see optical measurements
InGaAs 571
In$_x$Ga$_{1-x}$As 81
  –GaAs 235, 236
  –GaSb$_{1-y}$As$_y$ 364
  –In$_y$Al$_{1-y}$As 74, 79, 240, 241, 346, 403–405, 424, 434, 435, 592
  –InP 74, 86, 233–235, 241, 242, 320–322, 339, 343, 346, 347, 350–352, 358, 364, 366, 367, 380–395, 434, 439–442, 569, 571, 580, 588, 590

InGaAsP 439–441, 582, 590–592
$In_xGa_{1-x}As_{1-y}P_y$ 439–441
(In,Ga)(As,P)
  –InP 74, 342–344, 349, 380, 585, 586
$In_xGa_{1-x}P$ 81
  –GaAs 74
InN 89
InP 44, 81, 89, 90, 93, 100, 104
  –$Al_xIn_{1-x}As$ 73, 233–235
  –CdS 141, 142, 322
  –Ge 74, 78, 96, 97, 101, 143
  –$In_xGa_{1-x}As$ 74, 233–235, 241, 242, 320–322, 339, 343, 346, 347, 350–352, 358, 364, 366, 367, 380–395, 434, 439–442, 569, 571, 580, 588, 590
  –(In,Ga)(As,P) 74, 342–344, 349, 380
  –Si 74, 78, 96, 101, 143
InSb 44, 81, 89, 93, 101, 104
  –Ge 74, 78, 96, 101, 141, 142
  –Si 74, 78, 97, 102
interface defects 44, 62, 122, 154–157
interface dipole 10, 14, 19, 22–42, 49, 51, 53, 54, 62, 70, 82, 83, 90, 91, 95, 108, 109, 134, 140, 143, 144, 150, 170, 173, 181, 186–190, 194–198, 328, 348, 442–448
interface grading 352, 353
interface traps, *see* traps *and* interface defects
internal photoemission 68, 72–75, 117, 124, 254–257
intralayers 108–110, 157–160
ionization potential 17, 18, 22, 24, 85
I–V characteristics 65, 69, 72–75, 321, 354–372, 415–423, 470, 510–514, 517–532

$k \cdot p$ calculations 214–216, 252, 253

Landau levels, *see* magneto-optical measurements
lasers 453, 551–553, 567–568, 571–573, 577–592, 614–621
LCAO, *see* tight binding
light scattering 245–254
liquid phase epitaxy (LPE) 122, 292
local-density approximation 12, 14, 82

LPE, *see* liquid phase epitaxy

Magneto-optical measurements 287, 292–302, 308
MBE, *see* molecular beam epitaxy
metal-induced gap states, *see* gap states
metal-organic chemical vapor deposition, *see* MOCVD
MgTe 89
midgap-energy, *see* charge neutrality level
MIGS, *see* gap states
Mn 85
MOCVD 123, 235, 217, 237, 266, 292, 385
modulation doping 455–457, 460–462, 478–480, 513–523
molecular beam epitaxy (MBE) 123, 125, 144, 150, 155–157, 169, 175, 181, 190, 217, 242, 287, 291, 292, 343, 361, 385, 401–448
MnTe 44, 93
MQW, *see* quantum wells
multiple quantum wells, *see* quantum wells

Ni 85
neutrality level, *see* charge neutrality level
neutrality rule, *see* charge neutrality level
non-ohmic contacts 322, 323

optical absorption 217–236
optical measurements 68–70, 72–75, 119, 124, 209–262, 265–282, 302–305

P 343
PbTe 81, 104
  –Ge 75
photocurrent 354, 365–369, 434–435, 445–448, 609–612
photoelectron, *see* photoemission
photoemission 66–68, 70, 72–75, 106, 120, 124–132, 144–153, 162, 170–205, 307, 308
photoluminescence (*see also* optical measurements) 237–246, 265–280, 505–507
photoluminescence excitation 217–236
pinning strength 50
position-dependent band edges 94
pressure 266–280

ZnSe 44, 81, 86, 89, 90, 93, 95, 100, 104, 180, 191
 -GaAs 13, 75, 96, 97, 101, 139, 141, 142, 144, 154, 203
 -Ge 13, 24, 67, 75, 78, 84, 96, 97, 101, 106–109, 121, 128, 130, 136, 137, 139, 141, 144, 157, 161, 193, 203
 -Si 75, 78, 96, 97, 101, 139

ZnSnP$_2$ 81, 104
 -GaAs 75
ZnTe 44, 81, 89, 90, 93, 95, 100, 104
 -HgTe 308
 -Ge 75, 78, 97, 101, 139
 -Si 75, 78, 97, 101, 139, 142

Subject index

pseudopotential 12, 13–15, 34, 43, 90, 97, 98, 106, 120, 132, 133, 266

quantum confined Stark effect 606–613
quantum wells 68, 69, 210–259, 265–280, 313, 314, 343, 379–395, 418–424, 431–436, 438–442, 472–475, 501, 506, 513–524, 539–559, 567–592, 597–622
QW, *see* quantum wells

Raman scattering 252, 254, 305–307
real-space transfer 470–472, 483, 484, 513–539
reconstruction 145–153
reflection high-energy electron diffraction (RHEED) 225, 227, 228, 257
resonant Raman scattering, *see* Raman scattering
resonant tunneling 418–423, 472, 473, 539–551, 553–559
RHEED, *see* reflection high-energy electron diffraction

sawtooth structures 425–428
Schottky barrier 5, 7, 9, 13, 16, 20, 21, 28, 45–52, 61, 63, 66, 76, 82, 84, 94, 99–104, 108, 118, 121, 153–162, 341, 367, 379–395
Si 44, 81, 89, 90, 92, 93, 95, 100, 104, 341, 342
  p–n diodes 324
  –CdS 75, 78, 96, 97, 101, 108, 139
  –CdSe 75, 78, 96, 97, 101, 139
  –CdTe 75, 78, 96, 101, 139, 142
  –CuInSe$_2$ 75, 78
  –GaAs 73, 78, 96, 101, 139, 143
  –GaP 74, 78, 83, 96, 97, 101, 138–104, 143, 144, 161, 364
  –GaSb 74, 78, 96, 101, 139, 143
  –GaSe 75, 78, 97
  –Ge 13, 19, 72, 78, 90, 96, 97, 101, 121, 137, 139, 143, 322, 326–331, 358, 359, 364, 368, 369
  –InAs 74, 78, 96, 97, 139, 143
  –InP 74, 78, 96, 101, 139, 143
  –InSb 74, 78, 96, 101, 139
  –SiO$_2$ 79, 108
  –ZnSe 75, 78, 96, 97, 101, 139
  –ZnTe 75, 78, 97, 101, 139, 142

SiC 89
SiO$_2$
  –Si 79, 108
Sn 81, 89, 104
  –CdTe 75, 96
space-charge spectroscopy 379–395
staggered gaps 73–75, 275, 276
staircase structures 428–437
Stokes shift 237, 242, 305
superlattices (*see also* quantum wells) 285–310, 385–390, 420, 423, 424, 426–442, 485, 551–553
surface reconstruction, *see* reconstruction
surface stoichiometry 145–153
susceptibility 30, 32, 33
synchrotron radiation, *see* photoemission

thermionic emission 124, 470–471, 495, 513, 515
tight-binding 13, 25, 36, 47, 54, 55, 87–90, 96–98, 105, 120, 132–134, 141, 142, 169, 214–216, 285, 294
transistors
  ballistic 496–504
  charge injection 514, 515, 517, 518, 525–539
  graded base 407–413, 508, 509
  heterojunction bipolar 490–418, 508, 509
  high electron mobility (HEMT) 478–483
  hot electron 500–507
  induced base 504–506
  metal base 496–500
  negative differential resistance field effect 513–539
transitivity 19, 78, 135, 138, 139, 201–204
traps 323–331, 353, 354
tunneling 38, 290, 359, 360, 381, 424, 471, 473, 509–512, 551–553, 603, 606

V 85

XPS, *see* photoemission

ZnO 89
ZnS 81, 86, 89, 90
  –Cu$_2$S 75